D1536358

Bonebeds

# BONEBEDS

Genesis, Analysis, and Paleobiological Significance

*Edited by*

Raymond R. Rogers, David A. Eberth, and Anthony R. Fiorillo

The University of Chicago Press
Chicago and London

**Raymond R. Rogers** is professor in and chair of the geology department at Macalester College, St. Paul, Minnesota. **David A. Eberth** is a senior research scientist at the Royal Tyrrell Museum in Drumheller, Alberta, Canada. **Anthony R. Fiorillo** is a faculty member in the Department of Geological Sciences at Southern Methodist University and Curator of Paleontology at the Dallas Museum of Natural History, Dallas, Texas.

The University of Chicago Press, Chicago 60637
The University of Chicago Press, Ltd., London
© 2007 by The University of Chicago
All rights reserved. Published 2007
Printed in the United States of America

16 15 14 13 12 11 10 09 08 07     1 2 3 4 5

ISBN-13: 978-0-226-72370-9  (cloth)
ISBN-13: 978-0-226-72371-6  (paper)
ISBN-10: 0-226-72370-4      (cloth)
ISBN-10: 0-226-72371-2      (paper)

Library of Congress Cataloging-in-Publication Data

Bonebeds : genesis, analysis, and paleobiological significance / edited by Raymond R. Rogers, David A. Eberth, and Anthony R. Fiorillo.
    p.  cm.
Includes bibliographical references and index.
ISBN-13: 978-0-226-72370-9 (cloth : alk. paper)
ISBN-13: 978-0-226-72371-6 (pbk. : alk. paper)
ISBN-10: 0-226-72370-4 (cloth : alk. paper)
    ISBN-10: 0-226-72371-2 (pbk. : alk. paper)   1. Vertebrates, Fossil.   2. Paleontological excavations.   3. Paleobiology.   I. Rogers, Raymond R.   II. Eberth, David A.
III. Fiorillo, Anthony R.
    QE841.B675   2007
    560—dc22
                                                          2007003767


⊗ The paper used in this publication meets the minimum requirements of the American National Standard for Information Sciences—Permanence of Paper for Printed Library Materials, ANSI Z39.48-1992.

# CONTENTS

Preface    *vii*

Acknowledgments    *xi*


1   A Conceptual Framework for the Genesis and Analysis
    of Vertebrate Skeletal Concentrations                              *1*
    *Raymond R. Rogers and Susan M. Kidwell*

2   Bonebeds through Time                                              *65*
    *Anna K. Behrensmeyer*

3   A Bonebeds Database: Classification, Biases,
    and Patterns of Occurrence                                         *103*
    *David A. Eberth, Matthew Shannon, and Brent G. Noland*

4   From Bonebeds to Paleobiology: Applications of Bonebed Data        *221*
    *Donald B. Brinkman, David A. Eberth, and Philip J. Currie*

5   A Practical Approach to the Study of Bonebeds                      *265*
    *David A. Eberth, Raymond R. Rogers, and Anthony R. Fiorillo*

6   Numerical Methods for Bonebed Analysis                             *333*
    *Richard W. Blob and Catherine Badgley*

7   Trace Element Geochemistry of Bonebeds                             *397*
    *Clive Trueman*

**8** Stable Isotope Geochemistry of Bonebed Fossils:
Reconstructing Paleoenvironments, Paleoecology, and Paleobiology     *437*

*Henry Fricke*


Contributors    *491*

Index    *495*

# PREFACE

Bonebeds rank among the more notable features of the fossil record, and they tend to conjure up visions of expansive, densely packed animal graveyards rife with the scent of ancient catastrophe. Perhaps comparable visualizations gripped William Buckland, an early bonebed researcher, as he studied mammalian skeletal debris in 1821 from Kirkdale Cave, a Pleistocene locality in Yorkshire. Buckland pored over the amassed fossils and associated sediments and astutely concluded that the concentration of bones resulted from the activity of hyenas, with eventual burial of the skeletal remains by a diluvial flood event. Shortly thereafter, Sir Roderick Murchison published his classic description of the Silurian System, which included the first report of the widespread marine Ludlow Bonebed. Other workers soon followed the trail to the bones, working the Triassic fissure fills in the Bristol region and the Jurassic bone deposits of Lyme Regis, among other localities in and around the United Kingdom. The quest for bones soon extended throughout much of Europe and beyond, and by the beginning of the twentieth century a diverse array of fantastic bonebeds had been discovered and described from localities around the world. Today, thousands of bonebeds are known from Phanerozoic sedimentary rocks on every continent.

While the classic view of bonebeds as areally extensive deposits of bone heaped upon bone is still commonly held and, in fact, quite accurate in at least some spectacular cases (e.g., the classic Agate Springs mammal bonebed in the Miocene of Nebraska), it does not reflect the full spectrum of bone concentrations, and it is not a suitable basis for a working definition. Indeed, we struggled to clarify the precise meaning of the term "bonebed" as we wrote our respective chapters, and the reader is warned that some of the chapters provide slightly different versions of what exactly constitutes a bonebed (e.g., the 5% criterion of Behrensmeyer [Chapter 2] vs. the "greater than background" perspective of Eberth et al. [Chapter 3]). Nevertheless, we did reach general agreement on a basic definition that, while downplaying the dramatic side, serves to effectively characterize a bonebed (see Appendix 1 in Chapter 2 for a comparison of terms and definitions). The two most important criteria that distinguish a bonebed from other vertebrate occurrences are (1) whether a site (e.g., cave, bog, hollowed-out tree stump, etc.) or sedimentary stratum preserves the hardparts of only one individual (*not* a bonebed) or the hardparts of more than one individual in close spatial proximity (a bonebed), and (2) whether a site or stratum preserves hardparts in an abundance greater than the associated or "background" facies. The operational meaning of "close spatial proximity" varies with the context: consider, for example, the variable spatial scales of a tadpole bonebed and a sauropod bonebed (both are known from the rock record). Similarly, the relative abundance of bone in a bonebed can vary dramatically and is probably best judged by the individual researcher from a qualitative comparison with other fossil sites. The key point is that, in our opinion, any vertebrate locality that preserves the hardparts of two or more individuals in close association begs both careful scrutiny and, in most cases, an explanation that accommodates more than the singular demise of a lone individual.

In regard to indication of multi-individual mortality, we recognize that the vast majority of bonebeds documented in the literature do convincingly preserve the remains of two or more dead animals. In fact, we maintain that it is the very prospect of concentrated multi-individual mortality that piques the interest of bonebed researchers, be they taphonomists interested in reconstructing ancient death events or paleobiologists engaged in morphological or evolutionary studies. That said, we also recognize the possibility that some concentrations of vertebrate hardparts, such as shark-tooth lags embedded in marine strata, can develop due to processes entirely unrelated to localized mortality (sharks shed great numbers of teeth during life, and physical processes on the sea floor can concentrate them).

In *Bonebeds: Genesis, Analysis, and Paleobiological Significance*, we have attempted to distill the paleontological significance of bonebeds through a series of core papers that address current theoretical and practical treatments of bonebeds and their data. Our primary objectives are to (1) provide the reader with workable definitions and a framework for the consideration of bonebeds, and (2) supply the reader with an up-to-date compendium of current techniques of data collection and analysis with a specific focus on bonebeds. This edited compilation is not intended to be a comprehensive historical account of bonebed studies, and while we recognize that there is perhaps a rich historical story to be told, we have purposely kept the focus on a selection of more contemporary scientific issues. The chapters emphasize sites known from the rock record (true beds in a sedimentary basin sense), and this reflects the experiences and predominant interests of the authors. This emphasis is not meant to imply that cave-hosted bone accumulations are not bonebeds-*they most certainly are*! However, there are already several fine books and detailed professional reports devoted to the study and interpretation of bone accumulations preserved in caves. Finally, in a preemptive effort to avoid confusion, readers are alerted at the outset that throughout the book the term "bonebed" is often used synonymously with terms such as "skeletal concentration," "vertebrate accumulation," and "bone assemblage." It is also important to note that the term "assemblage" is frequently used to refer to subsets of bones recovered from a bonebed or a formation. For example, a bonebed researcher might refer to the "mammal assemblage" derived from a multitaxic bonebed that also yields the remains of fish and amphibians.

The first half of *Bonebeds: Genesis, Analysis, and Paleobiological Significance* delves into conceptual and interpretive aspects of the bonebed record. Chapter 1 considers bonebeds from a formative perspective, and an intuitive process-based classification is proposed that differentiates among various biogenic bone accumulation scenarios and physically generated concentrations. Chapter 1 draws upon numerous examples of modern phenomena in both terrestrial and marine settings that yield localized concentrations of vertebrate hardparts. A database approach to the study of bonebeds is developed in Chapters 2 and 3. Chapter 2 explores the Evolution of Terrestrial Ecosystems (ETE) Bonebed Database to provide an overview of bonebeds through time, with a focus on large-scale trends in taxonomic composition, ecological representation (carnivores vs. herbivores), and taphonomic quality (e.g., temporal trends in articulation patterns). The ETE Bonebed Database can be viewed online at www.press.uchicago.edu/books/rogers/. Chapter 3 utilizes a second "bonebeds database" (partially overlapping with the ETE Bonebed Database, available here in print form and online at the URL given above) to further elucidate patterns of origin and occurrence with specific reference to research biases and common facies associations. Both of these database-driven contributions examine terminology and classification. In

Chapter 4, bonebed datasets are reviewed in relation to paleobiology, with an emphasis on species characterization, paleobehavior, and paleocommunity reconstruction.

The second half of *Bonebeds* emphasizes practical approaches and serves as a "how-to" guide to the study of bonebeds. In Chapter 5, the reader is guided through preliminary site assessment, excavation and mapping techniques, and the collection of both geological and taphonomic data. Chapter 6 tackles bonebeds from a numerical perspective and probes the various quantitative methods that can be applied to bonebed assemblages. Protocols for counting specimens, evaluating taphonomic equivalence, comparing species richness among sites, estimating taxonomic abundance, and assessing faunal change (using bonebeds) are discussed, and multiple examples are provided. Chapters 7 and 8 provide a detailed review of the geochemistry of fossil bone and explore the phenomenon of fossilization in relation to vertebrate hardparts. Chapter 7 outlines specific examples of the utility of rare earth element concentrations in addressing bonebed-related questions (e.g., degrees of spatiotemporal mixing) and concludes with a useful primer on the analysis of trace metals in fossil bone. Chapter 8 tackles the topic of stable isotopes in relation to bonebeds and provides numerous examples of studies that utilize bonebed assemblages to address broad paleoenvironmental questions.

We envisage *Bonebeds: Genesis, Analysis, and Paleobiological Significance* as a useful resource for seasoned researchers, students, and amateurs interested in issues that pertain to vertebrate paleontology, archeology, paleoecology, and sedimentary geology. Through this book we demonstrate that bonebeds are a tremendous resource for the paleobiologist, geologist, and archeologist, and that the future of bonebed research is bright. We are confident that careful practice, in combination with new tools and methodologies, will lead to new and unique insights into the history of the vertebrates. Indeed, this seems destined to happen as new bonebeds are discovered and announced on a seemingly daily basis, and as classic bonebed localities are revisited and reinterpreted by a new generation of bonebed researchers with fresh eyes, innovative tools, and provocative new questions.

The idea for a book dedicated to bonebeds simmered on the back burner for years and only solidified into a concrete plan at the 1998 annual meeting of the Society of Vertebrate Paleontology in Snowbird, Utah. At this meeting the editors convened a symposium devoted to various types of bonebed studies, and it is from this symposium that the majority of contributions in this volume were drawn. We first thank all those who participated in the bonebeds symposium. The enthusiasm generated during the short span of the conference served as a litmus test and catalyst for the current project. We are also grateful to the program committee of the Society of Vertebrate Paleontology for approving and promoting our bonebed-centric efforts (and we of course forgive the official notice that welcomed attendees to the "boneheads" symposium). Considering how long it took to bring this book to fruition, the reference to "boneheads" may have in fact been prophetic and most appropriate. A sincere debt of gratitude is owed to all those who provided editorial comment, including Mara Brady, Kristi Curry Rogers, Cara Harwood, Francois Therrien, several anonymous readers, and of course the included authors, all of whom freely shared ideas and editorial suggestions. Christie Henry of the University of Chicago Press must also be acknowledged for her patience, guidance, and thoughtful encouragement. Finally, we thank our families, and especially our wives (Kristi, Marty, and Jessica), who graciously endured our ranting throughout the course of this prolonged but ultimately enriching endeavor.

# CHAPTER 1

# A Conceptual Framework for the Genesis
# and Analysis of Vertebrate Skeletal Concentrations

Raymond R. Rogers and Susan M. Kidwell

## INTRODUCTION

The record of vertebrate skeletal concentration begins with accumulations of agnathan dermal armor preserved in lower Paleozoic rocks (Behre and Johnson, 1933; Denison, 1967; Allulee and Holland, 2005). From these relatively modest beginnings, the record soon expands to include fossil deposits that yield the more diversified skeletal remains (calcified cartilage, endochondral elements, teeth) of marine and freshwater gnathostomes (Elles and Slater, 1906; Wells, 1944; Conkin et al., 1976; Antia, 1979; Adrain and Wilson, 1994). By the Late Devonian, tetrapods had ventured into marginal terrestrial ecosystems (Campbell and Bell, 1978; DiMichelle et al., 1992), and this major ecological foray marked a dramatic increase in the potential diversity of vertebrate taphonomic modes. Relative concentrations of vertebrate skeletal hardparts are found throughout the remainder of the Phanerozoic in a wide spectrum of marine and terrestrial depositional settings (Behrensmeyer et al., 1992; Behrensmeyer, Chapter 2 in this volume; Eberth et al., Chapter 3 in this volume).

Like macroinvertebrate shell beds, which are commonly studied for paleontological, paleoecological, paleoenvironmental, and stratigraphic information (Kidwell 1991a, 1993; Brett and Baird, 1993; Brett, 1995; Abbott, 1998; Kondo et al., 1998; del Rio et al., 2001; Mandic and Piller, 2001), vertebrate skeletal concentrations, or "bonebeds," provide a unique opportunity to explore an array of paleobiological and geological questions. These questions revolve around an array of fundamental "quality of data" issues, such as the degree of time averaging recorded by skeletal material and its fidelity to the spatial distribution and species or age-class composition of the source community. How do vertebrate paleoecology and behavior translate into bone-rich deposits? What are the genetic links between local sedimentary dynamics (event sedimentation, sediment starvation, erosion) and bonebed formation? To what extent are bonebeds associated with important stratigraphic intervals and discontinuity surfaces, such as well-developed paleosols, marine flooding surfaces, and sequence boundaries, and how faithfully does the nature of the skeletal material record the "time significance" of such features?

Clearly, understanding the diverse mechanisms of vertebrate hardpart concentration—and, needless to say, having the ability to recognize these in the fossil record—is vital to accurate paleoecological and paleoenvironmental reconstructions and is also essential for the development of productive collection strategies. Moreover, the stratigraphic distribution and taphonomic signatures of vertebrate skeletal concentrations have geological significance because of their potential to provide critical insights into sedimentary dynamics, local geochemical conditions, and basin-fill history in both marine and terrestrial settings (e.g., Badgley, 1986; Kidwell, 1986, 1993; Behrensmeyer, 1987, 1988; Bartels et al., 1992; Rogers and Kidwell, 2000; Straight and Eberth, 2002; Rogers, 2005; Eberth et al., 2006; Walsh and Martill, 2006).

Our aims here are first to clarify terminology by providing operational definitions, and then, by focusing on the dynamics of vertebrate hardpart accumulation, to distill a relatively simple intuitive scheme for categorizing bonebeds genetically. Like the benthic marine macroinvertebrate record, some concentrations of vertebrate skeletal elements reflect primarily biological agents or activities, such as gregarious nesting habits (yielding bone-strewn rookeries), predation (yielding bone-rich feces), and bone collecting (e.g., hyena dens and packrat middens), whereas others are the result of predominantly physical phenomena such as erosional exhumation (yielding bone lags) and sediment starvation (yielding time-averaged attritional accumulations). In addition, some concentrations reflect single, ecologically and geologically brief events, such as mass-kill deposits, whereas others have complex formative histories recording the interplay of multiple ecological and/or geological agents and events, generally over longer periods. Genetic scenarios are considered here both from a conceptual standpoint and from empirical observations (both actualistic and stratigraphic record-based studies), and the characteristic taphonomic signatures of different genetic themes are explored.

## OPERATIONAL DEFINITIONS

Here we essentially follow Behrensmeyer's (1991, Chapter 2 in this volume) and Eberth et al.'s (Chapter 3 in this volume) definitions of a "bonebed," which at its most basic is a "relative concentration" of vertebrate hardparts preserved in a localized area or stratigraphically limited sedimentary unit (e.g., a bed, horizon, stratum) and derived from more than one individual. Within this broad characterization, two distinct types of bone concentrations are commonly recognized by vertebrate paleontologists: (1) "macrofossil" bonebeds and (2) "microfossil" bonebeds (see Appendix 2.1 in Chapter 2 of this volume for a comparison of terminology [Behrensmeyer, this volume]).

Macrofossil bonebeds (sensu Eberth et al., Chapter 3 in this volume) are herein considered concentrated deposits of skeletal elements from two or more animals in which most bioclasts (>75%, be they isolated elements or entire skeletons) are >5 cm in maximum dimension. Macrofossil bonebeds are known from many different facies and depositional contexts, and they occur throughout the Phanerozoic history of the Vertebrata. Classic examples include the many Jurassic and Cretaceous dinosaur quarries of the western interior of North America (e.g., Hatcher, 1901; Brown, 1935; Sternberg, 1970; Lawton, 1977; Hunt, 1986; Rogers, 1990; Varricchio, 1995; Ryan et al., 2001; Gates, 2005). Mammalian counterparts also abound and include the spectacularly bone-rich Agate Spring locality and the Poison Ivy Quarry in the Miocene of Nebraska (Peterson, 1906; Matthew, 1923; Voorhies, 1981, 1985, 1992), among many others (e.g., Borsuk-Bialynicka, 1969; Voorhies, 1969; Barnosky, 1985; Voorhies et al., 1987; Fiorillo, 1988; Turnbull and Martill, 1988; Coombs and Coombs, 1997; Smith and Haarhoff, 2004).

Macrofossil bonebeds are also known to preserve aquatic and semi-aquatic animals. Examples include abundant amphibian-dominated assemblages from the Late Paleozoic of Texas and Oklahoma (Case, 1935; Dalquest and Mamay, 1963; Sander, 1987), and extensive fish-dominated bonebeds from Cretaceous deposits in Lebanon (Hückel, 1970), the Eocene Green River Formation of Utah and Wyoming (McGrew, 1975; Grande, 1980; Ferber and Wells, 1995), and elsewhere (e.g., Anderson, 1933; Pedley, 1978; Martill, 1988; Adrain and Wilson, 1994; Johanson, 1998; Davis and Martill, 1999; Fara et al., 2005). Interestingly, macrofossil bonebeds are seemingly less common in the marine realm—or at least less commonly analyzed and reported by vertebrate paleontologists—than might be expected given the great abundance and diversity of marine vertebrates and the potential for the addition of terrestrially derived skeletal material via bedload delivery and "bloat and float" (see Brongersma-Sanders, 1957; Schäfer, 1962, 1972). Well-documented examples of ancient marine macrofossil bonebeds include ichthyosaur lagerstätten from the Triassic of Nevada and the Jurassic of Europe (Ulrichs et al., 1979; Camp, 1980; Hogler, 1992). Less well documented but more common are shark-tooth beds and ecologically mixed assemblages of marine, estuarine, and terrestrial vertebrates associated with marine unconformities and surfaces of maximum transgression in Cretaceous to Neogene records (e.g., Barnes, 1977; Myrick, 1979; Norris, 1986; Kidwell, 1989; Schröder-Adams et al., 2001).

Microfossil bonebeds, which are commonly termed "vertebrate microsites" or "vertebrate microfossil assemblages" in the literature (McKenna, 1962; Estes, 1964; Sahni, 1972; Korth, 1979; Dodson, 1987; Brinkman, 1990; Eberth, 1990; Peng et al., 2001, among others), are sometimes construed as preserving the abundant remains of animals that have body masses on average 5 kg or less (e.g., Behrensmeyer, 1991). Instead of this overall body size criterion, Eberth et al. (Chapter 3 in this volume; see also Wood et al., 1988) propose that microfossil bonebeds be defined as relative concentrations of fossils where most component elements (>75%) are ≤5 cm in maximum dimension. This would include a variety of skeletal material (including entire carcasses) from small animals (such as frogs, salamanders, small snakes, fish, mammals, etc.) and small skeletal components or skeletal fragments from large animals (such as the teeth of crocodiles, dinosaurs, and sharks). In keeping with the operational definition of a bonebed, microfossil bonebeds should occur in a stratigraphically limited sedimentary unit, should demonstrably include the remains of at least two animals, and should preserve bones and teeth in considerably greater abundance than in surrounding strata (i.e., they should be "relatively enriched" with vertebrate bioclasts). Microfossil bonebeds have been described from many different facies (e.g., Nevo, 1968; Estes et al., 1978; Maas, 1985; Breithaupt and Duvall, 1986; Bell et al., 1989; Eberth, 1990; Henrici and Fiorillo, 1993; Khajuria and Prasad, 1998; Gau and Shubin, 2000; Rogers and Kidwell, 2000; Perea et al., 2001; Ralrick, 2004), and like macrofossil bonebeds, they occur throughout much of the Phanerozoic record in both terrestrial and marine depositional settings. Thoughts pertaining to the genesis of multitaxic microfossil bonebeds are presented later in this chapter.

Finally, "bone sands" are occasionally encountered in the stratigraphic literature and, as the name implies, consist mostly consist of sand-sized grains (0.0625–2 mm) to granules (2–4 mm) of bone. Particles are generally rounded fragments and are usually not identifiable beyond assignment to Vertebrata, but intact skeletal elements (e.g., teeth, vertebrae, phalanges, scales) and large bone pebbles are occasionally dispersed in the bone-sand matrix. Bone sands can vary in geometry from localized lenses to areally widespread sheets associated with unconformities (e.g., classic Rhaetic bone sand of Reif, 1982; SM-0 of Kidwell, 1989) and are most commonly found intercalated in marine strata.


## GENETIC FRAMEWORK OF SKELETAL CONCENTRATION

Vertebrates today inhabit virtually every depositional environment, from the deepest ocean basins to mountain lakes, and they exhibit tremendous variation in life strategies, ecological interactions, and body sizes (Pough et al., 2005). Given their current distribution (global) and diversity (∼50,000 extant species), the array of factors with potential to cause mass mortality or otherwise play a role in the concentration and preservation of vertebrate skeletal debris is staggering, and the possibilities only multiply when vertebrates are considered in an evolutionary context that spans more than 500 million years.

Nevertheless, a few general themes can be exploited in order to construct a genetic framework that is applicable across a broad spectrum of vertebrate occurrences, and we believe that the major formative scenarios explored here will provide general guidelines for the analysis of taphonomic history. This review is not exhaustive, however, and the reader is especially referred to the classic works of Weigelt (1927, 1989), Brongersma-Sanders (1957), and Schäfer (1962, 1972) for additional insights into an array of mortality and preservation scenarios in both terrestrial and marine settings. The reader is also referred to works by Behrensmeyer and Hill (1980), Shipman (1981), Behrensmeyer (1991), Martill (1991), Behrensmeyer and others (1992, 2000), and Lyman (1994) for in-depth considerations of vertebrate taphonomic modes and methodological approaches to the reconstruction of taphonomic history.

In keeping with the goal of developing a conceptual framework comparable to those already in existence for the macroinvertebrate fossil

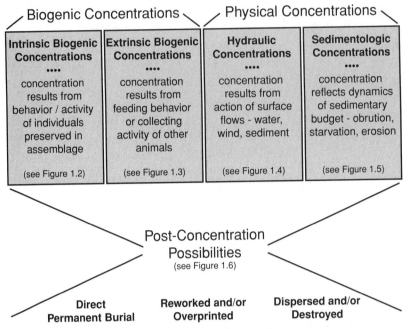

# Genetic Framework for
# Vertebrate Skeletal Concentration

**Biogenic Concentrations**  |  **Physical Concentrations**

| Intrinsic Biogenic Concentrations | Extrinsic Biogenic Concentrations | Hydraulic Concentrations | Sedimentologic Concentrations |
|---|---|---|---|
| **••••** | **••••** | **••••** | **••••** |
| concentration results from behavior / activity of individuals preserved in assemblage | concentration results from feeding behavior or collecting activity of other animals | concentration results from action of surface flows - water, wind, sediment | concentration reflects dynamics of sedimentary budget - obrution, starvation, erosion |
| (see Figure 1.2) | (see Figure 1.3) | (see Figure 1.4) | (see Figure 1.5) |

**Post-Concentration Possibilities**
(see Figure 1.6)

**Direct Permanent Burial**   **Reworked and/or Overprinted**   **Dispersed and/or Destroyed**

*Figure 1.1.* Vertebrate skeletal concentrations are grouped according to their inferred relations to biological and physical agents. Most vertebrate skeletal concentrations can be readily categorized as either biogenic or physical in origin. Biogenic concentrations are by definition produced by biological agents or events, and two general types are differentiated in relation to the formative role of intrinsic and extrinsic agents. Concentrations of vertebrate skeletal elements generated by physical processes are similarly subdivided into two general categories. Hydraulic concentrations result from the actions of transporting flows, whereas sedimentologic concentrations reflect the dynamics of the sedimentary budget. Postconcentration possibilities include (1) direct permanent burial, (2) reworking and/or overprinting, and (3) dispersal and/or destruction.

record (Johnson, 1960; Kidwell, 1986, 1991a; Kidwell et al., 1986), vertebrate skeletal concentrations are grouped here according to their inferred relations to biological and physical agents (Fig. 1.1). Most vertebrate skeletal concentrations can be readily categorized as either biogenic (intrinsic versus extrinsic) or physical (hydraulic versus sedimentologic) in origin, although the potential for mixed formative histories abounds. Mass mortality assemblages that have been reworked, sorted, and reconcentrated by fluvial processes are good examples of such mixed-origin concentrations (e.g., Voorhies, 1969; Wood et al., 1988; Eberth and Ryan, 1992). Contrary to the macroinvertebrate record, diagenetic processes such as compaction and pressure solution are deemed largely inconsequential with regard to

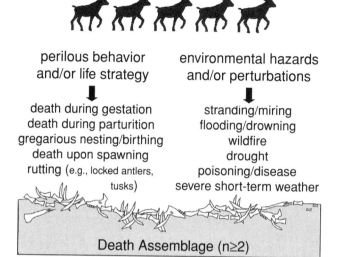

## Intrinsic Biogenic Concentrations

perilous behavior
and/or life strategy

environmental hazards
and/or perturbations

death during gestation
death during parturition
gregarious nesting/birthing
death upon spawning
rutting (e.g., locked antlers, tusks)

stranding/miring
flooding/drowning
wildfire
drought
poisoning/disease
severe short-term weather

## Death Assemblage (n≥2)

Figure 1.2. Intrinsic biogenic concentrations develop as a result of activity or behavior of the fossilized organism itself. Concentrations might reflect perilous behavior in life, such as colonial nesting habits, or gregarious behavior in death, such as fatal spawning events. Concentrations might also reflect behaviors forced by unusual environmental hazards or perturbations. The key factor is that concentration results ultimately from the behavior or activity of the animals represented in the death assemblage, and is not the result of direct action by other organisms or physical processes.

the primary concentration of vertebrate hardparts, although diagenesis is certainly a factor in long-term preservation potential, and early diagenetic prefossilization may play a role in the reworking of bones into some erosional lags (see below).

### Biogenic Concentrations

Biogenic concentrations of vertebrate hardparts are by definition produced by biological agents or events, and two general types are recognized here (Fig. 1.1). *Intrinsic biogenic concentrations* (Fig. 1.2) result from the "normal" activity or behavior of the vertebrates preserved in the death assemblage. Examples include concentrations that reflect gregarious behavior in life, such as colonial nesting habits, or in death, such as fatal spawning events. In some instances, the key behavior is forced by environmental conditions, specifically unusual hazards or perturbations that elicit behavior that results in a concentrated death assemblage. A prime example is drought, during which gregarious and nongregarious animals alike may modify their behavior so that their carcasses become concentrated in the

# Extrinsic Biogenic Concentrations

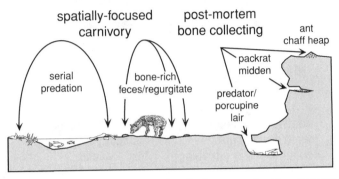

*Figure 1.3.* Extrinsic biogenic concentrations of vertebrate hardparts are produced by extrinsic biological agents, most notably predators. Nonpredatory animals such as porcupines and packrats also concentrate vertebrate hardparts due to habitual collecting. More rarely, intimate predator-prey associations are preserved, such as instances of fatal ingestion and dead carnivores with osseous gut contents.

vicinity of persistent food and water resources (Shipman, 1975; Haynes, 1988, 1991; Rogers, 1990; Dudley et al., 2001; among others). The key factor is that the formation of a death assemblage results ultimately from the behavior or activity of the hardpart producers and is not the result of direct action by other organisms or physical processes such as aqueous flows.

Biogenic concentrations of vertebrate hardparts also can be produced by the actions of other animals, acting on the hardpart producers (Fig. 1.3). Good examples of such *extrinsic biogenic concentrations* are predator-generated bone accumulations, such as bone-rich fecal masses and regurgitates (Mayhew, 1977; Dodson and Wexlar, 1979; Hoffman, 1988; Kusmer, 1990; Schmitt and Juell, 1994; Laudet and Selva, 2005; among others). Nonpredatory animals such as porcupines and packrats also concentrate vertebrate hardparts in appreciable quantities due to habitual collecting (e.g., Brain, 1980, 1981; Shipman, 1981; Betancourt et al., 1990). More rarely, intimate predator-prey associations are preserved, such as instances of fatal ingestion (or partial ingestion, see Grande, 1980; Davis and Martill, 1999) and dead carnivores with osseous gut contents (e.g., Eastman, 1911; Romer and Price, 1940; Eaton, 1964; Ostrom, 1978; Colbert, 1989; Charig and Milner, 1997; Chen et al., 1998; Varricchio, 2001; Hu et al., 2005; Nesbitt et al., 2006). This final category is somewhat of a hybrid mode of biogenic concentration (intrinsic plus extrinsic), in that it preserves the remains of the animal that generated the concentration via its typical behavior (the predator) and its ingested prey.

Raymond R. Rogers and Susan M. Kidwell

Animals are unlikely to engage in activities as part of their regular life strategies that result in mass mortality. However, some vertebrates, by virtue of their reproductive physiology or behavior, are predisposed to localized, multi-individual (n ≥ 2) mortality. At least three routine life-history events linked to reproduction predictably generate localized concentrations of bones and teeth.

The first reproduction-related scenario is the death of a pregnant viviparous or ovoviviparous female. Death of a pregnant female and one or more fetuses can result from a variety of causes (e.g., disease, breached birth, trauma) and can occur during gestation or during parturition. This scenario can transpire in any environmental context, and the only truly distinctive taphonomic signature is the intimate association of an adult female and a prenatal individual or individuals. Death during pregnancy or aberrant birth is a common phenomenon (e.g., Bergerund, 1971), and spectacular examples are known from the fossil record. For example, numerous specimens of the ichthyosaur *Stenopterygius* from the Early Jurassic Holtzmaden locality of southwestern Germany have been recovered with embryos both within the body cavity and partially expelled through the birth canal (Hauff, 1921; Böttcher, 1990).

A second reproduction-related scenario that can lead to accumulations of vertebrate hardparts relates to nesting in general, and gregarious nesting and birthing habits in particular. Given the definition of a bonebed, a single nest of any egg-laying vertebrate can in theory translate into a skeletal concentration if any combination of two or more embryos (assuming they have developed a mineralized and, thus, preservable skeleton) or hatchlings or nest-attending adults die and remain in close proximity. Death in and around a nest can readily result from intrinsic causes such as sickness or disease, abandonment, trampling/smothering, or siblicide (Burton, 1985; Anderson, 1990, 1995). Predators and inclement weather can also play a role in the buildup of bones in and around nesting sites. Mortality rates can be especially high in colonial nesting grounds, where overcrowding can lead to disease and agonistic behavior. Colonial nesting is especially common among seabirds (Furness and Monaghan, 1987), and aggressive behavior in seabird rookeries can take a particularly heavy toll on unattended chicks (Müller-Schwarze, 1984; Burton, 1985).

Pinnipeds also gather in rookeries in order to undergo synchronized mating and birthing. Mortality rates of seals, sea lions, and walruses in rookeries can be high, especially with regard to juveniles, and death can result from a variety of causes, including extreme or unusual weather, starvation, disease, and trauma (LeBoeuf and Briggs, 1977; Fay and Kelly, 1980; Johnson and Johnson, 1981; LeBoeuf and Condit, 1983; LeBoeuf and Laws, 1994; Ovsyanikov et al., 1994). Pups separated from their mothers are particularly vulnerable and often are crushed by breeding bulls or bitten by unrelated females in response to attempted suckling (LeBoeuf and Briggs, 1977). Huber et al. (1991) estimated that pup mortality in northern elephant seal colonies ranges from 10% to 40% on an annual basis. Mortality rates among elephant seals, and presumably other pinniped taxa, are density dependent and vary directly with the crowding of beach rookeries (e.g., Reiter et al., 1978).

From a taphonomic perspective, a concentration of vertebrate hardparts associated with a single nest or a colonial rookery should be dominated by a single taxon and by juveniles, which under most circumstances would be the most vulnerable age class (e.g., Emslie, 1995). Age spectra can vary, however, depending on the nature of the mortality event (Forcada et al., 1999; Meng et al., 2004). For example, frenzied episodes of mass mating in pinniped rookeries (e.g., "mobbing" by Hawaiian monk seals) can produce death assemblages that include both juveniles and adult females (Johnson and Johnson, 1981; Banish and Gilmartin, 1992). Bonebeds resulting from this scenario can be autochthonous (e.g., bones still in nest), parautochthonous (bones scattered within rookery), or allocththonous, given the propensity of seabirds and marine pinnipeds to gather in seaside rookeries where storms and marine transgression can readily rework and transport skeletal material. Site fidelity, combined with regular annual mortality, should in principle propagate time-averaged assemblages of skeletal elements. Bone assemblages derived from nesting animals (e.g., birds, nonavian dinosaurs) may be associated with additional evidence of colonial activity, including guano deposits, abundant fragmentary eggshell, remains of prey, and multiple nest structures (Horner and Makela, 1979; Horner, 1982, 1994; Baroni and Orombelli, 1994; Emslie, 1995; Chiappe et al., 1998, 2004; Emslie et al., 1998; McDaniel and Emslie, 2002).

Finally, several species of fish regularly undertake arduous and ultimately fatal journeys to spawning grounds. The classic example is the five species of Pacific salmon in the genus *Oncorhynchus*, which as sexually mature adults leave oceanic habitats and venture upstream into natal river systems to reproduce. This journey can stretch for more than 1000 km, and while in transit the salmon do not feed but rely instead on stores of fat (Curtis, 1949; Maxwell, 1995). Pacific salmon tend to die shortly after spawning, presumably a result of starvation and perhaps accelerated senescence, and their carcasses accumulate in vast numbers within and downstream from spawning grounds. A death assemblage of fishes linked to a fatal spawning event should be dominated by a single taxon and a single age class (reproductively capable adults) and should exhibit minimal levels of time averaging since spawning is typically synchronized (Wilson, 1996). However, given the generally poor preservation potential of fish carcasses (Schäfer, 1962, 1972; Nriagu, 1983; Smith et al., 1988), especially in slightly acidic freshwater systems, the likelihood of long-term preservation and entry into the fossil record is probably low (but see Wilson [1996]; and note that in the face of these taphonomic challenges, nitrogen isotopic signatures are proving to be extremely useful for reconstructing past fluctuations in salmon abundance [e.g., Finney et al., 2002]).

Biogenic concentrations of bones and teeth also develop when animals experience unusual or unpredictable circumstances in their local environment. Environmental hazards (such as the asphalt pools at Rancho La Brea [Stock, 1972]) and perturbations (such as drought [Shipman, 1975]) are envisioned as contributing factors related to events or behaviors that translate into local accumulations of vertebrate hardparts. A wide variety of environmental conditions can serve to bring or keep animals together prior to or during their death (Berger et al., 2001). The circumstances and taphonomic signatures of a selection of relatively common scenarios are explored here, namely stranding, miring, flooding, wildfire, severe short-term weather, drought, and disease.

With regard to strandings, Wilkinson and Worthy (1999) report that from 1989 through 1994, a total of 21,228 marine mammals (pinnipeds, cetaceans, sea otters, manatees) came ashore along the coastlines of the United States. Cetaceans, which comprise a significant fraction of this total (6768), are particularly susceptible to mass stranding because they are highly social and thus engage in coordinated group activity that leads on occasion to group mortality. On average, mass strandings of cetaceans number fewer than 15 animals, but some die-offs include hundreds of individuals (Smithers, 1938; Kellogg and Whitmore, 1957; Caldwell et al., 1970; Odell et al., 1980; Geraci and Lounsbury, 1993; Geraci et al., 1999).

The reasons behind cetacean mass strandings are often obscure. Schäfer (1962, 1972) proposed the possibility of "psychically induced" events, with panic or excitement perhaps coursing through a highly organized cetacean pod, ultimately leading to its subaerial demise. Panic attacks could conceivably result from predation pressures, or perhaps "extraordinary meteorological, volcanic, or oceanic events" (Schäfer, 1972,

p. 20). Local coastal effects, such as shore-parallel sand ridges, strong currents, or unusual tidal volumes could presumably also play a role (Schäfer, 1962, 1972; Brabyn and McLean, 1992). Some stranding events could potentially reflect a breakdown in navigation due to the impairment of echolocation in shallow waters, or geomagnetic anomalies (Dudok van Heel, 1966; Klinowska, 1985, 1986; Kirschvink et al., 1986). Resultant bonebeds should preserve a monospecific sample of a social unit (e.g., Camp, 1980; Hogler, 1992). The age spectrum would depend on the demographics of the affected pod. The coastal setting of mortality has great potential for reworking during storms, however, with bones likely to be incorporated into ravinement beds during marine transgression.

Miring as an agent of vertebrate mortality and skeletal concentration is perhaps best illustrated by the classic Rancho La Brea locality in the Pleistocene of southern California. Here, scores of animals big and small became trapped and perished in viscous pools of asphalt over tens of thousands of years (Stock, 1972; Spencer et al., 2003). A miring scenario has also been proposed for several low-diversity accumulations and isolated occurrences of prosauropod dinosaurs preserved in the Late Triassic of central Europe (Sander, 1992; Hungerbühler, 1998). Taphonomic features noted in support of miring include the prevalence of upright postures (including vertically oriented limbs), and a preservational bias in favor of the posterior/ventral body region. Sander (1992) further noted that juvenile prosauropods were absent in the bonebed assemblages and proposed that they may have escaped entrapment due to their smaller size and "lower foot pressure" (but see Berger [1983] for an account of juvenile horses trapped in a quagmire). Contorted strata indicative of extrication efforts (fugichnia) might also accompany carcasses mired in soft sediments.

From a conceptual standpoint, flooding in a terrestrial setting can lead to biogenic accumulations of vertebrate skeletal debris in at least three ways. First, when landscapes are inundated with floodwaters, terrestrial animals tend to move to higher ground, if there is any to be found. If animals succeed in securing an "island refuge," they may find themselves stranded for long periods. If the refuge lacks sufficient carrying capacity, animals may begin to starve. Close quarters may also lead to stress and agonistic behavior. If carnivores are present, predation could conceivably transpire at intensified levels. The resultant death assemblage would potentially include an assortment of animals of variable age classes that were mobile enough to reach high ground before drowning. The death assemblage should be relatively autochthonous with regard to the site of death, but the assemblage may not represent a typical ecological "community" because the animals arguably assembled under duress. The level of time averaging should be relatively low unless fatal flood events recur at sporadic intervals. Long-term preservation is not a given because burial potential is limited in "high-ground" locations. An apparent exception was described by Kormos (1911: cited in Weigelt, 1927, 1989), who invoked flooding (or perhaps fire) to explain a Pliocene mammal assemblage preserved in karst caves in a limestone crag. Burial potential would also be enhanced if flooding occurred in a large river system near sea level (e.g., Mississippi delta).

Aquatic animals also experience hardship during major flood events, potentially due to changes in aqueous chemistry (Whitfield and Patterson, 1995), or as the result of stranding as floodwaters recede. The latter scenario is particularly lethal to fish, which can become trapped in temporary back swamps and sloughs (Smith et al., 1988; Wilson, 1996; Johanson, 1998). The resultant death assemblage would presumably include fishes of variable age classes that ventured into flooded regions during high-water stage. The death assemblage should be autochthonous to parautochthonous with regard to the actual site of mortality. If fish bones survive scavengers and the rigors of subaerial weathering (Behrensmeyer, 1978; Smith et al., 1988), there is a reasonable chance of burial given their association with low ground on the floodplain.

Rivers and lakes in "high-water" or flood stage can also generate local concentrations of terrestrial vertebrate carcasses, with perhaps the most commonly cited example being the mass drowning of wildebeest during their overland migrations (Talbot and Talbot, 1963; McHugh, 1972; Schaller, 1972, 1973; Sinclair, 1979; Capaldo and Peters, 1995). Rivers need not be in flood stage, however, for mass drowning to occur (Dechant-Boaz, 1982). Apparently the pressure of the advancing herd is a killing agent in and of itself, as it forces animals in the lead to falter and drown as they pile up one upon another in swift-moving currents. The well-known *Centrosaurus* Bonebed (Quarry 143) in the Dinosaur Park Formation of Alberta (Currie and Dodson, 1984; Eberth and Ryan, 1992) has been interpreted to represent a dinosaur analog of the wildebeest stream-crossing scenario (but see Ryan et al. [2001] and Eberth and Getty [2005] for more recent interpretations). At this locality, more than 220 *Centrosaurus* individuals are preserved at the base of a broad, lenticular sandstone body interpreted as an ancient stream channel deposit. Several other fossil bonebeds (e.g., Turnbull and Martill, 1988; Wood et al., 1988) have also been attributed to mass drowning of vertebrates while crossing streams.

Bonebeds originating from fatal river or lake crossings should in principle preserve a monospecific sample comprising the herding taxon (e.g.,

Cole and Houston, 1969; Capaldo and Peters, 1995), although bones of other animals can be present at the accumulation site prior to the event and can also be added to the death assemblage subsequent to the event. If migratory herd structure is "fluid," as in the case of the wildebeest (Talbot and Talbot, 1963), a mixed age-class death assemblage is likely. If a mobile herd is socially organized, with perhaps adult animals leading the social unit, a more selective age spectrum could possibly be generated. With regard to depositional setting, bones should be preserved in fluvial or lacustrine deposits, although some reports (e.g., Turnbull and Martill, 1988) contend that floodwaters may actually transport carcasses out of the active channel belt and onto the adjacent floodplain. Preservation potential is generally good in fluvial and lacustrine settings (Behrensmeyer, 1988; Behrensmeyer et al., 1992).

Wildfires can also cause animals to congregate under duress and can ultimately lead to mass mortality. The likelihood of mass mortality depends to some extent on the abundance and mobility of animals in the path of the fire, and the dynamics of the fire advance (Lawrence, 1966; Singer et al., 1989). Mobile animals will often escape largely unscathed, but slow animals, such as turtles, and relatively immobile animals, such as nest-bound birds and newborns, can suffer dramatic losses. Aspects of the local terrain also play a role, especially with regard to the potential for concentration. If an obstruction is encountered, such as a cliff or a canyon, animals may gather and perish en masse by flames or by asphyxiation. Animals may also seek refuge at local water sources, where they may be overcome by heat or suffocate (Singer et al., 1989). The refuge scenario was invoked by Sander (1987) to explain a diverse assemblage of tetrapods preserved in lacustrine facies of Permian age. Bones derived from assemblages related to fire may show microscopic evidence of burning (Shipman, 1981; Shipman et al., 1984) and may be associated with charcoal deposits (fusain if fossilized [Jones, 1997; Falcon-Lang, 1998, 1999; Zeigler, 2002]). Preservation potential is generally good if animals congregate in or near water sources, which are often sites of sediment accumulation.

A wide range of mass mortality scenarios can be linked to extreme or unusual short-term weather events. Rapid spring thaws of snow pack or ice can lead to catastrophic flooding (see above) and can also leave animals stranded on islands or ice rafts, where starvation and possibly hyperpredation can lead to mass die-offs (Geraci and Lounsbury, 1993; Geraci et al., 1999). Conversely, rapid cooling events and severe winter stress can lead to hypothermia and starvation in both aquatic (Weigelt, 1927, 1989;

Storey and Gudger, 1936; Gunter, 1941, 1947; Brongersma-Sanders, 1957; Waldman, 1971; Economidas and Vogiatzis, 1992; McEachron et al., 1994; Marsh et al., 1999) and terrestrial (Berger, 1983; Borrero, 1990; Jehl, 1996) settings. If the drop in temperature is particularly severe, fatal ice-trapping can occur, as in the case of approximately 300 coots that were immobilized in ice and subsequently died of exposure in a small oxbow lake of the Illinois River in 1985 (Oliver and Graham, 1994). Marine mammals, such as beluga whales and narwhals, can also be trapped in ice by the thousands during unusual freezes (Freeman, 1968; Sergeant and Williams, 1983; Siegstad and Heide-Jørgensen, 1994). More exotic weather-related killing agents such as fog (Lubinski and O'Brien, 2001), large hail and lightning (Kuhk, 1956; NWHC, 1998, 2001), and unusually large oceanic waves (Bodkin et al., 1987) can also result in mass mortality and potentially generate localized concentrations of vertebrate skeletal debris. The long-term preservation potential and taphonomic signatures of a weather-related bone assemblage depend on taxon-specific and site-specific attributes and generally cannot be predicted for this diverse category of killing agents.

Drought is a very significant and recurrent killing agent in modern ecosystems (Tulloch, 1970; Corfield, 1973; Tramer, 1977; Coe, 1978; Haynes, 1988, 1991; Dudley et al., 2001), and numerous examples of drought-related mortality have been inferred from the fossil record (Matthew, 1924; Huene, 1928; Brown, 1935; Case, 1935; Dalquest and Mamay, 1963; Saunders, 1977; Hulbert, 1982; Rogers, 1990, 2005; Schwartz and Gillette, 1994; Fiorillo et al., 2000; Gates, 2005). During a drought, gregarious and nonsocial animals alike necessarily congregate in the vicinity of persistent reserves of food and water. If drought persists and resources are further depleted, fitness diminishes and animals begin to succumb to malnutrition and disease. Their plight may only worsen if drenching rains follow the drought and turn the terrain into a quagmire (Mellink and Martin, 2001). Carcass assemblages numbering in the hundreds may develop, and a variety of clues can be used to identify these drought-related skeletal concentrations in the fossil record (Shipman, 1975; Rogers, 1990; Falcon-Lang, 2003). Potential indicators include a suitable paleoclimate (indication of aridity and/or seasonality), an aqueous depositional setting, and age- and sex-specific mortality profiles (Voorhies, 1969; Hillman and Hillman, 1977; Conybeare and Haynes, 1984). The preservation potential of drought assemblages is good owing to the concentration of animals around water sources and the likelihood of flooding, erosion, and burial subsequent to drought (Weigelt, 1927, 1989; Kurtén, 1953; Shipman, 1975; Gillson, 2006).

Finally, under the general rubric of "sickness and disease," a host of biotoxins, abiotic chemical poisons, viruses, bacterial infections, and parasites can and often do trigger events of vertebrate mass mortality (Gunter et al., 1948; Brongersma-Sanders, 1957; Grindley and Taylor, 1962; Nyman, 1986; Pybus et al., 1986; Wurtsbaugh and Tapia, 1988; Worthylake and Hovingh, 1989; Thompson and Hall, 1993; Leonardos and Sinis, 1997; Berger et al., 1998; Swift et al., 2001; Braun and Pfeiffer, 2002; and see USGS National Wildlife Health Center Quarterly Mortality Reports for a wealth of additional examples). Whether a mortality event linked to sickness or disease translates into a concentration of vertebrate skeletal debris would of course depend on local conditions at the time of death and postmortem events. Likewise, the preservation potential and taphonomic attributes of a given assemblage would depend on the selectivity of the killing agent and local ecological and geological circumstances. Diagnosing sickness or disease in the fossil record, beyond the routine characterization of pathologic bone, is difficult at best, and often impossible (Baker and Brothwell, 1980; Rothschild and Martin, 1993; Hopley, 2001). A few potential examples include (1) lung failure in Miocene mammals, which was attributed to ash inhalation (Voorhies 1992), (2) botulism poisoning of dinosaurs (Varricchio, 1995), (3) red-tide-related poisoning of Pliocene cormorants (Emslie and Morgan, 1994) and (4) toxic cyanobacterial poisoning of Pleistocene mammals (Braun and Pfeiffer, 2002).

*Extrinsic Biogenic Concentrations*

Localized accumulations of vertebrate hardparts also result from the activities of other organisms, which can range from scavenging insects to large-bodied, voracious predators. Two general categories of accumulation are recognized, and these essentially relate to whether the bone assemblage reflects (1) spatially focused predatory activity or (2) purposeful postmortem bone collecting (Fig. 1.3).

One example of a predatory concentration is the direct ingestion of multiple prey items, which, depending on the intensity of gastric processing, can yield a bone-rich, multi-individual fecal mass or regurgitate. Numerous studies have explored the end results of avian, mammalian, and crocodilian digestion as it relates to the taphonomy of bones and teeth (e.g., Mayhew, 1977; Dodson and Wexlar, 1979; Fisher, 1981; Andrews and Nesbit Evans, 1983; Hoffman, 1988; Andrews, 1990; Kusmer, 1990; Denys et al., 1992; Denys and Mahboubi, 1992; Schmitt and Juell, 1994;

Terry, 2004; Laudet and Selva, 2005; among others), and the findings indicate that differentiating the feces and regurgitate of various bone-ingesting species is possible, even in the fossil record. For example, Mayhew (1977) compared the taphonomic characteristics of modern mammalian remains in pellets of diurnal and nocturnal birds of prey and identified characteristic breakage patterns and distinctive corrosion effects. These criteria were then applied to a Pleistocene fossil locality in order to identify the concentrating agent, which turned out to be a diurnal avian predator. Along these same lines, Schmitt and Juell (1994) and Denys et al. (1992) argued that skeletal representation, bone-fragment size, macroscopic and microscopic attributes of bone surfaces, and chemical composition could be used to distinguish bone assemblages of canid origin (specifically coyote and sand fox) from those of avian or human derivation. A multivariate analysis of new and published datasets demonstrates that small-mammal assemblages concentrated by owls are in fact consistently distinguishable from concentrations produced by diurnal raptors and mammalian predators on the basis of degree of fragmentation and skeletal-element representation (Terry, 2005). Distinctive taphonomic characteristics of digested bones and teeth, combined with the distinct morphologic characteristics of regurgitated pellets or fecal deposits, thus facilitate recognition of such accumulations (a coprocoenosis sensu Mellet [1974]) in the fossil record.

Another predatory scenario results from serial predation, which according to Haynes (1988, p. 219) "refers to regular and habitual killing of prey animals in the same loci, which are favoured by predators because of features of the terrain or localized abundance of prey." Haynes (1988) provided two examples of concentrations generated by serial predation. The first was a "cumulative bone site" in Wood Buffalo National Park in northern Alberta, Canada. The locus of bone accumulation was a small slough (~30 × 30 m in size) situated in an open grassland setting punctuated by small stands of trees. The bones of five bison—two adult males, two adult females, and one subadult—were found scattered around the edge of the slough. The monotaxic accumulation of bison elements was attributed to wolf predation that transpired over the span of a few years. Haynes' second example was from a water hole in Hwange National Park, Zimbabwe. This site contained the remains of at least 77 individuals, representing 14 vertebrate taxa (Haynes, 1988, p. 228), and was attributed to lion and hyena predation over a span of several years. Haynes (1988) argued that bone accumulations produced by serial predation are not straightforward to identify, and that distinguishing them from mass-death assemblages might be especially difficult. Criteria that might be useful in the fossil record include (1) evidence of time averaging, although mass-death sites can also show variable stages of bone weathering (Haynes, 1988), (2) indication of bone processing by carnivores, (3) a preponderance of potential prey animals, and (4) perhaps a depositional setting consistent with potential prey localization (e.g., stream margins, water holes).

The second category of extrinsic biogenic concentration reflects the intentional collection (and transport) by organisms of dead skeletal material. The habitual retrieval of bones and carcasses to cache sites or lairs for feeding or gnawing is a well-documented phenomenon. Carnivorous mammals notorious in this regard include hyenas, wolves, leopards, and a variety of mustelids (Yeager, 1943; Vander Wall, 1990). The bone-collecting habits of the hyenas have been particularly well studied (e.g., Buckland, 1823; Kruuk, 1972; Potts, 1986; Skinner et al., 1986; Horwitz and Smith, 1988; Skinner and van Aarde, 1991; Kerbis-Peterhans and Horwitz, 1992; Lam, 1992). The characteristic features of bone assemblages accrued by hyena include extensive evidence of crushing and gnawing (focused particularly on protruding features), an unusual abundance of bone chips (reflecting bone-crushing tendencies), and relatively rare evidence of distinct puncture marks. Avian taxa such as hawks and owls also regularly transport vertebrate prey, and they sometimes store their remains in multi-individual larders. Owls in particular make frequent use of prey stockpiles (Pitelka et al., 1955; Cope and Barber, 1978; Vander Wall, 1990), and the thick accumulations of pellets that develop under roost sites are effectively middens, combining the effects of serial predation with homing behavior and regurgitation of undigested remains (bones, teeth, and hair from small mammal prey).

Nonpredatory animals such as porcupines also collect bones, presumably to chew for dietary reasons ($Ca^{2+}$ uptake) and to keep ever-growing incisors in check. Bones collected by the African porcupine (*Hystrix africaeaustralis*) tend to show particularly heavy damage from gnawing (Brain, 1980, 1981) and may exhibit scoop-shaped excavations in the cancellous interiors of broken long bones in association with characteristic gnaw marks (Maguire et al., 1980). Packrats (*Neotoma* spp.) also transport and collect bones, although their motives are more difficult to ascertain. According to Elias (1990, p. 356), packrats "bring objects to their den site for a variety of reasons, including food, curiosity, and protection," and Hockett (1989) reported that relatively few bones recovered from a midden in the Nevada desert displayed gnaw marks that could be attributed to the resident packrat. The relatively small bones that are collected by packrats are found in association with an array of other small collected objects (e.g., shells, seeds, twigs), and would potentially be encased in the crystallized urine (amberat) of the packrat (Van Devender and Mead, 1978; Mead et al., 1983; Betancourt et al., 1990; Cole, 1990; Elias, 1990).

The curious category of ant-generated bone assemblages occupies the minuscule end of the bone-collecting spectrum. Shipman and Walker (1980) described the characteristics of modern bones collected by East African harvester ants (*Messor barbarus*), which apparently retrieve small bones from surrounding terrain to their chaff heaps in order to feed on adhering soft tissue (Shipman, 1981). Shipman and Walker (1980) found that ant-generated assemblages were superficially similar to owl pellet assemblages but could potentially be distinguished by a greater diversity of sampled taxa (predators are presumably more selective than scavenging ants), and preponderance of robust elements. A similar phenomenon occurs when ants encounter preexisting concentrations of fossil bones during excavation of their subterranean tunnels (Clark et al., 1967; RRR, pers. observ. 1993).

Finally, throughout much of their history, humans have interacted with vertebrates both as predators and as scavengers, producing extrinsic biogenic bone concentrations at sites of mass or serial killing (bison jump-offs, mammoth mires), consumption (middens), and tool making (postmortem utilization). A large anthropological and archeological literature is devoted to the phenomenology and recognition of "bonebeds" of these types (e.g., Frison, 1974; Graham et al., 1981; Grayson, 1984; Frison and Todd, 1986; Haynes, 1991; Holliday et al., 1994; Meltzer et al., 2002), and thus we do not discuss it here in further detail. Generally speaking, however, these present some of the same challenges as in diagnosing nonhominid biogenic concentrations. This includes the observation that humans were not necessarily agents of concentration, even where they are associated with bones via artifacts, and instead could have been agents of (minor) dispersion, for example removal of elements from drought-aggregated bison (e.g., Gadbury et al., 2000; Mandel and Hofman, 2003).

## Physical Processes of Concentration

Concentrations of vertebrate skeletal elements generated by physical processes are best conceptualized if vertebrates and their skeletal elements are viewed as sedimentary particles, or "bioclasts" (e.g., Behrensmeyer, 1975; Shipman, 1981). Like inorganic sedimentary particles, the behavior of vertebrate bioclasts in fluid flows is primarily a function of size, shape, and density, with bioclasts being entrained, transported, and deposited with moderate predictability. Whether accumulations of vertebrate hardparts develop as a result of physical hydraulic processes depends upon numerous factors, including the energy and persistence of the hydraulic medium, the threshold velocity of bioclasts relative to inorganic matrix, and the abundance of bioclastic material delivered to the system. A critical aspect of the concentration equation relates to the budget of nonbioclastic material (host matrix). If the system is starved of such sediment in relation to biological input, an attritional accumulation of vertebrate bioclasts could potentially accrue. If this sediment is supplied but reworked and winnowed from the system more readily than bioclasts, a residual hydraulic lag of relatively dense skeletal elements may result. If bioclast transport occurs at lower velocity than does the matrix, then a relative concentration of transported bioclasts could theoretically result "downstream," assuming the existence of some trap (but see discussion below). If an unusual amount of sediment is delivered to the system abruptly, an "obrution" event concentration of vertebrate bioclasts may be generated (sensu Seilacher et al., 1985; Brett, 1990). The key factor in all of these scenarios is that the *final concentration* of vertebrate skeletal components is predominantly the result of physical factors—either hydraulic processes or sedimentary budgets, rather than biological phenomena (Figs. 1.4 and 1.5).

### Fluvial Hydraulic Accumulations

In a study of vertebrate preservation in fluvial settings, Behrensmeyer (1988, p. 191) distinguished between assemblages that accumulated under the influence of "sustained active flow" (channel-lag assemblages) and those that accumulated after a channel is abandoned by flow (channel-fill assemblages). Bones and teeth preserved in channel-lag assemblages are generally considered parautochthonous to allochthonous. Component elements may exhibit a variety of taphonomic features generated by the multitude of biological and physical processes that can act upon vertebrate hardparts both before and during their interaction with fluvial processes. Bones may exhibit abraded edges and processes, indicating sustained interaction with abrasive sediment driven by currents, or may appear fresh and angular (Behrensmeyer, 1982, 1987, 1988). Accumulations may also show size and shape sorting and preferred orientations relative to prevailing currents (Voorhies, 1969; Behrensmeyer, 1975; Korth, 1979;

# Hydraulic Concentrations

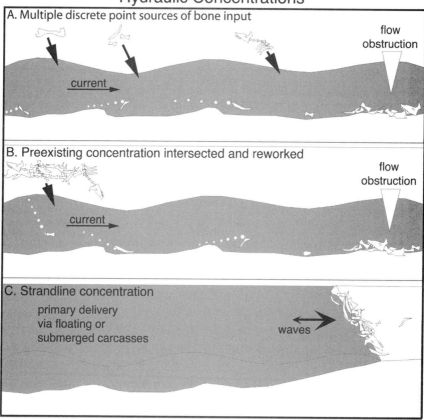

**A. Multiple discrete point sources of bone input**

flow
obstruction

current

**B. Preexisting concentration intersected and reworked**

flow
obstruction

current

**C. Strandline concentration**

primary delivery
via floating or
submerged carcasses

waves

*Figure 1.4.* Hydraulic concentrations of carcasses, parts of carcasses, or disarticulated bones and/or teeth form by the action of surface flows (wind, water, sediment) or wave activity. Transport of bioclastic material to the site of concentration is an integral part of this formative scenario. Numerous factors determine whether relative accumulations of vertebrate hardparts develop in these settings, including the energy and persistence of the hydraulic agent, the amount of bioclastic material delivered to the system, and the presence of trapping mechanisms (e.g., log jams).

Shipman, 1981). The degree of articulation and element association is generally low, especially for microfossil bonebeds (e.g., Rogers and Kidwell, 2000).

Fluvial transport and dispersal of vertebrate skeletal debris have received a great deal of attention (e.g., Dodson, 1971; Wolff, 1973; Lawton, 1977; Badgley, 1986; Behrensmeyer, 1988; Hook and Ferm, 1988), and numerous experiments have been conducted to learn how to recognize these phenomena (e.g., Voorhies, 1969; Dodson, 1973; Behrensmeyer,

# Sedimentologic Concentrations

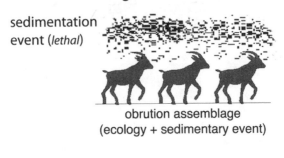

sedimentation event (*lethal*)

obrution assemblage
(ecology + sedimentary event)

sediment
starvation

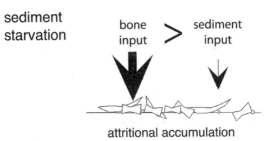

bone input $>$ sediment input

attritional accumulation

erosional
exhumation

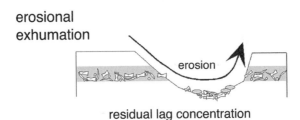

erosion

residual lag concentration

*Figure 1.5.* Sedimentologic concentrations, like hydraulic concentrations, are intimately linked to hydraulic processes. The key difference lies in their direct genetic link to the siliciclastic sedimentary budget, be it positive, zero, or negative. An obrution assemblage forms in response to, and is permanently buried by, a single unusual circumstance of sediment deposition. The obrution category is a composite mode of concentration that hinges upon both ecology (living animals initially congregate) and sedimentary geology (obrution triggers mortality and captures the concentration). An attritional assemblage of vertebrate hardparts initially forms during an episode of sediment starvation, with conditions of low sedimentary dilution fostering concentrations of vertebrate material by passive means. Residual lag concentrations form when erosion preferentially removes siliciclastic sedimentary matrix, leaving behind a hydraulic (transport-resistant) lag of larger and/or denser skeletal material exhumed from previous deposits.

1975; Korth, 1979; Hanson, 1980; Frison and Todd, 1986; Argast et al., 1987; Aslan and Behrensmeyer, 1996; Blob 1997; Trapani, 1998). For example, based on his seminal flume studies, Voorhies (1969) introduced the concept of fluvial transport groups (later referred to as "Voorhies groups" by Behrensmeyer, 1975), which reflect the transport potential and sorting behavior of disarticulated vertebrate elements (specifically those of sheep and coyote) with increasing current velocity. Using these guides, Voorhies (1969) interpreted the Pliocene Verdigre Quarry bone assemblage of Nebraska as hydraulic in origin. He argued that a preexisting assemblage of disarticulated and variably weathered bones and teeth had been concentrated during "a single flood of major proportions" at the base of an "off-channel sediment-trap."

Subsequent studies have extended the analysis to other vertebrate groups. Dodson (1973) focused on the flume behavior of disarticulated elements and component parts (e.g., skulls) of frogs, toads, and mice and identified the sequence of movement in unidirectional currents. He demonstrated that the bones of small vertebrates (microvertebrates) were particularly susceptible to transport (rather than "hiding" in the benthic boundary layer), and also identified the possibility of floating bones, which would of course have great dispersal potential in fluvial systems.

Behrensmeyer (1975) examined the consequences of bone size, shape, and density on transport potential for numerous different mammals. In addition to theoretical considerations of dispersal, she conducted experiments on bone settling velocities and was thus able to estimate the hydraulic equivalence of modern and fossil bones and their associated sedimentary matrix. Behrensmeyer (1975) also addressed the significant and often overlooked details of bottom morphology, current profile, and burial potential as they relate to bone transport and dispersal.

In a more recent analysis of hydrodynamic dispersal potentials, Blob (1997) returned to the question of Voorhies groups for nonmammalian material, specifically the soft-shelled turtle *Apalone spinifera*. He found that turtle elements generally fall into three groups that disperse in a relatively predictable fashion. However, the initial orientations of turtle elements on the flume bed played a significant role in their dispersal potential, and, contrary to expectation, threshold velocities for the entrainment of turtle elements correlated poorly with bone density. Thus, he cautioned against the uncritical application of mammalian hydrodynamic sorting patterns to the analyses of nonmammalian assemblages.

In one of very few actualistic studies of bones in natural fluvial systems, Aslan and Behrensmeyer (1996) surveyed a 2 km stretch of the East Fork River in western Wyoming to relocate 311 experimental bones and teeth placed in the river between 1974 and 1983 (Hanson, 1980; Behrensmeyer, 1982, 1991). One hundred and forty-two bones of the original experimental sample were relocated, and observations were taken on 372 additional naturally occurring bones between 1974 and 1987. Most of the experimental bones had been transported less than 1000 m from their point of origin, and virtually all showed minor evidence of abrasion. Interestingly, there was no correlation between degree of abrasion and distance of transport. Perhaps most important to the present discussion, however, was their discovery that both natural and experimental bones accumulated along the inside bends of sinuous channels with prominent point bars (Aslan and Behrensmeyer, 1996). They attributed these accumulations to flow resistance in sinuous channel bends, which is approximately twice as great as in straight reaches (Lisle, 1976; Dietrich, 1983). Aslan and Behrensmeyer (1996) did not, however, encounter multi-individual concentrations of naturally occurring bones on point bars in their study transect (Behrensmeyer, pers. comm., 2000).

In general, the dynamics of bone concentration under active flow in modern fluvial channels is poorly understood, at least in relation to our working knowledge of bone transport, bone degradation (abrasion and rounding), and bone burial in the same setting (but see Brady et al., 2005). This is unfortunate, because fossil bones and teeth are common in fluvial deposits (e.g., Korth, 1979; Behrensmeyer, 1987, 1988; Koster, 1987), and numerous channel-hosted concentrations of fossil bones have been reported in the literature (e.g., Lawton, 1977; Eberth, 1990; Fiorillo, 1991; Badgley et al., 1995; Rogers and Kidwell, 2000). Some of these ancient fluvial bone assemblages have been linked to mass-death events, and in these cases the concentrations of hardparts are envisioned as parautochthonous deposits derived from a nearby multi-individual source (e.g., Voorhies, 1969; Eberth and Ryan, 1992; Ryan et al., 2001). Others, however, have been interpreted as attritional accumulations that developed over time as bones and teeth of numerous animals were hydraulically transported within the fluvial system. Two trapping mechanisms are commonly invoked to explain these attritional accumulations: in-channel obstructions such as trees or large carcasses (Fiorillo, 1991; Le Rock, 2000), and drops in hydraulic competence associated with sinuous or otherwise compromised channel stretches (e.g., Lawton, 1977).

While the attritional scenarios outlined above are arguably possible, we find it conceptually difficult to accept the proposition that disarticulated bones and teeth of numerous animals delivered from widely separated point sources at different times would travel downstream through complex and hydraulically unstable channel belts and collectively accumulate on a regular basis (Fig. 1.4A). In fact, we find it much more likely that bones and teeth would tend to disperse over time as a function of differential transport. Moreover, as pointed out above, vertebrate elements released from point sources along the East Fork River tended to travel less than 1000 m from their origination point, despite years of residence time in the active channel (Aslan and Behrensmeyer, 1996). This suggests that bones and teeth derived from background mortality along the stretch of a river should, given time, diffuse downstream, resulting in dispersed occurrences relative to their respective sites of origin and individual sorting behaviors.

It strikes us as more likely that many, if not most, multi-individual concentrations of disarticulated vertebrate skeletal elements preserved in ancient fluvial channels were derived from a preexisting concentrated source (Fig. 1.4B). Eberth (1990) alluded to this possibility in an analysis of vertebrate microsites in the Judith River Formation of Alberta. He proposed that preexisting "floodplain concentrates" of vertebrate microfossils were the likely source for the numerous paleochannel-hosted microvertebrate assemblages in the Dinosaur Park Formation of Alberta. Badgley et al. (1998) also addressed the potential of preexisting sources and proposed that localized concentrations of vertebrate microfossils in fluvial facies of the Middle Miocene Ghinji Formation of Pakistan are the product of initial biological accumulation (probably linked to predator activity) with subsequent reworking, transport, and redeposition in fluvial channels. Rogers and Kidwell (2000) focused on the significant role of preexisting sources in their analysis of vertebrate skeletal concentrations and discontinuity surfaces in terrestrial and marine facies in the Cretaceous Two Medicine and Judith River Formations of Montana. They identified sites on the floodplain where physicochemically resistant vertebrate hardparts accumulated to concentrated levels via attritional processes (see below), and provided physical stratigraphic and taphonomic evidence for the fluvial reworking of preexisting floodplain concentrations into channel-hosted microvertebrate assemblages (see also Rogers, 1995). More information on the erosional exhumation and reconcentration of bone in both fluvial and shallow marine settings is presented below.

*Strandline Hydraulic Accumulations*

Vertebrate remains also accumulate hydraulically along strandlines, where they are subject to wave activity (Fig. 1.4C). Weigelt (1927, 1989) described the formation of strandline assemblages and provided examples from the fossil record. In most cases, bones and teeth are delivered to the strand via floating carcasses, which become anchored and may disarticulate in the swash zone. Onshore and longshore transport may also deliver isolated skeletal elements to the strand. Entire carcasses and individual skeletal components are usually driven by wave activity to lie tangential to the shoreline. However, if an element is irregular in shape such that one end serves as a pivot point, it may orient perpendicular to the shoreline. Vertebrate remains along strandlines are commonly associated with plant debris, especially along bodies of freshwater, such as the classic Smithers Lake locality of Weigelt (1927, 1989). Here, vertebrate carcasses accumulated amidst vast quantities of plant debris during a rare winter storm (a "norther"). In most cases, a multi-individual accumulation of skeletal debris along a strandline reflects a precursor event of heightened mortality.

Leggitt and Buchheim (1997) compared a modern strandline bone assemblage composed of thousands of disarticulated bird elements (predominantly derived from pelicans and cormorants that died of infectious disease) with an assemblage of *Presbyornis* bones from Eocene Fossil Lake (Wyoming) and found similar taphonomic features. Most striking was the strong alignment of elongate elements (humerus, radius, ulna, tibiotarsus, tarsometatarsus) with respective shorelines. In a study of much older vertebrate remains, Rogers et al. (2001) attributed an enigmatic concretion-hosted assemblage of partially articulated tetrapods in the Triassic of Argentina to a strandline scenario. Tetrapod fossils from the Los Chañares locality exhibit telltale evidence of strandline accumulation, including distribution in a narrow linear swath, and alignment of skeletal debris, including spectacular examples of parallel-arrayed carcasses.

*Sedimentary Budgets and Vertebrate Skeletal Accumulation*

Empirical data from the marine stratigraphic record demonstrate a strong association between benthic macroinvertebrate concentrations and discontinuity surfaces of many types, as well as congruence between taphonomic attributes and the inferred duration of the associated hiatus (Kidwell, 1989, 1991a, 1991b, 1993; Fürsich and Oschmann, 1993; Brett, 1995; Naish and Kamp, 1997; Abbott, 1998; Gillespie et al., 1998; Kondo et al., 1998; del Rio et al., 2001; Fernandez-Lopez et al., 2002; Fürsich and Pandey, 2003; Mandic et al., 2004; Cantalamessa et al., 2005; Parras and Casadio, 2005). It is thus reasonable to suspect that spatial and temporal variation in erosion, sedimentary omission, and deposition—that is, patterns in the accumulation of inorganic sediment, as opposed to active manipulation of hardparts—might also exert a strong influence on the distribution and quality of vertebrate fossil assemblages in marine and continental systems (see Rogers and Kidwell, 2000). Accumulations of vertebrate hardparts that ultimately result from budgets of sedimentation are herein categorized as sedimentologic concentrations (Fig. 1.5).

Sedimentologic concentrations overlap with the hydraulic concentrations described in the preceding section in that they too are intimately linked to hydraulic parameters (waves, currents, etc.), because these influence the capacity and competence of inorganic sediment removal and delivery, not simply bioclast mobility. The sedimentologic category, however, focuses on the nonbioclastic sedimentary budget, be it positive (aggradational), negative (erosional, degradational), or zero (starvation, dynamic or total bypassing). In the following analysis, we focus on mechanisms of sedimentologic concentration that operate in sedimentary basins that are fundamentally sinks for sediment and thus accumulating stratigraphic records, as opposed to regions undergoing widespread erosion or chemical degradation, such as karst terrains. In-depth considerations of bone concentrations that accumulate in caves and fissures can be found in works by Fraser and Walkden (1983), Brain (1980, 1981), Andrews (1990), and Benton et al. (1997), among others.

Theoretically, vertebrate hardparts may accumulate to concentrated levels during episodes of sediment starvation. Nondepositional hiatuses create conditions of low sedimentary dilution for the "rain" of skeletal material produced by contemporaneous vertebrate populations (Behrensmeyer and Chapman, 1993) and thus should foster relative concentrations of vertebrate material by passive means. The longer the sedimentary hiatus, the greater the potential quantity (and ecological mixture) of hardparts supplied, and thus the richer and laterally more continuous the ultimate skeletal concentration may be. On the other hand, conditions of low or zero net sedimentation also increase the period that any cohort of skeletal material is subject to destructive postmortem processes operating at or near the depositional interface. Thus, the advantages of lowered sedimentary dilution might be outweighed by taphonomic culling.

The stratigraphic record of marine settings does contain "time-rich" hiatal concentrations of vertebrate skeletal hardparts, generated during stratigraphically significant hiatuses in sedimentation and associated with significant discontinuities, such as parasequence-bounding flooding surfaces, transgressive surfaces, midcycle surfaces of maximum transgression, and third- or fourth-order sequence boundaries (e.g., most articles cited above for macroinvertebrate concentrations, plus Conkin et al., 1976, 1999; Sykes, 1977; Kidwell, 1989, 1993; Macquaker, 1994; Turner et al., 2001; Walsh and Naish, 2002; Allulee and Holland, 2005). Hiatal conditions can reflect a variety of short-term sedimentary dynamics and develop in an array of paleogeographic and historical contexts (e.g., sediment starvation of submarine paleohighs, distal portions of basins, and transgressive shelves; sediment bypassing of shallow-water environments at grade; erosional truncation associated with rapid transgression and baselevel lowering; Kidwell 1991a, b). Depending on the conditions, vertebrate material may be associated with either scarce or abundant shelly remains from macrobenthos and ranges in condition from articulated specimens of marine reptiles and mammals (most common in the context of anoxia and reduced sedimentation during maximum transgression, but see Brand et al. [2004]) to highly comminuted, abraded and polished teeth and fragmental bones from marine taxa (associated with erosional reworking of marine strata). All of these marine hiatal concentrations, however, are thinner than coeval, less fossiliferous strata, and they typically bear tangible evidence of prolonged low net sedimentation, such as ecological condensation and admixtures of hardparts with diverse taphonomic or diagenetic histories. In the marine realm, it thus appears that the production of new elements during the hiatus, both vertebrate and macroinvertebrate, compensates for the destructive aspects of hiatuses (e.g., retarded permanent burial, repeated small-scale burial-exhumation cycles, possible elevated attack from other taphonomic agents during bioclast residence at or near the sediment-water interface).

Taphonomically comparable hiatal skeletal concentrations are less common in the terrestrial record, and this is not entirely unexpected, given the generally dispersed nature of vertebrate populations and the harsh conditions that can accompany bone exposure on the land surface and pedogenesis during early burial (Behrensmeyer, 1978; Behrensmeyer and Chapman, 1993). These conditions include trampling, scavenging, weathering (UV exposure, oxidation, fungal and microbial attack, freeze-thaw), destructive soil processes (e.g., organic leaching of bases, wet/dry alteration of oxidation states), and exhumation-burial cycles (e.g., via bedform migration within channels).

Despite these negative factors in the accumulation process, some excellent examples of passive, hiatal concentrations of vertebrate hardparts are known from terrestrial systems. Paleosols are one plausible setting for an "attritional" accumulation. Soil formation takes time, and the duration of the hiatus is the key to accumulation: long-term landscape stability provides the opportunity for vertebrate skeletal debris to gradually accumulate in significant abundance, even from low-density living populations. If skeletal input from the local community exceeds the rate of recycling, bones and teeth may build up to concentrated levels. That said, actualistic data derived from land surface assemblages in the Amboseli Basin of Kenya indicate that even after 10,000 years of attritional input, bone densities would likely still be less than one bone per square meter (Behrensmeyer, 1982).

In a taphonomic study of the Eocene Willwood Formation of Wyoming, Bown and Kraus (1981) identified a specific type of paleosol that was apparently quite conducive to the concentration of vertebrate bioclasts. These authors described numerous concentrations of vertebrate fossils in bluish to greenish gray tabular mudstones that overlie variegated red, purple, and orange mudstones. This repetitive couplet was interpreted as A and B horizons, respectively, of podzolic spodosols (aquods: saturated soils). In the formative scenario, vertebrate elements "accumulated gradually as litter" on alluvial soil surfaces and became incorporated into A horizons presumably through incremental sedimentation and bioturbation. Bown and Kraus (1981) noted a preponderance of teeth and jaws in the Willwood paleosol accumulations, along with an abundance of small and compact postcranial elements such as vertebral centra, carpal and tarsal bones, phalanges, calcanea, and astragali. These relatively durable skeletal elements would presumably be most resistant to physical and chemical degradation at the ground surface and in soil matrices.

Floodbasin ponds and lakes provide another locale for the attritional accumulation of vertebrate bioclasts. Aquatic ecosystems typically support diverse communities of vertebrate animals, including abundant fish, crocodilians, and amphibians. They also tend to attract terrestrial animals to their shores and shallows for purposes of feeding, drinking, and wallowing. Over time, many generations of aquatic, semiaquatic , and terrestrial animals may perish in and around ponds and lakes for a plethora of reasons (senescence, disease, predation), and their skeletal hardparts may in turn contribute to cumulative death assemblages. Whether skeletal elements accrue to concentrated levels would depend on numerous factors, including the density and fecundity of vertebrate populations, and the intensity of biological recycling. The chemical nature of lake and pore waters would also play a role, with high pH and low Eh conditions being most conducive to long-term preservation. A final consideration is the rate at which sediment infills the aquatic basin. Under ideal circumstances for fossilization, sedimentary dilution is minimal (primarily organic), and the aquatic basin persists long enough for skeletal debris to accumulate to appreciable levels.

Rogers (1995) identified 21 microfossil bonebeds that presumably accumulated via such attritional processes within shallow lacustrine basins in the Upper Cretaceous Judith River and Two Medicine Formations of Montana. These subaqueous microfossil bonebeds are characterized by high-diversity assemblages of predominantly small fossils (<5 cm in maximum dimension) representing both aquatic (fish, amphibians, crocodilians, mollusks) and terrestrial (dinosaurs, mammals) animals (e.g., Sahni, 1972; Rogers, 1995). Beds that yield these vertebrate microfossil assemblages are generally tabular, organic rich, and laterally extensive, and skeletal debris can often be traced for tens to hundreds of meters along available exposures. Skeletal hardparts are abundant but typically disseminated throughout these bone-rich horizons (as opposed to densely concentrated in pockets or along bed contacts), and are also thoroughly disarticulated and dissociated, so that it is virtually impossible to conclude confidently that any two skeletal elements are from the same individual. Elements resistant to physical and chemical degradation, such as teeth, scales, scutes, and small dense bones such as vertebral centra and phalanges predominate. Most skeletal elements show only minimal effects of weathering, although the occasional bone does show advanced stages of surface degradation. Rare bones show surface pitting and etching suggestive of chemical dissolution. Ganoid scales and teeth devoid of ganoine and enamel, respectively, are occasionally recovered. This final condition is consistent with processing in crocodilian digestive tracts (Fisher, 1981).

From a taphonomic perspective, attritional accumulations of vertebrate skeletal hardparts, regardless of their sedimentary context, should be time averaged to a greater or lesser degree, depending on the duration of the hiatus, but should date to the period of the hiatus itself (i.e., be a true hiatal concentration, sensu Kidwell [1991], rather than including lag material reworked from older deposits). Local pockets of fossils might preserve census assemblages (e.g., mass mortalities from drowning or other ecologically brief mortality events within a lake), but in general the material should be time averaged. The longer the hiatus, the greater the opportunity for amassing ecologically heterogeneous assemblages (i.e., mixing of noncontemporaneous communities or community states, rather than simply mixing of generations from single communities). The assemblage is more or less autochthonous, with the accumulating bioclasts derived from attritional mortality in the local vertebrate community. The preservational quality of bones and teeth in the assemblage would likely be variable, as elements are added over time. The degree of articulation and element association will generally be low. Associated sediments may show evidence of the hiatus beyond the presence of concentrated vertebrate bioclasts. For example, in the marine realm attritional accumulations are sometimes associated with firmgrounds or hardgrounds and authigenic mineralization. In terrestrial settings, a prolonged hiatus might be marked by an anomalously mature paleosol profile.

Erosion (negative net sedimentation) can preferentially remove silici-clastic sedimentary matrix and leave behind a true lag of larger and/or denser skeletal material exhumed from previous deposits. This erosion-based scenario can occur anywhere that energy impinges upon sediment or lithified strata with a preexisting vertebrate component. Likely locales for the formation of vertebrate lag concentrations include channels (especially cutbanks) within fluvial systems and a variety of high-energy shallow marine settings, especially shorelines and various intertidal and subtidal channels (Wells, 1944; Behrensmeyer, 1982, 1988; Reif, 1982; Smith and Kitching, 1997; Rogers and Kidwell, 2000). Theoretically, the more deeply that erosion cuts into older sediment or strata, the greater the volume of previously deposited vertebrate material that can be intersected and concentrated into lag form. However, in this formative scenario, any cohort of skeletal material would probably undergo a larger number of reworking events than it would during purely aggradational conditions. Thus, erosion itself represents an additional agent of skeletal breakdown and potential transport out of the local system, and exhumation may have little net positive effect on hardpart supply. On the other hand, if the ex-humed material is particularly resilient (e.g., teeth), or is in a prefossilized state (permineralized, see below, and see Smith and Kitching [1997]), then it may endure the rigors of erosion and accumulate in lag form.

In a test of the association between vertebrate skeletal concentrations and erosional surfaces in the Upper Cretaceous Two Medicine and Judith River Formations of Montana, Rogers and Kidwell (2000) found no clear correlation between the abundance of vertebrate skeletal material and the inferred duration of the associated hiatus. The most significant erosional discontinuity in the section (the 80 Ma sequence boundary in the Two Medicine Formation) was virtually barren with regard to vertebrate skele-tal debris, whereas erosion surfaces of lesser extent and duration, ranging from marine flooding surfaces to scours within individual fluvial chan-nels, were frequently mantled by diverse skeletal concentrations. Also, the distribution of vertebrate lag concentrations closely tracked the abun-dance of vertebrate material in underlying and laterally disposed facies, providing evidence that skeletal material was supplied primarily through local erosional reworking with little subsequent lateral dispersion. For example, a prominent flooding surface in the marine portion of the Mon-tana study interval truncates several meters of underlying shoreface strata along its most fossiliferous segment, and the shark teeth, fish and ma-rine reptile bones, and scattered chert pebbles that mantle the surface can also be found, albeit in less concentrated form, in underlying beds. Additional evidence for exhumation consists of reworked steinkerns of burrows, which are found in abundance on the aforementioned ravine-ment surface and are of the shoreface ichnogenera commonly found in underlying beds. The taphonomy of the skeletal debris concentrated on the marine ravinement surface is also consistent with physical reworking, in that the majority of the vertebrate elements exhibit evidence of break-age and abrasion. Analogous examples are concentrations of fish plates along burrowed flooding surfaces within the Ordovician Harding Sand-stone (Allulee and Holland, 2005), and concentrations of turtle scutes, plant material, and fragments of allochthonous iguanodon, pterosaurs, and snakes in small intertidal channels associated with transgression (Neraudeau et al., 2003).

Rogers and Kidwell (2000) also provided evidence for the exhumation and concentration of vertebrate skeletal debris in the nonmarine portion of their study interval, where the recurrent skeletal lags that mantle fluvial erosion surfaces closely track the abundance of skeletal debris in surrounding floodplain facies. In fact, compelling evidence suggests that most, if not all, channel-hosted vertebrate skeletal lags (microfossil bonebeds) include material reworked from preexisting floodplain concentrations, specifically the lacustrine microfossil bonebeds of attritional origin described above. The reworked channel-lag assemblages contain the same array of taxa and the same assortment of elements as the lacustrine assemblages—teeth, vertebrae, phalanges, scutes, and scales—but some elements are polished. Polish is an enigmatic bone-modification feature (Morlan, 1984; Behrens-meyer et al., 1989), and its origin is less than clear. It can presumably be imparted to permineralized ("prefossilized") material upon exhuma-tion and physical abrasion, and an initial period of burial and diagenesis would seem to enhance the likelihood that abrasion will result in pol-ishing rather than rounding alone. The presence of polished bones in the vertebrate concentrations that drape fluvial scours suggests that at least a portion of the skeletal fraction was reworked in a prefossilized condition.

Erosion-generated vertebrate skeletal concentrations should be time averaged to a variable degree, depending on the depth of incision and the age profile of bioclasts preserved in underlying and laterally disposed strata (see Manning [1990] for an extreme example of erosion-related time averaging). In a purely exhumational scenario, the resultant lag concen-tration will be entirely older than the hiatal episode that formed the sur-face. However, skeletal debris can be added to the assemblage both during and after planation or incision. The more severe the downcutting, the greater the potential age differential between material in the lag and the hiatus that concentrated it, and the lower the relevance of its taxonomic composition to paleoenvironmental conditions during the hiatus. Data presented in Rogers and Kidwell (2000) suggest that residual lag concentrations of vertebrate skeletal debris should tend to be autochthonous to parautochthonous in a spatial sense, due to the apparent strong dependence of lag development on preexisting local sources of skeletal material. However, this does not imply autochthony or parautochthony in any ecological sense, because bioclasts can be preserved in a completely different sedimentary context than their in-life habitat.

Taphonomic artifacts of exhumation might include abrasion and rounding (although this modification feature is not necessarily diagnostic), polish (Rogers and Kidwell, 2000), angular as opposed to spiral breakage patterns (Morlan 1984), variable diagenetic signatures (Trueman and Benton, 1997; Trueman, Chapter 7 in this volume), and environmentally mixed assemblages. Sedimentological features consistent with exhumation include stratigraphic evidence of incision and exotic sedimentary matrices embedded within or adhering to exhumed skeletal debris. Care must be exercised with regard to this final potential indicator, however, because fine clays and silt particles moving within a fluvial system may infiltrate bones as draft fills, and contrast markedly with the coarser-grained fraction represented in the bedload.

*Obrution: Ecology and Sedimentology Combined*

The third end-member in the sedimentological bonebed spectrum is the obrution assemblage (sensu Seilacher et al., 1985; Brett, 1990; Bruton, 2001), which in our model is a concentration of vertebrate hardparts that initially formed as a carcass assemblage in response to, and permanently buried by, a single and generally unusual circumstance of sediment deposition. The scenario that most readily comes to mind is that of gregarious animals engulfed by a catastrophic sedimentation event, such as an ash fall, slip-face avalanche, or slump/bank collapse. In this category of bone concentration, the stage is certainly set by group activity (e.g., communal habitation, herding, predator-prey interaction), but geology effectively sets taphonomy in motion by triggering the event of concentrated mortality and permanently burying the result. The obrution scenario is thus a special category of concentration that hinges upon both ecology and sedimentary geology.

Fossorial animals should be particularly susceptible to this mode of mortality because their burrows would serve as sediment sinks, and their aestivation chambers could potentially become permanent tombs if the surface were blanketed by an anomalously thick bed of sediment.

However, there are relatively few reports of mortality horizons comprised of clustered aestivation chambers (e.g., Olson, 1939; Carlson, 1968; Olson and Bolles, 1975; Wood, 1988), and even fewer documented examples of multiple individuals entombed in subterranean dwelling chambers (Voorhies, 1975; Hunt et al., 1983; Abdala et al., 2006). A notable exception was provided by Smith (1993), who described pairs of articulated therapsid skeletons preserved in burrow casts in the Upper Permian Beaufort Group of the Karoo Basin in South Africa. Smith postulated that rare occurrences of intertwined skeletons of the genus *Diictidon* were buried alive when crevasse-splay sands plugged their burrows. Potts (1989) and Potts and others (1999) described a similar occurrence in Pleistocene deposits of the Olorgelailie Formation of Kenya, where four intact hyena skeletons were discovered in an ancient burrow system in association with the scattered bones of other animals.

Catastrophic sedimentation can entomb assemblages of gregarious but nonburrowing animals. One dramatic example is the abundant articulated skeletons of theropod, ankylosaurian, and protoceratopsian dinosaurs, lizards, and mammals entombed within structureless beds of sandstone at the Upper Cretaceous Ukhaa Tolgod locality of the Gobi Desert, Mongolia (Dashzeveg et al., 1995). In a recent analysis of this spectacular locality, Loope et al. (1998, 1999) implicated lethal sandslides triggered by heavy rainfall as the depositional events that both captured and entombed the Late Cretaceous biota in a sand dune landscape. Another classic example of vertebrate mass mortality linked to catastrophic sedimentation is the Poison Ivy Quarry of Nebraska (Voorhies, 1985, 1992). Here, many tens of individuals of the gregarious rhinoceros *Teleoceras* are buried alongside horses, camels, and other animals mostly in articulated and three-dimensional condition, owing apparently to a catastrophic volcanic eruption that blanketed the Miocene landscape with a thick ash. This pyroclastic event apparently killed and buried many or all of the smaller animals almost immediately, while larger animals such as *Teleoceras* died shortly after the onset of the event due to related complications. All specimens in the Poison Ivy Quarry are entombed within either primary or reworked volcanic ash that was blown into a local hollow very shortly after death.

Concentrations of vertebrate skeletal debris that enter the fossil record in response to unusual sedimentation events should show minimal evidence of time averaging, because all individuals in the assemblage presumably perished and were buried simultaneously (e.g., Finch et al., 1972; but see Voorhies, 1985, 1992, for a slight variation on this theme). Under most circumstances the assemblage should also be authochthonous, although some depositional events may transport carcasses a short distance prior to final burial. The preservational quality of specimens should be excellent (e.g., Chiappe et al., 1998; Grellet-Tinner, 2005) unless entombed materials are degraded diagenetically. The degree of articulation and element association should generally be high. The animals represented arguably should be prone to gregarious behavior or at least be likely to interact (e.g., predator-prey associations). Finally, the vertebrate assemblage should be within or immediately beneath a thick or otherwise anomalous sedimentation unit, such as an ash bed or debris flow deposit. Whether this event bed is readily discernible would of course depend on the nature of surrounding sediments, and the extent of postdepositional modification (e.g., bioturbation and pedogenesis).

## CAVEATS AND COMPLICATIONS

The general mechanisms of bone concentration presented here are based on conceptual arguments and empirical observations, and the formative scenarios conform to expected and observed physical and biological parameters—they are intended to be intuitively reasonable. More importantly, every category of biogenic and physical concentration distinguished in our genetic framework can be linked to modern or fossil record-based examples (Table 1.1). Arguably, our conceptual framework of skeletal concentration will accommodate many, if not most, of the carcass/bone concentrations found on modern landscapes (or under the sea) and in the fossil record.

However, we recognize that relative concentrations of vertebrate skeletal debris can develop under circumstances that might not readily relate to our scheme. For example, complex circumstances that simultaneously superpose two or more formative scenarios could arise, such as bone collectors (e.g., packrats, hyenas) trapped and entombed within their bone-laden lair by an event of obrution (e.g., Potts, 1989; Potts et al., 1999). Events can also transpire after initial concentration that obfuscate or compound the original signal. Figure 1.6 outlines a selection of hypothetical pathways that could follow after an intrinsic biogenic concentration is generated. One outcome (path A) leads to permanent burial. In this "simple case" scenario, the sample can be somewhat degraded by taphonomic processes prior to final burial, but the mortality and accumulation signal remain relatively uncomplicated. A compound concentration (path B) might develop if the mass mortality event transpires under conditions of sediment starvation, and the resulting bone assemblage

Table 1.1. Modern and ancient examples of group/focused mortality and bone concentrating scenarios.

| Group/Focused Mortality | Modern Example | Potential Ancient Example |
| --- | --- | --- |
| *Reproduction related* | Bergerund, 1971; Burton, 1985; Anderson, 1990; Maxwell, 1995 | Jordan, 1920; Horner, 1982, 1994; Böttcher, 1990; Baroni and Orombelli, 1994; Wilson, 1996; Chiappe et al., 1998 |
| *Stranding* | Wilkinson and Worthy, 1999; Kellogg and Whitmore, 1957; Geraci et al., 1999 | Camp, 1980; Hogler, 1992 |
| *Miring* | Chamberlin, 1971; Berger, 1983; Mellink and Martin, 2001 | Coope and Lister, 1987; Haynes, 1991; Sander, 1992; Hungerbühler, 1998; Spencer et al., 2003 |
| *Flooding/drowning* | Whitfield and Patterson, 1995; Johanson, 1998; Varricchio et al., 2005 | Kormos, 1911; Parrish, 1978; Hunt et al., 1983; Turnbull and Martill, 1988; Wood et al., 1988; Kahlke and Gaudzinski, 2005 |
| *Wildfire* | Lawrence, 1966; Nelson, 1973, p. 139; Singer et al., 1989 | Sander, 1987; Falcon-Lang, 1998; Zeigler, 2002; Zeigler et al., 2005 |
| *Extreme or unusual weather (rapid thaws, rapid cooling, severe storms)* | Geraci et al., 1999; Berger, 1983; Borrero, 1990; Oliver and Graham, 1994; Jehl, 1996; Sergeant and Williams, 1983; Siegstad and Heide-Jørgensen, 1994 | Voorhies, 1969; Waldman, 1971; Parrish, 1978; Ferber and Wells, 1995 |
| *Drought* | Tulloch, 1970; Corfield, 1973; Tramer, 1977; Coe, 1978; Conybeare and Haynes, 1984; Haynes, 1988; Dudley et al., 2001; Mellink and Martin, 2001 | Matthew, 1924; Huene, 1928; Brown, 1935; Case, 1935; Dalquest and Mamay, 1963; Saunders, 1977; Hulbert, 1982; Rogers, 1990, 2005; Schwartz and Gillette, 1994; Fiorillo et al., 2000 |
| *Sickness, disease, poisoning* | Gunter et al., 1948; Grindley and Taylor, 1962; Brongersma-Sanders et al., 1980; Nyman, 1986; Pybus et al., 1986; Wurtsbaugh and Tapia, 1988; Worthylake and Hovingh, 1989; Thompson and Hall, 1993; Leonardos and Sinis, 1997; Berger et al., 1998 | Martill, 1988; Bell et al., 1989; Leckie et al., 1992; Henrici and Fiorillo, 1993; Emslie and Morgan, 1994; Varricchio, 1995; Emslie et al., 1996; Smith, 2000; Braun and Pfeiffer, 2002 |

| | |
|---|---|
| *Volcanism related* | Brongersma-Sanders, 1957; Taber et al., 1982; Kling et al., 1987; Cotel, 1999 | Anderson, 1933; Voorhies, 1985, 1992; Rolf et al., 1990; Anderson et al., 1995; Brand et al., 2000; Rogers et al., 2001 |
| *Predation* | Mellett, 1974; Andrews and Nesbit Evans, 1983; Haynes, 1988; Hoffman, 1988; Schmitt and Juell, 1994; Terry, 2004; Laudet and Selva, 2005 | Mayhew, 1977; Grande, 1980; Maas, 1985; Wilson, 1987; Pratt, 1989; Murphey, 1996; Davis and Martill, 1999; Northwood, 2005; Nesbitt et al., 2006 |
| **Bone/Carcass Concentration Scenario** | | |
| *Biogenic carcass or bone collection* | Yeager, 1943; Vander Wall, 1990; Kruuk, 1972; Brain, 1980, 1981; Shipman, 1981; Potts, 1986; Skinner et al., 1986; Horwitz and Smith, 1988; Skinner and van Aarde, 1991; Kerbis-Peterhans and Horwitz, 1992 | Hunt, 1990; Haynes, 1991; Sundell, 1999; Palmqvist and Arribas, 2001 |
| *Fluvial hydraulics* | Aslan and Behrensmeyer, 1986; Varricchio et al., 2005 | Voorhies, 1969; Lawton, 1977; Fiorillo, 1991; Eberth and Ryan, 1992; Bandyopadhyay et al., 2002 |
| *Strandline hydraulics* | Weigelt, 1927, 1989; Leggitt and Buchheim, 1997 | Weigelt, 1927, 1989; Leggitt and Buchheim, 1997; Rogers et al., 2001; Zonneveld et al., 2001 |
| *Hiatus/sediment starvation* | Behrensmeyer, 1982, Copenhagen, 1953 | Conkin et al., 1976; Bown and Kraus, 1981; Rogers, 1995; Schröder-Adams et al., 2001 |
| *Erosion/reworking* | Popova, 2004; deflation lags in fossiliferous terrain | Manning, 1990; Smith and Kitching, 1997; Rogers and Kidwell, 2000; Walsh and Martill, 2006 |
| *Ecology plus obrution* | Hayward et al., 1982 | Finch et al., 1972; Voorhies, 1985, 1992; Potts, 1989; Smith, 1993; Chiappe et al., 1998; Loope et al., 1998, 1999; Potts et al., 1999; Grellet-Tinner, 2005 |

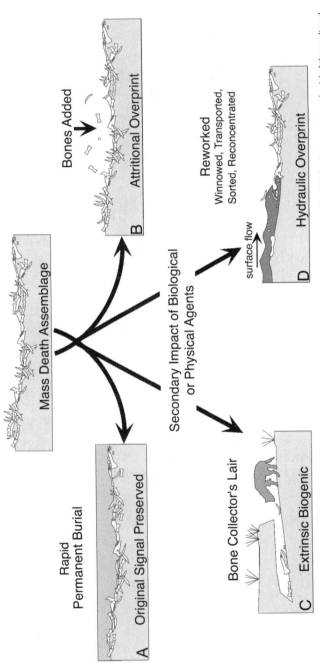

*Figure 1.6.* Hypothetical pathways that could be followed after a mass-death assemblage is generated. One potential outcome (path A) leads to permanent burial of the unaltered assemblage. A compound concentration (path B) might develop if the mass mortality event transpires under conditions of sediment starvation, and the resulting bone assemblage receives an attritional overprint. A preexisting concentration can also be reworked, transported, sorted, and ultimately reconcentrated by either biological (path C) or physical (path D) agents.

receives an attritional overprint. A preexisting concentration can also be reworked, transported, sorted, and ultimately reconcentrated by either biological (path C) or physical (path D) agents. If a relative concentration of vertebrate skeletal debris exhibits a taphonomic signature that does not readily correspond to one of our major formative categories, it should be viewed as a clear warning to consider a suite of more complex formative scenarios.

Finally, we appreciate the fact that a bonebed, at its most basic expression (remains of two individuals in close association), can form by pure happenstance. Two carcasses can certainly come to rest in close proximity for entirely unrelated, unpredictable, and perhaps ultimately undecipherable reasons. However, it is probably unlikely that this type of chance occurrence is common either in modern settings or in the vertebrate fossil record. The framework for vertebrate skeletal concentration developed here is intended to address a straightforward suite of formative scenarios more pertinent to the central theme of skeletal accumulation. Our scenarios should be viewed as general and recurrent mechanisms or agents of vertebrate hardpart concentration (bonebed generators), and we provide criteria that facilitate their recognition.

## SUMMARY AND CONCLUSION

Bonebeds are remarkable features of the vertebrate fossil record. They can embody ancient environmental catastrophes and, thus, reveal some of the most dramatic aspects of past ecosystems. They can also serve as less sensational but equally informative gauges of sedimentary dynamics and biological recycling. Regardless of their particular mode(s) of origin, bonebeds, in their many forms, provide exceptional opportunities to investigate a variety of paleobiological and geological questions.

Here we have proposed a system for the categorization and analysis of bonebeds analogous to the genetic classification scheme proposed by Kidwell et al. (1986) for marine macrobenthic concentrations. Like shell bed counterparts, most vertebrate skeletal concentrations can be readily classed as either biogenic or physical (hydraulic/sedimentologic) in terms of the primary driving process(es), although mixed taphonomic histories are certainly possible (Figs. 1.1 and 1.6).

Two types of biogenic concentrations are recognized in our conceptual treatment. (1) *Intrinsic biogenic concentrations* are the product of the activity or behavior of the hardpart-producing organism(s) itself and are predictable by products of the vertebrate life cycle. In some instances intrinsic biogenic concentrations reflect behaviors that are forced by environmental conditions, such as unusual hazards or perturbations that result in concentrated death assemblages. (2) *Extrinsic biogenic concentrations* are produced by extraneous biological agents, most notably predators.

Two modes of physical concentration are distinguished in our genetic consideration of bonebeds. (1) *Hydraulic concentrations* of carcasses, parts of carcasses, or disarticulated bones and/or teeth form by the action of surface flows (wind, water, sediment) or wave activity. Transport of bioclastic material to the eventual point of concentration is an integral part of this formative scenario. Two primary settings are fluvial channels (including estuarine channels) and strandlines. Numerous factors determine whether relative accumulations of vertebrate hardparts develop in these settings, including the energy and persistence of the hydraulic agent, the amount of bioclastic material delivered to the system, and the presence of trapping mechanisms (e.g., log jams). Notably, many, if not most, multi-individual concentrations that are preserved in ancient fluvial channels or along ancient shorelines were apparently derived from preexisting concentrated sources.

(2) *Sedimentologic concentrations* are the second major category of physical concentration and are the consequences of largely hydraulically controlled variation in rates of inorganic matrix accumulation, that is concentration of bioclasts via conditions of sediment omission (starvation, bypassing), erosion, or sudden intense deposition. An attritional assemblage of vertebrate hardparts initially forms during an episode of sediment starvation or otherwise zero net sediment accumulation (e.g., bypassing), with conditions of low sedimentary dilution fostering relative concentrations of vertebrate material by passive means. Lag concentrations form when erosion (negative sedimentation) preferentially removes sedimentary matrix, leaving behind a true hydraulic (transport-resistant) lag of larger and/or denser skeletal material exhumed from previous deposits. Finally, an obrution concentration forms in response to, and is permanently buried by, a single, generally unusual circumstance of sediment deposition, such as an ash fall, a slip-face avalanche, or a bank collapse acting upon an ecologically aggregated group of individuals. The obrution category is a composite mode of concentration that hinges upon both ecology (the living animals concentrated themselves) and sedimentary geology (the obrution event triggers mortality and ultimately captures the concentration).

The conceptual framework for vertebrate skeletal concentration developed here is intended to facilitate the classification and study of a wide variety of bonebeds. Our categories are intentionally broad in scope, with the hope that they will be applicable under most biostratinomic circumstances. Generalizations that pertain to genesis and expected taphonomic signatures (damage states) and bias levels (degrees of time averaging, fidelity to life habitat) are based on natural history studies, actualistic studies, and stratigraphic investigations, and here much work remains to be done—we certainly do not suggest that our community's understanding is now complete. Our goal is to organize the current understanding of bonebeds and to stimulate a broader conception of their origins and significance from a paleobiological perspective. Ideally, the proposed system of categorization and analysis will serve as a workable template for the study of both modern and ancient bonebeds and will thereby promote a better understanding of the spectacular skeletal concentrations that punctuate the long and diverse history of the vertebrates.

## REFERENCES

Abbott, S.T. 1998. Transgressive systems tracts and onlap shellbeds from mid-Pleistocene sequences, Wanganui Basin, New Zealand. Journal of Sedimentary Research 68:253–268.

Abdala, F., J.C. Cisneros, and R.M.H. Smith. 2006. Faunal aggregation in the Early Triassic Karoo Basin: Earliest evidence of shelter-sharing behavior among tetrapods? Palaios 21:507–512.

Adrain, J.M., and M.V.H. Wilson. 1994. Early Devonian cephalaspids (Vertebrata: Cornuata) from the southern Mackenzie Mountains, N.W.T., Canada. Journal of Vertebrate Paleontology 14:301–319.

Allulee, J.L., and S.M. Holland. 2005. The sequence stratigraphic and environmental context of primitive vertebrates: Harding Sandstone, Upper Ordovician, Colorado, USA. Palaios 20:518–533.

Anderson, D.J. 1990. Evolution of obligate siblicide in boobies. 1. A test of the insurance-egg hypothesis. American Naturalist 135:334–350.

Anderson, D.J. 1995. The role of parents in sibilicidal brood reduction of two booby species. The Auk 112:860–869.

Anderson, D.K., J. Damuth, and T.M. Bown. 1995. Rapid morphological change in Miocene marsupials and rodents associated with a volcanic catastrophe in Argentina. Journal of Vertebrate Paleontology 15:640–649.

Anderson, R.V.V. 1933. The diatomaceous and fish-bearing Beida Stage of Algeria. Journal of Geology 41:673–698.

Andrews, P. 1990. Owls, caves, and fossils. University of Chicago Press, Chicago.

Andrews, P., and E. M. Nesbit Evans. 1983. Small mammal bone accumulations produced by mammalian carnivores. Paleobiology 9:289–307.

Antia, D.D.J. 1979. Bone-beds: A review of their classification, occurrence, genesis, diagenesis, geochemistry, palaeoecology, weathering, and microbiotas. Mercian Geologist 7:93–174.

Argast, S., J.O. Farlow, R.M. Gabet, and D.L. Brinkman. 1987. Transport-induced abrasion of fossil reptilian teeth: Implications for the existence of Tertiary dinosaurs in the Hell Creek Formation, Montana. Geology 15:927–930.

Aslan, A., and A.K. Behrensmeyer. 1996. Taphonomy and time resolution of bone assemblages in a contemporary fluvial system: The East Fork River, Wyoming. Palaios 11:411–421.

Badgley, C. 1986. Taphonomy of mammalian remains from Siwalik rocks of Pakistan. Paleobiology 12:119–142.

Badgley, C., W.S. Bartels, M.E. Morgan, A.K. Behrensmeyer, and S.M. Raza. 1995. Taphonomy of vertebrate assemblages from the Paleogene of northwestern Wyoming and the Neogene of northern Pakistan. Palaeogeography, Palaeoclimatology, Palaeoecology 115:157–180.

Badgley, C., W. Downs, and L.J. Flynn. 1998. Taphonomy of small-mammal fossil assemblages from the Middle Miocene Chinji Formation, Siwalik Group, Pakistan. Pp. 125–166 in Advances in vertebrate paleontology and geochronology. Y. Tomida, L.J. Flynn, and L.L. Jacobs, eds. National Science Museum Monographs, Tokyo.

Baker, J., and D. Brothwell. 1980. Animal diseases in archaeology. Academic Press, London.

Bandyopadhyay, S., T.K. RoyChowdhury, and D.P. Sengupta. 2002. Taphonomy of some Gondwana vertebrate assemblages of India. Sedimentary Geology 147: 219–245.

Banish, L.D., and W.G. Gilmartin. 1992. Pathological findings in the Hawaiian monk seal. Journal of Wildlife Diseases 28:428–434.

Barnes, L.G. 1977. Outline of eastern North Pacific fossil cetacean assemblages. Systematic Zoology 25:321–343.

Barnosky, A.D. 1985. Taphonomy and herd structure of the extinct Irish elk, *Megaloceras giganteus*. Science 228:340–344.

Baroni, C., and G. Orombelli. 1994. Abandoned penguin rookeries as Holocene paleoclimatic indicators in Antarctica. Geology 22:23–26.

Bartels, W.S., T.M. Bown, C. Badgley, A.K. Behrensmeyer, M. Morgan, and S.M. Raza. 1992. Taphonomy of Paleogene and Neogene vertebrate assemblages. Fifth North American Paleontological Convention, Abstracts and Program 6:19.

Behre, C.H. Jr., and J.H. Johnson. 1933. Ordovician and Devonian fish horizons in Colorado. American Journal of Science 25:477–486.

Behrensmeyer, A.K. 1975. The taphonomy and paleoecology of Plio-Pleistocene vertebrate assemblages east of Lake Rudolph, Kenya. Bulletin of the Museum of Comparative Zoology 146:473–578.

Behrensmeyer, A.K. 1978. Taphonomic and ecologic information from bone weathering. Paleobiology 4:150–162.

Behrensmeyer, A.K. 1982. Time resolution in fluvial vertebrate assemblages. Paleobiology 8:211–227.

Behrensmeyer, A.K. 1987. Miocene fluvial facies and vertebrate taphonomy in northern Pakistan. Pp. 169–176 in Recent developments in fluvial sedimentology. Special publication 39. F.G. Ethridge, R.M. Flores, and M.D. Harvey, eds. Society of Economic Paleontology and Mineralogy.

Behrensmeyer, A.K. 1988. Vertebrate preservation in fluvial channels. Palaeogeography, Palaeoclimatology, Palaeoecology 63:183–199.

Behrensmeyer, A.K. 1991. Terrestrial vertebrate accumulations. Pp. 291–335 in Taphonomy: Releasing the data locked in the fossil record. P.A. Allison and D.E.G. Briggs, eds. Plenum Press, New York.

Behrensmeyer, A.K. This volume. Bonebeds through time. Chapter 2 in Bonebeds: Genesis, analysis, and paleobiological significance. R.R. Rogers, D.A. Eberth, and A.R. Fiorillo, eds. University of Chicago Press, Chicago.

Behrensmeyer, A.K., and R.E. Chapman. 1993. Models and simulations of time-averaging in terrestrial vertebrate accumulations. Pp. 125–149 in Taphonomic approaches to time resolution in fossil assemblages. Paleontological Society short courses in paleontology no. 6. S.M. Kidwell and A.K. Behrensmeyer, eds. University of Tennessee, Knoxville.

Behrensmeyer, A.K., and A.P. Hill. 1980. Fossils in the making: Vertebrate taphonomy and paleoecology. University of Chicago Press, Chicago.

Behrensmeyer, A.K., K.D. Gordon, and G.T. Yanagi. 1989. Non-human bone modification in Miocene fossils from Pakistan. Pp. 99–120 in Bone modification. Proceedings of the first international conference on bone modification. R. Bonnichsen and M. Sorg, eds. Center for the Study of the First Americans, Orono, Maine.

Behrensmeyer, A.K., R.S. Hook, C.E. Badgley, J.A. Boy, R.E. Chapman, P. Dodson, R.A. Gastaldo, R.W. Graham, L.D. Martin, P.E. Olsen, R.A. Spicer, R.E. Taggart, and M.V.H. Wilson. 1992. Paleoenvironmental contexts and taphonomic modes. Pp. 15–136 in Terrestrial ecosystems through time: The evolutionary paleoecology of terrestrial plants and animals. A.K. Behrensmeyer, J.D. Damuth, W.A. DiMichelle, R. Potts, H.-D. Sues, and S.L. Wing, eds. University of Chicago Press, Chicago.

Behrensmeyer, A.K., S.M. Kidwell, and R.A. Gastaldo. 2000. Taphonomy and paleobiology. Paleobiology 26:103–144.

Bell, M.A., C.E. Wells, and J.A. Marshall. 1989. Mass mortality of layers of fossil stickleback fish: Catastrophic kills of polymorphic schools. Evolution 43:607–619.

Benton, M.J., E. Cook, D. Grigorescu, E. Popa, and E. Tallódi. 1997. Dinosaurs and other tetrapods in an Early Cretaceous bauxite-filled fissure, northwestern Romania. Palaeogeography, Palaeoclimatology, Palaeoecology 130:275–292.

Berger, J. 1983. Ecology and catastrophic mortality in wild horses: Implications for interpreting fossil assemblages. Science 220:1403–1404.

Berger, L., R. Speare, P. Daszak, D.E. Green, A.A. Cunningham, C.L. Goggin, R. Slocombe, M.A. Ragan, A.D. Hyatt, K.R. McDonald, H.B. Hines, K.R. Lips, G. Marantelli, and H. Parkes. 1998. Chytridiomycosis causes amphibian mortality associated with population declines in the rain forests of Australia and Central America. Proceedings of the National Academy of Sciences 95:9031–9036.

Berger, J., S. Dulamtseren, S. Cain, D. Enkkhbileg, P. Lichtman, Z. Namshir, G. Wingard, and R. Reading. 2001. Back-casting sociality in extinct species: New perspectives using mass death assemblages and sex ratios. Proceedings of the Royal Society of London B 268:131–139.

Bergerund, A.T. 1971. The population dynamics of Newfoundland caribou. Wildlife Monographs 25. Wildlife Society, Louisville, Kentucky.

Betancourt, J. L., T. R. V. Devender, and P. S. Martin. 1990. Packrat middens: The last 40,000 years of biotic change. University of Arizona Press, Tucson.

Blob, R.W. 1997. Relative hydrodynamic dispersal potentials of soft-shelled turtle elements; implications for interpreting skeletal sorting in assemblages of non-mammalian terrestrial vertebrates. Palaios 12:151–164.

Bodkin, J.L., G.R. VanBlaricom, and R.J. Jameson. 1987. Mortalities of kelp-forest fishes associated with large oceanic waves off central California, 1982–1983. Environmental Biology of Fishes 18:73–76.

Borrero, L.A. 1990. Taphonomy of guanaco bones in Tierra del Fuego. Quaternary Research 34:361–371.

Borsuk-Bialynicka, M. 1969. Lower Pliocene rhinocerotids from Altan Teli, western Mongolia. Palaeontologia Polonica 21:73–92

Böttcher, R. 1990. Neue Erkenntnisse über die Fortpflanzungsbiologie der Ichthyosaurier (Reptilia). Stuttgarter Beiträge zur Naturkunde, Serie B (Geologie und Paläontologie) 164:1–51.

Bown, T. M., and M.J. Kraus. 1981. Vertebrate fossil-bearing paleosol units (Willwood Formation, Lower Eocene, Northwest Wyoming, U.S.A.): Implications for taphonomy, biostratigraphy, and assemblage analysis. Palaeogeography, Palaeoclimatology, Palaeoecology 34:31–56.

Brabyn, M.W., and I.G. McLean. 1992. Oceanography and coastal topography of herd-stranding sites for whales in New Zealand. Journal of Mammalogy 73:469–476.

Brady, M., R. Rogers, and B. Sheets. 2005. An experimental and field-based approach to microvertebrate bonebed taphonomy in the Judith River Formation of north-central Montana. Geological Society of America, North-Central Section, Abstracts with Programs 37(5):24.

Brain, C.K. 1980. Some criteria for the recognition of bone-collecting agencies in African caves. Pp. 107–130 in Fossils in the making. A.K. Behrensmeyer and A. Hill, eds. University of Chicago Press, Chicago.

Brain, C.K. 1981. The Hunters or the Hunted? An introduction to African cave taphonomy. University of Chicago Press, Chicago.

Brand, L.R., H.T. Goodwin, P.D. Ambrose, and H.P. Buchheim. 2000. Taphonomy of turtles in the Middle Eocene Bridger Formation, SW Wyoming. Palaeogeography, Palaeoclimatology, Palaeoecology 162:171–189.

Brand, L.R., R. Esperante, A.V. Chadwick, O.P. Porras, and M. Alomia,2004. Fossil whale preservation implies high diatom accumulation rate in the Miocene-Pliocene Pisco Formation of Peru. Geology 32:165–168.

Braun, A., and T. Pfeiffer. 2002. Cyanobacterial blooms as the cause of a Pleistocene large mammal assemblage. Paleobiology 28:139–154.

Breithaupt, B.H., and D. Duvall. 1986. The oldest record of serpent aggregation. Lethaia 19:181–185.

Brett, C.E. 1990. Obrution deposits. Pp. 239–243 in Paleobiology: A synthesis. D.E.G. Briggs and P.R. Crowther, eds. Blackwell Scientific Publications, London.

Brett, C.E. 1995. Sequence stratigraphy, biostratigraphy, and taphonomy in shallow marine environments. Palaios 10:597–616.

Brett, C.E., and G.C. Baird. 1993. Taphonomic approaches to temporal resolution in stratigraphy: Examples from Paleozoic marine mudrocks. Pp. 250–274 in Taphonomic approaches to time resolution in fossil assemblages. Paleontological

Society short courses in paleontology no. 6. S.M. Kidwell and A.K. Behrensmeyer, eds. University of Tennessee, Knoxville.

Brinkman, D.B. 1990. Paleoecology of the Judith River Formation (Campanian) of Dinosaur Provincial Park, Alberta, Canada; evidence from vertebrate microfossil localities. Palaeogeography, Palaeoclimatology, Palaeoecology 7–8:37–54

Brongersma-Sanders, M. 1957. Mass mortality in the sea. Pp. 941–1010 *in* Treatise on marine ecology and paleoecology. Geological Society of America memoir 67. J.W. Hedgepath, ed. Geological Society of America, New York.

Brongersma-Sanders, M., K.M. Stephan, T.G. Kwee, and M. DeBruin. 1980. Distribution of minor elements in cores from the southwest Africa shelf with notes on plankton and fish mortality. Marine Geology 37:91–132.

Brown, B. 1935. Sinclair dinosaur expedition, 1934. Natural History 36:3–15.

Bruton, D.L. 2001. A death assemblage of priapulid worms from the Middle Cambrian Burgess Shale. Lethaia 34:163–167.

Buckland, W. 1823. Reliquiae diluvianae; or, observations on the organic remains contained in caves, fissures, and diluvial gravel, and on other geological phenomena, attesting the action of an universal deluge. John Murray, London.

Burton, R.,1985. Bird behavior. Alfred A. Knopf, New York.

Caldwell, D.K., Caldwell, M.C., and J. Walker. 1970. Mass and individual strandings of False Killer Whales, *Pseudorca crassidens*, in Florida. Journal of Mammalogy 51:634–635.

Camp, C.L. 1980. Large ichthyosaurs from the Upper Triassic of Nevada. Palaeontographica. Abteilung A: Palaeozoologie-Stratigraphie 170:139–200.

Campbell, K.S.W., and M.W. Bell. 1978. A primitive amphibian from the Late Devonian of New South Wales. Alcheringa 1:369–381.

Cantalamessa, G. C. Di Celma, and L Ragaini. 2005. Sequence stratigraphy of the Punta Ballena Member of the Jama Formation (Earlty Pleistocene, Eduador): Insights from integrated sedimentologic, taphonomic, and paleoecologic analysis of molluscan shell concentrations. Palaeogeography, Palaeoclimatology, Palaeoecology 216:1–25.

Capaldo, S.D., and C.R. Peters. 1995. Skeletal inventories from wildebeest drownings at Lakes Masek and Ndutu in the Serengeti ecosystem of Tanzania. Journal of Archaeological Science 22:385–408.

Carlson, K.J. 1968. The skull morphology and aestivation burrows of the Permian lungfish, *Gnathorhiza serrata*. Journal of Geology 76:641–643.

Case, E.C. 1935. Description of a collection of associated skeletons of *Trimerorhachis*. Contributions from the Museum of Paleontology, University of Michigan 4:227–274.

Chamberlin, L.C. 1971. The fabled moose graveyard? Alces 11:1–22.

Charig, A.J., and A.C. Milner. 1997. *Baryonix walkeri*, a fish-eating dinosaur from the Wealden of Surrey. Bulletin of the Natural History Museum 53:11–70.

Chen, P., Z. Dong, and S. Zhen. 1998. An exceptionally well-preserved theropod dinosaur from the Yixian Formation of China. Nature 391:147–152.

Chiappe, L.M., R.A. Coria, L. Dingus, F. Jackson, A. Chinsamy, and M. Fox. 1998. Sauropod dinosaur embryos from the Late Cretaceous of Patagonia. Nature 396: 258–261.

Chiappe, L.M., J.G. Schmitt, G.D. Jackson, A. Garrido, L. Dingus, G. Grellet-Tinner. 2004. Nest structures for sauropods: Sedimentary criteria for recognition of dinosaur nesting traces. Palaios 19:89–95.

Clark, J., J.R. Beerbower, and K.K. Kietzke. 1967. Oligocene sedimentation, stratigraphy and paleoclimatology in the Big Badlands of South Dakota. Fieldiana geology memoir 5. Field Museum of Natural History, Chicago.

Coe, M. 1978. The decomposition of elephant carcasses in the Tsavo (East) National Park, Kenya. Journal of Arid Environments 1:71–86.

Colbert, E.H. 1989. The Triassic dinosaur *Coelophysis*. Bulletin series 57. Museum of Northern Arizona Press, Flagstaff.

Cole, G.F., and D.B. Houston. 1969. An incidence of mass elk drowning. Journal of Mammalogy 50:640–641.

Cole, K.L. 1990. Reconstruction of past desert vegetation along the Colorado River using packrat middens. Palaeogeography, Palaeoclimatology, Palaeoecology 76: 349–366.

Conkin, J.E., B.M. Conkin, and Z.L. Lipchinsky. 1976. Middle Devonian (Hamiltonian) stratigraphy and bone beds on the east side of the Cincinnati Arch in Kentucky. 2. The Kidds Store section, Casey County. Studies in paleontology and stratigraphy 6. University of Louisville, Louisville, Kentucky.

Conkin, J.E., B.M. Conkin, and M.R. Dasari. 1999. Sequential disconformities in the Devonian succession of southern Indiana and northwestern Kentucky. American Association of Petroleum Geologists Bulletin 83:1367.

Conybeare, A., and G. Haynes. 1984. Observations on elephant mortality and bones in water holes. Quaternary Research 22:189–200.

Coombs, M.C., and W.P. Coombs Jr. 1997. Analysis of the geology, fauna, and taphonomy of Morava Ranch Quarry, Early Miocene of northwest Nebraska. Palaios 12:165–187.

Coope, G.R., and A.M. Lister. 1987. Late-glacial mammoth skeletons from Condover, Shropshire, England. Nature 330:472–474.

Cope, J.B., and J.C. Barber. 1978. Caching behavior of screech owls in Indiana. Wilson Bulletin 90:450.

Corfield, T.F. 1973. Elephant mortality in Tsavo National Park, Kenya. East African Wildlife Journal 11:339–368.

Cotel, A.J. 1999. A trigger mechanism for the Lake Nyos disaster. Journal of Volcanology and Geothermal Research 88:343–347

Currie, P.J., and P. Dodson. 1984. Mass death of a herd of ceratopsian dinosaurs. Pp. 61–66 *in* Third symposium on Mesozoic terrestrial ecosystems, short papers. W.E. Reif and F. Westphal, eds. Attempto, Tübingen, Germany.

Curtis, B. 1949. The life story of the fish. Harcourt, Brace and Company, New York.

Dalquest, W.W., and S.H. Mamay. 1963. A remarkable concentration of Permian amphibian remains in Haskell County, Texas. Journal of Geology 71:641–644.

Dashzeveg, D., M.J. Novacek, M.A. Norell, J.M. Clark, L.M. Chiappe, A. Davidson, M.C. McKenna, L. Dingus, C. Swisher, and P. Altangerel. 1995. Extraordinary preservation in a new vertebrate assemblage from the Late Cretaceous of Mongolia. Nature 374:446–449.

Davis, S.P., and D.M. Martill. 1999. The gonorynchiform fish *Dastilbe* from the Lower Cretaceous of Brazil. Palaeontology 42:715–740.

Dechant-Boaz, D. 1982. Modern riverine taphonomy: Its relevance to the interpretation of Plio-Pleistocene hominid paleoecology in the Omo Basin, Ethiopia. Unpublished Ph.D. dissertation, University of California, Berkeley.

del Rio, C.J., S.A. Martinez, and R.A. Scasso. 2001. Nature and origin of spectacular Miocene shell beds of northeastern Patagonia (Argentina); paleoecological and bathymetric significance. Palaios 16:3–25.

Denison, R.H. 1967. Ordovician vertebrates from the western United States. Fieldiana Geology 16:131–192.

Denys, C., and M. Mahboubi. 1992. Altérations structurales et chimiques des éléments squelettiques de pelotes de régurgitation d'un rapace diurne. Bulletin Musée Nationale d'Histoire Naturelle, Paris, 4ᵉ série, 14:229–249.

Denys, C., K. Kowalski, and Y. Dauphin. 1992. Mechanical and chemical alterations of skeletal tissues in a recent Saharian accumulation of faeces from *Vulpes rueppelli* (Carnivora, Mammalia). Acta Zoologica Cracov 35:265–283.

Dietrich, W.E. 1983. Boundary shear stress, sediment transport and bed morphology in a sand-bedded meander during high and low flow. Pp. 632–639 *in* River meandering: Proceedings of the conference on rivers 1983. C.M. Elliott, ed. American Society of Civil Engineers, New York.

DiMichelle, W.A., R.W. Hook, R. Beerbower, J.A. Boy, R.A. Gastaldo, N. Hotton III, T.L. Phillips, S.E. Scheckler, W.A. Shear, and H.-D. Sues. 1992. Paleozoic terrestrial ecosystems, Pp. 205–325 *in* Terrestrial ecosystems through time: The evolutionary paleoecology of terrestrial plants and animals. A.K. Behrensmeyer, J.D. Damuth, W.A. DiMichelle, R. Potts, H.-D. Sues, and S.L. Wing, eds. University of Chicago Press, Chicago.

Dodson, P. 1971. Sedimentology and taphonomy of the Oldman Formation (Campanian), Dinosaur Provincial Park, Alberta (Canada). Palaeogeography, Palaeoclimatology, Palaeoecology 10:21–74.

Dodson, P. 1973. The significance of small bones in paleoecological interpretation. Contributions to Geology, University of Wyoming 12:15–19.

Dodson, P. 1987. Microfaunal studies of dinosaur paleoecology, Judith River Formation of southern Alberta. Pp. 70–75 *in* Fourth symposium on Mesozoic terrestrial ecosystems, short papers. P.M. Currie and E.H. Koster, eds. Royal Tyrrell Museum of Paleontology, Drumheller, Alberta.

Dodson, P., and D. Wexlar. 1979. Taphonomic investigation of owl pellets. Paleobiology 5:275–284.

Dudley, J.P., G.C. Craig, D.S.C. Gibson, G. Haynes, and J. Klimowicz. 2001. Drought mortality of bush elephants in Hwange National Park, Zimbabwe. African Journal of Ecology 39:187–194.

Dudok van Heel, W.H. 1966. Navigation in Cetacea, Pp. 597–606 *in* Whales, dolphins and porpoises. K.S. Norris, ed. University of California Press, Berkeley.

Eastman, C.R. 1911. Jurassic saurian remains ingested within fish. Annals of the Carnegie Museum 8:182–187.

Eaton, T.H. 1964. A captorhinomorph predator and its prey (Cotylosauria). American Museum Novitates 2169:1–3.

Eberth, D.A. 1990. Stratigraphy and sedimentology of vertebrate microfossil sites in the uppermost Judith River Formation (Campanian), Dinosaur Provincial Park, Alberta, Canada. Palaeogeography, Palaeoclimatology, Palaeoecology 7–8:1–36.

Eberth, D.A., and M.A. Getty. 2005. Ceratopsian bonebeds at Dinosaur Provincial Park: Stratigraphy, geology, taphonomy, origins, and significance. Pp. 501–536 in Dinosaur Provincial Park, a spectacular ancient ecosystem revealed. P.J. Currie and E. Kopplehus, eds. Indiana University Press, Bloomington.

Eberth, D.A., and M. Ryan. 1992. Stratigraphy, depositional environments and paleontology of the Judith River and Horseshoe Canyon formations (Upper Cretaceous), southern Alberta, Canada. Field trip guidebook, AAPG/SEPM annual convention, Calgary, Alberta.

Eberth, D.A., B.B. Britt, R. Scheetz, K.L. Stadtman, and D.B. Brinkman. 2006. Dalton Wells: Geology and significance of debris-flow-hosted dinosaur bonebeds in the Cedar Mountain Formation (Lower Cretaceous) of eastern Utah, USA. Palaeogeography, Palaeoclimatology, Palaeoecology 236:217–245.

Eberth, D.A., M. Shannon, and B.G. Noland. This volume. A bonebeds database: Classification, biases, and patterns of occurrence. Chapter 3 in Bonebeds: Genesis, analysis, and paleobiological significance. R.R. Rogers, D.A. Eberth, and A.R. Fiorillo, eds. University of Chicago Press, Chicago.

Economidis, P.S., and V.P. Vogiatzis. 1992. Mass mortality of *Sardinlla aurita* Valenciennes (Pisces, Clupeidae) in Thessaloniki Bay (Macedonia, Greece). Journal of Fish Biology 41:147–149.

Elias, S.A. 1990. Observations on the taphonomy of late Quaternary insect fossil remains in packrat middens of the Chihuahuan Desert. Palaios 5:356–363.

Elles, G.L., and I.L. Slater,1906. The highest Silurian rocks of the Ludlow district. Quarterly Journal of the Geological Society of London 62:195–222.

Emslie, S.D. 1995. Age and taphonomy of abandoned penguin colonies in the Antarctic Peninsula region. Polar Record 31:409–418.

Emslie, S.D., and G.S. Morgan. 1994. A catastrophic death assemblage and paleoclimatic implications of Pliocene seabirds of Florida. Science 264:684–685.

Emslie, S.D., W.D. Allmon, F.J. Rich, J.H. Wrenn, and S.D. de France. 1996. Integrated taphonomy of an avian death assemblage in marine sediments from the late Pliocene of Florida. Palaeogeography, Palaeoclimatology, Palaeoecology 124:107–136.

Emslie, S.D., W. Fraser, R.C. Smith, and W. Walker. 1998. Abandoned penguin colonies and environmental change in the Palmer Station area, Anvers Island, Antarctic Peninsula. Antarctic Science 10:257–268.

Estes, R. 1964. Fossil vertebrates from the Late Cretaceous Lance Formation, eastern Wyoming. University of California Publications in Geological Science 49:1–187.

Estes, R., Z.V. Spinar, and E. Nevo. 1978. Early Cretaceous pipid tadpoles from Israel (Amphibia: Anura). Herpetologica 34:374–393.

Falcon-Lang, H.J. 1998. The impact of wildfire on an Early Carboniferous coastal environment, North Mayo, Ireland. Palaeogeography, Palaeoclimatology, Palaeoecology 139:121–138.

Falcon-Lang, H.J. 1999. Fire ecology of a Late Carboniferous floodplain, Joggins, Nova Scotia. Journal of the Geological Society of London 156:137–148.

Falcon-Lang, H.J. 2003. Growth interruptions in silicified conifer woods from the Upper Cretaceous Two Medicine Formation, Montana, USA: Implications for palaeoclimate and dinosaur palaeoecology. Palaeogeography, Palaeoclimatology, Palaeoecology 199:299–314.

Fara, E., A.Á.F. Saraiva, D. de Almeida Campos, J.K.R. Moreira, D. de Carvalho Siebra, and A.W.A. Kellner. 2005. Controlled excavations in the Romualdo Member of the Santana Formation (Early Cretaceous, Araripe Basin, northeastern Brazil): Stratigraphic, palaeoenvironmental, and palaeoecological implications. Palaeogeography, Palaeoclimatology, Palaeoecology 218:145–160.

Fay, F.H., and B.P. Kelly. 1980. Mass natural mortality of walruses (*Odobenus rosmarus*) at St. Lawrence Island, Bering Sea, autumn 1978. Arctic 33:226–245.

Ferber, C.T., and N.A. Wells. 1995. Paleolimnology and taphonomy of some fish deposits in "Fossil" and "Uinta" Lakes of the Eocene Green River Formation, Utah and Wyoming. Palaeogeography, Palaeoclimatology, Palaeoecology 117:185–210.

Fernandez-Lopez, S.R., M.H. Henriques, and L.V. Duarte. 2002. Taphonomy of ammonite condensed associations—Jurassic examples from carbonate platforms of Iberia. Abhandlungen Geologischen Bundesanstalt 57:423–430.

Finch, W.I., F.C. Whitmore Jr., and J.D. Sims. 1972. Stratigraphy, morphology, and paleoecology of a fossil peccary herd from western Kentucky. United States Geological Survey professional paper 790.

Finney, B.P., I. Gregory-Eaves, M.S.V. Douglas, and J.P. Smol. 2002. Fisheries productivity in the northeastern Pacific Ocean over the past 2,200 years. Nature 416:729–733.

Fiorillo, A.R. 1988. Taphonomy of Hazard Homestead Quarry (Ogallala Group), Hitchcock County, Nebraska. Contributions to Geology, University of Wyoming 26:57–97.

Fiorillo, A.R. 1991. Taphonomy and depositional setting of Careless Creek Quarry (Judith River Formation), Wheatland County, Montana, U.S.A. Palaeogeography, Palaeoclimatology, Palaeoecology 81:281–311.

Fiorillo, A.R., K. Padian, and C. Musikasinthorn,2000. Taphonomy and depositional setting of the *Placerias* Quarry (Chinle Formation: Late Triassic, Arizona). Palaios 15:373–386.

Fisher, D. 1981. Crocodilian scatology, microvertebrate concentrations, and enamelless teeth. Paleobiology 7:262–275.

Forcada, J., P.S. Hammond, and A. Aguilar. 1999. Status of the Mediterranean monk seal *Monachus monachus* in the western Sahara and the implications of a mass mortality event. Marine Ecology Progress Series 188:249–261.

Fraser, N.C., and G.M. Walkden. 1983. The ecology of a Late Triassic reptile assemblage from Gloucestershire, England. Palaeogeography, Palaeoclimatology, Palaeoecology 42:341–365.

Freeman, M.M.R. 1968. Winter observations on beluga whales (*Delphinapterus leucas*) in Jones Sound, N.W.T. Canadian Field Naturalist 82:276–286.

Frison, G.C. 1974. The Casper Site. Academic Press, New York.

Frison, G.C., and L.C. Todd. 1986. The Colby Mammoth Site: Taphonomy and archaeology of a Clovis kill in northern Wyoming. University of New Mexico Press, Albuquerque.

Furness, R.W., and P. Monaghan. 1987. Seabird ecology. Chapman and Hall, New York.

Fürsich F.T., and W. Oschmann. 1993. Shell beds as tools in basin analysis: The Jurassic of Kachchh, western India. Journal of the Geological Society of London 150:169–185.

Fürsich, F.T., and D.K. Pandey. 2003. Sequence stratigraphic significance of sedimentary cycles and shell concentrations in the Upper Jurasic–Lower Cretaceous of Kachchh, western India. Palaeogeography, Palaeoclimatology, Palaeoecology 193: 285–309.

Gadbury, C., L. Todd, A.H. Jahren, and R. Amundson. 2000. Spatial and temporal variations in the isotopic composition of bison tooth enamel from the Early Holocene Hudson-Meng Bone Bed, Nebraska. Palaeogeography Palaeoclimatology Palaeoecology 157:79–93

Gates, T.A. 2005. The Late Jurassic Cleveland-Lloyd dinosaur quarry as a drought-induced assemblage. Palaios 20:363–375.

Gau, K., and N. Shubin. 2000. Late Jurassic salamanders from northern China: Phylogenetic and biogeographic implications. Journal of Vertebrate Paleontology 20(supplement to 3):43A.

Geraci, J.R., and V.J. Lounsbury. 1993. Marine mammals ashore: A field guide for strandings. Texas A&M University Sea Grant College Program, Galveston.

Geraci, J.R., J. Harwood, and V.J. Lounsbury. 1999. Marine mammal die-offs. Pp. 367–395 in Conservation and management of marine mammals. J.R. Twiss Jr. and R.R. Reeves, eds. Smithsonian Institution Press, Washington, D.C.

Gillespie, J.L., C.S. Nelson, and S.D. Nodder. 1998. Post-glacial sea-level control and sequence stratigraphy of carbonate-terrigenous sediments, Wanganui Shelf, New Zealand. Sedimentary Geology 122:245–266.

Gillson, L. 2006. A "large infrequent disturbance" in an East African savanna. African Journal of Ecology 44:458–467.

Graham, R.W., V.V. Haynes, D.L. Johnson, and M. Kay. 1981. Kimmswick: A Clovis-mastodon association in eastern Missouri. Science 213:1115–1117.

Grande, L. 1980. Paleontology of the Green River Formation with a review of the fish fauna. Bulletin of the Geological Survey of Wyoming 63:1–333.

Grayson, D.K. 1984. Quantitative zooarchaeology: Topics in the analysis of archaeological faunas. Academic Press, Orlando.

Grellet-Tinner, G. 2005. Membrana testacea of titanosaurid dinosaur eggs from Auca Mahuevo (Argentina): Implications for exceptional preservation of soft tissue in lagerstätten. Journal of Vertebrate Paleontology 25:99–106.

Grindley, J.R., and F.J.R. Taylor. 1962. Red water and mass mortality of fish near Cape Town. Nature 195:1324.

Gunter, G. 1941. Death of fishes due to cold on the Texas coast, January, 1940. Ecology 22:203–208.

Gunter, G. 1947. Catastrophism in the sea and its paleontological significance, with special reference to the Gulf of Mexico. American Journal of Science 245:46–67.

Gunter, G., R.H. Williams, C.C. Davis, and F.G.W. Smith. 1948. Catastrophic mass mortality of marine animals and coincident phytoplankton bloom on the west coast of Florida, November 1946 to August 1947. Ecological Monographs 18: 309–324.

Hanson, C.B. 1980. Fluvial taphonomic processes: Models and experiments. Pp. 156–181 *in* Fossils in the making. A.K. Behrensmeyer and A.P. Hill, eds. University of Chicago Press, Chicago.

Hatcher, J.B. 1901. The Jurassic dinosaur deposits near Canyon City, Colorado. Annals of the Carnegie Museum 1:327–341.

Hauff, B. 1921. Untersuching der Fossilifundstätten von Holzmaden im Posidonien-schiefer des oberen Lias Württembergs. Palaeontographica 64:1–42.

Haynes, G. 1988. Mass deaths and serial predation: Comparative taphonomic studies of modern large mammal death sites. Journal of Archaeological Science 15: 219–235.

Haynes, G. 1991. Mammoths, mastodons and elephants: Biology, behavior, and the fossil record. Cambridge University Press, Cambridge.

Hayward, J.L., D.E. Miller, and C.R. Hill. 1982. Mount St. Helens ash: Its impact on breeding ring-billed and California gulls. The Auk 99:623–631.

Henrici, A.C., and A.R. Fiorillo. 1993. Catastrophic death assemblage of *Chelomophrynus bayi* (Anura: Rhinophrynidae) from the middle Eocene Wagon Bed Formation of central Wyoming. Journal of Paleontology 67:1016–1026.

Hillman, J.C., and A.K.K. Hillman. 1977. Mortality of wildlife in Nairobi National Park during the drought of 1973–1974. East African Wildlife Journal 15:1–18.

Hocket, B.S. 1989. The concept of "carrying range": A method for determining the role played by woodrats in contributing bones to archaeological site. Nevada Archaeologist 7:28–35.

Hoffman, R. 1988. The contribution of raptorial birds to patterning in small mammal assemblages. Paleobiology 14:81–90.

Hogler, J.A. 1992. Taphonomy and paleoecology of *Shonisaurus popularis* (Reptilia: Ichthyosauria). Palaios 7:108–117.

Holliday, V.T., C.V. Haynes Jr., J.L. Hoffman, and D.J. Meltzer. 1994. Geoarchaeology and geochronology of the Miami (Clovis) Site, southern High Plains of Texas. Quaternary Research 41:234–244.

Hook, R., and J.C. Ferm. 1988. Paleoenvironmental controls on vertebrate-bearing abandoned channels in the Upper Cretaceous. Palaeogeography, Palaeoclimatology, Palaeoecology 63:159–181.

Hopley, P.J. 2001. Plesiosaur spinal pathology: The first fossil occurrence of Schmorl's nodes. Journal of Vertebrate Paleontology 21:253–260.

Horner, J.R. 1982. Evidence of colonial nesting and site fidelity among orhithischian dinosaurs. Nature 297:675–676.

Horner, J.R. 1994. Comparative taphonomy of some dinosaur and extant bird colonial nesting grounds. Pp. 116–123 *in* Dinosaur eggs and babies. K. Carpenter, K.F. Hirsch, and J.R. Horner, eds. Cambridge University Press, Cambridge, U.K.

Horner, J.R., and R. Makela. 1979. Nest of juveniles provides evidence of family structure among dinosaurs. Nature 282:296–298.

Horwitz, L.K., and P. Smith. 1988. The effects of striped hyaena activity on human remains. Journal of Archaeological Science 15:471–481.

Hu, Y., J. Meng, Y. Wang, and C. Li. 2005. Large Mesozoic mammal fed on young dinosaurs. Nature 433:149–152.

Huber, H.R., A.C. Rovetta, L.A. Fry, and S. Johnston. 1991. Age-specific natality of northern elephant seals at the South Farallon Islands, California. Journal of Mammalogy 72:525–534.

Hückel, U. 1970. Fossil-Lagerstaetten, Nr. 7; Die Fischschiefer von Haqel und Hjoula in der Oberkreide des Libanon. Neues Jahrbuch fuer Geologie und Palaeontologie. Abhandlungen 135:113–149.

Huene, F. von,1928. Lebensbild des Saurischier-Vorkommens im obersten Keuper von Trossingen in Württemberg. Palaeobiologica 1:103–116.

Hulbert, R.C. Jr. 1982. Population dynamics of the three-toed horse *Neohipparion* from the Late Miocene of Florida. Paleobiology 8:159–167.

Hungerbühler, A. 1998. Taphonomy of the prosauropod dinosaur *Sellosaurus*, and its implications for carnivore faunas and feeding habits in the Late Triassic. Palaeogeography, Palaeoclimatology, Palaeoecology 143:1–29.

Hunt, A.P. 1986. Taphonomy of the Cleveland-Lloyd Quarry, Morrison Formation (Late Jurassic), Emery County, Utah: A preliminary report. Fourth North American Paleontological Convention, Boulder, Colorado, A21.

Hunt, R.M. Jr. 1990. Taphonomy and sedimentology of Arikaree (lower Miocene) fluvial, eolian, and lacustrine paleoenvironments, Nebraska and Wyoming: A paleobiota entombed in fine-grained volaniclastic rocks. Pp. 69–11 *in* Volcanism and fossil biotas. Special paper 244. M.G. Lockley and A. Rice, eds. Geological Society of America, Boulder, Colorado.

Hunt, R.M. Jr., X.-X. Xu, and J. Kaufman. 1983. Miocene burrows of extinct bear dogs: Indication of early denning behavior of large mammalian carnivores. Science 221:364–366.

Jehl, J.R. Jr. 1996. Mass mortality of eared grebes in North America. Journal of Field Ornithology 67:471–476.

Johanson, Z. 1998. The Upper Devonian fish *Bothriolepis* (Placodermi, Antiarchi) from near Canowindra, New South Wales, Australia. Records of the Australian Museum 50:315–348.

Johnson, R.G. 1960. Models and methods for analysis of the mode of formation of fossil assemblages. Bulletin of the Geological Society of America 71:1075–1085.

Johnson, B.W., and P.A. Johnson. 1981. The Hawaiian monk seal on Laysan Island: 1978. Report no. MMC-78/15, Marine Mammal Commission, U.S. Department of Commerce.

Jones, T.P. 1997. Fusain in Late Jurassic sediments from the Witch Grond Graben, North Sea, U.K. Mededelingen Nederlands Instituut voor Toegepaste Geowetenschappen TNO 58:93–103.

Jordan, D.S. 1920. A Miocene catastrophe. Natural History 20:18–22.

Kahlke, R.-D., and S. Gaudzinski. 2005. The blessing of a great flood: Differentiation of mortality patterns in the large mammal record of the Lower Pleistocene fluvial site of Untermassfeld (Germany) and its relevance for the interpretation of faunal assemblages from archaeological sites. Journal of Archaeological Science 32:1202–1222.

Kellogg, R., and F.C. Whitmore Jr. 1957. Marine mammals; annotated bibliography. Pp. 1021–1024 *in* Ecology, vol. 1 of Treatise on marine ecology and paleoecology.

Geological Society of America memoir 67. J.W. Hedgpeth, ed. Memoir, Geological Society of America, New York.

Kerbis-Peterhans, J.C., and L.K. Horwitz. 1992. A bone assemblage from a striped hyaena (*Hyaena hyaena*) den in the Negev Desert, Israel. Israel Journal of Zoology 37:225–245.

Khajuria, C.K., and G.V.R. Prasad. 1998. Taphonomy of a Late Cretaceous mammal bearing microvertebrate assemblage from the Deccan inter-trappean beds of Naskal, peninsular India. Palaeogeography, Palaeoclimatology, Palaeoecology 137:153–172.

Kidwell, S.M. 1986. Models for fossil concentrations: Paleobiologic implications. Paleobiology 12:6–24.

Kidwell, S.M. 1989. Stratigraphic condensation of marine transgressive records: Origin of major shell deposits in the Miocene of Maryland. Journal of Geology 97:1–24.

Kidwell, S.M. 1991a. The stratigraphy of shell concentrations. Pp. 211–290 *in* Taphonomy: Releasing the data locked in the fossil record. Topics in Geobiology 9. P.A. Allison and D.E.G. Briggs, eds. Plenum Publishing, New York.

Kidwell, S.M. 1991b. Condensed deposits in siliciclastic sequences: Expected and observed features. Pp. 682–695 *in* Cycles and events in stratigraphy. G. Einsele, W. Ricken, and A. Seilacher, eds. Springer Verlag, Berlin.

Kidwell, S.M. 1993. Taphonomic expressions of sedimentary hiatuses: Field observations on bioclastic concentrations and sequence anatomy in low, moderate and high subsidence settings. Geologische Rundschau 82:189–202.

Kidwell, S.M., F.T. Fürsich, and T. Aigner. 1986. Conceptual framework for the analysis and classification of fossil concentrations. Palaios 1:228–238.

Kirschvink, J.L., A.E. Dizon, and J.A. Westphal. 1986. Evidence from strandings for geomagnetic sensitivity in cetaceans. Journal of Experimental Biology 120:1–24.

Kling, G.W., M. Clark, H.R. Compton, J.D. Devine, W.C. Evans, A.M. Humphrey, J.P. Lockwood, and M.L. Tuttle. 1987. The 1986 Lake Nyos gas disaster, Cameroon, West Africa. Science 236:169–175.

Klinowska, M. 1985. Cetacean live stranding sites relate to geomagnetic topography. Aquatic Mammals 1:27–32.

Klinowska, M. 1986. Cetacean live stranding dates relate to geomagnetic disturbances. Aquatic Mammals 11:109–119.

Kondo Y., S.T. Abbott, A. Kitamura, P.J.J. Kamp, T.R. Naish, T. Kamataki, and G.S. Saul. 1998. The relationship between shellbed type and sequence architecture: Examples from Japan and New Zealand. Sedimentary Geology 122:109–128.

Kormos, T. 1911. Der Pliozäne Knochenfund bei Polgardi. Földtani Közlöni 41:48–64

Korth, W.W. 1979. Taphonomy of microvertebrate fossil assemblages. Annals of the Carnegie Museum 48:235–285.

Koster, E.H. 1987. Vertebrate taphonomy applied to the analysis of ancient fluvial systems. Pp. 159–168 *in* Recent developments in fluvial sedimentology. Special publication 39. F.G. Ethridge, R.M. Flores, and M.D. Harvey, eds. Society of Economic Paleontologists and Mineralogists.

Kruuk, H. 1972. The spotted hyena: A study of predation and social behavior. University of Chicago Press, Chicago.

Kuhk, R. 1956. Hagelunwetter als Verlustursache bei Störchen und anderen Vögeln. Vogelwarte 18:180–182.

Kurtén, B. 1953. On the variation and population dynamics of fossil and recent mammal populations. Acta Zoologica Fennica 76:5–122.

Kusmer, K.D. 1990. Taphonomy of owl pellet deposition. Journal of Paleontology 64:629–637.

Lam, Y.M. 1992. Variability in the behavior of spotted hyaenas as taphonomic agents. Journal of Archaeological Science 19:389–406.

Laudet, F., and N. Selva. 2005. Ravens as small mammal bone accumulators: First taphonomic study on mammal remains in raven pellets. Palaeogeography, Palaeoclimatology, Palaeoecology 226:272–286.

Lawrence, G.E. 1966. Ecology of vertebrate animals in relation to chaparral fire on the Sierra Nevada foothills. Ecology 47:278–291

Lawton, R. 1977. Taphonomy of the dinosaur quarry, Dinosaur National Monument. Contributions to Geology, University of Wyoming 15:119–126.

Le Boeuf, B.J., and K.T. Briggs. 1977. The cost of living in a seal harem. Mammalia 41:167–195.

Le Boeuf, B.J., and R.S. Condit. 1983. The high cost of living on the beach. Pacific Discovery 36:12–14.

Le Boeuf, B.J., and R.M. Laws. 1994. Elephant seals: Population ecology, behavior, and physiology. University of California Press, Berkeley.

Leckie, D.A., C. Singh, J. Bloch, M. Wilson, and J. Wall. 1992. An anoxic event at the Albian-Cenomonian boundary: The Fish Scale Marker Bed, northern Alberta, Canada. Palaeogeography, Palaeoclimatology, Palaeoecology 92:139–166.

Le Rock, J.W. 2000. Sedimentology and taphonomy of a dinosaur bonebed from the Upper Cretaceous (Campanian) Judith River Formation of north central Montana. Unpublished masters thesis, Montana State University, Bozeman, Montana.

Leggitt, V.L., and H.P. Buchheim. 1997. Bird bone taphonomic data from Recent lake margin strandlines compared with an Eocene *Presbyornis* (Aves: Anseriformes) bone strandline. Geological Society of America, Abstracts with Programs 29:105.

Leonardos, I., and A. Sinis. 1997. Fish mass mortality in the Etolikon Lagoon, Greece: The role of local geology. Cybium 21:201–206.

Lisle, T.E. 1976. Components of flow resistance in a natural channel. Unpublished Ph.D. thesis, University of California, Berkeley.

Loope, D.B., L. Dingus, C.C. Swisher III, and C. Minjin. 1998. Life and death in a Late Cretaceous dune field, Nemegt basin, Mongolia. Geology 26:27–30.

Loope, D.B., J.A. Mason, and L. Dingus. 1999. Lethal sandslides from eolian dunes. Journal of Geology 107:707–713.

Lubinski, P.M., and C.J. O'Brien. 2001. Observations on seasonality and mortality from a Recent catastrophic death assemblage. Journal of Archaeological Science 28:833–842.

Lyman, R.L. 1994. Vertebrate taphonomy. Cambridge University Press, Cambridge.

Maas, M.C. 1985. Taphonomy of a Late Eocene microvertebrate locality, Wind River Basin, Wyoming (U.S.A.). Palaeogeography, Palaeoclimatology, Palaeoecology 52:123–142.

Macquaker, J.H.S. 1994. Palaeoenvironmental significance of "bone-beds" in organic-rich mudstone successions: An example from the Upper Triassic of south-west Britain. Zoological Journal of the Linnean Society 112:285–308.

Maguire, J.M., D. Pemberton, and M.H. Collett. 1980. The Makapansgat Limeworks Grey Breccia: Hominids, hyaenas, hystricids or hillwash? Paleontologia Africana 23:75–98.

Mandel, R.D., and J.L. Hofman. 2003.Geoarchaeological investigations at the Winger site: A late Paleoindian bison bonebed in southwestern Kansas, USA. Geoarchaeology, an International Journal 18:129–144

Mandic, O., and W.E. Piller. 2001. Pectinid coquinas and their palaeoenvironmental implications; Examples from the early Miocene of northeastern Egypt. Palaeogeography, Palaeoclimatology, Palaeoecology 172:171–191.

Mandic, O., M. Harzhauser, and R. Roetzel. 2004. Taphomomy and sequence stratigraphy of spectacular shell accumulations form the type stratum of the Central Paratethys stage Eggenburgian (Lower Miocene, NE Austria). Courier Forschungs-Institut Senckenberg (Frankfurt) 246:69–88.

Manning, E. 1990. The late early Miocene Sabine River. Transactions Gulf Coast Association of Geological Societies 40:531–549.

Marsh, R., B. Petrie, C.R. Weidman, R.R. Dickson, J.W. Loder, C.G. Hannah, K. Frank, and K. Drinkwater. 1999. The 1882 tilefish kill: A cold event in shelf waters off the north-eastern United States? Fisheries Oceanography 8:39–49.

Martill, D.M. 1988. Preservation of fish in the Cretaceous Santana Formation of Brazil. Palaeontology 31:1–18.

Martill, D.M. 1991. Bones as stones: The contribution of vertebrate remains to the lithologic record. Pp. 270–292 in The processes of fossilization. S.K. Donovan, ed. Columbia University Press, New York.

Matthew, W.D. 1923. Fossil bones in the rock. Natural History 23:358–369.

Matthew, W.D. 1924. Third contribution to the Snake Creek fauna. Bulletin of the American Museum of Natural History 50:61–210.

Maxwell, J. 1995. Swimming with the salmon. Natural History (September):26–39.

Mayhew, D.F. 1977. Avian predators as accumulators of fossil material. Boreas 6:25–31.

McDaniel, J.D., and S.D. Emslie. 2002. Fluctuations in Adélie penguin prey size in the mid to late Holocene, northern Marguerite Bay, Antarctic Peninsula. Polar Biology 25:618–623.

McEachron, L.W., G.C. Matlock, C.E. Bryan, P. Unger, T.J. Cody, and J.H. Martin. 1994. Winter mass mortality of animals in Texas bays. Northeast Gulf Science 13:121–138.

McGrew, P.O. 1975. Taphonomy of Eocene fish from Fossil Basin, Wyoming. Fieldiana Geology 33:257–270.

McHugh, T. 1972. The time of the buffalo. University of Nebraska Press, Lincoln.

McKenna, M.C. 1962. Collecting small fossils by washing and screening. Curator 5:221–235.

Mead, J.I., T.R. Van Devender, and K.L. Cole. 1983. Late Quaternary small mammals from Sonoran Desert packrat middens, Arizona and California. Journal of Mammalogy 64:173–180.

Mellet, J.S. 1974. Scatological origins of microvertebrate fossils. Science 185:349–350.

Mellink, E., and P.S. Martin. 2001. Mortality of cattle on a desert range: Paleobiological implications. Journal of Arid Environments 49:671–675.

Meltzer, D.J., L.C. Todd, and V.T. Holliday. 2002. The Folsom (Paleoindian) type site: Past investigations, current studies. American Antiquity 67:5–36.

Meng, Q., J. Liu, D.J. Varricchio, T. Huang, and C. Gao. 2004. Paleontology: Parental care in an ornithischian dinosaur. Nature 431:145–146.

Morlan, R.E. 1984. Toward the definition of criteria for the recognition of artificial bone alterations. Quaternary Research 22:160–171.

Müller-Schwarze, D. 1984. The behavior of penguins. State University of New York Press, Albany.

Murphey, P.C. 1996. Depositional setting and fauna on the *Omomys* Quarry, a possible owl site in the Bridger Formation (Middle Eocene) of Southwestern Wyoming. Journal of Vertebrate Paleontology 16(supplement to 3):55A.

Myrick, A.C. 1979. Variation, taphonomy, and adaptation of the Rhabdosteidae (Eurinodelphidae) (Odontoceti, Mammalia) from the Calvert Formation of Maryland and Virginia. Unpublished Ph.D. dissertation, University of California, Los Angeles.

Naish, T., and P.J.J. Kamp. 1997. Sequence stratigraphy of sixth-order (41 k.y.) Pliocene-Pleistocene cyclothems, Wanganui Basin, New Zealand; a case for the regressive systems tract. Geological Society of America Bulletin 109:978–999.

Nelson, J.G. 1973. The last refuge. Harvest House, Montreal.

Neraudeau D., R. Allain, V. Perrichot, B. Videt, F.L. de Broin, F. Guillochaeu, M. Philippe, J.-C. Rage, and R. Vullo. 2003. A new Early-Cenomanian paralic deposit with fossil wood, amber with insects and Iguanodontidae (Dinosauria, Ornithopoda) at Fouras (Charente-Maritime, southwestern France). Comptes Rendus Palevol 2(3): 221–230.

Nesbitt, S.J., A.H. Turner, G.M. Erickson, and M.A. Norell. 2006. Prey choice and cannibalistic behaviour in the theropod *Coelophysis*. Biology Letters 2(4):611–614.

Nevo, E. 1968. Pipid frogs from the Early Cretaceous of Israel and pipid evolution. Bulletin of the Museum of Comparative Zoology 136:255–318.

Norris, R.D. 1986. Taphonomic gradients in shelf fossil assemblages. Pliocene Purisma Formation, California. Palaios 1:252–266.

Northwood, C. 2005. Early Triassic coprolites from Australia and their palaeobiological significance. Palaeontology 48:49–68.

Nriagu, J.O. 1983. Rapid decomposition of fish bones in Lake Erie sediments. Hydrobiologica 106:217–222.

NWHC. 1998. USGS National Wildlife Health Center quarterly mortality reports: Fourth quarter.

NWHC. 2001. USGS National Wildlife Health Center quarterly mortality reports: Fourth quarter.

Nyman, S. 1986. Mass mortality of larval *Rana sylvatica* attributable to the bacterium, *Aeromonas hydrophila*. Journal of Herpetology 20:196–201.

Odell, D.K., E.D. Asper, J. Baucom, and L.H. Cornell. 1980. A recurrent mass stranding of the False Killer Whale, *Pseudorca crassidens*, in Florida. Fishery Bulletin 78:171–177.

Oliver, J.S., and R.W. Graham. 1994. A catastrophic kill of ice-trapped coots: Time averaged versus scavenger-specific disarticulation patterns. Paleobiology 20:229–244.

Olson, E.C. 1939. The fauna of the *Lysorophus* pockets in the Clear Fork Permian, Baylor County, Texas. Journal of Geology 46:389–397.

Olson, E.C., and K. Bolles. 1975. Permo-Carboniferous freshwater burrows. Fieldiana Geology 33:271–290.

Ostrom, J.H. 1978. The osteology of *Compsognathus longipes*. Wagner Zitteliana 4:73–118.

Ovsyanikov, N.G., L.L. Bove, and A.A. Kochnev. 1994. Walrus mortality causes on hauling-out grounds. Zoologicheskij Zhurnal 73:80–87.

Palmqvist, P., and A. Arribas, A. 2001. Taphonomic decoding of the paleobiological information locked in a lower Pleistocene assemblage of large mammals. Paleobiology 27:512–530.

Parras, A., and S. Casadio. 2005. Taphonomy and sequence stratigraphic significance of oyster-dominated concentrations from the San Julian formation, Oligocene of Patagonia, Argentina. Palaeogeography, Palaeoclimatology, Palaeoecology 217:47–66.

Parrish, W.C. 1978. Palaeoenvironmental analysis of a Lower Permian bonebed and adjacent sediments, Wichita County, Texas. Palaeogeography, Palaeoclimatology, Palaeoecology 24:209–237.

Pedley, H.M. 1978. A new fish horizon from the Maltese Miocene and its palaeoecological significance. Palaeogeography, Palaeoclimatology, Palaeoecology 24:73–83.

Peng, J., A.P. Russell, and D.B. Brinkman. 2001. Vertebrate microsite assemblages (exclusive of mammals) from the Foremost and Oldman Formations of the Judith River Group (Campanian) of southeastern Alberta: An illustrated guide. Natural history occasional paper no. 25. Provincial Museum of Alberta, Edmonton.

Perea, D., M. Ubilla, A. Rojas, and C.A. Goso. 2001. The West Gondwanan occurrence of the hybodontid shark Priohybodus, and the Late Jurassic-Early Cretaceous age of the Tacuarembo Formation, Uruguay. Palaeontology 44:1227–1235.

Peterson, O.A. 1906. The Agate Spring fossil quarry. Annals of the Carnegie Museum 3:487–494.

Pitelka, F.A., P.Q. Tomich, and G.W. Treichel. 1955. Ecological relations of jaegers and owls as lemming predators near Barrow, Alaska. Ecological Monographs 25:85–117.

Popova, L.V. 2004. The micromammal fauna of the Dnieper modern channel alluvium: Taphonomic and biostratigraphic implications. Quaternaire 15:233–242.

Potts, R. 1986. Temporal span of bone accumulation at Olduvai Gorge and implications for early hominid foraging behavior. Paleobiology 12:25–31.

Potts, R. 1989. Olorgesailie: New excavations and findings in Early and Middle Pleistocene contexts, southern Kenya rift valley. Journal of Human Evolution 18:477–484.

Potts, R., A.K. Behrensmeyer, and P. Ditchfield. 1999. Paleolandscape variation and early Pleistocene hominid activities: Members 1 and 7, Olorgesailie Formation. Journal of Human Evolution 37:747–788.

Pough, F.H., C.M. Janis, and J.B. Heiser. 2005. Vertebrate life, seventh edition. Prentice-Hall, Upper Saddle River, New Jersey.

Pratt, A.E. 1989. Taphonomy of the microvertebrate fauna from the early Miocene Thomas Farm locality, Florida (U.S.A.). Palaeogeography, Palaeoclimatology, Palaeoecology 76:125–151.

Pybus, M.J., D.P. Hobson, and D.K. Onderka. 1986. Mass mortality of bats due to probable blue-green algal toxicity. Journal of Wildlife Diseases 22:449–450.

Ralrick, P.E. 2004. Subfossil vertebrate mass kill event at Little Fish Lake (Alberta, Canada): Preliminary observations. Pp. 45–50 *in* Alberta Palaeontological Society, eighth annual symposium. Alberta Paleontological Society, Calgary.

Reif, W.-E. 1982. Muschelkalk/Keuper bone-beds (Middle Triassic, SW-Germany): Storm condensation in a regressive cycle. Pp. 299–325 *in* Cyclic and event stratification. G. Einsele and A. Seilacher, eds. Springer-Verlag, Berlin.

Reiter, J., N.L. Stinson, and B.J. Le Boeuf. 1978. Northern elephant seal development: The transition from weaning to nutritional independence. Behavioral Ecology and Sociobiology 3:337–367.

Rogers, R.R. 1990. Taphonomy of three bone beds in the Upper Cretaceous Two Medicine Formation of northwestern Montana: Evidence for drought-related mortality. Palaios 5:394–413.

Rogers, R.R. 1995. Sequence stratigraphy and vertebrate taphonomy of the Upper Cretaceous Two Medicine and Judith River formations, Montana. Unpublished Ph.D. dissertation, University of Chicago.

Rogers, R.R. 2005. Fine-grained debris flows and extraordinary vertebrate burials in the Late Cretaceous of Madagascar. Geology 33:297–300.

Rogers, R.R., and S.M. Kidwell. 2000. Associations of vertebrate skeletal concentrations and discontinuity surfaces in terrestrial and shallow marine records: A test in the Cretaceous of Montana. Journal of Geology 108:131–154.

Rogers, R.R., A.B. Arcucci, F. Abdala, P.C. Sereno, C.A. Forster, and C.L. May. 2001. Paleoenvironment and taphonomy of the Chañares Formation tetrapod assemblage (Middle Triassic), northwestern Argentina: Spectacular preservation in volcanogenic concretions. Palaios 16:461–481.

Rolf, W.D.I., G.P. Durant, A.E. Fallick, A.J. Hall, D.J. Large, A.C. Scott, T.R. Smithson, and G.M. Walkden. 1990. An early terrestrial biota preserved by Visean vulcanicity in Scotland. Pp. 13–24 *in* Volcanism and fossil biotas. Special paper 244. M.G. Lockley and A. Rice, eds. Geological Society of America, Boulder, Colorado.

Romer, A.S., and L.W. Price. 1940. Review of the Pelycosauria. Geological Society of America Special Papers 28:1–538.

Rothschild, B.M., and L.D. Martin. 1993. Paleopathology: Disease in the fossil record. CRC Press, London.

Ryan, M.J., A.P. Russell, D.A. Eberth, and P.J. Currie. 2001. The taphonomy of a *Centrosaurus* (Ornithischia: Ceratopsidae) bone beds from the Dinosaur Park Formation (Upper Campanian), Alberta, Canada, with comments on cranial ontogeny. Palaios 16:482–506.

Sahni, A. 1972. The vertebrate fauna of the Judith River Formation of Montana. American Museum of Natural History Museum Bulletin 147:321–412.

Sander, P.M. 1987. Taphonomy of the Lower Permian Geraldine Bonebed in Archer County, Texas. Palaeogeography, Palaeoclimatology, Palaeoecology 61:221–236.

Sander, P.M. 1992. The Norian *Plateosaurus* bonebeds of central Europe and their taphonomy. Palaeogeography, Palaeoclimatology, Palaeoecology 93:255–299.

Saunders, J.J. 1977. Late Pleistocene vertebrates of the Western Ozark highland, Missouri. Reports of investigations, no. 33. Illinois State Museum, Springfield.

Schäfer, W. 1962. Aktuo-paläontologie nach studien in der Nordsee. W. Kramer, Frankfurt.

Schäfer, W. 1972. Ecology and paleoecology of marine environments. University of Chicago Press, Chicago.

Schaller, G.B. 1972. The Serengeti lion. University of Chicago Press, Chicago.

Schaller, G.B. 1973. Golden shadows, flying hooves. Alfred A. Knopf, New York.

Schmitt, D.N., and K.E. Juell. 1994. Toward the identification of coyote scatological faunal accumulations in archaeological contexts. Journal of Archaeological Science 21:249–262.

Schröder-Adams, C.J., S.L. Cumbaa, J. Bloch, D.A. Leckie, J. Craig, S.A. Seif El-Dein, D.-J.H.A.E. Simons, and F. Kenig. 2001. Late Cretaceous (Cenomanian to Campanian) paleoenvironmental history of the Eastern Canadian margin of the Western Interior Seaway: Bonebeds and anoxic events. Palaeogeography, Palaeoclimatology, Palaeoecology 170:261–289.

Schwartz, H.L., and D.D. Gillette. 1994. Geology and taphonomy of the *Coelophysis* quarry, Upper Triassic Chinle Formation, Ghost Ranch, New Mexico. Journal of Paleontology 68:1118–1140.

Seilacher, A., W.E. Reif, and F. Westphal. 1985. Sedimentological, ecological, and temporal patterns of fossil Lagerstätten. Philosophical Transactions of the Royal Society of London B 311:5–23.

Sergeant, D.E., and G.A. Williams. 1983. Two recent entrapments of narwhals, *Monodon monoceros*, in arctic Canada. Canadian Field Naturalist 97:459–460.

Shipman, P. 1975. Implications of drought for vertebrate fossil assemblages. Nature 257:667–668.

Shipman, P.G. 1981. Life history of a fossil: An introduction to taphonomy and paleoecology. Harvard University Press, Cambridge.

Shipman, P., and A. Walker. 1980. Bone-collecting by harvesting ants. Paleobiology 6:496–502.

Shipman, P., G. Foster, and M. Schoeninger. 1984. Burnt bones and teeth: An experimental study of color, morphology, crystal structure and shrinkage. Journal of Archaeological Science 11:307–325.

Siegstad, H., and M.P. Heide-Jørgensen. 1994. Ice entrapments of narwhals (*Monodon monoceros*) and white whales (*Delphinapterus leucas*) in Greenland. Meddelelser om Grønland, Bioscience 39:151–160.

Sinclair, A.R.E. 1979. The eruption of the ruminants. Pp. 82–103 *in* Serengeti: Dynamics of an ecosystem. A.R.E. Sinclair and M. Norton-Griffiths, eds. University of Chicago Press, Chicago.

Singer, F.J., W. Schreier, J. Oppenheim, and E.O. Garton. 1989. Drought, fire, and large mammals. BioScience 39:716–722.

Skinner, J.D., and R.J. van Aarde. 1991. Bone collecting by brown hyaenas *Hyaena brunnea* in the Central Namib Desert, Namibia. Journal of Archaeological Science 18:513–523.

Skinner, J.D., J.R. Henschel, and A.S. van Jaarsveld. 1986. Bone-collecting habits of spotted hyaenas *Crocuta crocuta* in the Kruger National Park. South African Tydskrif Dierk 21:303–308.

Smith, G.R., R.F. Stearley, and C.E. Badgley. 1988. Taphonomic bias in fish diversity from Cenozoic floodplain environments. Palaeogeography, Palaeoclimatology, Palaeoecology 63:263–273.

Smith, R.M.H. 1993. Vertebrate taphonomy of Late Permian floodplain deposits in the southwestern Karoo Basin of South Africa. Palaios 8:45–67.

Smith, R.M.H. 2000. Sedimentology and taphonomy of Late Permian vertebrate fossil localities in southwestern Madagascar. Palaeontologia Africana 36:25–41.

Smith, R., and P. Haarhoff. 2004. Taphonomy of an Early Pliocene Sivathere bonebed at Lange-Baanweg, Cape Province, South Africa. Journal of Vertebrate Paleontology 24(supplement to 3):115A.

Smith, R., and J. Kitching. 1997. Sedimentology and vertebrate taphonomy of the *Trity-lodon* Acme Zone: A reworked paleosol in the Lower Jurassic Elliot Formation, Karoo Supergroup, South Africa. Palaeogeography, Palaeoclimatology, Palaeoecology 131:29–50.

Smithers, R.H.N. 1938. Notes on the stranding of a school of *Pseudorca crassidens* at the Berg River mouth. Transactions of the Royal Society of South Africa 25:403–411.

Spencer, L.M., B. Van Valkenburgh, and J.M. Harris. 2003. Taphonomic analysis of large mammals recovered from the Pleistocene Rancho La Brea tar seeps. Paleobiology 29:561–575

Sternberg, C.M. 1970. Comments on dinosaurian preservation in the Cretaceous of Alberta and Wyoming. Publications in Palaeontology 4. National Museums of Canada, Ottawa.

Stock, C. 1972. Rancho La Brea: A record of Pleistocene life in California. Science series no. 20, paleontology no. 11, Los Angeles County Museum of Natural History, Los Angeles.

Storey, M., and E.W. Gudger. 1936. Mortality of fishes due to cold at Sanibel Island, Florida, 1886–1936. Ecology 17:640–648.

Straight, W.H., and D.A. Eberth. 2002. Testing the utility of vertebrate remains in recognizing patterns in fluvial deposits: An example from the lower Horseshoe Canyon Formation, Alberta. Palaios 17:472–490.

Sundell, K.A. 1999. Taphonomy of a multiple *Peobrotherium* kill site: An archaeotherium meat cache. Journal of Vertebrate Paleontology 19(supplement to 3):79A.

Swift, P.K., J.D. Wehausen, H.B. Ernset, R.S. Singer, A.M. Pauli, H. Kinde, T.E. Rocke, and V.C. Bleich. 2001. Desert bighorn sheep mortality due to presumptive type C botulism. Journal of Wildlife Diseases 36:184–189.

Sykes, J.H. 1977. British Rhaetian bone-beds. Mercian Geologist 5:39–48.

Taber, R.D., K.J. Radeke, and D.K. Paige. 1982. Wildlife-forest interactions in the Mt. St. Helens blast zone. Final Report US National Forest, contribution no. 0m-40-055K3-1-2572. United States Forest Service, Vancouver, Washington.

Talbot, L.M., and M.H. Talbot. 1963. The wildebeest in western Masailand, East Africa. Wildlife monographs, no. 12. National Academy of Sciences and National Research Council, Washington, D.C.

Terry, R.C. 2004. Owl pellet taphonomy: A preliminary study of the post-regurgitation taphonomic history of pellets in a temperate forest. Palaios 19:497–506.

Terry, R.C. 2005. Raptors, rodents, and paleoecology: Testing paleoecological assumptions using taphonomic damage patterns at Homestead Cave, Utah. Geological Society of America, Abstracts with Programs 37(7):528.

Thompson, P.M., and A.J. Hall. 1993. Seals and epizootics: What factors might affect the severity of mass mortalities? Mammal Reviews 23:149–154.

Tramer, E.J. 1977. Catastrophic mortality of stream fishes trapped in shrinking pools. The American Midland Naturalist 97:469–478.

Trapani, J. 1998. Hydrodynamic sorting of avian skeletal remains. Journal of Archaeological Science 25:477–487.

Trueman, C. This volume. Trace element geochemistry of bonebeds. Chapter 7 *in* Bonebeds: Genesis, analysis, and paleobiological significance. R.R. Rogers, D.A. Eberth, and A.R. Fiorillo, eds. University of Chicago Press, Chicago.

Trueman, C.N., and M.J. Benton. 1997. A geochemical method to trace the taphonomic history of reworked bones in sedimentary settings. Geology 25:263–266.

Tulloch, D.G. 1970. Seasonal movements and distribution of the sexes in the water buffalo, *Bubalus bubalis*, in the Northern Territory. Australian Journal of Zoology 18:399–414.

Turnbull, W.D., and D.M. Martill. 1988. Taphonomy and preservation of a monospecific titanothere assemblage from the Washakie Formation (Late Eocene), southern Wyoming: An ecological accident in the fossil record. Palaeogeography, Palaeoclimatology, Palaeoecology 63:91–108.

Turner, A.H., C.E. Brett, P.I. McLaughlin, D.J. Over, and G.W. Storrs. 2001. Middle-Upper Devonian (Givetian-Famennian) bone/conodont beds from central Kentucky; reworking and event condensation in the distal Acadian foreland basin. Geological Society of America, North-Central Section, Abstracts with Programs 33(4):8.

Ulrichs, M., R. Wild, and B. Ziegler. 1979. Fossilien aus Holzmaden. Stuttgarter Beiträge zur Naturkunde, Serie C 11:1–34.

Vander Wall, S.B. 1990. Food hoarding in animals. University of Chicago Press, Chicago.

Van Devender, T.R., and J.I. Mead. 1978. Early Holocene and Late Pleistocene amphibians and reptiles in Sonoran Desert packrat middens. Copeia 1978(3):464–475.

Varricchio, D.J. 1995. Taphonomy of Jack's Birthday Site, a diverse dinosaur bonebed from the Upper Cretaceous Two Medicine Formation of Montana. Palaeogeography, Palaeoclimatology, Palaeoecology 114:297–323.

Varricchio, D.J. 2001. Gut contents from a Cretaceous tyrannosaurid; implications for theropod dinosaur digestive tracts. Journal of Paleontology 75:401–406.

Varricchio, D., F. Jackson, B. Scherzeer, and J. Shelton. 2005. Don't have a cow, man! It's only actualistic taphonomy on the Yellowstone River of Montana. Journal of Vertebrate Paleontology 25(supplement to 3):126A

Voorhies, M.R. 1969. Taphonomy and population dynamics of an early Pliocene vertebrate fauna, Knox County, Nebraska. Contributions to geology, special paper 1. University of Wyoming, Laramie.

Voorhies, M.R. 1975. Vertebrate burrows. Pp. 325–350 *in* The study of trace fossils. R.W. Frey, ed. Springer-Verlag, New York.

Voorhies, M.R. 1981. Ancient skyfall creates Pompeii of prehistoric animals. National Geographic 159:66–75.

Voorhies, M.R. 1985. A Miocene rhinoceros herd buried in volcanic ash. National Geographic Research Reports 19:671–688.

Voorhies, M.R. 1992. Ashfall: Life and death at a Nebraska waterhole ten million years ago. Museum Notes no. 81. University of Nebraska State Museum, Lincoln.

Voorhies, M.R., J.A. Holman, and X. Xiang-Xu. 1987. The Hottell Ranch rhino quarries (basal Ogallala: medial Barstovian), Banner County, Nebraska. 1. Geologic setting, faunal lists, lower vertebrates. Contributions to Geology, University of Wyoming 25:55–69.

Waldman, M. 1971. Fish from the freshwater Lower Cretaceous of Victoria, Australia with comments on the palaeoenvironment. Special papers in palaeontology 9. The Palaeontological Association, London.

Walsh, S.A., and D. Martill. 2006. A possible earthquake-triggered mega-boulder slide in a Chilean Mio-Pliocene marine sequence: Evidence for rapid uplift and bonebed genesis. Journal of the Geological Society, London 163:697–705.

Walsh, S., and D. Naish. 2002. Fossil seals from Late Neogene deposits in South America: A new pinniped (Carnivora, Mammalia) assemblage from Chile. Palaeontology 45:821–842

Weigelt, J. 1927. Rezente wirbeltierleichen und ihre paläobiologische bedeutung. Verlag von Max Weg, Leipzig.

Weigelt, J. 1989. Recent vertebrate carcasses and their paleobiological implications, J. Schaefer, trans. University of Chicago Press, Chicago.

Wells, J.W. 1944. Middle Devonian bone beds of Ohio. Bulletin of the Geological Society of America 55:273–302.

Whitfield, A.K., and A.W. Paterson. 1995. Flood-associated mass mortality of fishes in the Sundays Estuary. Water SA 21:385–389.

Wilkinson, D., and G.A.J. Worthy. 1999. Marine mammal stranding networks. Pp. 396–411 in Conservation and management of marine mammals. J. R. Twiss Jr. and R. R. Reeves, eds. Smithsonian Institution Press, Washington, D.C.

Wilson, M.V.H. 1987. Predation as a source of fish fossils in Eocene lake sediments. Palaios 2:497–504.

Wilson, M.V.H. 1996. Taphonomy of a mass-death layer of fishes in the Paleocene Paskapoo Formation at Joffre Bridge, Alberta, Canada. Canadian Journal of Earth Sciences 33:1487–1498.

Wolff, R.G. 1973. Hydrodynamic sorting and ecology of a Pleistocene mammalian assemblage from California (U.S.A.). Palaeogeography, Palaeoclimatology, Palaeoecology 13:91–101.

Wood, J., R. Thomas, and J. Visser. 1988. Fluvial processes and vertebrate taphonomy: The Upper Cretaceous Judith River Formation, south-central Dinosaur Provincial Park, Alberta, Canada. Palaeogeography, Palaeoclimatology, Palaeoecology 66:127–143.

Wood, R.C. 1988. A monospecific death assemblage of fossil side-necked turtles from the Cretaceous of Brazil. International Symposium on Vertebrate Behavior as Derived from the Fossil Record, occasional paper 22. Museum of the Rockies, Bozeman, Montana.

Worthylake, K.M., and P. Hovingh. 1989. Mass mortality of salamanders (*Ambystoma tigrinum*) by bacteria (*Acinetobacter*) in an oligotrophic seepage mountain lake. Great Basin Naturalist 49:364–372.

Wurtsbaugh, W.A., and R.A. Tapia. 1988. Mass mortality of fishes in Lake Titicaca (Peru-Bolivia) associated with the protozoan parasite *Ichthyophthirius multifiliis*. Transactions of the American Fisheries Society 117:213–217.

Yeager, L.E. 1943. The storing of muskrat and other food by minks. Journal of Mammalogy 24:100–110.

Zeigler, K.E. 2002. A taphonomic analysis of a fire-related Upper Triassic fossil assemblage. Unpublished masters thesis, University of New Mexico, Albuquerque.

Zeigler, K.E., L.H. Tanner, and S.G. Lucas. 2005. Mass mortality and post-paleowildfire erosion: Evidence from a Late Triassic death assemblage, Chama Basin, north-central New Mexico. Geological Society of America, Rocky Mountain Section, Abstracts with Programs 37(6):7.

Zonneveld, J.P., W.S. Bartels, S.D. Wolfe, G.F. Gunnell, and J.I. Bloch. 2001. The occurrence of fossil vertebrates along strandlines of paleo-Lake Gosiute, Wasatch Formation, South Pass, Wyoming. Journal of Vertebrate Paleontology 21(supplement to 3):117A.

# CHAPTER 2

# Bonebeds through Time

*Anna K. Behrensmeyer*

## INTRODUCTION

Bonebeds have attracted great interest for centuries and remain a source of many important specimens that document the history of the vertebrates. A quote from W. S. Symonds in an 1858 article "On Bone Beds and Their Characteristic Fossils" indicates that they contributed to debates about catastrophism: "And whereas we formerly held that "bone beds" were boundaries of Creation—universal phenomena—and that they marked out a certain horizon where whole tribes and races of animals were extinguished, never to be renewed, it may be useful to Geologists to sum up shortly the evidence as it *now* stands, respecting these most interesting platforms of death." With the growth of understanding of stratigraphy and the history of the Earth, it became clear to early workers that bonebeds occurred at many different times and involved a wide range of vertebrate taxa. Perhaps in part because of the shear diversity of bonebed types and ages, there has been little previous synthetic work on their taphonomy and importance in paleontology.

The goals of this chapter are to provide an overview of bonebeds through time, based on the analysis of a bonebed database derived from publications on the Phanerozoic terrestrial record, and to develop hypotheses for any significant patterning observed in this record. With standardized data, it is possible to provide some preliminary tests of hypotheses such as the following: (1) How do bonebed frequency and type correlate with broad climatic and other environmental trends in Phanerozic history, and are there more bonebeds during environmental "bad times" for vertebrates? (2) Were some ecomorphs or taxonomic groups more susceptible to mass deaths? (3) Do changes in the bonebed record relate to evolutionary trends in skeletonization, body size, habitat utilization, and bone-ingesting capabilities of vertebrates? (4) Since bonebeds provide a significant portion of the vertebrate fossil record, how might secular changes in taphonomic processes that create bonebeds affect our view of vertebrate history?

## WHAT IS A BONEBED?

The working definition for the purpose of this chapter is that *a bonebed is a single sedimentary stratum with a bone concentration that is unusually dense (often but not necessarily exceeding 5% bone by volume), relative to adjacent lateral and vertical deposits.* It is generally understood that more than one individual is represented. The sedimentary stratum can be a homogeneous lithological unit or a suite of variable lithologies and multiple layers, all of which were generated by the event(s) that formed the bonebed. The above definition differs from many other uses of the term over the past centuries; for instance, in the paleontological literature of Germany, bonebeds referred to "placer" type concentrations of bones that had been reworked into dense, thin, but widespread deposits (Weigelt 1986; Seilacher et al. 1985). In general, the definition used in this chapter is more inclusive than most past usages and is meant to focus attention on a range of processes that can generate unusual bone concentrations while also excluding the more "normal" dispersed and isolated occurrences of vertebrate fossil remains. Definitions of terms differ somewhat among the chapters in this volume; for example, Eberth et al. (Chapter 3 in this volume) define it "as consisting of the complete or partial remains of more than one vertebrate animal in notable concentration along a bedding plane, erosional surface, or throughout a single bed"; for a comparison of terminologies see Appendix 2.1.

Recognized subcategories of bonebeds are based on objective criteria such as number of taxa (modified from Rogers, 1990):

- A *monotaxic* bonebed consists of multiple elements of two or more individuals of the same species. Most bonebeds of this type include minor remains from other species ("background") unrelated to the event that formed the concentration of bones from the single species. Such incidental remains are unevenly reported in the literature, and they are not used here as a criterion for excluding bonebeds from the "monospecific" category.

- A *paucitaxic* bonebed consists of multiple elements from two or three different species that form the bulk of the assemblage and may include minor remains from other species.
- A *multitaxic* bonebed consists of skeletal remains from multiple species (more than three).

There are many other ways to classify bonebeds, some of which will be elaborated in other chapters in this volume (see Chapters 1, 3, and 4). These represent a range from strictly descriptive criteria (size of bone particles or elements, such as macro- versus micro-bone accumulations) to inferences regarding different causes of death (mass versus attritional mortality). All such schemes serve different but overlapping purposes in providing frameworks that can aid in planning collecting strategies, facilitating taphonomic and paleoecological comparisons among different bonebeds, interpreting causal processes, and examining long-term trends in processes that cause bonebed formation. There can and should be no single "right" way to classify bonebeds, because the questions at hand determine the categories that will be most useful for organizing the data for hypothesis testing. However, we also need to make sense of the massive amount of available information about bonebeds in order to look for patterns that can support inferences about process, and some classification schemes may be more appropriate for this than others. In this chapter, I examine the ETE Bonebed Database, compiled in conjunction with the Evolution of Terrestrial Ecosytems Program (ETE), National Museum of Natural History (Appendices 2.2 and 2.3, available online at www.press. uchicago/edu/books/rogers/; see also www.nmnh.si.edu/ete and the Paleobiology Database at http://paleodb.org/cgi-bin/bridge.pl), for patterns relating to geography, distribution through time, lithological context, and taphonomic features, as well as published inferences concerning causality. I then provide an overview based on my own classification of bonebed types, using information provided in the published records as well as personal experience with bonebed localities in late Paleozoic, Mesozoic, and Cenozoic deposits. This empirical approach offers some interesting parallels and contrasts with the more deductive scheme discussed by Rogers and Kidwell in Chapter 1 in this volume.

## THE ETE BONEBED DATABASE

This ETE Bonebed Database has been assembled from the published literature and in 2002 consisted of 315 localities in Filemaker Pro 8.0 with

A.

Biryokwonin Tuff

Lacustrine Clay

K089A Bonebed

B.

C.

basic locality and age information, species lists, and any available data on taphonomy, paleoenvironmental context, and the cause of bonebed formation, as inferred by the author(s) (Tables 2.1 and 2.2; Appendices 2.2 and 2.3). It includes many of the same references as the database used by Eberth et al. (Chapter 3 in this volume) but differs in the inclusion of more Cenozoic localities and in the categories used to organize bonebed data.

Basic information on the locality name, age, formation, place, and dominant taxon is unambiguous and provides a solid dataset for examining frequency through time, geographic distribution, etc. As is typical of any literature survey, other types of information are variably reported and often subject to qualitative judgments upon data entry; these were standardized as much as possible through repeated cross-checking during the growth of the database. The composition of the fauna (taxonomic lists) is usually provided, although in some cases only at family level or above. Species lists may be complete for some groups in the deposit, while others are reported as "reptile indet.," etc. Assessment of geological context, skeletal part representation, numbers of specimens, and cause of bonebed formation require careful reading prior to data entry, and information of this kind is uneven from publication to publication. Many references to bonebeds are in abstracts and represent preliminary investigations; these are included in the database when the reported information adds substantively to specific fields. Some of the entries are unpublished and were included with the permission of the researchers responsible for the ongoing studies; these are clearly marked as personal communications (Appendix 2.2).

It should be noted that the following results do not distinguish between macro- and micro-accumulations, although important differences between these will be discussed later and are an important theme in Chapter 3 in this volume by Eberth et al. The decision to include a locality in the ETE Bonebed Database (Appendix 2.2) was determined solely by whether it fit the definition of a bonebed presented at the beginning of this chapter.

---

*Figure 2.1.* The middle Miocene bonebed KO89A at Kipsaramon, Muruyur Formation, Tugen Hills, Kenya (Hill et al. 1991; Behrensmeyer et al. 2002). A. Stratigraphic context of the bonebed, showing that it is a single, ~20 cm thick unit within a sequence of lacustrine deposits, capped by a green clay and an ignimbritic tuff. B. View of the excavation surface with many fragmentary remains of proboscidean, rhinoceros, suid, and smaller mammals in an area of about 5 m$^2$ (average of ~30 bones/m$^2$); white, plaster-covered block was removed for laboratory preparation. C. Close-up view of part of the excavation surface, showing the poor sorting and variable quality of preservation of the bones, which ranged from unabraded and unweathered to rounded and highly weathered. The bonebed is conservatively estimated to cover about 0.09 km$^2$ and appears to represent a long-term accumulation of vertebrate remains (e.g., on a delta topset) that was rapidly emplaced into a lake, perhaps by an unusual flooding event or a turbidity flow.

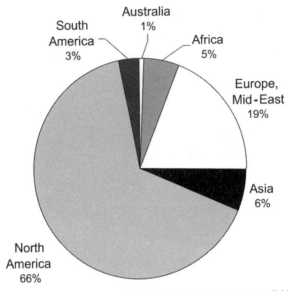

**Bonebed Sample by Continent**

N = 315 Localities

South America 3%

Australia 1%

Africa 5%

Europe, Mid-East 19%

Asia 6%

North America 66%

*Figure 2.2.* Continental distribution of bonebed localities in the ETE Bonebed Database (Table 2.1A). This database is drawn primarily from the published literature, in English, and for this and other reasons has a strong bias toward North American and European localities. Results in this chapter therefore pertain mainly to the northern and western hemispheres.

## CHARACTERISTICS OF BONEBEDS THROUGH TIME

### Geographic Distribution of Bonebed Sample

The literature search for bonebed localities resulted in a sample that is heavily skewed toward North America (67%) and Europe (19%) (Fig. 2.2). This is an expected consequence of the long-term focus of western science on the fossil record of these two continental areas, plus the bias toward English-language publications and listings in the standard search programs (e.g., GeoRef). For the purposes of this chapter, we assume that the patterns through time based mostly on the North American and European record probably reflect general global trends. However, it is clear that the less-developed paleontological resources of the other continents could significantly alter the results reported here as they become known at an equivalent level of detail. Thus, the caveat provided by the geographic bias in the ETE Bonebed Database should be kept in mind throughout the following discussions.

## Overall Phanerozoic Patterns

The raw numbers of bonebed localities through the Phanerozoic show significant peaks in the Cretaceous and the Pleistocene (Table 2.1B, Fig. 2.3). When corrected for the variable lengths of the geological periods and epochs (Gradstein and Ogg, 1996), the Paleocene rather than the Cretaceous becomes the high point in bonebed frequency prior to the Miocene. Within the Cretaceous record, however, 53/68 (78%) of the records occur in the Upper Cretaceous (99–65 Ma, Cenomanian–Maastrichtian), for a score of about 1.6 bonebeds per Ma., which is comparable to the Paleocene. The Miocene through Pleistocene interval provides a significant proportion of the total bonebed sample (32%), which undoubtedly reflects a bias imposed by the widespread availability of late Cenozoic deposits in North America and Europe. The relatively low number of bonebeds in the Oligocene ($N = 8$) is anomalous, and it is not possible to determine at present whether this is a consequence of underreporting, deficiencies in data acquisition, or a temporary shift in the frequency of bonebed-generating taphonomic processes.

## Lithology and Environmental Context

About 82% of the localities included in the database had information pertaining to depositional environment (Table 2.2), permitting classification into the general categories shown in Figure 2.4A. It is obvious that bonebeds occur over a wide range of terrestrial and marine settings, wherever there are animals or remains to entomb and depositional processes to do the job. Approximately half of the bonebeds in the database occur in fluvial environments, which include various types of channel and overbank subenvironments. This speaks to the availability of fluvial deposits on all continents and throughout the Phanerozoic, the suitability of riverine habitats for vertebrate habitation, and the frequency of favorable circumstances for burial and preservation of vertebrate remains. There is considerable variability in the proportion of bonebeds in fluvial deposits through time (Table 2.1B), although it is difficult to assess the significance of this variation because of uneven sampling for the different time intervals. However, with some exceptions, the post-Carboniferous proportion of fluvial bonebeds is on the order of 60%–70% for most time intervals.

Lithologies associated with bonebeds are predominantly sandstone and siltstone or mudstone (Fig. 2.4B), with finer grained, volcanogenic and chemically precipitated sediments making up only about one-quarter

Table 2.1. A. Geographic distribution of localities in the Bonebed Database, by continent

| Continent | # Bonebed Localities | % |
|---|---|---|
| Africa | 16 | 5 |
| Asia | 19 | 6 |
| Australia | 3 | 1 |
| Europe, Middle East | 61 | 19 |
| North America | 206 | 65 |
| South America | 10 | 3 |
| Total | 315 | |

Table 2.1. B. Sample of bonebed localities by geological period and epoch through the Phanerozoic

| | Max. Age | Min. Age | Time Span (Ma) | No. Bonebeds | BB/Ma | Fluvial Only |
|---|---|---|---|---|---|---|
| Pleistocene | 1.8 | 0 | 1.8 | 57 | 31.67 | 0.30 |
| Pliocene | 5.3 | 1.8 | 3.5 | 11 | 3.14 | 0.64 |
| Miocene | 23 | 5.3 | 23 | 32 | 1.39 | 0.47 |
| Oligocene | 33.4 | 23 | 33 | 8 | 0.24 | 0.67 |
| Eocene | 55.5 | 33.4 | 22.1 | 13 | 0.59 | 0.46 |
| Paleocene | 65 | 55.5 | 9.5 | 16 | 1.68 | 0.69 |
| Cretaceous | 142 | 65 | 77 | 68 | 0.88 | 0.62 |
| Jurassic | 206 | 142 | 64 | 26 | 0.41 | 0.50 |
| Triassic | 248 | 206 | 42 | 30 | 0.71 | 0.69 |
| Permian | 290 | 248 | 42 | 26 | 0.62 | 0.60 |
| Carboniferous | 354 | 290 | 64 | 10 | 0.16 | 0.13 |
| Devonian | 417 | 354 | 63 | 9 | 0.14 | 0.22 |
| Silurian | 443 | 417 | 26 | 8 | 0.31 | 0.13 |
| Ordovician | 495 | 443 | 52 | 1 | 0.02 | 0.00 |
| N | | | | 315 | | 135/265 |

Note: Boundary ages and age assignments for the localities are based on the D.T.S. (Digital Time Scale, Purdue University; Gradstein and Ogg, 1996), the International Stratigraphic Chart (IUGS 2000), and Janis et al. 1998. Holocene is included with Pleistocene. The "fluvial only" sample gives the proportion of localities in fluvial contexts, based on a total of 265 localities with adequate information to assign depositional environment.

A.

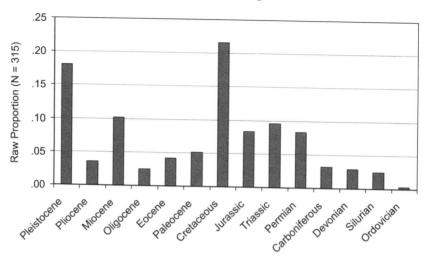

B.

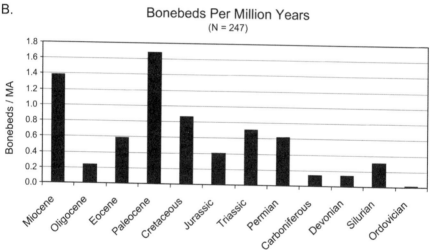

*Figure 2.3.* A. Distribution of bonebed occurrences through the Phanerozoic, based on the total sample of 315 localities in the ETE Bonebed Database. B. Distribution of bonebed localities per million years, corrected for the varying lengths of the geological periods and epochs. The Pliocene and Pleistocene were omitted from this diagram because of their much higher numbers of localities relative to time span (Table 2.1B).

Table 2.2. Environmental context of a subset of 265 localities in the Bonebed Database

|  | Environment | | | | | | | | |
|---|---|---|---|---|---|---|---|---|---|
| Period/Epoch | Volcanic | Karst | Eolian | Marine | Coastal | Deltaic | Fluvial | Lacustrine | Totals |
| Pleistocene | 2 | 5 | 5 | 2 | 3 |  | 10 | 6 | 33 |
| Pliocene |  |  |  | 1 | 2 |  | 7 | 1 | 11 |
| Miocene | 2 | 1 |  | 1 | 2 | 2 | 15 | 9 | 32 |
| Oligocene |  | 1 |  |  |  |  | 2 |  | 3 |
| Eocene |  |  |  |  |  |  | 6 | 7 | 13 |
| Paleocene |  |  |  |  |  |  | 9 | 4 | 13 |
| Cretaceous |  | 1 | 5 | 4 | 2 | 4 | 38 | 7 | 61 |
| Jurassic |  |  |  |  | 8 |  | 12 | 4 | 24 |
| Triassic | 1 |  |  | 3 | 3 |  | 20 | 2 | 29 |
| Permian |  | 1 |  |  | 1 | 4 | 12 | 2 | 20 |
| Carboniferous | 1 | 2 |  | 2 | 1 | 1 | 1 |  | 8 |
| Devonian |  |  |  | 7 |  |  | 2 |  | 9 |
| Silurian |  |  |  | 6 | 1 |  | 1 |  | 8 |
| Ordovician |  |  |  | 1 |  |  |  |  | 1 |
| Total | 6 | 11 | 10 | 27 | 23 | 11 | 135 | 42 | 265 |

*Note:* Localities that could be clearly assigned based on published evidence. "Coastal" includes estuarine, "deltaic" includes fluvial-deltaic, and "lacustrine" includes fluvial-lacustrine.

## A. Environmental Context
N = 265

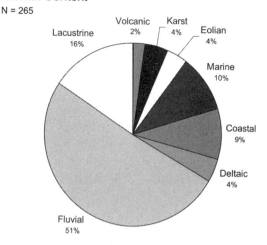

## B. Lithologies of Bonebeds
N = 234

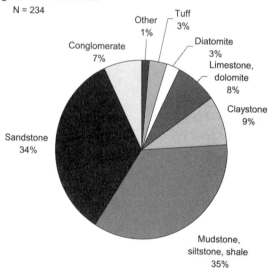

*Figure 2.4.* A. Dominant lithologies associated with bonebeds, for a sample of 234 localities that reported unambiguous information. B. Inferred paleoenvironmental context of bonebed localities, for a sample of 265. Note that more localities had information on environmental context than specific lithologies.

of the sample. This is consistent with the strong role of clastic sediments in burying and preserving vertebrate remains, although clearly bonebeds occur in a diverse array of sediment types. There is no way of knowing whether the pattern of lithological association for bonebeds reflects the proportions of available lithologies in a broader sense, although one might expect that bonebed context might be skewed toward finer-grained lithofacies (i.e., clays, silts, sands, and gravels rather than boulder or cobble conglomerates).

### Taxonomic Representation

Three aspects of taxonomic representation in bonebeds are considered here: vertebrate group, species richness (i.e., mono- and paucispecific versus multispecific), and carnivores versus herbivores. Localities were classified according to which major vertebrate group was dominant in the bone assemblage, and this was straightforward based on the published information (Fig. 2.5A, Table 2.3). All categories are at the level of vertebrate class (fish, amphibian, reptile, bird, mammal) except for dinosaurs, which were considered as a distinct group for the purpose of this analysis. Mammal-dominated bonebeds make up over a third of the sample, but significant numbers are primarily fish, amphibian, reptile, or dinosaur. Birds seem to be infrequent bonebed formers, with only two localities in the database.

The ETE Bonebed Database includes 303 localities that can be categorized in terms of species richness as monospecific (MO) (21%), paucispecific (PA) (27%), or multispecific (MU) (52%). These data can be applied to two different but related questions: (1) Do the major vertebrate groups show any differences in their tendency to form mono-, pauci-, or multispecific bonebeds? (2) Are there changes in taxonomic representation (MO, PA, MU) in bonebeds through geological time?

The tallies in Figure 2.5B and Table 2.3 show that fish are most likely to form MU bonebeds, but among the other four major groups, there are only minor differences in the proportions of the three types of taxonomic representation. There is a suggestion that dinosaurs and amphibians form MO assemblages less commonly (or PA assemblages more commonly) than non-dinosaurian reptiles and mammals. However, the overall similarity of the patterns for amphibians, reptiles, dinosaurs and mammals is the most notable result of this analysis. This indicates that processes or circumstances largely independent of taxonomic influence control the

A.

### Bonebeds: Dominant Vertebrate Group
N = 315

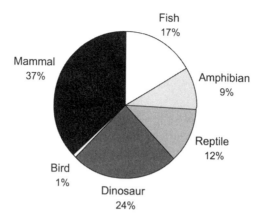

Fish
17%

Mammal
37%

Amphibian
9%

Reptile
12%

Bird
1%

Dinosaur
24%

B.

### Bonebeds: Taxonomic Category for Different Groups
N = 299

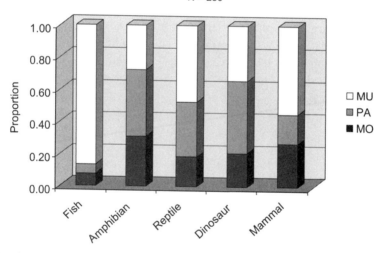

*Figure 2.5.* A. Dominant vertebrate groups in the total sample of bonebed localities, showing that the total for dinosaurs plus other reptiles is approximately equal to the total for mammals. B. Patterns of representation in mono- (MO), pauci- (PA), and multispecific (MU) bonebeds for the major vertebrate groups, based on a sample of 299 localities that could be classified with these criteria (Table 2.3). Birds are omitted from this diagram because of their small sample (N = 2, both paucispecific).

Table 2.3. Major vertebrate groups that are dominant in the bonebed localities, with numbers of mono-, pauci- and multispecific bonebeds

| Dominant Group | No. Bonebed Localities | Monospecific | Paucispecific | Multispecific |
|---|---|---|---|---|
| Fish | 52 | 4 | 3 | 45 |
| Amphibian | 29 | 9 | 12 | 8 |
| Reptile | 38 | 7 | 13 | 18 |
| Dinosaur | 77 | 16 | 34 | 26 |
| Bird | 2 | 0 | 2 | 0 |
| Mammal | 117 | 28 | 19 | 57 |
| Total[a] | 315 | 64 | 83 | 146 |

*Notes:* [a] Totals represent the subset of localities that could be classified according to the taxonomic abundance categories.

occurrence of MO versus PA versus MU types of bonebeds. It is possible that body size is one of these circumstances because large animals tend to occur together in social groups and clusters of their carcasses have a higher probability of discovery than those of small animals. This can be tested when adequate body size information is incorporated into the ETE Bonebed Database.

The patterns of MO-PA-MU bonebed occurrences through the Phanerozoic (Fig. 2.6) show several distinct divisions, with fish-dominated, hydraulically concentrated MU assemblages up through the Devonian, a relatively stable combination of the three types from the Carboniferous through the Cretaceous, a sharp change to MU-dominated assemblages in the Paleocene, and then a return to a balance of MO-PA-MU assemblages through the rest of the Cenozoic. Superimposed upon this overall pattern is a tendency for PA bonebeds to be more common in the late Paleozoic and Mesozoic, and MO bonebeds to increase in the Pleistocene. The greater number of PA bonebeds could indicate that environmental circumstances, rather than social preferences, were dictating the co-occurrence of different species in mass-death situations (e.g., amphibians trapped together in a dried-up pond versus herd-oriented mammals). In the Pleistocene, there may be a collecting bias against MU assemblages, which tend to be fragmentary, because of the availability of fossiliferous localities with more complete skeletal material. However, the greater number of MO assemblages also reflects the inclusion in the ETE Bonebed Database of

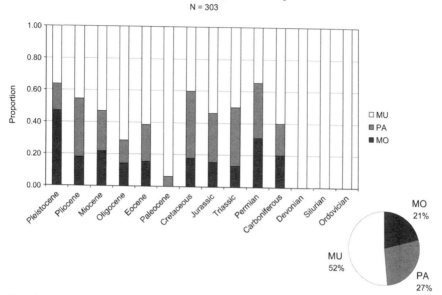

Figure 2.6. Mono-, pauci-, and multitaxic bonebeds through the Phanerozoic, based on a total of 303 localities that could be classified according to these criteria (includes birds and a few localities that did not have a dominant vertebrate group).

assemblages of large mammals that were naturally trapped in MO groups or killed by early humans. The lack of MO bonebeds in the Paleocene indicates a strong focus on screen-washing of hydraulically concentrated microsites, but this collecting bias also is partly a result of the scarcity of other types of bone concentrations during this time interval. This could in turn reflect the dominantly small body sizes of Paleocene vertebrates, limited modes of biological carcass concentration (owls, other predatory birds?) in the aftermath of the K/T extinction, and/or a difference in social behaviors of mammals that made MO or PA death and burial situations uncommon.

### Carnivores versus Herbivores

Another way to look at the patterns of bonebed occurrence through the Phanerozoic is to classify localities by an ecomorphic criterion, in this case whether the dominant species preserved in the assemblage is herbivore

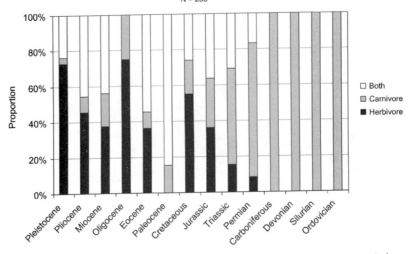

*Figure 2.7.* The pattern of carnivore- versus herbivore-dominated bonebeds through the Phanerozoic, for a sample of 283 localities that could be unambiguously classified according to these criteria (Table 2.4). The trend through the Cretaceous likely reflects the increasing diversity of herbivorous vertebrates; proportions of herbivores and carnivores after the Paleocene are relatively stable, except for the Oligocene (small sample size) and the Pleistocene (possible impact of localities reflecting human predation).

or carnivore. This was possible for 199 of the localities, and an additional 84 had both herbivores and carnivores without clear dominance (Fig. 2.7, Table 2.4). In the Paleozoic–Mesozoic, the proportion of herbivores is concurrent with the rise of terrestrial herbivores in the Permian, continuing through the increase of large-bodied herbivorous dinosaurs in the Jurassic and Cretaceous. The Paleocene represents a break point in this trend, but much of the rest of the Cenozoic has frequencies of carnivore versus herbivore dominance similar to the later Mesozoic. The small Oligocene sample ($N = 4$) does not provide a fair representation for this time interval. In the Pleistocene, the higher proportion of herbivore-dominated bonebeds suggests human predation combined with an increased mortality of groups of large herbivores near the end of the Pleistocene, which is also supported by the pattern of increased MO bonebeds discussed in the previous section. Of possible significance is the observation that carnivore-dominated assemblages become less common after the Jurassic; this conceivably could relate to an underlying change in predator behavior and/or energy transfer through terrestrial food webs (i.e., lower numbers of vertebrate predators relative to their prey).

Table 2.4. Numbers of bonebed localities dominated by herbivores or carnivores

| Period/Epoch | Herbivore | Carnivore | Both | Total |
|---|---|---|---|---|
| Pleistocene | 40 | 2 | 13 | 55 |
| Pliocene | 5 | 1 | 5 | 11 |
| Miocene | 12 | 6 | 14 | 32 |
| Oligocene | 3 | 1 | | 4 |
| Eocene | 4 | 1 | 6 | 11 |
| Paleocene | | 2 | 11 | 13 |
| Cretaceous | 32 | 11 | 15 | 58 |
| Jurassic | 8 | 6 | 8 | 22 |
| Triassic | 4 | 14 | 8 | 26 |
| Permian | 2 | 18 | 4 | 24 |
| Carboniferous | | 9 | | 9 |
| Devonian | | 9 | | 9 |
| Silurian | | 8 | | 8 |
| Ordovician | | 1 | | 1 |
| Total[a] | 110 | 89 | 84 | 283 |

Notes: [a] Totals represent the subset of localities that could be classified according to this criterion.

## Taphonomic Features

There are many taphonomic features of bonebeds that would be interesting to examine through Phanerozoic time, including patterns of breakage, bone surface modification (weathering, tooth marks), differential preservation of body parts, degrees of articulation, sorting by body size, etc. (Behrensmeyer, 1991). Such evidence could be used to test hypotheses about changes in biotic versus abiotic processes that have affected the vertebrate record, such as the proposal that the Cenozoic shows more evidence of bone modification than the Mesozoic because of the evolution of bone-crunching predators and scavengers (Van Valkenburgh 1999; Behrensmeyer et al. 2000). However, information in the published literature is highly uneven with regard to most taphonomic variables, and datasets are too small to document trends through long spans of geological time. An exception to this is the degree of articulation of skeletal parts, which frequently is reported or obvious from quarry maps. Thus, it was possible to analyze degrees of articulation for a set of 216 bonebed localities where remains could be classified as "articulated," "associated,"

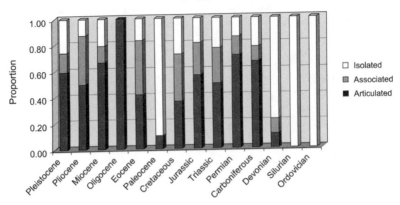

*Figure 2.8.* Patterns of skeletal articulation versus dispersion through the Phanerozoic, based on the presence or absence of articulated or associated skeletal remains in a sample of 216 localities (Table 2.5). "Associated" includes disarticulated but still closely associated skeletal parts from single individuals. The small number of Oligocene localities ($N = 3$) do not provide an adequate representation of degrees of articulation for this time interval.

or "isolated." If an assemblage had any bones articulated, even if most were not, this was counted as articulated in the tallies, likewise for "associated." This approach provides an assessment that favors any evidence for preservation of portions of individual bodies as opposed to dispersed parts. Even two articulated bones indicate a lower level of taphonomic processing than complete dispersion, mixing, and reworking, hence the results should be sensitive to a wide range of assemblage histories. (This approach could be further refined with a measure of the proportion of articulated versus associated versus isolated remains.) Results indicate that there is no notable difference in the proportion of articulated skeletal parts in any time interval since the Devonian (Fig. 2.8, Table 2.5), with the exception of the Paleocene and the Oligocene. The latter suffers from small sample size ($N = 3$), and the former reflects the high proportion of isolated remains associated with microsites and perhaps also the screen-washing methodology, which does not usually recover associated or articulated remains even if they are present in the original assemblage. There is a possible slight decline in the proportion of articulated remains from the Permian through the Cretaceous, which could reflect increased processing of carcasses by large predators.

The hypothesis that the evolution of bone-crushing mammalian predators in the Cenozoic increased the amount of carcass processing compared with the Mesozoic is not supported by the articulation evidence, as

Table 2.5. Numbers of localities with articulated, associated, or isolated skeletal elements by time interval

| Period/Epoch | Articulated[a] | Associated[a] | Isolated | Total |
|---|---|---|---|---|
| Pleistocene | 16 | 4 | 7 | 27 |
| Pliocene | 4 | 3 | 1 | 8 |
| Miocene | 16 | 3 | 5 | 24 |
| Oligocene | 3 | | | 3 |
| Eocene | 5 | 5 | 2 | 12 |
| Paleocene | 1 | | 9 | 10 |
| Cretaceous | 17 | 17 | 13 | 47 |
| Jurassic | 9 | 4 | 3 | 16 |
| Triassic | 13 | 7 | 6 | 26 |
| Permian | 15 | 3 | 3 | 21 |
| Carboniferous | 6 | 1 | 2 | 9 |
| Devonian | 1 | 1 | 7 | 9 |
| Silurian | | | 3 | 3 |
| Ordovician | | | 1 | 1 |
| Total | 106 | 48 | 62 | 216 |

Notes: [a] If at least some bones were reported as articulated, the locality was counted in the "articulated" column; likewise for "associated."

there is no significant difference overall in the relative proportions of articulated, associated, and isolated bones in bonebed assemblages in the Mesozoic and Cenozoic. However, bonebeds may not be a fair dataset for testing this hypothesis, as they often represent mass-death events that give carnivores and scavengers a surfeit of carcasses. In such cases, carcass processing tends to be minimized relative to predation and scavenging on single kills (Lyman 1994).

## CAUSES OF DENSE BONE CONCENTRATIONS

Publications on bonebeds generally focus on descriptions of taxonomic, taphonomic, and sedimentological attributes of the assemblage, but many also include interpretations or speculations on the cause of bonebed formation. It is natural to ask "Why?" when confronted with a concentrated mass of skeletons or isolated bones, although earlier workers had a limited amount of comparative taphonomic information that could help them answer this question for any particular assemblage. This chapter uses two approaches to the issue of causality, the first being a simple tabulation

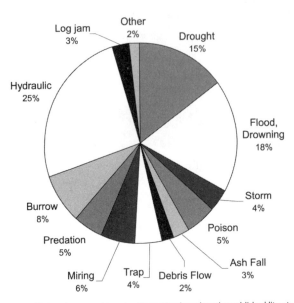

Published Causes of Bonebed Formation

N = 185 Localities

*Figure 2.9.* Overview of inferred causes of bonebed formation based on the published literature, for localities where this information was available and unambiguous. Many reports were uncertain or noncommittal about causality, and only 185 of the 315 localities in the ETE Bonebed Database are included in this diagram. "Other" includes winter cold, fire, and disease. "Hydraulic" primarily includes reworked and winnowed fluvial or nearshore concentrations.

of interpretations that were available in the published literature used for the ETE Bonebed Database, the second an attempt to organize taphonomic and contextual information in these publications and interpret it in terms of general categories representing processes and circumstances that affect bonebed assemblages. It is clear that any given bonebed represents a series of causes, from the events resulting in death to the processes leading to final burial, but the approach in this chapter focuses on the primary cause of the concentration of bodies or body parts. Without a concentrating circumstance or process, there may be fossils, but there will be no bonebed as defined at the beginning of this chapter.

## Inferred Causes of Bonebed Formation

The most frequently inferred causes of bonebed formation relate to drought, flooding, and hydraulic sorting (Fig. 2.9). Drought causes animals to congregate in or around shrinking waterholes and may decimate populations of water-dependent animals, leaving abundant, sometimes mummified carcasses that are minimally scavenged. Droughts can be followed by flooding that strips sediment from devegetated landscapes, thereby enhancing opportunities for burial. Flooding can trap and drown land-adapted animals, either as they attempt to cross swollen rivers or by overtaking them in floodplain or other alluvial or coastal habitats. In both drought and flood events, herd species may be particularly susceptible to mass-death events. Hydraulic causes represent a wide range of water-related processes that may concentrate carcasses or bones and overlap with flooding and drowning. The totals reported in Figure 2.9 depend on the published assessments; if the author specifically mentioned "flood" or "drowning" the locality was counted as such; if only hydraulic processes were specified, it was counted as hydraulic. Reports tend to assign multispecies concentrations dominated by fragmentary bones and teeth to hydraulic winnowing or sorting, if these occur in a fluvial or nearshore marine context. In contrast, bonebeds with one to three taxa and relatively complete skeletons are more commonly assigned to drought or flooding. This reflects a general tendency to infer more specific causes for bone concentrations that have suffered lower degrees postmortem taphonomic alteration.

Other inferred causes of bonebed formation typically relate to particular circumstances or events that are indicated by the taphonomic and contextual features of the deposit. Log jam assemblages are formed by an obstruction to hydraulic flow that is directly responsible for concentrating organic debris (carcasses, bones, plant material) in one place, usually in a river bed. Bone concentrations attributed to storms usually are associated with flooding (storm surges) or a temporarily lowered wave base but may also involve hypothermia, freezing, or severe sand or dust storms. Burrow collapse or infill, volcanic ash falls, and debris flows suffocate animals, and the classic examples usually involve herds or groups that were caught by unexpected catastrophic events. Traps involve pitfalls, sinkholes, and fissures that animals fall into and cannot escape, and miring is similar but occurs in mud or tar. Bonebeds attributed to predation can be formed by the killing of numbers of individuals at one time or, more commonly, by serial predation and/or bone collecting that accumulates deposits in denning, caching, or (in the case of humans) feeding/camping areas. Poison is invoked as a cause of bonebed formation, especially for aquatic organisms, when there is evidence that animals were subject to fatal biologically or geologically driven changes in water chemistry (e.g., red tides, anoxia, excess $CO_2$). Many of these causes involve death by suffocation, in one form or another, which is less destructive with respect to skeletal completeness than predation and often involves sedimentary processes that inter the animals during or shortly after death. Other causes include unusual winter cold spells, fire, and disease, which are based on the presence of circumstantial evidence such as associated sedimentary features, charcoal, carcass positioning, and bone pathologies.

### Abiotic and Biotic Processes in Bonebed Formation

The ETE Bonebed Database provides examples of bonebeds that range from monospecific, skeletally complete, single-event mass-death assemblages to time-averaged, fragmentary, multispecific lag, or "placer" bone fragment concentrations associated with widespread erosional surfaces (Appendix 2.2). One way to make sense of the variability in these data is to focus on the processes or circumstances that caused the *final concentration* of skeletal remains. From this viewpoint, "biotic" mass-death assemblages would result primarily from herding, schooling, or other types of behavior that lead to the clustering together of individuals in life, even though the cause of death may be a physical event such as a flood or volcanic ash fall. On the other hand, when the processes of erosion and winnowing during stratigraphic condensation or fluvial reworking of floodplain deposits create a bone concentration from initially dispersed remains, these bonebeds result primarily from physical processes and are here termed "abiotic." A third distinct set of processes concentrates bones when sedimentation is low relative to long-term bone input from many different individual death events, such as in traps formed by cave deadfalls or sinkholes, paleosols (stable land surfaces), abandoned channels, and deep lacustrine or marine situations where dense, autochthonous accumulations may form over long periods of time with minimal influence of transport or other active physical processes. This type of bone concentration is here termed "passive attritional." Based on the reasoning outlined above, we can examine patterns of bonebed formation through the Phanerozoic, in terms of inferences about the final concentration mechanism, as further detailed below:

> *Abiotic.* Concentrations of disarticulated bones or bone fragments bearing evidence that physical processes such as hydraulic transport or variations in sediment supply (hydraulic versus sedimentologic concentrations of Rogers and Kidwell, Chapter 1 in this volume) were primarily responsible

for forming the bonebed. Transport includes surface flow processes (water or wind) that hydraulically concentrate remains spatially, and such concentrations may also involve flow obstructions such as log jams or even stabilized vertebrate remains. Variations in sediment supply cause stratigraphic (e.g., hiatal) concentrations and time averaging of remains through sediment starvation and stratigraphic condensation. Positive net sedimentation may bury preexisting concentrations of vertebrate remains, and catastrophic events can form obrution deposits by overwhelming and burying groups of live animals. However, in this classification scheme, obrution bonebeds are included under "biotic" (see below). Fluvial reworking of floodplain deposits combines hydraulic surface processes and low net sedimentation, as well as concentrations formed by flow obstructions. These bonebeds are usually multispecific, and clustering of remains due to current action, erosion, sediment starvation, and so forth, indicates a primary abiotic cause for the bonebed.

*Biotic.* Concentrations of associated to articulated carcasses or skeletal parts that bear evidence for mass- or clustered death events (e.g., death in an aestivation burrow, clustering behavior during a drought, group miring or drowning, burial of a herd in a volcanic ash fall). The obrution assemblages of Rogers and Kidwell (Chapter 1 in this volume) fall under this category because of the importance of a biological or ecological cause for the clustering of individuals prior to a catastrophic death event caused by unusual sedimentation. Most biotic assemblages formed over short time intervals and are mono- or paucispecific. However, this category also includes concentrations of remains formed by other organisms, such as predators and bone collectors (e.g., owls, hyenas), which may be time averaged over long intervals and have a large number of species. In all cases described above, the clustering of individuals immediately prior to or shortly after death indicates a primary biotic cause for the bonebed. This also means that there may be strong taxonomic, body size, or other biases (relative to the contemporaneous paleocommunity) resulting from species herd structure or predator selectivity.

*Passive attritional.* Concentrations of autochthonous, articulated, disassociated, and/or fragmentary remains that occur in situations where sediment accumulation is slow relative to bone input (e.g., paleosols, low-energy channel fills, water holes, some lakes, fissure or sinkhole fills). These situations can be characterized as sediment-starved, relatively passive (hiatal) contexts for concentration that accumulate time-averaged, multispecific attritional assemblages. They represent low sedimentation rates in an overall aggrading context favorable to bone burial combined

Figure 2.10. Example of a fluvial bonebed, Omo Locality 33, a channel lag bonebed in the base of Tuff F, Shungura Formation, Omo Group, southern Ethiopia (Johanson et al. 1976). A. French O.33 excavation in 1973 under the direction of Claude Guillemot. The bones (dark brown) are distributed within a tuffaceous sand at the base of the tuff-filled channel. B. Closer view of the bonebed, showing variable preservation of the mammalian skeletal elements. Circled object is a maxilla of a large suid, one of the better preserved fossils in this excavation. American excavation L.398 sampled the same bonebed several kilometers to the north (Johanson et al. 1976), indicating that the channel lag bone concentration extended over a very large, linear area coinciding with the paleochannel thalweg.

with a sustained input of vertebrate remains, which may be further concentrated by behavior that results in greater numbers of carcasses in certain areas (e.g., multispecies clustering around waterholes).

*Mixed.* Concentrations that clearly represent mixtures of at least two of the other categories, such as channel bar assemblages that include both abundant disarticulated, multispecific remains and partial carcasses or disassociated remains of one or two species indicating a mass death (e.g., Fig. 2.10).

The relative frequencies of these different bonebed categories through the Phanerozoic (Fig. 2.11, Table 2.6) indicate dominance of abiotic processes in the early part of the vertebrate record and in the Paleocene. Otherwise, biotic bonebed-forming processes are dominant for most of the time intervals. The Permian has an especially large number of bonebed localities attributed to biotic processes; these typically are mass-death assemblages of amphibians or reptiles associated with drought conditions. The passive attritional category appears to be most important in the Carboniferous, where bonebeds are associated with sapropels in abandoned channels in coal-bearing strata.

The relationship of the three categories of bonebeds to skeletal completeness highlights differences that affect the quality of biological or ecological information in bonebed assemblages (Fig. 2.12). Abiotic processes typically result in less complete, size-sorted skeletal material but offer more complete censuses of the species present over relatively large geographic areas or through ecologically long time spans. Biotic processes often result in short-term (snapshot) samples with relatively complete skeletal preservation and, hence, provide a wealth of anatomical information, but these assemblages are usually limited to a few species or subject to possible selectivity by bone-concentrating predators/scavengers. The passive attritional category, with its combination of time averaging and minimal transport or reworking of bones, provides a different type of sample with more complete skeletal material as well as the potential for recovery of multiple species that contributed to the assemblage through time, or at different times. Although there is overlap in the number of taxa that can be present in localities formed by these three types of processes (Fig. 2.13), it is clear that biotic assemblages tend to provide limited taxonomic samples compared with abiotic and passive attritional assemblages.

A.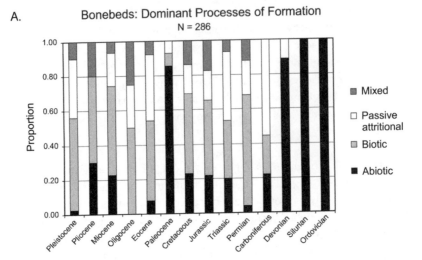

Bonebeds: Dominant Processes of Formation
N = 286

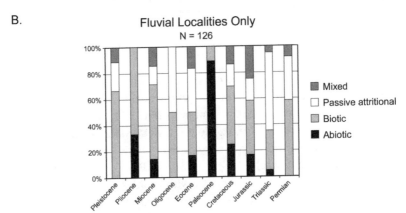

B. Fluvial Localities Only
N = 126

*Figure 2.11.* A. The pattern through the Phanerozoic of bonebeds resulting from different categories of bonebed-forming processes, based on information from 286 localities in the ETE Bonebed Database and the assessment of the author (Table 2.6). The Oligocene sample is small (Table 2.1B) and may not give a fair representation of the proportions for this time interval. However, the anomalous dominance of abiotic localities (mostly winnowed fluvial concentrations) in the Paleocene indicates a significant break in the otherwise similar patterns in dinosaur- versus mammal-dominated assemblages in the Mesozoic and Cenozoic. B. The pattern of bonebed forming processes for a subset of fluvial localities from Permian to Pleistocene (126 that could be categorized as biotic, abiotic, passive attritional or mixed), showing that the pattern in A is driven largely by bonebeds in the fluvial paleoenvironmental context.

## DISCUSSION

The above results explore patterns in the data captured for the ETE Bonebed Database, and much more could be done with multivariate analysis to test for additional patterns. Moreover, with continuing efforts to include

Table 2.6. Distribution through the Phanerozoic of bonebeds attributed to different types of bonebed-forming processes

| Period/Epoch | Biotic | Abiotic | Passive Attritional | Mixed | Total |
|---|---|---|---|---|---|
| Pleistocene | 22 | 1 | 14 | 4 | 41 |
| Pliocene | 5 | 3 | | 2 | 10 |
| Miocene | 16 | 7 | 6 | 2 | 31 |
| Oligocene | 4 | | 2 | 2 | 8 |
| Eocene | 6 | 1 | 5 | 1 | 13 |
| Paleocene | 1 | 12 | 1 | | 13 |
| Cretaceous | 30 | 15 | 11 | 9 | 65 |
| Jurassic | 10 | 5 | 4 | 4 | 23 |
| Triassic | 10 | 6 | 12 | 2 | 30 |
| Permian | 16 | 1 | 5 | 3 | 25 |
| Carboniferous | 2 | 2 | 5 | | 9 |
| Devonian | | 8 | 1 | | 9 |
| Silurian | | 8 | | | 8 |
| Ordovician | | 1 | | | 1 |
| Total | 122 | 70 | 66 | 29 | 286 |

bonebed information from different continents, it should be possible to capture a sample that more fairly represents global-scale trends. Based on the results available at present, however, it is clear that processes that cause bonebeds have operated throughout the span of vertebrate history. All major groups of vertebrates have been affected by mass-mortality events and other mechanisms or circumstances that concentrate their remains. The questions posed at the beginning of this chapter are considered in light of the analysis of the sample provided by the ETE Bonebed Database.

1. *How does bonebed frequency and type correlate with broad climatic and other environmental trends in Phanerozoic history, and are there more bonebeds during environmental "bad times" for vertebrates?* Overall, there is no strong signal in the patterns of bonebed occurrence through the Phanerozoic (Figs. 2.3A, 2.11A) that would support this idea. This said, however, there is a hint of increased numbers of biotic bonebed-forming events in the Permian, Cretaceous, and Pleistocene. Most of the Cretaceous localities occur in the last 20 million years of the Cretaceous, suggesting an unusual tendency for large land vertebrates to end up in mass graves during this time. The Pleistocene record is similar, although in both cases there are so many possibly confounding variables (especially continent and outcrop

A.

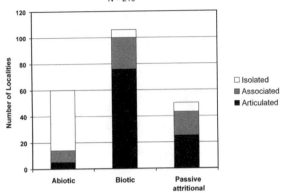

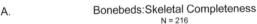

B.

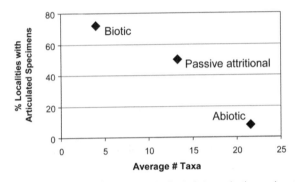

*Figure 2.12.* A. Relationship of different degrees of skeletal articulation to the three major categories of processes responsible for bonebed formation (abiotic, biotic, and passive attritional), showing that articulated specimens are preserved in all three categories but are more commonly associated with biotic processes. B. Inverse relationship of taxonomic richness and skeletal completeness/articulation, showing the differences in species representation for bonebeds attributed to abiotic, biotic, and passive-attritional processes.

biases favoring documentation of localities in these time intervals) that it is not possible to assess the significance of a link to "bad times" for vertebrates without further analysis. On the other hand, the Permian, with its assemblages of water-dependent vertebrates that appear to have been concentrated in lakes and ponds prior to being killed by drought conditions, provides a more credible signal that could be tied to the highly continental climatic extremes of Pangaea. Examples from several different continents and paleolatitudes support this pattern.

2. *Were some ecomorphs or taxonomic groups more susceptible to mass deaths?* Further work is needed on possible correlations between body size

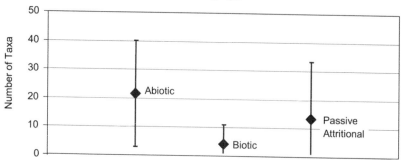

*Figure 2.13.* The mean and standard deviation of numbers of taxa preserved in each of three major categories of bonebed-forming processes. Although all three categories overlap, bonebeds inferred to result primarily from biotic processes have a narrower range (2–10) of taxa.

and tendency to form different types of bonebeds (MO, PA, MU, also biotic versus abiotic), but the present data indicate that all body sizes and major groups have been susceptible to mass deaths leading to bonebed formation. Sociality likely plays a key role in making a species susceptible to biotic bonebed-forming processes, but social behavior such as herding or schooling is not necessarily linked with body size or other ecomorphic characters. In the Mesozoic, there is evidence that predators as well as herbivores congregated in large groups that ended up in bonebeds; this pattern does not seem to carry over into the mammal-dominated record of the Cenozoic. Bird-dominated bonebeds are rare in the ETE Bonebed Database, but this may simply follow birds' general tendency to be uncommon in the fossil record relative to other types of vertebrates. The large percentages of bonebed localities dominated by mammals and dinosaurs (Fig. 2.5, Table 2.3) undoubtedly is affected by the paleontological "interest factor"; mass kills of fish and amphibians are probably much more common in the record than indicated by the published literature.

3. *Do changes in the bonebed record relate to evolutionary trends in skeletonization, body size, habitat utilization, and bone-ingesting capabilities of vertebrates?* The bonebed record clearly tracks the vertebrate expansion from aquatic (marine and freshwater) to land habitats and also indicates the ecological shift in water dependency from Permian to Triassic. There is no obvious signal relating to degrees of skeletonization, except for the paucity of bird bonebeds, but body size may play an important role in controlling the frequency of biotic bonebeds during times when large vertebrates

(e.g., >200 kg) were common in terrestrial and marine ecosystems. The shift to abiotic assemblages dominated by teeth and jaws in the Paleocene is an exception to the overall consistency of bonebed forming processes through the Phanerozoic. This anomaly suggests a fundamentally different "taphosystem" during a time when most land vertebrates were small. Concerning the changing effects of evolving predator/scavenger bone-consuming capabilities, patterns of skeletal completeness, as measured by articulation in bonebed localities, do not indicate a shift between the bone-swallowing strategies of the Mesozoic and the bone-crushing strategies of the Cenozoic. There may be other indicators of changes in patterns of bone modification, but these have yet to be documented well enough to permit through-time analysis.

4. *Bonebeds provide a significant portion of the vertebrate fossil record; how might changes in taphonomic processes that create bonebeds affect our view of vertebrate history?* There are two issues raised by this question: the effects of a changing balance of different types of bonebed-forming processes, and the potential effects of "oversampling" bonebeds in fluvial environments.

The trends through time highlighted in this study suggest taphonomic "domains" that apply to broad portions of the vertebrate record, with a dominance of abiotic assemblages of aquatic organisms from the Ordovician through Carboniferous, and a balance of abiotic and biotic assemblages from the Permian through Mesozoic and again in the Cenozoic, but divided by the anomaly of abiotic dominance in the Paleocene. This indicates that the sum of information derived from bonebed assemblages will be more or less equivalent for much of the post-Carboniferous record, in terms of representation of taxonomic and ecological variables such as species richness. However, vertebrate diversity tallies in the Paleocene may be artificially elevated because of the temporary shift to abiotic processes of concentration and the focus on microsite collection strategies. This is somewhat mitigated by efforts to collect and document microsites in the Cretaceous and Eocene, but it would be worth taking a look at the micro- versus macrosite records to determine the extent of this potential bias in Phanerozoic vertebrate diversity.

It is clear that if ≥50% of the bonebed record derives from fluvial contexts, then our view of vertebrate history, including our understanding of the anatomy of well-preserved species, is biased with respect to the range of vertebrate ecomorphs that dwelled in fluvial habitats versus other types of continental and marine environments. However, since fluvial environments are represented throughout the Phanerozoic (Tables 2.1B, 2.2), in essence we have, by default, a more or less "isotaphonomic" sample of vertebrate history in the fluvial ecosystem. Occasionally the record provides glimpses of other habitats, such as the dunes of the Cretaceous Gobi Desert (e.g., Jerzykiewicz et al. 1993; Loope et al. 1998) and the lacustrine deposits Liaoning Province, China (e.g., Zhou et al. 2003), reminding us of the existence of other ecologies, vertebrate communities, and taphonomic circumstances that occasionally result in spectacular bonebeds. It would be interesting, in future analyses, to examine the degree of taxonomic and ecomorphic novelty provided by such unusual preservational circumstances and compare this with the fluvial bonebed record.

## CONCLUSION

This study is a initial attempt to take a broad view of the bonebed record, and although it has revealed some general patterns, probably its most important contribution is to open the door to a wealth of interesting questions to pursue in the future. We won't really know which aspects of the patterns are real and which are sampling artifacts until we further investigate the "odd times," such as the Paleocene, Oligocene, and Pleistocene, and expand the sample to more adequately represent the different continents. However, the ETE Bonebed Database, even as it now stands, provides opportunities to take the same type of bonebeds and do comparative study through time and/or space of the taphonomic features and paleoecology. For example, one could do an isotaphonomic investigation of miring events (8 localities in Mesozoic, Cenozoic), burrow collapses (12, mainly in Permian and some in Cenozoic), drought (25, being careful about varying levels of taphonomic equivalence), hydraulic sorting or reworking (26), or stratigraphic condensation (19). Alternatively, one could examine assemblages representing biotic or passive attritional concentration mechanisms through time. This book provides the increased understanding of bonebeds—classification, examples, and methodologies—that will make comparative research of this kind an exciting and accessible goal for the future.

## ACKNOWLEDGMENTS

I owe the volume editors, especially Ray Rogers, a debt of thanks for their patience and persistence in making this book a reality. Ray Rogers and Sue Kidwell provided very helpful discussion of the ideas and terminology relating to bonebeds, resulting in a more united perspective with regard to initially different classification systems. I thank Christine O'Reilly, who was responsible for much of the input of bonebed data to the ETE Database. I thank Alan Cutler and my husband, Bill Keyser, for contributing to fieldwork and other research relating to this study, and the many individuals who assisted with our bonebed field odyssey in 1994. I also thank Bill and my daughters, Kristina and Sarah, for their help and patience. Long-term support for the ETE Program from the National Museum of Natural History and the Smithsonian Institution is gratefully acknowledged. This is ETE Publication no. 155.

## REFERENCES

Behrensmeyer, A.K., S.M. Kidwell, and R.A. Gastaldo. 2000. Taphonomy and paleobiology. Paleobiology 26(4, supplement):103–144.

Behrensmeyer, A.K. 1991. Terrestrial vertebrate accumulations. Pp. 291–335 in Taphonomy: Releasing the data locked in the fossil record. P. Allison and D.E.G. Briggs, eds. Plenum, New York.

Behrensmeyer, A.K., A.L. Deino, A. Hill, J. Kingston and J.J. Saunders. 2002. Geology and chronology of the middle Miocene Kipsaramon fossil site complex, Muruyur Beds, Tugen Hills, Kenya. Journal of Human Evolution 42:11–38.

Eberth, D.A., M. Shannon, and B.G. Nolan. This volume. A bonebeds database: Classification, biases, and patterns of occurrence. Chapter 3 in Bonebeds: Genesis, analysis, and paleobiological significance. R.R. Rogers, D.A. Eberth, and A.R. Fiorillo, eds. University of Chicago Press, Chicago.

Gradstein, F.M., and J.G. Ogg. 1996. A geologic time scale. Episode 9 (1 and 2); see also www.geosociety.org/science/timescale/timescl.pdf.

Hill, A., A.K. Behrensmeyer, B. Brown, A. Deino, M. Rose, J. Saunders, S. Ward, and A. Winkler. 1991. Kipsaramon: A lower Miocene hominoid site in the Tugen Hills, Baringo District, Kenya. Journal of Human Evolution 20:67–75.

Janis, C.M., K.M. Scott, and L.L. Jacobs. 1998. Evolution of Tertiary mammals of North America. Cambridge University Press, New York.

Jerzykiewicz, T., P.J. Currie, D.A. Eberth, P.A. Johnston, E.H. Koster, and Jia-Jian Zheng. 1993. Djadokha Formation correlative strata in Chinese Inner Mongolia: An overview of the stratigraphy, sedimentary geology, and paleontology and comparisons with the type locality in the pre-Altai Gobi. Canadian Journal of Earth Sciences 30:2180–2195.

Johanson, D. C., M. Splingaer, and N. T. Boaz,1976. Paleontological excavations in the Shungura Formation, Lower Omo Basin, 1969-73. Pp. 402-420 in Earliest Man and Environments in the Lake Rudolf Basin: Stratigraphy, Paleoecology, and Evolution. Chicago University Press, Chicago.

Loope, D.B., L. Dingus, C.C. Swisher III, and C. Minjin,1998. Life and death in a Late Cretaceous dune field, Nemegt Basin, Mongolia. Geology 26:27–30.

Lyman, R.L. 1994. Vertebrate taphonomy. Cambridge manuals in archaeology. Cambridge University Press, Cambridge, U.K.

Rogers, R.R. 1990. Taphonomy of three dinosaur bone beds in the Upper Cretaceous Two Medicine Formation of northwestern Montana: Evidence for drought-related mortality. Palaios 5:394–413.

Rogers, R.R., and S.M. Kidwell. This volume. A conceptual framework for the genesis and analysis of vertebrate skeletal concentrations. Chapter 1 *in* Bonebeds: Genesis, analysis, and paleobiological significance. R.R. Rogers, D.A. Eberth, and A.R. Fiorillo, eds. University of Chicago Press, Chicago.

Seilacher, A., W.-E. Reif, and F. Westphal. 1985. Sedimentological, ecological and temporal patterns of fossil Lagerstätten. Philosophical Transactions of the Royal Society of London B 311:5–23.

Symonds, W.S. 1858. On bone beds and their characteristic fossils. Geologist 1:15–18.

Van Valkenburgh, B. 1999. Major patterns in the history of carnivorous mammals. Annual Review of Earth and Planetary Sciences 27:463–93.

Weigelt, J. 1986 (1927). Recent vertebrate carcasses and their paleobiological implications J. Schaefer, transl. University of Chicago Press, Chicago.

Zhou, Z., P.M. Barrett, and J. Hilton. 2003. An exceptionally preserved Lower Cretaceous ecosystem. Nature 421:807–814.

**APPENDIX 2.1. COMPARISONS OF TERMS AND DEFINITIONS**

Comparison of terms and definitions regarding bonebed descriptive classification; see other papers in this volume

| | | Rogers and Kidwell | Behrensmeyer | Eberth et al. |
|---|---|---|---|---|
| **Description** (definitions) | **Bonebeds** | A "relative concentration" of vertebrate hardparts preserved in a localized area or stratigraphically limited sedimentary unit (e.g., bed, horizon, stratum) and derived from more than one individual | A single sedimentary stratum with a bone concentration that is unusually dense (often but not necessarily exceeding 5% bone by volume), relative to adjacent lateral and vertical deposits, and consisting of remains from more than one individual | An accumulation of vertebrate fossils consisting of the complete or partial remains of more than one vertebrate animal in notable concentration along a bedding plane, erosional surface, or throughout a single bed; can be identified where a concentration of fossil bone exceeds the background concentration in the host formation or facies |
| **Size (animals or elements)** | | Follows Eberth et al. (Chapter 3 in this volume) in differentiation of microfossil and macrofossil bonebeds<br><br>Bone sands: particles are rounded bone fragments of sand grade | Not differentiated | Macrofossil: elements >5 cm maximum dimension, may be disarticulated to articulated; includes most described bonebeds (75% or more of the number of identifiable specimens (NISP))<br><br>Microfossil: elements ≤5 cm maximum dimension (75% or more of NISP), may be disarticulated to articulated. |

| | | |
|---|---|---|
| **Diversity (taxonomic richness)** | | Mixed: macro- and micro-components both present. (each represented by at least 25% of NISP)<br><br>Low: 2–9 taxa; in the case of 6–9 taxa, assignment must be considered relative to host formation<br><br>High: ≥6 taxa; in the case of 6–9 taxa, assignment must be considered relative to host formation |
| **Relative abundance** | Monospecific: bonebed consists of remains of two or more individuals of the same species; may include minor remains from other species ("background")<br><br>Paucispecific: bonebed dominated by remains of two or three different species; may include minor remains from other species<br><br>Multispecific: skeletal remains from multiple species (>3) | Monotaxic: one taxon, most often articulated or associated skeletal elements<br><br>Monodominant: multiple taxa (high or low diversity), with one taxon >50% MNI (minimum number of individuals or NISP)<br><br>Multidominant: multiple taxa (high or low diversity) with no taxon >50% of MNI or NISP |

(Continued)

**APPENDIX 2.1.** (continued)

Comparison of terms and definitions regarding causes of bonebed formation; see other papers in this volume

| | | Rogers and Kidwell | Behrensmeyer | Eberth et al. |
|---|---|---|---|---|
| **Genesis** (cause) | **Biogenic concentrations** | Intrinsic: resulting from "normal" activity or behavior of the vertebrates that are preserved in the assemblage Extrinsic: resulting from the behavior or activity of other organisms (e.g., predators) | Biotic: concentrations caused primarily by biological processes or agents, including gregarious behavior that makes some species vulnerable to catastrophic death of large numbers of individuals; or also predator arenas or dens, roosts, and lairs where carcass parts are concentrated by carnivores resulting in many remains available for burial as a dense bone concentration | |

| **Physical concentrations** | Hydraulic concentrations result from the action of surface flows or wave action (fluvial channels and strandlines) Sedimentologic concentrations result from physical patterns of sedimentation linked to sediment budget, i.e., obrution, sediment starvation (hiatal), stratigraphic condensation (lag)<br><br>Mixed: clear combination of any of the above (see their Fig. 1.6) | Abiotic: physical concentrations caused by hydraulic (and sometimes eolian) processes that winnow and sort bones and teeth by size, density, and shape, including stratigraphic condensation, fluvial channel lags, strandline accumulations, bone-rich ravinements<br><br>Passive attritional: combination of sediment starvation and sustained biologically mediated input of carcasses or bones from many different, unrelated individual death events, such as around water holes or in traps, fissures, some lakes, paleosols<br><br>Mixed: clear combinations of any of the above |

# CHAPTER 3

# A Bonebeds Database: Classification, Biases, and Patterns of Occurrence

*David A. Eberth, Matthew Shannon, and Brent G. Noland*

## INTRODUCTION

In this chapter we present and review a database of 383 citations that refer to more than 1,000 bonebeds (Appendix 3.1; also available online at www.press.uchicago.edu/books/rogers/). Our goals in compiling and assessing this database are to

- identify, characterize, and document different types of bonebeds;
- identify historical and scientific biases in the treatment of bonebeds;
- demonstrate the relative frequency of different types of bonebeds in the database;
- identify and quantify patterns of occurrence for the different types of bonebeds; and
- identify and quantify patterns of association between bonebed type and paleoenvironment.

Ultimately, our intention is to provide additional insight into how bonebeds formed in the past, and to better understand what they can tell us about the ancient environments within which they formed.

The citations in our database are drawn almost exclusively from the scientific literature. Accordingly, the database reflects biases in our search and compilation methods, as well as biases in how original researchers selected, worked, and collected data. In this context, the patterns and conclusions presented here should be regarded only as testable hypotheses, rather than statements of ultimate fact.

Data were entered into Microsoft Excel (Appendix 3.1), and the data and calculations shown in Figures 3.1–3.4 and Tables 3.1 and 3.2 were all drawn from the final data set. The "pivot table" function was employed both as a check on entry format consistency and as a means of counting and comparing different entries.

## DATABASE TERMINOLOGY

Each citation in the database includes some or all of the following information:

- Author(s)
- Title/publication date/source
- Taxonomic content
- Site/bonebed name(s)
- Number of sites
- Geographic location(s)
- Stratigraphy
- Element size (macrofossil, microfossil, or mixed)
- Taxonomic diversity (high versus low)
- Relative taxonomic abundance (monotaxic, monodominant, multidominant)
- Depositional system (setting)
- Depositional environment
- Host facies
- Depositional system code

Under "taxonomic content" we employ abbreviations (Table 3.1) that pertain to higher-order taxa, many of which are paraphyletic (e.g., fish, amphibian, and reptile). Paraphyletic taxa are deemed useful here because they provide broad paleoecologic insights. For example, bonebeds that include assemblages of "fish" and "amphibians" in association with non-marine facies are indicative of aquatic paleoenvironments that occurred either at the bonebed site or nearby. We use the term "dinosaur" (d) in reference to nonavian dinosaurs, and we use the term "nondinosaur reptile" (ndr) in reference to amniotes exclusive of dinosaurs, birds, and mammals.

The terms under the heading "depositional system" (e.g., alluvial, marine) are interpretive and are intended to characterize the regional

Table 3.1. Database statistics and abbreviations

| | n | % | | n | % |
|---|---|---|---|---|---|
| **Citations** | | | **Taxonomic Content** | | |
| # of citations | 383 | | Dinosaur | 183 | 33% |
| # of bonebeds included | 1084 | | Non-Dinosaur Reptile | 91 | 16% |
| Publication dates | 1856–2006 | | Large Mammal | 78 | 14% |
| pre-1900 | 5 | 1% | Fish | 79 | 14% |
| 1901–1950 | 17 | 4% | Amphibian | 60 | 11% |
| 1951–1975 | 28 | 7% | Small Mammal | 51 | 9% |
| 1976–2005 | 333 | 87% | Bird | 20 | 4% |
| English | 377 | 98% | **Classification (size)** | | |
| Abstracts | 94 | 25% | Macrofossil | 254 | 59% |
| Manuscripts and other | 290 | 76% | Microfossil | 116 | 27% |
| **Stratigraphy** | | | Mixed | 57 | 13% |
| Mesozoic | 220 | 56% | **Classification (diversity)** | | |
| Tertiary | 89 | 23% | High | 237 | 58% |
| Paleozoic | 48 | 13% | Low | 144 | 35% |
| Quaternary | 32 | 8% | Unknown | 30 | 7% |
| **Continental Distribution** | | | **Classification (relative abundance)** | | |
| North America | 276 | 72% | Montotaxic | 32 | 8% |
| Europe | 39 | 10% | Multitaxic | 288 | 74% |
| Asia | 26 | 7% | Unknown | 68 | 18% |
| Africa | 19 | 5% | **Depositional System Code** | | |
| South America | 13 | 3% | | | |
| Australia | 5 | 1% | | | |
| Antarctica | 4 | 1% | | | |

**Depositional System Code**

| | n | % | | n | % |
|---|---|---|---|---|---|
| CHLSP | 139 | 27% | MAR | 30 | 6% |
| WTLND | 132 | 26% | DBF | 17 | 3% |
| PSOL | 54 | 10% | EOL | 7 | 1% |
| PESH | 40 | 8% | O | 30 | 6% |
| LAC | 31 | 6% | U | 36 | 7% |

**Database Abbreviations**

**Taxa**

| a | amphibian |
|---|---|
| b | bird |
| d | dinosaur |
| f | fish |
| lm | large mammal |
| ndr | nondinosaur reptile |
| sm | small mammal |

**Depositional System Codes**

| CHLSP | channel, levee, splay |
|---|---|
| DBF | debris flow, alluvial fan deposit, sheet flood |
| EOL | eolian |
| LAC | lacustrine |
| MAR | marine |
| O | other |
| PESH | paralic setting, estuarine, shoreline |
| PSOL | paleosol, well drained alluvial plain |
| WTLND | wetland, poorly drained alluvial plain |
| U | unknown |

paleogeographic context of a bonebed. The terms used under the heading "depositional environment" are also interpretive but refer to the environment of deposition of the bonebed's host sediments. The terms used under the heading "facies" are descriptive and are standard in most sedimentological studies. They refer to lithologies and sedimentary structures that are present at the bonebed site.

Entries under the heading "depositional system code" consist of one of 10 acronyms (Table 3.1), each of which refers to a depositional system that is made up of a group of genetically related depositional environments (Davis, 1983). For example, the acronym CHLSP refers to a depositional system made up of channel and channel-margin deposits (e.g., levees and splays). Because the quality of paleoenvironmental data retrievable from the literature is highly variable, we employ this broadly inclusive coding system to facilitate searches for patterns of association between different kinds of bonebeds and paleoenvironments (see below).

## BONEBEDS DEFINED

Before bonebeds can be classified or compared, there must be agreement as to the definition of a bonebed. Here, a bonebed is defined as consisting of the complete or partial remains of more than one vertebrate animal in notable concentration along a bedding plane or erosional surface, or throughout a single bed. This definition is similar to Behrensmeyer's (1991, p. 293) definition of "vertebrate accumulations" but does not limit the assemblage to a "well-defined area." Indeed, some marine bonebeds are traceable for tens of kilometers and are useful stratigraphic markers (e.g., Rogers and Kidwell, 2000; Schröder-Adams et al., 2001).

In our use, a "notable concentration" of vertebrate fossils means that the accumulation exceeds the normal (background) occurrence of fossils in the host formation or local strata. Clearly, this approach avoids using a strict fossil/matrix ratio or phosphate content (e.g., minimum 5.0% fossil/matrix by volume [Behrensmeyer, 1999]). Instead, it relies on the opinions and experience of the investigators as to what is normal versus unusual on a case by case basis (e.g., Smith's "fossil-rich occurrence" of *Diictodon* [1993]; Eberth's "microfossil sites" that consist of 0.03%–0.50% fossils by weight [1990, Table 3]).

## WHY CLASSIFY BONEBEDS?

A bonebed classification provides a framework in which to assess whether a bonebed is "typical" or "unusual" in the context of its age, host formation, and relation to other bonebeds. In turn, this assessment can be a time-saving tool for establishing hypotheses for bonebed origins and taphonomic histories (e.g., Rogers and Kidwell, Chapter 1 in this volume). A practical classification also provides opportunities to conduct comparative taphonomic studies and to examine trends in the kinds of bonebeds that occur within a given stratigraphic unit or geographic area (e.g., Behrensmeyer, 1975; Eberth, 1990; White et al., 1998; Brinkman et al., Chapter 4 in this volume). Although published information about macroscale stratigraphic trends in bonebeds are rare, identifying these trends is important because of the potential insights they can provide with regard to the evolution of vertebrate ecosystems and taphonomic processes through time (e.g., Carpenter, 1988; Behrensmeyer and Cutler, 1994; Chure et al., 2000; Eberth et al., 2001; Behrensmeyer, Chapter 2 in this volume).

Classifying a bonebed prior to excavation can provide insight into the potential significance and origins of a site. Accordingly, the working hypotheses that follow can be extremely helpful in refining the design of a bonebed-based research project, potentially saving money and time. Establishing working hypotheses early on can also help a researcher characterize aspects of the depositional system and paleoecology of the host formation or unit, especially during surveys in "new" areas (e.g., White et al., 1998; Eberth et al., 2000).

Most bonebeds in the accompanying database are easily classified using one or more of the following three measures: (1) element size; (2) taxonomic diversity; and (3) relative taxonomic abundance (Fig. 3.1; Table 3.2). Element size, taxonomic diversity, and relative taxonomic abundance can often be assessed directly from the literature or can be quantified in the field during an examination of fossil bone surface concentrates. These measures relate exclusively to the fossils in the bonebed and require no geologic background, which is clearly advantageous when a paleontologist is unfamiliar with geologic terminology (contrast this with the more geologically oriented approaches of Reif [1971, 1982], Sykes [1977], Antia [1979], and Eberth and Getty [2005]).

## A BONEBED CLASSIFICATION

### Element Size: Microfossil, Macrofossil, and Mixed Bonebeds

The dominant size of elements comprising a bonebed can be used to effectively distinguish between vertebrate microfossil, macrofossil, and mixed (microfossil and macrofossil) bonebeds. Although these categories (and terms) are not ubiquitously employed, their use beyond this volume is encouraged because they succinctly refer to the spectrum of all bonebeds in nature, and, more importantly, they reflect a commonly recognized dichotomy among bonebeds: vertebrate microfossil bonebeds versus all others. The term "vertebrate microfossil bonebed" (Eberth, 1990; Peng

## Macrofossil Dominated

### Low-Diversity
Monotaxic
> *Camp, 1980; Berman et al., 1987; Coria, 1994*

Multitaxic
> Monodominant
> *Turnbull & Martill, 1988; Eberth and Getty, 2005*
> Multidominant
> *Badam, et al., 1986; Eberth et al., 2000*

### High Diversity
Multitaxic
> Monodominant
> *Ryan et al., 2001; Emslie et al., 1996*
> Multidominant
> *Andrews & Ersoy, 1990; Coombs & Coombs, 1997*

## Mixed (macrofossil & microfossil)

### Low-Diversity
Multitaxic
> Monodominant
> *Maxwell and Ostrom, 1995; Bell and Padian, 1995*
> Multidominant
> *Benton et al., 1997*

### High Diversity
Multitaxic
> Monodominant
> *Sander, 1992; Lehman, 1982*
> Multidominant
> *Voorhies, 1969; Varricchio, 1995*

## Microfossil Dominated

### Low-Diversity
Monotaxic
> *Henrici & Fiorillo, 1994; Breithaupt & Duvall, 1986*

Multitaxic
> Monodominant
> *Unknown*
> Multidominant
> *Antia, 1981; Sanz et al., 2001*

### High-Diversity
Multitaxic
> Monodominant
> *Emslie & Morgan, 1994; Ryan et al., 2000*
> Multidominant
> *Brinkman, 1990; Badgley et al., 1998*

*Figure 3.1.* The bonebed classification system used here (Table 3.2) and highlighted (italics) with examples from our bonebeds database (Appendix 3.1).

Table 3.2. Bonebed classsification matrix. Number of database citations in parentheses.

| Taxon. Diversity & Abundance. / Element Size | | | Macrofossil Dominated (254) | Mixed Macro/Micro (57) | Microfossil Dominated (116) |
|---|---|---|---|---|---|
| Monotaxic (32) | | | Common | ✕ | Rare |
| Multitaxic (288) | Low-Diversity | Mono-dominant | Abundant | Rare | Unknown |
| | | Multi-dominant | Rare | Rare | Rare |
| | High-Diversity | Mono-dominant | Common | Rare | Rare |
| | | Multi-dominant | Abundant | Abundant | Abundant |

et al., 2001; Brinkman et al., 2005) equates broadly with the intended meaning of the more often used terms: microsite, microvertebrate locality, and microvertebrate site. Unfortunately, however, there exists little clarity or consensus as to what, exactly, constitutes a vertebrate microfossil bonebed. Some workers have characterized them as containing the remains of vertebrates of small body mass (1 kg or less [Pratt, 1989]; 5 kg or less [Behrensmeyer, 1991]), which, we believe, excludes the potentially significant contribution of small elements from large vertebrates (e.g. crocodiles, dinosaurs). Many other workers have described these bonebeds as concentrations of disarticulated, small, physicochemically resistant vertebrate hardparts (Korth, 1979; Maas, 1985; Brinkman, 1990; Blob and Fiorillo, 1996; Fiorillo and Rogers, 1999; Peng et al., 2001). Although we regard this latter definition as a truer reflection of the broadly intended use of the terminology, we also believe that it requires further clarification. Specifically, must vertebrates in a microfossil bonebed always be disarticulated? And, what are the size limits for vertebrate microfossils? In an effort to standardize practice and provide a workable means of differentiating vertebrate microfossil assemblages from vertebrate macrofossil and mixed (microfossil-macrofossil) assemblages, we propose the following simple and, hopefully, useful criteria.

We regard a microfossil bonebed as containing an assemblage of elements, bone fragments, and bone pebbles where >75% of the identifiable specimens (NISP) are smaller than 5 cm in maximum dimension (Wood et al., 1988; Fiorillo and Rogers, 1999). Conversely, we regard a macrofossil bonebed as containing an assemblage of disarticulated to articulated elements, where >75% of the NISP are larger than 5 cm in maximum dimension. Sites containing significant mixtures of macrofossils and microfossils—more than 25% NISP each—are referred to here as "mixed bonebeds."

Setting the size distinction at 5 cm to distinguish between macrofossils and microfossils reflects both our personal experience and historical procedures (e.g., Brinkman, 1990; Eberth, 1990; Blob and Fiorillo, 1996) and, thus, avoids a reclassification of many sites that are already referenced in the scientific literature as "microsites" (e.g., Peng et al., 2001; Sankey, 2001).

Setting the content limits for macrofossil and microfossil bonebeds at 75% or more, rather than making microfossil and macrofossil assemblages completely exclusive, allows for the realistic possibility that some elements have entered an assemblage by processes that played only a minor role in the formation of the bonebed. Accordingly, this approach allows for the presence of a small number of exotic or background elements in either a microfossil or macrofossil bonebed without having to designate it as "mixed."

Our classification is based on NISP rather than N because of the potential problem caused by bone fragments. For example, at Dalton Wells, a stacked succession of Lower Cretaceous dinosaur bonebeds in Eastern Utah (Eberth et al., 2006), tiny bone chips and fragments that formed due to syndepositional trampling dominate the site. However, almost every identifiable element is larger than 5 cm (B. Britt, pers. comm., 2000).

The three "element size" categories in the classification matrix in Table 3.2 are not intended to imply a continuum of decreasing or increasing fossil size. Instead, we regard these as discrete and, often, easily recognized categories. A review of the literature and our unpublished data suggests that mixed bonebeds often show a multimodal size-frequency distribution, with discrete modes in both the macrofossil and microfossil size range, rather than a single monomodal size-frequency distribution (e.g., Voorhies, 1969, p. 52; Lehman, 1982, WPA #1, p. 46; Fiorillo, 1991). A multimodal size-frequency distribution strongly suggests the influence of a mixture of multiple bone accumulating and bone concentrating influences, and indicates a complex history. Accordingly, it is not surprising that researchers who have worked with mixed bonebeds have tended to emphasize either the macrofossil or microfossil component (e.g., Camp and Welles, 1956; Kaye and Padian, 1994; Fastovsky et al., 1995; Varricchio, 1995; Eberth et al., 2000; Fiorillo et al., 2000).

Assessing taxonomic diversity in a fossil assemblage provides a simple and yet powerful means of further discriminating between bonebed types, especially in nonmarine depositional settings (e.g., Koster, 1987; Rogers, 1990; Varricchio and Horner, 1993; Fiorillo and Rogers, 1999). Furthermore, an assessment of taxonomic diversity combined with a facies analysis can provide significant insight into the degree of paleoenvironmental mixing, and time averaging (e.g., Behrensmeyer, 1982; Badgley, 1986; Behrensmeyer, 1988; Eberth, 1990; Rogers and Kidwell, 2000). In our use of this measure, we classify bonebeds as monotaxic, multitaxic high diversity, and multitaxic low diversity (Fig. 3.1; Table 3.2).

*Monotaxic bonebeds.* All macrofossil, microfossil, and mixed bonebeds are relatively easily classified as either monotaxic (only one species or genus present) or multitaxic (more than one species or genus present; Table 3.2). We employ the term "monotaxic" rather than "monospecific" or "monogeneric" because it does not imply a taxonomic understanding that may not be retrievable from the assemblage. In fact, it is rarely possible to classify with confidence all fossil remains in a bonebed to the species level. The term "monotaxic" is also preferred because, historically, "monospecific" and "monogeneric" have been used to refer to bonebeds that also contain significant numbers of other taxa (e.g., Turnbull and Martill, 1988; Varricchio and Horner, 1993; Ryan et al., 2001). Examples of monotaxic bonebeds (our usage) include those of Camp (1980), Breithaupt and Duvall (1986), Berman et al. (1987), Coria (1994), and Henrici and Fiorillo (1993).

*Multitaxic bonebeds.* Most bonebeds are multitaxic and can be further classified as high or low diversity (Fig. 3.1; Table 3.2). An approximation of taxonomic diversity can help clarify paleoecological and paleobiological influences in a bonebed's formation and can also provide insights into the ecosystem of the host formation (Eberth et al., 2000; Brinkman et al., Chapter 4 in this volume).

Two theoretical approaches can be used to determine whether a multitaxic bonebed is of high or low taxonomic diversity. The first is a relativistic approach that assesses diversity in the context of the host stratigraphic unit. The second is an absolute approach that employs a numerical cutoff, regardless of the taxonomic diversity within the host formation. The relativistic approach requires a good understanding of the taxonomic composition of the host stratigraphic unit, and data are often difficult to compare consistently between bonebeds and stratigraphic units. For example, the vertebrate diversity in the Upper Cretaceous Dinosaur Park Formation of

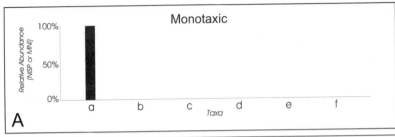

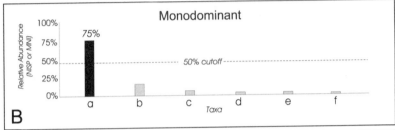

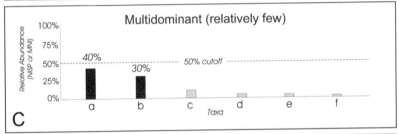

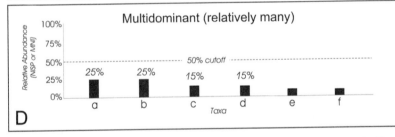

*Figure 3.2.* Theoretical distributions that exemplify the relative abundance of taxa in (A) monotaxic, (B) monodominant, and (C, D) multidominant bonebeds.

southern Alberta is high by any measure, consisting of 72 families, 116 genera, and more than 130 species (Eberth et al., 2001; Currie and Koppelhus, 2005). In contrast, the richest bonebeds in the formation have yielded 30 taxa, which represent only 15% of the known formational diversity (e.g., Brinkman, 1990; Getty et al., 1997; Tumarkin et al., 1997; Ryan et al., 2001; Eberth and Getty, 2005). Accordingly, all the bonebeds in the Dinosaur Park Formation justifiably could be regarded as having a low diversity compared to the host formation. Conversely, Lehman's (1982) study of bonebeds from the nonmarine portion of the Aguja Formation, Big Bend National Park, Texas, showed that even though only 5 to 8 taxa are present, these bonebed assemblages represent 50%–80% of the known formational diversity (as of 1982).

Because of these potential relativistic problems, we propose that an absolute approach that employs a numerical cutoff is the best means of assessing taxonomic diversity in a multitaxic bonebed. The following empirically derived ranges for high- and low-diversity bonebeds are based on a review of the bonebeds in our database (Appendix 3.1; e.g., Turnbull and Martill, 1988; Rogers, 1990; Varricchio and Horner, 1993).

- Monotaxic: one taxon, which presumably represents one species
- Low diversity: 2 to 5 taxa
- High diversity: 10 or more taxa
- High or low diversity: 6 to 9 taxa, where assignment depends on relative abundance of taxa, and a researcher's experience or opinions about the formation, depositional system, paleoecology, and taphonomic processes

Bonebeds with 6 to 9 taxa are often classified as having low diversity, especially where the relative abundance of most taxa is very low (e.g., Fig. 3.2B and 3.2C; Lehman, 1982; Turnbull and Martill, 1988; Rogers, 1990). The terms "paucitaxic" and "paucispecific" previously have been used in reference to low-diversity macrofossil bonebeds dominated by one or two taxa (e.g., Rogers, 1990; Christians, 1991; Fiorillo and Rogers, 1999). However, because these terms mix the concepts of diversity and relative taxonomic abundance (see below), we regard them as vague and do not recommend their further use.

### Relative Taxonomic Abundance: Monodominant and Multidominant Bonebeds

Most high-diversity and low-diversity multitaxic bonebeds can be further classified on the basis of relative taxonomic abundance. In this context we recognize "monodominant" and "multidominant" bonebeds (Figs. 3.1 and 3.2; Table 3.2). Relative taxonomic abundance is often assessed when behavioral, ecological, and community-composition aspects of the life assemblage (biocoenose) are being considered (e.g., Shotwell, 1955; Voorhies, 1969; Currie and Dodson, 1984; Rogers, 1990; Varricchio, 1995; Coombs and Coombs, 1997; Ryan et al., 2001).

We define a monodominant bonebed as an assemblage of animals where one taxon dominates and accounts for 50% or more of the NISP or the minimum number of individuals (MNI) (Fig. 3.2B; Table 3.2). The 50% cutoff is empirically derived from a review of our bonebeds database (Appendix 3.1) and reflects consistently the intended meaning of the terms monospecific, monospecies, monspecific, monogeneric, paucispecific, and paucitaxic in bonebed studies, specifically where there is an overwhelming and quantifiable dominance of one taxon in the assemblage (e.g., Dalquest and Mamay, 1963; Currie and Dodson, 1984; Voorhies, 1985; Visser, 1986; Norman, 1987; Winkler et al., 1988; Rogers, 1990; Sander, 1992; Varricchio and Horner, 1993; Schwartz and Gillete, 1994; Schmitt et al., 1998; Ryan et al., 2001; Eberth and Getty, 2005).

In contrast, we designate multidominant bonebeds as those in which two or more taxa account for 50% or more of the NISP or MNI (Fig. 3.2C and 3.2D; e.g., Hunt, 1978; Hunt et al., 1983; Brinkman, 1990; Stucky et al., 1990; Fiorillo, 1991; Kaye and Padian, 1994; Varricchio, 1995; Coombs and Coombs, 1997; Fiorillo et al., 2000). Multidominant bonebeds can show variation in the relative number of dominant taxa, ranging from relatively few (Fig. 3.2C; e.g., Badam et al., 1986; Visser, 1986; Brinkman, 1990; Fiorillo, 1991; Coombs and Coombs, 1997; Eberth et al., 2000; Fiorillo et al., 2000) to relatively many (Fig. 3.2D, e.g., Hunt, 1978). Most vertebrate microfossil assemblages are multidominant bonebeds.

## PATTERNS OF OCCURRENCE

After the bonebeds in our database were classified using the scheme described above, patterns of occurrence were identified in relation to (1) the relative abundance of different types of bonebeds; (2) inferred mechanisms of origin; and (3) recurrent associations of bonebed type and paleoenvironment. However, because our bonebeds database is biased in a number of ways, those biases must be explored before patterns of occurrence can be interpreted meaningfully.

### Database Biases

The database is organized by citation rather than by individual bonebed because of the uneven scientific treatment that many sites have received in the literature. In short, we were unable to adequately describe in sufficient detail each of the more than 1000 bonebeds that are referred to in

David A. Eberth, Matthew Shannon, and Brent G. Noland

the 383 included citations. Some citations in the database include descriptions of numerous bonebeds that vary broadly in element size, taxonomic composition and diversity, facies associations, and depositional setting (e.g., Dodson et al., 1980), or refer to groups of bonebeds that are not described in any detail (e.g., Fiorillo, 1998). Although we included data from these studies in the univariate statistics shown in Table 3.1, we excluded many of these citations from our analysis of patterns of association, especially where we were attempting to identify bivariate associations (e.g., element size *and* diversity). Accordingly, the number of citations (and bonebeds) employed to assess patterns of association is a subset of the entire database, which generally consists of studies focusing on individual, well-understood bonebeds (e.g., Varricchio, 1995; Coombs and Coombs, 1997), or studies that focus on a select group of similar, well-understood bonebeds (e.g., Rogers, 1990; Eberth and Getty, 2005).

The database was compiled using standard North American geological search tools (e.g., Georef), references in previously published compendium-style articles, and citations known to contributors of this volume. However, reports of bonebeds and data from bonebed studies are frequently presented in unpublished theses, semiscientific-to-popular media articles, and abstracts at scientific conferences (Fiorillo and Eberth, 2005). Because it can be difficult to identify and compile these kinds of references, we do not regard the database as exhaustive, and the reader should regard it as only a sample of the literature. Table 3.1 provides an at-a-glance summary of statistics from our database and underscores that the citations include the following significant biases:

- Predominantly in English (98%)
- Mostly postdate 1975 (87%)
- Emphasize the Mesozoic (56%)
- Emphasize North America (72%)
- Emphasize bonebeds containing dinosaurs (33%)

Other scientific biases are also revealed by the database. For example, Figure 3.3 shows that among the references to macrofossil bonebeds—where taxonomic diversity and relative taxonomic abundance can be determined—53% refer to sites with a low taxonomic diversity (Fig. 3.3A), and, more interestingly, 59% refer to sites whose taxonomic composition is dominated by one taxon (monotaxic and monodominant, Fig. 3.3B). Based on our review of the literature, anecdotal data, and personal experience, we regard this pattern as reflecting a preferred scientific interest in monotaxic and monodominant macrofossil bonebeds, and not a

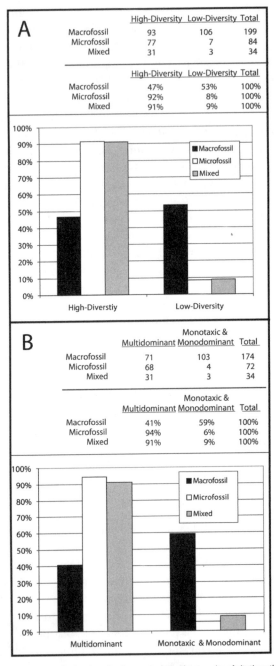

**A**

| Macrofossil | High-Diversity | Low-Diversity | Total |
|---|---|---|---|
| Macrofossil | 93 | 106 | 199 |
| Microfossil | 77 | 7 | 84 |
| Mixed | 31 | 3 | 34 |

| | High-Diversity | Low-Diversity | Total |
|---|---|---|---|
| Macrofossil | 47% | 53% | 100% |
| Microfossil | 92% | 8% | 100% |
| Mixed | 91% | 9% | 100% |

**B**

| | Multidominant | Monotaxic & Monodominant | Total |
|---|---|---|---|
| Macrofossil | 71 | 103 | 174 |
| Microfossil | 68 | 4 | 72 |
| Mixed | 31 | 3 | 34 |

| | Multidominant | Monotaxic & Monodominant | Total |
|---|---|---|---|
| Macrofossil | 41% | 59% | 100% |
| Microfossil | 94% | 6% | 100% |
| Mixed | 91% | 9% | 100% |

*Figure 3.3.* Tables and histograms showing absolute and relative frequencies of citations that describe or refer to (A) high- and low-diversity macrofossil, microfossil, and mixed bonebeds, and (B) monotaxic, monodominant, and multidominant bonebeds (also as a function of element size). All data from the bonebeds database (Appendix 3.1).

true reflection of the relative abundance of different kinds of macrofossil bonebeds in the stratigraphic record. For example, Eberth and Currie (2005) reported that there are more than 104 documented multitaxic (multidominant) bonebeds in the Oldman and Dinosaur Park formations at Dinosaur Provincial Park, Canada, and Eberth and Getty (2005) reported that there are only 20 monodominant bonebeds (ceratopsian) known from these same strata. However, a review of the literature on the bonebeds at Dinosaur Provincial Park shows eight popular articles, abstracts, theses, and published refereed articles that focus on the monodominant bonebeds, and only one master's thesis and a single published abstract that focus exclusively on the multitaxic bonebeds. The apparent preferred scientific interest in monodominant bonebeds at Dinosaur Provincial Park undoubtedly derives from the perceived and real opportunities they afford to explore the paleobiology and paleoecology of ceratopsid dinosaurs (e,g., Currie, 1981; Eberth and Getty, 2005; Brinkman et al., Chapter 4 in this volume). We propose that this pattern obtains in general, and that in any given fossiliferous stratigraphic succession, monodominant and monotaxic macrofossil bonebeds are likely to receive more scientific attention than multidominant bonebeds.

Another significant scientific bias in the study of bonebeds revealed by examining the citations in this database is that the origins of macrofossil bonebeds are most often studied in more detail than microfossil bonebeds, and publications that focus on macrofossil bonebeds typically refer to lower numbers of sites per publication (a minimum of 435 sites cited in 230 publications = 1.89 sites/study; Appendix 3.1) than those that focus on microfossil bonebeds (474 sites cited in 92 publications = 5.15 sites/study; Appendix 3.1). Thus, although the total number of citations in our database that focus on microfossil bonebeds is less than one-half that for macrofossil bonebeds (92 versus 230, respectively), the number of microfossil sites referred to in those citations (474) exceeds the number of macrofossil sites (435). These data underscore the fundamental difference in the way that macrofossil and microfossil bonebeds are studied. Whereas any given macrofossil site often requires large expenditures of resources and time to expose and collect sufficient sample sizes (Eberth et al., Chapter 5 in this volume), numerous microfossil bonebeds can be sampled during a field season with the same amount of financial resources and labor. Given the differences in time and effort expended at these two kinds of sites, it is not surprising that far more detailed information is usually included in citations that focus on macrofossil bonebeds.

Given the foregoing, it is not surprising that the quality and reliability of paleoenvironmental interpretations reported in our database citations are highly variable. The associated depositional environment(s) for some of these bonebeds can only be inferred. We have attempted to address this variation by assigning all possible depositional environments to 1 of the following 10 broad depositional systems (sensu Davis, 1983), which are defined by the presence of genetically related paleoenvironments (Table 3.1):

- Alluvial fan/debris flow/mass sediment deposit (DBF)
- Eolian (EOL)
- Channel/levee/splay (CHLSP)
- Overbank-paleosol (PSOL)
- Overbank-wetland/marginal-pond/watering-hole (WTLND)
- Lacustrine (LAC)
- Paralic/estuarine/shoreline (PESH)
- Marine shoreline to open marine (MAR)
- Trap/seep/fissure-fill/cave/other (O)
- Unknown (U)

However, the boundaries between these categories are clearly arbitrary. For example, the differences between PSOL, WTLND, and LAC are based on an interpretation of sustained saturation versus subaerial exposure at a site, as well as evidence for rooting, plant growth, and soil development. Similarly, although we recognize that the PESH setting is quite broad and may include channels, splays, lagoonal ponds, marshes, and bogs, we regard coastal processes and influences as the overriding control on the depositional processes (Eberth, 1997). Lastly, we make no claims that the published depositional interpretations are accurate.

We quantified and compared paleoenvironmental associations among different kinds of bonebeds by counting the maximum number of different paleoenvironmental associations per citation. For example, Langston (1953) described nine macrofossil bonebed sites in the Lower Permian redbeds of New Mexico. In our review of his paleoenvironmental data, the sites occur in association with channels (CHLSP), splays (CHLSP), ponds (WTLND), wetland overbank (WTLND), and caliche-rich paleosol (PSOL) settings, which together represent a maximum of three different paleosystem associations (CHLSP, WTLND, and PSOL). Similarly, Fiorillo (1998) refers to 108 microfossil bonebeds from the Hell Creek Formation of Montana where each site was apparently associated with a channel deposit. In this case, we have counted only one channel deposit (CHLSP) as the maximum number of associations. Several other biases occur in our paleoenvironmental data:

- Paleoenvironmental associations in citations that include reference to multiple sites may be only broadly described or interpreted (e.g., Olson, 1967).
- There may be multiple citations that refer to the same bonebed (e.g., Peterson, 1906; Matthew, 1923; Cook, 1926).
- Different citations may refer to the same bonebed but may include different paleoenvironmental interpretations (e.g., Bilbey, 1999, versus Gates, 2005).

Although there are clearly biases in the content, sources, and selection of entries in our database, we believe that a commonsense approach can be used to identify where these biases have skewed our results. For example, the fact that 72% of the entries in our database are from sites in North America obviously does not indicate that 72% of all existing bonebeds occur in North America. Furthermore, because our bonebeds database contains a significant number of studies covering a comprehensive range of taxa, stratigraphic units, geographic locations, and paleoenvironmental associations, we regard the patterns described below as general hypotheses that can be tested with additional data from the literature and field studies. We encourage future workers to challenge and test our interpretations with additional data.

### Pattern 1: Rarity of Low-Diversity and Monodominant Microfossil Bonebeds

Figure 3.3A illustrates that high-diversity microfossil bonebeds far outnumber low-diversity microfossil bonebeds in the database (92% and 8%, respectively). In contrast, references to high- and low-diversity macrofossil bonebeds are roughly subequal in number (47% and 53%, respectively). Similarly, microfossil bonebeds dominated by multiple taxa (multidominant, Fig. 3.3B) grossly outnumber (94%) microfossil bonebeds dominated by a single taxon (6%). In contrast, multidominant and monodominant macrofossil bonebeds are both abundant (41% and 59%, respectively, Fig. 3.3B).

In the recent, low-diversity death assemblages consisting of small vertebrates are common (see Rogers and Kidwell, Chapter 1 in this volume) and form in response to a wide variety of factors, including (1) mass death among members of a taxon that are naturally gregarious during some portion of their lives (e.g., Weigelt, 1989); (2) attritional accumulation within a geographically or ecologically limited habitat or where sediment input is low (e.g., Colley, 1990); and (3) fecal or regurgitated concentrates (e.g., Korth, 1979; Andrews and Nesbitt Evans, 1983). Therefore, in theory low-diversity microfossil assemblages should be well represented in the fossil record, and their absence in the literature—as reflected by our database—suggests that there is some degree of bias against their preservation or recognition. Given the importance of microfossil bonebeds in biostratigraphic and paleoecological studies, we regard it as unrealistic that researchers are overlooking or ignoring them. Instead, we speculate that physicotaphonomic processes that transport, sort, and mix elements—especially in marine and alluvial-channel paleoenvironments—naturally favor the formation and preservation of high-diversity rather than low-diversity microfossil assemblages (Kidwell, 1986; Behrensmeyer, 1988; Brinkman, 1990; Eberth, 1990; Rogers and Kidwell, 2000; Schröder-Adams et al., 2001; Rogers and Kidwell, Chapter 1 in this volume).

The preferred study of monotaxic and monodominant bonebeds described above further supports our interpretation that monotaxic and monodominant microfossil bonebeds are truly rare in the rock record. Were they more common, it is likely that they would be the objects of considerable scientific interest. Furthermore, mixed microfossil/macrofossil assemblages in our database exhibit a pattern similar to that for microfossil assemblages (91% high diversity versus 9% low diversity; 91% multidominant versus 9% monotaxic/monodominant). Given this pattern, it is not surprising that microfossil assemblages are generally regarded as consisting of hydraulically sorted and mixed vertebrate remains (Olson, 1977; Murray and Johnson, 1987; Stewart, 1994; Martin and Foster, 1998).

### Pattern 2: Different Origins For Monotaxic/Monodominant versus Multidominant Bonebeds

Further review of the database reveals that monotaxic and monodominant bonebeds (microfossil and macrofossil) are most often interpreted as resulting from

1. catastrophic or short-term mass-death events (e.g., Voorhies, 1985; Rogers, 1990; Schwartz and Gillette, 1994; Ryan et al., 2001); and
2. multiple death events (attrition), but at a specific site possibly due to predation, trapping, disease, drought, and so forth (e.g., Carroll et al., 1972; Laury, 1980; Hunt et al., 1983; Agenbroad, 1984; Conybeare and Haynes, 1984; Norman, 1987; Sander, 1992; Sundell, 1999).

In contrast, multidominant bonebeds are typically interpreted as time-averaged and reworked assemblages of elements that were concentrated by a variety of events and processes including

1. attrition in low-sedimentation-rate paleoenvironments (e.g., Bown and Kraus, 1981; Badgley, 1986; Schröder-Adams et al., 2001);
2. hydraulic reworking (e.g., Wood et al., 1988);
3. mass-death events (e.g., Fiorillo, 1991);
4. surface deflation (Rodrigues–de la Rosa and Cevallos-Ferriz, 1998); and
5. complex combinations of the above phenomena (Hunt, 1978; Hook and Baird, 1986; Pratt, 1989; Coombs and Coombs, 1997; Britt et al., 1997; Badgley et al., 1998).

These contrasting sets of interpretations emphasize that biological and ecological factors dominate the origins of monotaxic and monodominant bonebeds, whereas postdepositional physical factors dominate the origins of multitaxic assemblages (Rogers and Kidwell, Chapter 1 in this volume; Behrensmeyer, Chapter 2 in this volume; Brinkman et al., Chapter 4 in this volume).

**Pattern 3: Paleoenvironmental Associations of Macrofossil versus Microfossil Bonebeds**

Figure 3.4A shows published and inferred paleoenvironmental interpretations for the macrofossil versus microfossil assemblages cited in our database. Three paleoenvironmental associations stand out and are highlighted by arrows in the associated histograms:

**1.** Microfossil assemblages occur most frequently (33%) in paleochannel depositional systems (CHLSP), supporting the widely accepted interpretation that vertebrate microfossil assemblages form largely under the influence of physical processes that mix, sort, and concentrate microfossils. This further suggests that significant numbers of microfossil assemblages are to some degree reworked and time averaged.

**2.** Although wetland (WTLND) depositional settings commonly host macrofossil bonebeds (31%), they much less frequently host microfossil assemblages (13%). Wetlands are frequented regularly by large vertebrates in the modern (Shipman, 1975; Conybeare and Hayes, 1984) and are the location for a wide range of mortality events (disease, predation, attrition, stress, trapping) that can lead to bonebed formation (see Rogers and

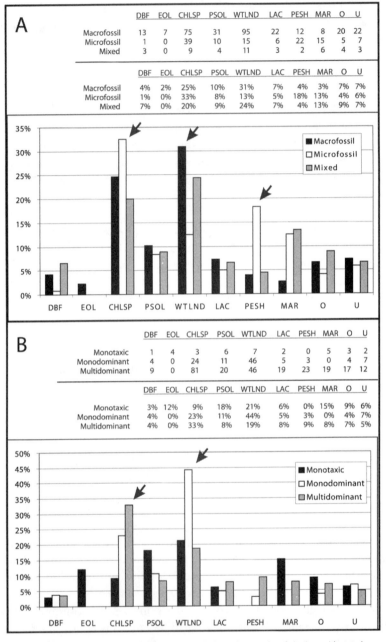

**A**

| | DBF | EOL | CHLSP | PSOL | WTLND | LAC | PESH | MAR | O | U |
|---|---|---|---|---|---|---|---|---|---|---|
| Macrofossil | 13 | 7 | 75 | 31 | 95 | 22 | 12 | 8 | 20 | 22 |
| Microfossil | 1 | 0 | 39 | 10 | 15 | 6 | 22 | 15 | 5 | 7 |
| Mixed | 3 | 0 | 9 | 4 | 11 | 3 | 2 | 6 | 4 | 3 |

| | DBF | EOL | CHLSP | PSOL | WTLND | LAC | PESH | MAR | O | U |
|---|---|---|---|---|---|---|---|---|---|---|
| Macrofossil | 4% | 2% | 25% | 10% | 31% | 7% | 4% | 3% | 7% | 7% |
| Microfossil | 1% | 0% | 33% | 8% | 13% | 5% | 18% | 13% | 4% | 6% |
| Mixed | 7% | 0% | 20% | 9% | 24% | 7% | 4% | 13% | 9% | 7% |

**B**

| | DBF | EOL | CHLSP | PSOL | WTLND | LAC | PESH | MAR | O | U |
|---|---|---|---|---|---|---|---|---|---|---|
| Monotaxic | 1 | 4 | 3 | 6 | 7 | 2 | 0 | 5 | 3 | 2 |
| Monodominant | 4 | 0 | 24 | 11 | 46 | 5 | 3 | 0 | 4 | 7 |
| Multidominant | 9 | 0 | 81 | 20 | 46 | 19 | 23 | 19 | 17 | 12 |

| | DBF | EOL | CHLSP | PSOL | WTLND | LAC | PESH | MAR | O | U |
|---|---|---|---|---|---|---|---|---|---|---|
| Monotaxic | 3% | 12% | 9% | 18% | 21% | 6% | 0% | 15% | 9% | 6% |
| Monodominant | 4% | 0% | 23% | 11% | 44% | 5% | 3% | 0% | 4% | 7% |
| Multidominant | 4% | 0% | 33% | 8% | 19% | 8% | 9% | 8% | 7% | 5% |

*Figure 3.4.* Tables and histograms showing absolute and relative frequencies of citations with stated or inferred depositional setting associations for (A) macrofossil, microfossil, and mixed bonebeds, and (B) monotaxic, monodominant, and multidominant bonebeds. All data from the bonebeds database (Appendix 3.1).

Kidwell, Chapter 1 in this volume). More importantly, the potential for physical transport, sorting, and reworking are generally at a minimum in these settings compared to channel, channel margin, alluvial fan, and shorelines. Thus, we conclude that the high frequency of occurrence for macrofossil assemblages in WTLND settings more likely reflects biologic preferences than physical concentrating mechanisms.

**3.** Microfossil assemblages are more abundant in paralic settings (PESH) than are macrofossil assemblages (18% versus 4%, respectively). The reason for the relatively greater abundance of microfossil assemblages in paralic settings is not obvious to us. Accordingly, we speculate that this pattern might result from the interplay of a number of factors including (1) high organic productivity, which is especially influential in the development of small, aquatic vertebrates such as fish (Rogers and Kidwell, 2000); and (2) the intensity and relatively high frequency of high-energy wind and flood events that can kill, or accumulate, sort, and preserve vertebrate skeletal elements (e.g., Eberth and Brinkman, 1997; Eberth and Currie, 2005).

### Pattern 4: Monodominant-Wetland and Multidominant-Paleochannel Associations

In the database, 44% of monodominant bonebeds occur in association with WTLND settings (Fig. 3.4B), whereas only 23% occur in association with CHLSP settings. In contrast, multidominant bonebeds are most frequently associated with CHLSP settings (33%) and less frequently with WTLND settings (19%). These patterns suggest that monodominant bonebeds most frequently form and develop in settings with low transport and reworking potential (see above). This supports the widely accepted view that gregariousness or natural species-specific groupings are the primary influence in the formation of monodominant bonebeds (Currie, 2000). In contrast, the relatively high proportion of multidominant bonebeds associated with CHLSP settings (33%) suggests that physical sorting and concentrating factors are at least as important as biologic factors in their formation and preservation.

### SUMMARY

Bonebeds described in the literature can be readily classified using element size, measures of taxonomic diversity, and estimates of relative abundance.

Accordingly, we can recognize macrofossil, microfossil, and mixed bonebeds that exhibit a monotaxic or multitaxic composition. Multitaxic bonebeds are further divisible into high- and low-diversity types that exhibit either a monodominant or multidominant relative taxonomic abundance. Although paleoenvironmental associations of different bonebed classes are generally more difficult to assess, a broad depositional systems approach is recommended in order to approximate patterns of paleoenvironmental association.

By employing a large database and assessing relative frequencies of different types of bonebeds and their paleoenvironmental associations, it is possible to gain insight into the formation and preservation of all bonebeds. Although a variety of biases are inevitable in any bonebeds database compiled from the literature, a common sense approach can help in the framing and testing of patterns of occurrence. Accordingly, we recognize the following patterns as worthy of further study and testing:

- Citations that include reference to low-diversity microfossil bonebeds are rare in the database. We interpret this as reflecting the influences of physical taphonomic processes that transport, sort, and mix microfossil elements, resulting in the preferential formation and preservation of high-diversity microfossil bonebeds.
- The literature on the origins of monotaxic and monodominant bonebeds emphasizes catastrophic and short-term mass-death events as the primary influences on their formation. In contrast, multidominant bonebeds are more frequently interpreted as time averaged and physically reworked.
- Historically, there has been more scientific interest (bias) in monotaxic and monodominant bonebeds than multitaxic bonebeds, probably because of the species-level paleobiologic information that the former are perceived as yielding.
- Microfossil bonebeds are most frequently hosted by paleochannel, levee, and splay facies, which is compatible with the premise that they form principally as a result of hydraulic sorting and concentrating mechanisms.
- Bonebeds containing monodominant assemblages are frequently interpreted as having been preserved in low-energy, wetland settings where ecological and biological concentrating mechanisms dominate.
- For unknown reasons, microfossil bonebeds appear to be more abundant than macrofossil bonebeds in paralic settings.

### REFERENCES

Agenbroad, L.D. 1984. Hot springs, South Dakota: Entrapment and taphonomy of Colombian Mammoth. Pp. 113–127 *in* Quaternary extinctions, a prehistoric revolution. P.S. Martin and R.G. Klein, eds. University of Arizona Press, Tucson.

Andrews, P., and E.M. Nesbitt Evans. 1983. Small mammal bone accumulations produced by mammalian carnivores. Paleobiology 9:289–307.

Antia, D.D.J. 1979. Bone-beds: A review of their classification, occurence, genesis, diagenesis, geochemistry, palaeoecology, weathering, and microbiotas. Mercian Geologist 7:93–174.

Badam, G.L., R.K. Ganjoo, and R.K.G. Salahuddin. 1986. Preliminary taphonomical studies of some Pleistocene fauna from the central Narmada Valley, Madhya Pradesh, India. Palaeogeography, Palaeoclimatology, Palaeocology 53:335–348.

Badgley, C. 1986. Taphonomy of mammalian fossil remains from Siwalik rocks of Pakistan. Paleobiology 12:119–142.

Badgley, C., W. Downs, and L.J. Flynn. 1998. Taphonomy of small-mammal fossil assemblages from the Middle Miocene Chinji Formation, Siwalik Group, Pakistan. Pp. 145–166 in Advances in vertebrate paleontology and geochronology. National Science Museum Monographs 14. Y. Tomida, L.J. Flynn, and L.L. Jacobs, eds. National Science Museum, Tokyo.

Behrensmeyer, A.K. 1975. The taphonomy and paleoecology of Plio-Pleistocene vertebrate assemblages east of Lake Rudolf, Kenya. Bulletin of the Museum of Comparative Zoology 146:474–578.

Behrensmeyer, A.K. 1982. Time sampling intervals in the vertebrate fossil record. Third North American Paleontological Convention Proceedings 1:41–45.

Behrensmeyer, A.K. 1988. Vertebrate preservation in fluvial channels. Palaeogeography, Palaeoclimatology, Palaeoecology 63:183–199.

Behrensmeyer, A.K. 1991. Terrestrial vertebrate accumulations. Pp. 291–335 in Taphonomy: Releasing the data locked in the fossil record. P.A. Allison and D.E.G. Briggs, eds. Plenum, New York.

Behrensmeyer, A.K. 1999. Bonebeds through geologic time. Journal of Vertebrate Paleontology 19(supplement to 3):31A–32A.

Behrensmeyer, A.K. This volume. Bonebeds through time. Chapter 2 in Bonebeds: Genesis, analysis, and paleobiological significance. R.R. Rogers, D.A. Eberth, and A.R. Fiorillo, eds. University of Chicago Press, Chicago.

Berman, D.S, R.R. Reisz, and D.A. Eberth. 1987. Seymoria sanjuanensis (Amphibia, Batrachosauria) from the Lower Permian Cutler Formation of north-central New Mexico and the occurrence of sexual dimorphism in that genus questioned. Canadian Journal of Earth Sciences 24:1769–1784.

Bilbey, S.A. 1999. Taphonomy of the Cleveland-Lloyd dinosaur quarry in the Morrison Formation, Central Utah: A lethal spring-fed pond. Pp. 121–133 in Vertebrate paleontology in Utah. Miscellaneous publication 99-1. D.D. Gillette, ed. Utah Geological Survey, Salt Lake City.

Blob, R.W., and A.R. Fiorillo. 1996. The significance of vertebrate microfossil size and shape distributions for faunal abundance reconstructions: A Late Cretaceous example. Paleobiology 22:422–435.

Bown, T.M., and M.J. Kraus. 1981. Vertebrate fossil-bearing paleosol units (Willwood Formation, Lower Eocene, Northwest Wyoming, U.S.A.): Applications for taphonomy, biostratigraphy, and assemblage analysis. Palaeogeography, Palaeoclimatology, Palaeoecology 34:31–56.

Breithaupt, B.H., and D. Duvall. 1986. The oldest record of serpent aggregation. Lethaia 19:181–185.

Brinkman, D.B. 1990. Paleoecology of the Judith River Formation (Campanian) of Dinosaur Provincial Park, Alberta, Canada: Evidence from microfossils localities. Palaeogeography, Palaeoclimatology, Palaeoecology 78:37–54.

Brinkman, D.B., D.A. Eberth, and P.J. Currie. This volume. From bonebeds to paleobiology: Applications of bonebed data. Chapter 4 in Bonebeds: Genesis, analysis, and paleobiological significance. R.R. Rogers, D.A. Eberth, and A.R. Fiorillo. University of Chicago Press, Chicago.

Brinkman, D.B., A.P. Russell, and Jiang-Hua Peng. 2005. Vertebrate microfossil sites and their contribution to studies of paleoecology. Pp. 88–98 in Dinosaur Provincial Park: A spectacular ecosystem revealed. P.J. Currie and E.B. Koppelhus, eds. Indiana University Press, Bloomington.

Britt, B.B., D.A. Eberth, D. Brinkman, R. Scheetz, K.L. Stadtman, and J.S. McIntosh. 1997. The Dalton Wells fauna (Cedar Mountain Formation): A rare window into the little-known world of Early Cretaceous dinosaurs in North America. Geological Society of America, Abstracts with Programs 29:A–104.

Camp, C.L. 1980. Large ichthyosaurs from the Upper Triassic of Nevada. Palaeontographica A 170:139–200

Camp, C.L., and S.P. Welles. 1956. Triassic dicynodont reptiles. 1. The North American genus Placerias. Memoirs of the University of California 13:255–348.

Carpenter, K. 1988. Dinosaur bone beds and mass mortality: Implications for the K-T extinction. Global catastrophes in Earth history: An interdisciplinary conference on impacts, volcanism and mass mortality. Lunar and Planetary Institute Contributions 673:24–25.

Carroll, R.L., E.S. Belt, D.L. Dineley, D. Baird, and D.C. McGregor. 1972. Vertebrate paleontology of Eastern Canada. 24th International Geological Conference, Ottawa, Excursion A59:1–113.

Christians, J.P. 1991. Taphonomic review of the Mason Dinosaur Quarry: Hell Creek Formation, Upper Cretaceous, South Dakota. Journal of Vertebrate Paleontology 11:22A.

Chure, D.J., Fiorillo, A.R., and Jacobsen, A. 2000. Prey bone utilization by predatory dinosaurs in the late Jurassic of North America, with comments on prey bone use by dinosaurs throughout the Mesozoic. Gaia 15: 227–232.

Colley, S.M. 1990. Humans as taphonomic agents. Pp. 50–64 in Problem solving in Taphonomy. S. Solomon, I. Davidson, and D. Watson, eds. Tempus 2:50–64.

Conybeare, A., and G. Haynes. 1984. Observations on elephant mortality and bones in water holes. Quaternary Research 22:189–200.

Cook, L.M. 1926. A rhinoceros bone bed. Black Hills Engineer 14:95–99.

Coombs, M.C., and W.P. Coombs Jr. 1997. Analysis of the geology, fauna, and taphonomy of Morava Ranch Quarry, Early Miocene of Northwest Nebraska. Palaios 12:165–187.

Coria, R.A. 1994. On monospecific assemblage of sauropod dinosaurs from Patagonia: Implications for gregarious behavior. Gaia 10:209–213.

Currie, P.J. 1981. Hunting dinosaurs in Alberta's great bonebed. Canadian Geographic 10(4):34–39.

Currie, P.J. 2000. Possible evidence of gregarious behavior in tyrannosaurids. Gaia 15:123–133.

Currie, P.J., and P. Dodson. 1984. Mass death of a herd of ceratopsian dinosaurs. Pp. 52–60 *in* Third symposium of Mesozoic terrestrial ecosystems. W.E. Reif and F. Westphal, eds. Attempto Verlag, Tubingen.

Currie, P.J., and E.B. Koppelhus, eds. 2005. Dinosaur Provincial Park: A spectacular ecosystem revealed. Indiana University Press, Bloomington.

Dalquest, W.W., and S.H. Mamay. 1963. A remarkable concentration of Permian amphibian remains in Haskall County, Texas. Journal of Geology 71:641–644.

Davis, R.C. 1983. Depositional Systems. Prentice Hall, New York.

Dodson, P., A.K. Behrensmeyer, R.T. Bakker, and J.S. Mcintosh. 1980. Taphonomy and paleoecology of the dinosaur beds of the Jurassic Morrison Formation. Paleobiology 6:208–232.

Eberth, D.A. 1990. Stratigraphy and sedimentology of vertebrate microfossil sites in the uppermost Judith River Formation (Campanian), Dinosaur Provincial Park, Alberta, Canada. Palaeogeography, Palaeoclimatology, Palaeoecology 78:1–36.

Eberth, D.A. 1997. Non-marine vertebrate taphonomy in estuarine paleoenvironments. Journal of Vertebrate Paleontology, 17(supplement to 3):44A.

Eberth, D.A., and P.J. Currie. 2005. Vertebrate taphonomy and taphonomic modes. Pp. 453–577 *in* Dinosaur Provincial Park. P.J. Currie and E.B. Koppelhus, eds. Indiana University Press, Bloomington, Indiana.

Eberth, D.A., and M.A. Getty. 2005. Ceratopsian bonebeds at Dinosaur Provincial Park: Occurrence, origin, and significance. Pp. 501–536 *in* Dinosaur Provincial Park: A spectacular ecosystem revealed. P.J. Currie and E.B. Koppelhus, eds. Indiana University Press, Bloomington, Indiana.

Eberth, D.A., D.S Berman, S.S. Sumida, and H. Hopf. 2000. Lower Permian terrestrial paleoenvironments and vertebrate paleoecology of the Tambach Basin (Thuringia, Central Germany): The upland holy grail. Palaios 15:293–313.

Eberth, D.A., P.J. Currie, D.B. Brinkman, M.J. Ryan, D.R. Braman, J.D. Gardner, V.D. Lam, D.N. Spivak, and A.G. Neuman. 2001. Alberta's dinosaurs and other fossil vertebrates: Judith River and Edmonton Groups (Campanian-Maastrichtian). Pp. 47–75 *in* Mesozoic and Cenozoic paleontology in the western plains and Rocky Mountains. Museum of the Rockies Occasional Paper 3. C.L. Hill, ed. Museum of the Rockies, Bozeman, Montana.

Eberth, D.A., B.B. Britt, R. Scheetz, K.L. Stadtman, and D.B. Brinkman. 2006. Dalton Wells: Geology and significance of debris-flow-hosted dinosaur bonebeds in the Cedar Mountain Formation (Lower Cretaceous) of eastern Utah, USA. Palaeogeography, Palaeoclimatology, Palaeoecology 236:217–245.

Eberth, D.A., R.R. Rogers, and A.R. Fiorillo. This volume. A practical approach to the study of bonebeds. Chapter 5 *in* Bonebeds: Genesis, analysis, and paleobiological significance. R.R. Rogers, D.A. Eberth, and A.R. Fiorillo. University of Chicago Press, Chicago.

Fastovsky, D.E., J.M. Clark, N.H. Strater, M. Montellano, R. Hernandez, and J.A. Hopson. 1995. Depositional environments of a Middle Jurassic terrestrial vertebrate assemblage, Huizachal Canyon, Mexico. Journal of Vertebrate Paleontology 15:561–575.

Fiorillo, A.R. 1991. Taphonomy and depositional setting of Careless Creek Quarry (Judith River Formation), Wheatland County, Montana, U.S.A. Palaeogeography, Palaeoclimatology, Palaeoecology 81:281–311.

Fiorillo, A.R. 1998. Measuring fossil reworking within a fluvial system: An example from the Hell Creek Formation (Upper Cretaceous) of Eastern Montana. Pp. 243–251 *in* Advances in vertebrate paleontology and geochronology. National Science Museum Monograph 14. Y. Tomida, L.J. Flynn, and L.L. Jacobs, eds. National Science Museum, Tokyo.

Fiorillo, A.R., and D.A. Eberth. 2005. Taphonomy. Pp. 607–613 *in* The Dinosauria. D. Weishampel, P. Dodson, and M. Osmolska, eds. University of California Press, Berkeley.

Fiorillo, A.R., and R.R. Rogers. 1999. Bonebeds: A nonmenclatural and historical overview. Journal of Vertebrate Paleontology 19(supplement to 3):44A.

Fiorillo, A.R., K. Padian, and C. Musikasinthorn. 2000. Taphonomy and depositional setting of the Placerias Quarry (Chinle Formation: Late Triassic, Arizona). Palaios 15:373–386.

Gates, T.A. 2005. The Late Jurassic Cleveland-Lloyd Dinosaur Quarry as a drought-induced assemblage. Palaios 20:363–375.

Getty, M., D.A. Eberth, D.B. Brinkman, D. Tanke, M.J. Ryan, and M. Vickaryous. 1997. Taphonomy of two *Centrosaurus* bonebeds in the Dinosaur Park Formation, Alberta, Canada. Journal of Vertebrate Paleontology 17(supplement to 3):48A–49A.

Henrici, A.C., and A.R. Fiorillo. 1993. Catastrophic death assemblage of *Chelomophrynus bayi* (Anura, Rhinophrynidae) from the Middle Eocene Wagon Bed Formation of Central Wyoming. Journal of Paleontology 7:1016–1026.

Hook, R.W., and D. Baird. 1986. The Diamond Coal Mine of Linton, Ohio, and its Pennsylvanian age vertebrates. Journal of Vertebrate Paleontology 6:174–190.

Hunt, R.M. Jr. 1978. Depositional setting of a Miocene mammal assemblage, Sioux County, Nebraska (U.S.A.). Palaeogeography, Palaeoclimatology, Palaeoecology 24:1–52.

Hunt, R.M. Jr., X. Xiang-Xu, and J. Kaufman. 1983. Miocene burrows of extinct bear dogs: Indication of early denning behavior of large mammalian carnivores. Science 221:364–366.

Kaye, F.T., and K. Padian. 1994. Microvertebrates from the *Placerias* Quarry: A window on Late Triassic vertebrate diversity in the American Southwest. Pp. 171–196 *in* The shadow of the dinosaurs: Early Mesozoic tetrapods. N.C. Fraser and H.-D. Sues, eds. Cambridge University Press, Cambridge.

Kidwell, S.M. 1986. Models for fossil concentrations: Paleobiologic implications. Paleobiology 12:6–24.

Korth, W.W. 1979. Taphonomy of microvertebrate fossil assemblages. Annals of the Carnegie Museum 48:235–284.

Koster, E.H. 1987. Vertebrate taphonomy applied to the analysis of ancient fluvial systems. Recent developments in fluvial sedimentology. SEPM Special Publications 39:159–168.

Langston, W. Jr. 1953. Permian Amphibians from New Mexico. University of California Publications in Geological Sciences 29:349–416.

Laury, R.L. 1980. Paleoenvironment of a Late Quaternary mammoth-bearing sinkhole deposit, Hot Springs, South Dakota. Geological Society of America Bulletin 91:465–475.

Lehman, T.M. 1982. A ceratopsian bone bed from the Aguja Formation (Upper Cretaceous) Big Bend National Park, Texas. Unpublished masters thesis, University of Texas at Austin.

Maas, M.C. 1985. Taphonomy of a late Eocene microvertebrate locality, Wind River Basin, Wyoming (U.S.A.). Palaeogeography, Palaeoclimatology, Palaeoecology 52:123–142.

Martin, J.E., and J.R. Foster. 1998. First Jurassic mammals from the Black Hills, Northeastern Wyoming. Modern Geology 23:381–392.

Matthew, W.D. 1923. Fossil bones in the rock. Natural History 23:359–369.

Murry, P.A., and G.D. Johnson. 1987. Clear Fork vertebrates and environments for the Lower Permian of north-central Texas. The Texas Journal of Science 39:253–266.

Norman, D.B. 1987. A mass-accumulation of vertebrates from the Lower Cretaceous of Nehden (Sauerland), West Germany. Proceedings of the Royal Society of London 230:215–255.

Olson, E.C. 1967. Early Permian vertebrates. Oklahoma Geological Survey Circular 74:1–111.

Olson, E.C. 1977. Permian lake faunas: A study in community evolution. Journal of the Palaeontological Society of India 20:146–163.

Peng, J.-H., A.P. Russell, and D.B. Brinkman. 2001. Vertebrate microsite assemblages (exclusive of mammals) from the Foremost and Oldman formations of the Judith River Group (Campanian) of southeastern Alberta: An illustrated guide. Provincial Museum of Alberta, Natural History Occasional Papers 25:1–54.

Peterson, O.A. 1906. The Agate Spring fossil quarry. Annals of the Carnegie Museum 3:487–494.

Pratt, A.E. 1989. Taphonomy of the microvertebrate fauna from the early Miocene Thomas Farm locality, Florida (U.S.A.). Palaeogeography, Palaeoclimatology, Palaeoecology 76:125–151.

Reif, W.E. 1971. On the genesis of the bone bed at the Muschelkalk-Keuper-Boundary "Grenzbonebed" in S.W. Germany. Neues Jahrbuch für Geologie und Palaeontologie 139:82–98.

Reif, W.E. 1982. Muschelkalk/Keuper bone-beds (Middle Triassic, SW Germany): Storm condensation in a regressive cycle. Pp. 299–325 in Cyclic and event stratification. G. Einsele and A. Seilacher, eds. Springer, Berlin.

Rodrigues-De La Rosa, R.A., and S.R.S. Cevallos-Ferriz. 1998. Vertebrates of the El Pelillal locality (Campanian, Cerro Del Pueblo Formation), Southeastern Coahuila, Mexico. Journal of Vertebrate Paleontology 18:751–764.

Rogers, R.R. 1990. Taphonomy of three dinosaur bone beds in the Upper Cretaceous Two Medicine Formation of northwestern Montana: Evidence for drought-related mortality. Palaios 5:394–413.

Rogers, R.R., and S.M. Kidwell. 2000. Associations of vertebrate skeletal concentrations and discontinuity surfaces in terrestrial and shallow marine records: A test in the Cretaceous of Montana. Journal of Geology 108:131–154.

Rogers, R.R., and S.M. Kidwell. This volume. A conceptual framework for the genesis and analysis of vertebrate skeletal concentrations. Chapter 1 in Bonebeds: Genesis, analysis, and paleobiological significance. R.R. Rogers, D.A. Eberth, and A.R. Fiorillo. University of Chicago Press, Chicago.

Ryan, M.J., A.P. Russell, D.A. Eberth, and P.J. Currie. 2001. The taphonomy of a *Centrosaurus* (Ornithischia: Certopsidae [sic]) bone bed from the Dinosaur Park Formation (Upper Campanian), Alberta, Canada, with comments on cranial ontogeny. Palaios 16:482–506.

Sander, P.M. 1992. The Norian *Plateosaurus* bonebeds of central Europe and their taphonomy. Palaeogeography, Palaeoclimatology, Palaeoecology 93:255–299.

Sankey, J.T. 2001. Late Campanian southern dinosaurs, Aguja Formation, Big Bend, Texas. Journal of Paleontology 75:208–215.

Schmitt, J.G., J.R. Horner, R.R. Laws, and F. Jackson. 1998. Debris-flow deposition of a hadrosaur-bearing bone bed, Upper Cretaceous Two Medicine Formation, Northwest Montana. Journal of Vertebrate Paleontology 18(supplement to 3):76A.

Schröder-Adams, C.J., S.L. Cumbaa, J. Bloch, D.A. Leckie, J. Craig, S.A. Seif El-Dein, D.-J.H.A.E. Simons, and F. Kenig. 2001. Late Cretaceous (Cenomanian to Campanian) paleoenvironmental history of the eastern Canadian margin of the Western Interior Seaway: bonebeds and anoxic events. Palaeogeography, Palaeoclimatology, Palaeoecology 170:261–289.

Schwartz, H.L., and D.D. Gillette. 1994. Geology and taphonomy of the *Coelophysis* quarry, Upper Triassic Chinle Formation, Ghost Ranch, New Mexico. Journal of Paleontology 68:1118–1130.

Shipman, P. 1975. Implications of drought for vertebrate fossil assemblages. Nature 257:667–668.

Shotwell, J.A. 1955. An approach to the paleoecology of mammals. Ecology 36:327–337.

Smith, R.M.H. 1993. Vertebrate taphonomy of Late Permian floodplain deposits in the Southwestern Karoo Basin of South Africa. Palaios 8:45–67.

Stewart, K.M. 1994. A Late Miocene fish fauna from Lothagam, Kenya. Journal of Vertebrate Paleontology 14:592–594.

Stucky, R.K., L. Krishtalka, and A.D. Redline. 1990. Geology, vertebrate fauna, and paleoecology of the Buck Spring Quarries (early Eocene, Wind River Formation), Wyoming. Pp. 169–186 *in* Dawn of the age of mammals in the northern part of the Rocky Mountain interior, North America. Special paper 243. T.M. Bown and K.D. Rose, eds. Geological Society of America, Boulder, Colorado.

Sundell, K.A. 1999. Taphonomy of a multiple *Poebrotherium* kill site: An *Archeotherium* meat cache. Journal of Vertebrate Paleontology 19(supplement to 3):79A.

Sykes, J.H. 1977. British Rhaetian bone-beds. Mercian Geologist 6:197–240.

Tumarkin, A.R., D.H. Tanke, and L.L. Malinconico Jr. 1997. Sedimentology, taphonomy, and faunal review of a multigeneric bonebed (bone bed 47) in the Dinosaur Park Formation (Campanian) of Southern Alberta, Canada. Geological Society of America, Abstracts with Programs 29:86.

Turnbull, W.D., and D.M. Martill. 1988. Taphonomy and preservation of a monospecific titanothere assemblage from the Washakie Formation (Late Eocene), southern Wyoming: An ecological accident in the fossil record. Palaeogeography, Palaeoclimatology, Palaeoecology 63:91–108.

Varricchio, D.J. 1995. Taphonomy of Jack's Birthday Site, a diverse dinosaur bonebed from the Upper Cretaceous Two Medicine Formation of Montana. Palaeogeography, Palaeoclimatology, Palaeoecology 114:297–323.

Varricchio, D.J. and J.R. Horner. 1993. Hadrosaurid and lambeosaurid bone beds from the Upper Cretaceous Two Medicine Formation of Montana: Taphonomic and biologic implications. Canadian Journal of Earth Sciences 30:997–1006.

Visser, J. 1986. Sedimentology and taphonomy of a *Styracosaurus* bonebed in the Late Cretaceous Judith River Formation, Dinosaur Provincial Park, Alberta. Unpublished masters thesis, University of Alberta, Calgary.

Voorhies, M.R. 1969. Taphonomy and population dynamics of an early Pliocene vertebrate fauna, Knox County, Nebraska. Contributions to geology, special paper 1. University of Wyoming,Laramie.

Voorhies, M.R. 1985. A Miocene rhinoceros herd buried in volcanic ash. National Geographic Research Reports, Projects 1978:671–688.

Weigelt, J. 1989. Recent vertebrate carcasses and their paleobiological implications. J. Schaefer, transl. University of Chicago Press, Chicago.

White, P.D., Fastovsky, D.E., and Sheehan, P.M. 1998. Taphonomy and suggested structure of the dinosaurian assemblage of the Hell Creek Formation (Maastrichtian) Eastern Montana, and Western North Dakota. Palaios 13:41–51.

Winkler, D.A., P.A. Murry, L.L. Jacobs, W.R. Downs, J.R. Branch, and P. Trundel. 1988. The Proctor Lake dinosaur locality, Lower Cretaceous of Texas. Hunteria 2:1–8.

Wood, J.M., R.G. Thomas, and J. Visser. 1988. Fluvial processes and vertebrate taphonomy: The Upper Cretaceous Judith River Formation, south-central Dinosaur Provincial Park, Alberta, Canada. Palaeogeography, Palaeoclimatology, Palaeoecology 66:127–143.

# APPENDIX 3.1. BONEBEDS DATABASE

| Author | Date | Title | Citation | Taxa | Source | Bonebed Name |
|---|---|---|---|---|---|---|
| Adrain, J.M., and M.V.H. Wilson. | 1994 | Early Devonian cephalaspids (Vertebrata: Osteostraci: Cornuata) from the Southern Mackenzie Mountains, N.W.T., Canada. | Journal of Vertebrate Paleontology 14:301–319. | f | Ms | Man on the Hill (MOTH) 180m |
| Agenbroad, L.D. | 1984 | Hot springs, South Dakota: Entrapment and taphonomy of Colombian Mammoth. | Pp. 113–127 *in* Quaternary extinctions: A prehistoric revolution. P.S. Martin and R.G. Klein, eds. University of Arizona Press, Tuscon. | lm | Ms | Hot Springs Mammoth Site |
| Akersten, W.A., C.A. Shaw, and G.C. Jefferson. | 1983 | Rancho La Brea: Status and future. | Paleobiology 9:211–217. | b, f, lm, sm | Ms | La Brea Tar Pits |
| Albright, L.B. | 1990 | A submerged Early Miocene site in the Fleming Formation near Toledo Bend Dam, Texas. | Journal of Vertebrate Paleontology 10(supplement to 3):12A. | lm | Ab | Toledo Bend |
| Albright, L.B. | 1996 | Insectivores, rodents, and carnivores of the Toledo Bend local fauna: An Arikareean (Earliest Miocene) assemblage from the Texas coastal plain. | Journal of Vertebrate Paleontology 16:458–473. | sm | Ms | Toledo Bend |
| Allen, P. | 1949 | Notes on Wealden bonebeds. | Proceedings of the Geological Association 60:275–283. | d, f, ndr | Ms | Brede Bone-Bed; Telham Bone-Bed; Mountfield Bone-Bed |
| Allison, P.A. | 1988 | Taphonomy of the Eocene London Clay. | Palaeontology 31:1079–1100. | b, f, ndr, sm | Ms | Sheppey Locality |
| Ambrose, P.D. | 1992 | Taphonomy and paleoenvironments of a turtle-bearing unit of the Bridger Fm. | Journal of Vertebrate Paleontology 12(supplement to 3):16A. | ndr | Ab | ? |
| Anderson, B.G., R.E. Barrick, and K.L. Stadtman. | 1995 | Paleoenviornmental and diagentic controls on the preservation of hadrosaur integument: Evidence from the Cretaceous Western interior USA. | Geological Society of America, Abstracts with Programs 27:318. | d | Ab | ? |
| Anderson, C., and H.O. Fletcher. | 1934 | The Cuddie Springs Bone Bed. | Australian Museum Magazine 1934:24–25. | lm | Ms | Cuddie Springs Bone Bed |

| Location | Period / Epoch | Stratigraphy | Element Size | Diversity | Relative Abundance | Depositional Setting | Depositional Environment | Facies | Paleoenv Code |
|---|---|---|---|---|---|---|---|---|---|
| North America, Canada, Northwest Territories, Mackenzie Mountains | Paleozoic; Devonian | Delorme Fm. | Mixed | High | Multidominant | Paralic | Estuarine, marine | Limestone, light and dark laminae | MAR, PESH |
| North America, USA, South Dakota, Hot Springs | Cenozoic; Quaternary, Pleistocene | ? | Macrofossil | Low | Monodominant | Sinkhole | Sinkhole | Sand, gravel, clay | O |
| North America, USA, California, Los Angles Basin, Rancho La Brea | Cenozoic; Quaternary, Pleistocene | Palos Verdes Sand | Macrofossil, microfossil | High | Multidominant | Tar seep | Channel, sheet-flood, tar seep | Fine to coarse sand, tar | CHLSP, O |
| North America, USA, Texas, Newton County | Cenozoic; Tertiary, Miocene | Fleming Fm. | Microfossil | High | Multidominant | Alluvial | Channel | Sand | CHLSP |
| North America, USA, Texas, Newton County | Cenozoic; Tertiary, Miocene | Fleming Fm. | Microfossil | High | Multidominant | Alluvial | Channel | Conglomerate | CHLSP |
| Europe, United Kingdom, England, Sussex | Mesozoic; Cretaceous | Lower Wealden | Microfossil | High | Multidominant | Paralic | ? | ? | PESH |
| Europe, United Kingdom, England, Kent | Cenozoic; Tertiary, Eocene | London Clay | Mixed | High | Multidominant | Marine | Offshore | Claystone, concretions | MAR |
| North America, USA, Wyoming | Cenozoic; Tertiary, Eocene, Bridgerian | Bridger Fm. | Macrofossil | Low | Monodominant | Alluvial | Paludal wetlands | ? | WTLND |
| North America, USA, Utah, Wyoming, Book Cliffs | Mesozoic; Cretaceous, Campanian | Nelson Fm. | Macrofossil | ? | ? | Paralic | Estuarine (fluvio-Estuarine) | Sandstone (calcareous) | PESH |
| Australia | Cenozoic; Quaternary, Pleistocene | ? | Macrofossil | ? | ? | Paralic | ? | ? | PESH |

*(Continued)*

| Author | Date | Title | Citation | Taxa | Source | Bonebed Name | # of Sites |
|--------|------|-------|----------|------|--------|--------------|------------|
| Andrews, P., and B. Alpagut. | 1990 | Description of the fossiliferous units at Pasalar, Turkey. | Journal of Human Evolution 19:343–361. | lm | Ms | ? | 1 |
| Andrews, P., and A. Ersoy. | 1990 | Taphonomy of the Miocene bone accumulations at Pasalar, Turkey. | Journal of Human Evolution 19:379–396. | lm | Ms | ? | 1 |
| Anonymous | 1936 | Notice of a new bone bed in the Early Pleistocene of Morrill County, Nebraska. | Bulletin of the University of Nebraska State Museum. | lm | Ab | Broadwater Quarries | 1 |
| Antia, D.D.J. | 1979 | Bone-beds: A review of their classification, occurrence, genesis, diagenesis, geochemistry, paleoecology, weathering, and microbiotas. | Mercian Geologist 7:93–174. | f | Ms | ? | 2+ |
| Antia, D.D.J. | 1981 | The Temeside bone bed and associated sediments from Wales and the Welsh Borderland. | Mercian Geologist 8:163–215. | f | Ms | Temeside Bone Bed | 2+ |
| Antia, D.D.J., and J.H. Sykes. | 1979 | The surface textures of quartz grains from a Rhaetian bone-bed, Blue Anchor Bay Somerset. | Mercian Geologist 7:205–210. | f | Ms | Basal Bone-Bed; The Clough; The Bone-Bed | 3 |
| Antia, D.D.J., and J.H. McD Whitaker. | 1978 | Scanning electron microscope study of the genesis of the Upper Silurian Ludlow bone bed. | Pp. 119–136 in Scanning electron microscopy in the study of sediments. W.B. Whalley, ed. Geoabstracts, Norwich. | f | Ms | Ludlow Bone Bed | 8 |
| Anton, M., M.J. Salesa, J. Morales, and A. Turner. | 2004 | First known complete skulls of the scimitar-toothed cat Machairodus aphanistus (Felidae, Carnivora) from the Spanish late Miocene site of Batallones-1. | Journal of Vertebrate Paleontology 24:957–969. | lm | Ms | Batallones-1 | 1 |
| Ayer, J. | 1999 | The Howe Ranch dinosaurs. | Sauriermuseum Aathal, Zürich. | d | Bk | Howe Quarry; Howe Stephens Quarry; Siber Site; Spring Hill | 4 |
| Azevedo, S.A.K., D.A. Campos, and A.W.A Kellner. | 1999 | A new Notosuchidae from the Late Cretaceous of Minas Gerais, Brazil. | Journal of Vertebrate Paleontology 19(supplement to 3):30A. | ndr | Ab | ? | 1 |

| Location | Period / Epoch | Stratigraphy | Element Size | Diversity | Relative Abundance | Depositional Setting | Depositional Environment | Facies | Paleoenv Code |
|---|---|---|---|---|---|---|---|---|---|
| Europe, Turkey, Pasalar | Cenozoic; Tertiary, Miocene | ? | Macrofossil | High | Multidominant | Alluvial | Sheet flood | Sandstone, gravelly | DBF |
| Europe, Turkey, Pasalar | Cenozoic; Tertiary, Miocene | ? | Macrofossil | High | Multidominant | Alluvial | Sheet flood | Sandstone, gravelly | DBF |
| North America, USA, Nebraska, Broadwater | Cenozoic; Quaternary, Pleistocene | ? | Mixed | High | Multidominant | Alluvial, lacustrine | ? | Diatomaceous earth | U |
| Europe | Phanerozoic | ? | Microfossil | High | Multidominant | Paralic | ? | Sandstone, mudstone, conglomerate | MAR, PESH |
| Europe, United Kingdom, Wales | Paleozoic; Silurian, Devonian | Temeside Fm., Ledbury Fm., Downton Castle Fm. & Overton Fm. | Microfossil | Low | Multidominant | Paralic | Marine; intertidal; back barrier | Sandstone | MAR, PESH |
| Europe, United Kingdom, England, Somerset, Blue Anchor Bay | Mesozoic; Triassic, Rhaetian | Brassington Fm. ? | Microfossil | High | Multidominant | Paralic | ? | Sandstone, mudstone | MAR, PESH |
| Europe, United Kingdom, Wales, Ludlow | Paleozoic; Silurian, Downtonian | Temeside Fm.; Ledbury Fm.; Downton Castle Fm.; Overton Fm. | Microfossil | Low | Multidominant | Paralic | Marine; intertidal; back barrier | Sandstone (thelodont Sand) | MAR, PESH |
| Europe, Spain | Cenozoic; Tertiary, Miocene | ? | Macrofossil | ? | ? | ? | ? | ? | U |
| North America, USA, Wyoming, Shell | Mesozoic; Jurassic | Morrison Fm. | Macrofossil | High | Monodominant | Alluvial | Channel, watering hole | Sandstone | CHLSP, WTLND |
| South America, Brazil, Minas Gerais | Mesozoic; Cretaceous | Bauru Grp. | Mixed | High | Multidominant | ? | ? | Calcareous sandstone | U |

*(Continued)*

| Author | Date | Title | Citation | Taxa | Source | Bonebed Name | # of Sites |
|---|---|---|---|---|---|---|---|
| Badam, G.L., R.K. Ganjoo, and R.K.G. Salahuddin | 1986 | Preliminary taphonomical studies of some Pleistocene fauna from the central Narmada Valley, Madhya Pradesh, India. | Palaeogeography, Palaeoclimatology, Palaeoecology, 53:335–348. | lm | Ms | Devakachar; Talayaghat; Daubaro; Chhindaghat; Gauri | 5 |
| Badgley, C. | 1986 | Taphonomy of mammalian fossil remains from Siwalik rocks of Pakistan. | Paleobiology 12:119–142. | lm, sm | Ms | Taphonomic Assemblages III and IV | 2+ |
| Badgley, C., W. Downs, and L.J. Flynn. | 1998 | Taphonomy of small-mammal fossil assemblages from the Middle Miocene Chinji Formation, Siwalik Group, Pakistan. | Pp. 145–166 in Advances in vertebrate paleontology and geochronology. Monograph 14. Y. Tomida, L.J. Flynn, and L.L. Jacobs, eds. National Science Museum, Tokyo. | sm | Ms | multiple sites | 13 |
| Bakker, R.T. | 1975 | Dinosaur renaissance. | Scientific American 232:58–78. | d, ndr | Ms | multiple sites | 1+ |
| Bakker, R.T. | 1982 | Juvenile-adult habitat shift in Permian fossil reptiles and amphibians. | Science 217:53–55. | a, ndr | Ms | multiple sites | 4 |
| Bakker, R.T. | 1997 | Raptor family values: Allosaur parents brought giant carcasses into their lair to feed their young. | Pp. 51–63 in Dinofest International, proceedings of a symposium sponsored by Arizona State University. D.L. Wolberg, E. Stump, and G.D. Rosenberg, eds. Academy of Natural Sciences, Philadelphia. | d | Ms | multiple sites | 6+ |
| Barnosky, A.A. | 1985 | Taphonomy and herd structure of the extinct Irish Elk, *Megaloceros giganteus*. | Science 228:340–344. | lm | Ms | Ballybetagh Bog | 1 |
| Bartlett, J. | 1999 | Taphonomy and characteristics of a diverse Maastrichtian assemblage: The Sandy Site, Hell Creek Formation, South Dakota. | Journal of Vertebrate Paleontology 19(supplement to 3):31A. | d | Ab | Sandy site | 1 |
| Barton, G.B., and M.V.H. Wilson. | 1999 | Microstratigraphic study of meristic variation in an Eocene fish from a 10, 000-year varved interval at Horsefly, British Columbia. | Canadian Journal of Earth Sciences 36:2059–2072. | f | Ms | Horsefly locality | 1 |

| Location | Period / Epoch | Stratigraphy | Element Size | Diversity | Relative Abundance | | Depositional Setting | Depositional Environment | Facies | Paleoenv Code |
|---|---|---|---|---|---|---|---|---|---|---|
| Asia, India, Central Narmada Valley | Cenozoic; Quaternary, Pleistocene | "Sandy Pebbly Gravel" unit | Macrofossil | Low | Multidominant | Alluvial | | Channel-splay | Mudstone, sandstone, conglomerate | CHLSP |
| Asia, Pakistan, Khuar | Cenozoic; Tertiary, Miocene | Siwalik Grp. | Macrofossil | High? | Multidominant? | Alluvial | | Pond, channel-splay, overbank | Conglomerate, sandstone, mudstone | CHLSP, PSOL, WTLND |
| Asia, Pakistan, Potwar Plateau | Cenozoic; Tertiary, Miocene | Chinji Fm. | Microfossil | High | Multidominant | Alluvial | | Channel, channel-splay, overbank, pond, sheetwash | Conglomerate, sandstone, mudstone | DBF, CHLSP, PSOL, WTLND |
| North America, USA, Texas | Paleozoic, Mesozoic; Permian, Triassic, Jurassic, Cretaceous | | Macrofossil | High | Multidominant | Alluvial | | Channels, overbank, pond | Sandstone, mudstone | CHLSP, WTLND, LAC |
| North America, USA, Texas | Paleozoic; Permian | Admiral Fm.; Belle Plains Fm. | Macrofossil | High | Multidominant | Alluvial, lacustrine | | Channel sandstone, mudstone | Sandstone, shale | CHLSP, WTLND, LAC |
| North America, USA, Wyoming, Como Bluff | Mesozoic; Jurassic | Morrison Fm. | Mixed | High | Multidominant | Alluvial | | Splay, overbank | Mudstone, siltstone, sandstone | CHLSP, PSOL |
| Europe, United Kingdom, Ireland, Enniskerry | Cenozoic; Quaternary, Pleistocene | | Macrofossil | Low | Monodominant | Alluvial | | Bog | Clay | WTLND |
| North America, USA, South Dakota | Mesozoic; Cretaceous | Hell Creek Fm. | Mixed | High | Multidominant | Alluvial | | Channel, splay | Sandstone | CHLSP |
| North America, Canada, British Columbia, Horsefly | Cenozoic; Tertiary, Eocene | H3 interval | Macrofossil | Low | Monodominant | Lacustrine | | Hypoxic lake | Varved silt | LAC |

*(Continued)*

| Author | Date | Title | Citation | Taxa | Source | Bonebed Name | # of Sites |
|--------|------|-------|----------|------|--------|--------------|------------|
| Barton, G.B., M.V.H. Wilson. | 2005 | Taphonomic variations in Eocene fish-bearing varves at Horsefly, British Columbia, reveal 10, 000 years of environmental change. | Canadian Journal of Earth Sciences 42:137–149. | f | Ms | Horsefly locality | 1 |
| Baszio, S. | 1997 | Investigations on Canadian dinosaurs. | Courier Forschungsinstitut, Senckenberg. | d | Ms | KUA-2; others | 2+ |
| Bayrock, L.A. | 1964 | Fossil *Scaphiopus* and *Bufo* in Alberta. | Journal of Paleontology 38:1111–1112. | a | Ms | ? | 1 |
| Behrensmeyer, A.K. | 1975 | The taphonomy and paleoecology of Plio-Pleistocene vertebrate assemblages east of Lake Rudolf, Kenya. | Bulletin of the Museum of Comparative Zoology 146:474–578. | lm, sm | Ms | seven sites | 7 |
| Bell, C.M., and K. Padian. | 1995 | Pterosaur fossils from the Cretaceous of Chile: Evidence for a pterosaur colony on an inland desert plain. | Geology Magazine 132:31–38. Also Journal of Vertebrate Paleontology 14(supplement to 3):16A. | ndr | Ms | ? | 1 |
| Benton, M.J., S. Bouaziz, E. Buffetaut, D. Martill, M. Ouaja, M. Soussi, and C. Tureman. | 2000 | Dinosaurs and other fossil vertebrates from fluvial deposits in the Lower Cretaceous of southern Tunisia. | Palaeogeography, Palaeoclimatology, Palaeoecology 157:227–246. | d | Ms | Chenini Fm. Bone Beds | 2+ |
| Benton, M.J., E. Cook, D. Grigorescu, E. Popa, and E. Tallódi. | 1997 | Dinosaurs and other tetrapods in an Early Cretaceous bauxite-filled fissure, northwestern Romania. | Palaeogeography, Palaeoclimatology, Palaeoecology 130:275–292. | d | Ms | Lens 204 | 1 |
| Berman, D.S, D.A. Eberth, and D.B. Brinkman. | 1988 | *Stegotretus agyrus* a new genus and species of microsaur (amphibian) from the Permo-Pennsylvanian of New Mexico. | Annals of the Carnegie Museum 57:293–323. | a | Ms | ? | 1 |

| Location | Period / Epoch | Stratigraphy | Element Size | Diversity | Relative Abundance | Depositional Setting | Depositional Environment | Facies | Paleoenv Code |
|---|---|---|---|---|---|---|---|---|---|
| North America, Canada, British Columbia, Horsefly | Cenozoic; Tertiary, Eocene | H3 interval | Macrofossil | Low | Monodominant | Lacustrine | Hypoxic lake | Varved silt | LAC |
| North America, Canada, Alberta, Southern | Mesozoic; Cretaceous | Scollard Fm.; Horseshoe Canyon Fm.; Dinosaur Park Fm.; Oldman Fm. | Microfossil | High | Multidominant | Alluvial | In-channel, splay, overbank | Sandstone, siltstone, mudstone | CHLSP, PSOL, WTLND |
| North America, Canada, Alberta, Killam | Cenozoic; Quaternary | Quaternary | Macrofossil | Low | ? | Alluvial | Alluvial fan | Sand | DBF |
| Africa, Kenya, Lake Rudolf | Cenozoic; Tertiary, Pliocene, Quaternary, Pleistocene | Koobi Fora Fm. | Macrofossil | High | Multidominant | Alluvial, paralic | Channel, channel-splay, pond, overbank, lacustrine deltaic, alluvial plain | Sandstone, mudstone | CHLSP, PSOL, WTLND |
| South America, Chile, Atacama region, Cerro La Isla | Mesozoic; Cretaceous | Quebrada Monardes Fm. | Mixed | Low | Monodominant | Alluvial | Debris flow | Sandstone, conglomerate | DBF |
| Africa, Tunisia, Tataouine region | Mesozoic; Cretaceous | Chenini Fm. | Mixed | High | Multidominant | Alluvial | Channel lag | Conglomeratic sandstone | CHLSP |
| Europe, Romania, Judetul Bihor, Cornet | Mesozoic; Cretaceous | | Mixed | Low | Multidominant | karst | Fissure fill | Conglomerate, (bauxite) | O |
| North America, USA, New Mexico, Arroyo del Agua | Paleozoic; Permo-Pennsylvanian | Cutler Fm. | Macrofossil | Low | Monotaxic | Alluvial | Burrows? | Caliches, mudstone | O |

*(Continued)*

| Author | Date | Title | Citation | Taxa | Source | Bonebed Name | # of Sites |
|---|---|---|---|---|---|---|---|
| Berman, D.S, A.C. Henrici, S.S. Sumida, and T. Martens. | 2000 | Redescription of *Seymouria sanjuanensis* (Seymouriamorpha) from the Lower Permian of Germany based on complete, mature specimens with a discussion of paleoecology of the Bromacker locality assemblage. | Journal of Vertebrate Paleontology 20:253–268. | a | Ms | Bromacker Locality | 1 |
| Berman, D.S, R.R. Reisz, and D.A. Eberth. | 1987 | *Seymouria sanjuanensis* (Amphibia, Batrachosauria) from the Lower Permian Cutler Formation of north-central New Mexico and the occurrence of sexual dimorphism in that genus questioned. | Canadian Journal of Earth Sciences 24:1769–1784. | a | Ms | ? | 1 |
| Bilbey, S.A. | 1998 | Cleveland-Lloyd dinosaur quarry: Age, stratigraphy and depositional environments. | Modern Geology 22:87–120. | d | Ms | Cleveland-Lloyd Quarry | 1 |
| Bilbey, S.A. | 1999 | Taphonomy of the Cleveland-Lloyd dinosaur quarry in the Morrison Formation, Central Utah: A lethal spring-fed pond. | Pp. 121–133 *in* Vertebrate paleontology in Utah. D.D. Gillette, ed. Miscellaneous publication 99–1. Utah Geological Survey, Salt Lake City. | d | Ms | Cleveland-Lloyd Quarry | 1 |
| Bjork, P.R. | 1985 | Preliminary report on the Ruby Site bone bed, Upper Cretaceous South Dakota. | Geological Society of America, North Central and South Central Sections, Abstracts with Programs 17:4. | d | Ab | Ruby Site Bone Bed | 1 |
| Blob, R.W. | 1994 | The significance of microfossil size distributions in faunal abundance reconstructions: A Late Cretaceous example. | Journal of Vertebrate Paleontology 14(supplement to 3):17A. | d | Ab | ? | 3 |
| Blob, R.W., and A.R. Fiorillo. | 1996 | The significance of vertebrate microfossil size and shape distributions for faunal abundance reconstructions: A Late Cretaceous example. | Paleobiology 22:422–435. | ndr | Ms | Blackbird's Ridge, Emily's Ankle | 2 |
| Bolt, J.R., R.M. McKay, B.J. Witzke, and M.P. McAdams. | 1988 | A new Lower Carboniferous tetrapod locality in Iowa. | Nature 333:768–770. | a | Ms | Delta Site | 1 |

| Location | Period / Epoch | Stratigraphy | Element Size | Diversity | Relative Abundance | Depositional Setting | Depositional Environment | Facies | Paleoenv Code |
|---|---|---|---|---|---|---|---|---|---|
| Europe, Germany, Gotha, Thurigian Forest | Paleozoic; Permian | Tambach Fm. | Macrofossil | Low | Multidominant | Alluvial | Overbank, debris flow, pond | Mudstone | DBF, WTLND |
| North America, USA, New Mexico | Paleozoic; Permian | Cutler Fm. | Macrofossil | Low | Monotaxic | Alluvial | Burrows? | Mudstone (siltstone) | WTLND |
| North America, USA, Utah | Mesozoic; Jurassic | Morrison Fm. | Macrofossil | High | Multidominant | Alluvial | Shallow lake | Calcareous mudstone | WTLND, LAC |
| North America, USA, Utah, Emery County | Mesozoic; Jurassic | Morrison Fm. | Macrofossil | High | Monodominant | Alluvial | Pond, quick mud, predator trap | Mudstone (smetitic claystone) | WTLND |
| North America, USA, South Dakota, Butte County | Mesozoic; Cretaceous | Hell Creek Fm. | Macrofossil | Low | ? | Alluvial | Channel, splay | Sandstone | CHLSP |
| North America, USA, Montana | Mesozoic; Cretaceous | Judith River Fm. | Microfossil | High | Multidominant | Alluvial | Channel, splay | Sandstone | CHLSP |
| North America, USA, Montana | Mesozoic; Cretaceous | Judith River Fm. | Microfossil | High | Monodominant, multidominant | Alluvial | Channel, splay | Sandstone | CHLSP |
| North America, USA, Iowa | Paleozoic; Carboniferous | St. Louis Fm. | Mixed | High | Multidominant | Karst, paralic | Fissure-fill | Limestone conglomerate | O |

*(Continued)*

| Author | Date | Title | Citation | Taxa | Source | Bonebed Name | # of Sites |
|---|---|---|---|---|---|---|---|
| Bown, T.M., and M.J. Kraus. | 1981 | Vertebrate fossil-bearing paleosol units (Willwood Formation, Lower Eocene, Northwest Wyoming, U.S.A.): Applications for taphonomy, biostratigraphy, and assemblage analysis. | Palaeogeography, Palaeoclimatology, Palaeoecology 34:31–56. | sm | Ms | ? | 2+ |
| Boyd, M.J. | 1984 | The Upper Carboniferous tetrapod assemblage from Newsham, Northumberland. | Palaeontology 27:367–392. | a, f | Ms | Hannah Pit | 1 |
| Braun, A., and T. Pfeiffer. | 2002 | Cyanobacterial blooms as the cause of a Pleistocene large mammal assemblage. | Paleobiology 28:139–154. | lm | Ms | Nuemark Nord | 1 |
| Bray, E.S., and K.F. Hirsch. | 1998 | Eggshell from the Upper Jurassic Morrison Formation. | Modern Geology 23:219–240. | d | Ms | multiple sites | 9+ |
| Breithaupt, B.H., and D. Duvall. | 1986 | The oldest record of serpent aggregation. | Lethaia 19:181–185. | ndr | Ms | ? | 1 |
| Brinkman, D.B. | 1990 | Paleoecology of the Judith River Formation (Campanian) of Dinosaur Provincial Park, Alberta, Canada: Evidence from microfossils localities. | Palaeogeography, Palaeoclimatology, Palaeoecology 78:37–54. | d | Ms | ? | 24 |
| Brinkman, D.B., D.R. Braman, A.G. Neuman, P.E. Ralrick, and T. Sato. | 2005 | A vertebrate assemblage from marine shales of the Lethbridge coal zone. | Pp. 486–500 in Dinosaur Park: A spectacular ancient ecosystem revealed. P.J. Currie and E.B. Koppelhus, eds. Indiana University Press, Bloomington. | b, d, f, ndr, sm | Ms | BB96, L2377 | 2 |
| Brinkman, D.B., M.J. Ryan, and D.A. Eberth. | 1998 | The palegeographic and stratigraphic distribution of ceratopsids (Ornithischia) in the Upper Judith River Group of western Canada. | Palaios 13:160–169. | d | Ms | ? | 2+ |
| Britt, B.B. | 1991 | Theropods of Dry Mesa Quarry (Morrison Formation, Late Jurassic), Colorado, with emphasis on the osteology of Torvosaurus tanneri. | Brigham Young University Geology Studies 37:1–72. | d | Ms | Dry Mesa Quarry | 1 |

| Location | Period / Epoch | Stratigraphy | Element Size | Diversity | Relative Abundance | Depositional Setting | Depositional Environment | Facies | Paleoenv Code |
|---|---|---|---|---|---|---|---|---|---|
| North America, USA, Wyoming | Cenozoic; Tertiary, Eocene | Willwood Fm. | Macrofossil, microfossil, mixed | High | Multidominant | Alluvial | Overbank, paleosol | Mudstone | PSOL |
| Europe, United Kingdom, England, Northumberland, Newsham | Paleozoic; Carboniferous, Westphalian B | | Macrofossil | High | ? | Lacustrine | Lacustrine | Mudstone (shale) | LAC |
| Europe, Germany, Sachsen-Anhalt | Cenozoic; Quaternary, Pleistocene | Pleistocene | Macrofossil | High | Multidominant | Lacustrine | Laminae, algal | Shale | LAC |
| North America, USA, Utah & Colorado | Mesozoic; Jurassic | Morrison Fm. | Microfossil, mixed | High, low | Monodominant, multidominant | Alluvial | Overbank | Mudstone | PSOL, WTLND |
| North America, USA, Wyoming, Converse County | Cenozoic; Tertiary, Oligocene, Orellan | White River Fm. | Microfossil | Low | Monotaxic | Alluvial | Overbank | Mudstone | PSOL |
| North America, Canada, Alberta, Dinosaur Provincial Park | Mesozoic; Cretaceous | Judith River Fm. | Microfossil | High | Multidominant | Alluvial | Channel, splay | Sandstone, mudstone | CHLSP |
| North America, Canada, Dinosaur Provincial Park | Mesozoic; Cretaceous, Campanian | Belly River Group, Dinosaur Park Formation | Microfossil | High | Multidominant | Marine | Transgressive lag | Intraclst-rich mudstone | MAR |
| North America, Canada, Alberta & Saskatchewan | Mesozoic; Cretaceous | Judith River Grp. | Microfossil | High, low | Multidominant | Alluvial, paralic | Channel, splay | ? | CHLSP, PESH |
| North America, USA, Colorado | Mesozoic; Jurassic | Morrison Fm., Brushy Basin Mbr. | Macrofossil | High | ? | Alluvial | Channel, splay | Sandstone (conglomeratic) | CHLSP |

*(Continued)*

| Author | Date | Title | Citation | Taxa | Source | Bonebed Name | # of Sites |
|---|---|---|---|---|---|---|---|
| Britt, B.B., and K.L. Stadtman. | 1996 | The Early Cretaceous Dalton Wells Dinosaur fauna and the earliest North American titanosaurid sauropod. | Journal of Vertebrate Paleontology 16(supplement to 3):24A. | d | Ab | Dalton Wells | 1 |
| Britt, B.B., D. Eberth, B. Brinkman, R. Scheetz, K.L. Stadtmen, and J.S. McIntosh. | 1997 | The Dalton Wells fauna (Cedar Mountain Formation): A rare window into the little-known world of Early Cretaceous dinosaurs in North America. | Geological Society of America, Abstracts with Programs 29:A–104. | d | Ab | Dalton Wells | 1 |
| Brouwers, E.M., W.A. Clemens, R.A. Spicer, T.A. Ager, D. Carter, and W.V. Sliter. | 1987 | Dinosaurs on the North Slope, Alaska: High latitude, latest Cretaceous environments. | Science 237:1608–1610. | d | Ms | Liscomb Bonebed; other sites | 6 |
| Burge, D.L., J.H. Bird, B.B. Britt, D.J. Chure, and R.L. Scheetz. | 2000 | A brachiosaurid from the Ruby Ranch Mbr. (Cedar Mountain Fm.) near Price, Utah, and sauropod faunal change across the Jurassic-Cretaceous boundary of North America. | Journal of Vertebrate Paleontology 20(supplement to 3):32A. | d | Ab | Price River II Quarry | 1 |
| Camp, C.L. | 1980 | Large ichthyosaurs from the Upper Triassic of Nevada. | Palaeontographica A 170:139–200. | ndr | Ms | Quarry 5 | 1 |
| Camp, C.L., and S.P. Welles. | 1956 | Triassic dicynodont reptiles. 1. The North American genus Placerias. | Memoirs of the University of California 13:255–348. | ndr | Ms | Placerias Quarry | 1 |
| Campbell, K.E. Jr. | 1979 | The non-passerine Pleistocene avifauna of the Talara Tar Seeps, Northwestern Peru. | Life Science Contribution, Royal Ontario Museum 118:1–203. | b | Ms | ? | 1 |
| Carpenter, K. | 1988 | Dinosaur bone beds and mass mortality: Implications for the K-T extinction. | Global catastrophes in Earth history: An interdisciplinary conference on impacts, volcanism and mass mortality. Lunar and Planetary Institute Contribution 673:24–25. | d | Ab | Bernissart, Revuelto Creek Quarry, Ashfall | 2+ |
| Carpenter, K. | 1998 | Evidence of predatory behavior by carnivorous dinosaurs. | Gaia 15:135–144. | d | Ms | Albertosaurus Bonebed | 1 |

| Location | Period / Epoch | Stratigraphy | Element Size | Diversity | Relative Abundance | Depositional Setting | Depositional Environment | Facies | Paleoenv Code |
|---|---|---|---|---|---|---|---|---|---|
| North America, USA, Utah, Moab | Mesozoic; Cretaceous | Cedar Mountain Fm. | Macrofossil | High | Multidominant | Alluvial | Channel | Altered ash, sandstone | CHLSP |
| North America, USA, Utah, Moab | Mesozoic; Cretaceous | Cedar Mountain Fm. | Macrofossil | High | Multidominant | Alluvial, lacustrine | Lacustrine-shoreline | Mudstone (siltstone) | WTLND |
| North America, USA, Alaska, Colville River | Mesozoic; Cretaceous | Colville Grp., Prince Creek Fm. | Macrofossil | High | Multidominant | Alluvial | Overbank, pond, paleosol | Mudstone, organic rich | PSOL, WTLND |
| North America, USA, Utah, Proce | Mesozoic; Cretaceous | Cedar Mountain Fm. | Macrofossil | High | Multidominant | Alluvial | Channel | Sandstone | CHLSP |
| North America, USA, Nevada, Nye County | Mesozoic; Triassic | Luning Fm. | Macrofossil | Low | Monotaxic | Marine | Nearshore | Limestone | MAR |
| North America, USA, Arizona, St. Johns | Mesozoic; Triassic | Chinle Fm. | Mixed | High | Multidominant | Alluvial | Overbank, paludal | Mudstone | WTLND |
| South America, Peru | Cenozoic; Quaternary, Pleistocene, Sangamon | Talara Tar Seeps | Microfossil | High | Multidominant | Seep | Seep | Tar | O |
| Europe & NA, Belgium & USA, New Mexico, Montana, Nebraska | Mesozoic; Cenozoic; Triassic, Cretaceous, Tertiary, Miocene | Hell Creek Fm.; Dockum Grp.; Wealden | Macrofossil | High, low | Monodominant, multidominant | Alluvial | Overbank, channel | Sandstone, mudstone, ash | CHLSP, PSOL, WTLND |
| North America, Canada, Alberta | Mesozoic; Cretaceous | Edmonton Fm. | Macrofossil | Low | Monodominant | Alluvial | Channel | Sandstone | CHLSP |

(Continued)

| Author | Date | Title | Citation | Taxa | Source | Bonebed Name | # of Sites |
|---|---|---|---|---|---|---|---|
| Carroll, R.L. | 1994 | Evaluation of geological age and environmental factors in changing aspects of the terrestrial vertebrate fauna during the Carboniferous. | Transactions of the Royal Society of Edinbourgh, Earth Sciences 84:427–431. | a, f | Ms | East Kirkton, Joggins | 2 |
| Carroll, R.L., E.S. Belt, D.L. Dineley, D. Baird, and D.C. McGregor. | 1972 | Vertebrate paleontology of eastern Canada. | 24th International Geological Conference, Ottawa, Excursion A59:1–113. | a, f, ndr | Ms | Joggins | 3+ |
| Carvalho, I.S., and E. Pedrao. | 1998 | Brazilian theropods from the equatorial Atlantic margin: Behavior and environmental setting. | Gaia 15:369–378. | d, ndr | Ms | multiple sites | 6 |
| Case, E.C. | 1900 | The vertebrates from the Permian bone bed of Vermillion County, Illinois. | Journal of Geology 8:698–729. | a, ndr | Ms | Vermillion Bone Bed | 1 |
| Case, E.C. | 1932 | A collection of stegocephalians from Scurry County, Texas. | Contributions to the Museum of Paleontology, University of Michigan 4:1–56. | a | Ms | Elkins Place | 1 |
| Chaffee, R.G. | 1952 | The Deseadan vertebrate fauna of the Scarritt Pocket, Patagonia. | Bulletin of the American Museum of Natural History 98:507–562. | a, lm, sm | Ms | Scarritt Pocket | 1+ |
| Chiappe, L., F. Jackson, L. Dingus, G. Grellet-Tinner, and R. Coria. | 1999 | Auca Mahuevo: An extraordinary dinosaur nesting ground from the Late Cretaceous of Patagonia. | Journal of Vertebrate Paleontology 19(supplement to 3):37A. | d | Ab | Auca Mahuevo | 1 |
| Chiappe, L.M., R.A. Coria, L. Dingus, F. Jackson, A. Chinsamy, and M. Fox. | 1998 | Sauropod dinosaur embryos from the Late Cretaceous of Patagonia. | Nature 396:258–261. | d | Ms | Auca Mahuevo | 1 |
| Chinnery, B. | 2004 | Description of *Prenoceratops pieganensis* gen.et sp. nov. (*Dinosauria, Neoceratopsia*) from the Two Medicine Formation of Montana. | Journal of Vertebrate Paleontology 24:572–590. | d | Ms | Neoceratops Bonebed | 1 |

| Location | Period / Epoch | Stratigraphy | Element Size | Diversity | Relative Abundance | Depositional Setting | Depositional Environment | Facies | Paleoenv Code |
|---|---|---|---|---|---|---|---|---|---|
| Europe, United Kingdom, Scotland | Paleozoic; Carboniferous | | Mixed | High | Multidominant | Paralic | Ponds, pools | Black shale | WTLND, PESH |
| North America, Canada, Nova Scotia, Joggins | Paleozoic; Carboniferous | Cumberland Grp. | Microfossil | High, low | Monodominant, multidominant, monotaxic | Alluvial, paralic | Tree-stump trap | Carbonaceous siltstone | O |
| South America, Brazil, Sao Luis Basin, Sao Luis & Alcantara counties, Sao Luis Basin | Mesozoic; Cretaceous, Aptian-Cenomanian | Itapecuru Fm. | Macrofossil | High | Multidominant | Coastal plain | Tidal sand flats, tidal channels | Sandstones, siltstones, shales, mudstones, and carbonates | CHLSP, PSOL, PESH |
| North America, USA, Illinois | Paleozoic; Permian | | Macrofossil | High | Multidominant | ? | ? | ? | U |
| North America, USA, Texas, Scurry County | Mesozoic; Triassic | | Macrofossil | Low | Monodominant | Alluvial, lacustrine | Pond | Sandstone | WTLND |
| South America, Argentina | Cenozoic; Tertiary, Oligocene, Deseadan | Sarmientan Grp, Crater Lake Beds, Scarritt Pocket | Macrofossil | High | Multidominant | Volcanic crater basin | Lacustrine, volcanic | Breccia, tuffs, sandstone | LAC, O |
| South America, Argentina, Patagonia | Mesozoic; Cretaceous | Río Colorado Fm. | Macrofossil | Low | Monotaxic? | Alluvial | Overbank | Mudstone | PSOL |
| South America, Argentina, Patagonia | Mesozoic; Cretaceous | Río Colorado Fm. | Macrofossil | Low | Monotaxic | Alluvial | Overbank | Mudstone | PSOL |
| North America, USA, Montana, Pondera County | Mesozoic; Cretaceous, Campanian | Two Medicine Fm. | Macrofossil | Low | Monotaxic | Alluvial | ? | ? | U |

*(Continued)*

| Author | Date | Title | Citation | Taxa | Source | Bonebed Name | # of Sites |
|---|---|---|---|---|---|---|---|
| Chinnery, B.J., and D. Trexler. | 1999 | The first bonebed of a basal neoceratopsian, with new information on the skull morphology of *Leptoceratops*. | Journal of Vertebrate Paleontology 19(supplement to 3):38A. | d | Ab | Neoceratops Bonebed | 1 |
| Christians, J.P. | 1991 | Taphonomic review of the Mason Dinosaur Quarry: Hell Creek Formation, Upper Cretaceous, South Dakota. | Journal of Vertebrate Paleontology 11(supplement to 3):22A. | d | Ab | Mason Dinosaur Quarry | 1 |
| Chure, D.J., K. Carpenter, R. Litwin, S. Hasiotis, and E. Evanoff. | 1998 | The fauna and flora of the Morrison Formation. | Modern Geology 23:507–537. | d | Ms | multiple sites | 1+ |
| Chure, D.J., A.R. Fiorillo, and A. Jacobsen. | 2000 | Prey bone utilization by predatory dinosaurs in the late Jurassic of North America, with comments on prey bone use by dinosaurs throughout the Mesozoic. | Gaia 15:227–232. | d | Ms | Reed's Quarry, Cleveland-Lloyd Qarry | 2 |
| Clark, J., J.R. Beerbower, and K.K. Kietzke. | 1967 | Oligocene sedimentation, stratigraphy, paleoecology, and paleoclimatology in the Big Badlands of South Dakota. | Fieldiana, Geology Memoirs 5:158. | lm, sm | Ms | ? | 2+ |
| Clemens, W.A. | 1963 | Fossil mammals of the type Lance Formation, Wyoming. I. Introduction and multituberculata. | Publications in the Geological Sciences 48. University of California, Berkeley. | sm | Ms | multiple sites | 1+ |
| Colbert, E.H. | 1989 | The Triassic dinosaur *Coelophysis*. | Museum of Northern Arizona Bulletin 57:1–160. | d | Ms | Coelophysis Quarry | 1 |
| Colson, M.C., R.O. Colson, and R. Nellermoe. | 2004 | Stratigraphy and depositional environments of the upper Fox Hills and lower Hell Creek formations at the Concordia Hadrosaur Site in northwestern South Dakota. | Rocky Mountain Geology 39:93–111. | d | Ms | Concordia Hadrosaur Site | 1 |
| Cook, E. | 1995 | Taphonomy of two non-marine Lower Cretaceous bone accumulations from southeastern England. | Palaeogeography, Palaeoclimatology, Palaeoecology 116:263–270. | d | Ms | Cliff End Bone Bed, Keymer Tile Works | 2 |
| Cook, L.M. | 1926 | A rhinoceros bone bed. | Black Hills Engineer 14:95–99. | lm | Ms | Agate Springs Quarry | 1 |

| Location | Period / Epoch | Stratigraphy | Element Size | Diversity | Relative Abundance | Depositional Setting | Depositional Environment | Facies | Paleoenv Code |
|---|---|---|---|---|---|---|---|---|---|
| North America, USA, Montana, Pondera County | Mesozoic; Cretaceous, Campanian | Two Medicine Fm. | Macrofossil | Low | monotaxic | Alluvial | ? | ? | U |
| North America, USA, South Dakota | Mesozoic; Cretaceous | Hell Creek Fm. | Macrofossil | Low | ? | Alluvial | Debris flow | ? | DBF |
| North America, USA, Colorado & New Mexico | Mesozoic; Jurassic | Morrison Fm. | Macrofossil, microfossil, mixed | ? | ? | Alluvial | ? | many | U |
| North America, USA, Wyoming, Utah | Mesozoic; Jurassic | Morrison Fm. | Macrofossil | High | ? | Alluvial | ? | ? | U |
| North America, USA, South Dakota | Cenozoic; Tertiary, Oligocene | Chadron Fm.; Brule Fm. | Macrofossil | Low | Monodominant? | Alluvial | Channel, splay, overbank, pond | ? | CHLSP, PSOL, WTLND |
| North America, USA, Montana | Mesozoic; Cretaceous | Lance Fm. | Microfossil | High | Multidominant | Alluvial | Channel sandstone | Sandstone | CHLSP |
| North America, USA, New Mexico, Abiquiu | Mesozoic; Triassic | Chinle Fm. | Macrofossil | Low | Monodominant | Alluvial | Pond | Mudstone (siltstone) | CHLSP, WTLND |
| North America, USA, South Dakota | Mesozoic, Cretaceous, Maastrichtian | Hell Creek Fm., Little Beaver Creek Mbr. | Macrofossil | Low | Monotaxic | Paralic | Coastal swamp | Organic mudstone | WTLND |
| Europe, United Kingdom, England | Mesozoic; Cretaceous | Wealden Grp. | Microfossil | High | Multidominant | Alluvial | Channel, splay, overbank | ? | CHLSP, WTLND |
| North America, USA, Nebraska, Agate | Cenozoic; Tertiary, Miocene | | Macrofossil | Low | Monodominant | Alluvial | Channel, splay | Sandstone | CHLSP |

*(Continued)*

| Author | Date | Title | Citation | Taxa | Source | Bonebed Name | # of Sites |
|---|---|---|---|---|---|---|---|
| Cooley, J.T., and J.G. Schmitt. | 1998 | An anastomosed fluvial system in the Morrison Formation (Upper Jurassic) of southwest Montana. | Modern Geology 22:171–208. | d | Ms | Strickland Creek Site | 1 |
| Coombs, M.C., and W.P. Coombs Jr. | 1997 | Analysis of the geology, fauna, and taphonomy of Morava Ranch quarry, Early Miocene of Northwest Nebraska. | Palaios 12:165–187. | lm | Ms | Morava Ranch Quarry | 1 |
| Cope, E.D. | 1877 | On the vertebrata of the bone bed in eastern Illinois. | Read before the American Philosophical Society, May 20, 1877. | a, f | Ms | ? | 1 |
| Coria, R.A. | 1994 | On monospecific assemblage of sauropod dinosaurs from Patagonia: Implications for gregarious behavior. | Gaia 10: 209–213. | d | Ms | Cerro Condor Norte (area) | 1 |
| Coria, R.A., and P.J. Curie. | 2006 | A new carcharodontosaurid (Dinosauria, Theropoda) from the Upper Cretaceous of Argentina. | Geodiversitas 28:71–118. | d | Ms | Canadon del Gato Bonebed | 1 |
| Cumbaa, S.L., T.T. Tokaryk, C. Collum, J.D. Stewart, T.S. Ercit, and R.G. Day. | 1997 | A Cenomanian age bone bed of marine origin, Saskatchewan, Canada. | Journal of Vertebrate Paleontology 17(supplement to 3):40A. | f | Ab | ? | 1 |
| Currie, P.J. | 1981 | Hunting dinosaurs in Alberta's great bonebed. | Canadian Geographic 101:34–39. | d | Ms | Alberta's Great Bonebed | 2+ |
| Currie, P.J. | 2000 | Possible evidence of gregarious behavior in tyrannosaurids. | Gaia 15:123–133. | d | Ms | *Albertosaurus* Bonebed | 1 |
| Currie, P.J., and P. Dodson. | 1984 | Mass death of a herd of ceratopsian dinosaurs. | Pp. 52–60 *in* Third symposium of Mesozoic terrestrial ecosystems. W.E. Reif and F.Westphal, eds. Attempto Verlag, Tubingen. | d | Ms | Bone Bed 143 | 1 |
| Currie, P.J., and D.A. Eberth. | 1993 | Paleontology, sedimentology and paleoecology of the Iren Dabasu Formation (Upper Cretaceous), Inner Mongolia, People's Republic of China. | Cretaceous Research 14:127–144. | d | Ms | multiple sites | 3+ |

| Location | Period / Epoch | Stratigraphy | Element Size | Diversity | Relative Abundance | Depositional Setting | Depositional Environment | Facies | Paleoenv Code |
|---|---|---|---|---|---|---|---|---|---|
| North America, USA, Montana, southwestern | Mesozoic; Jurassic | Morrison Fm. | Macrofossil | Low | Multidominant | Alluvial | Channel | Sandstone | CHLSP |
| North America, USA, Nebraska | Cenozoic; Tertiary, Miocene | Harrison Fm., Upper Harrison Fm. | Macrofossil | High | Multidominant | Alluvial | Pond | ? | WTLND |
| North America, USA, Illinois | Paleozoic; Permian | | Microfossil | Low | ? | ? | ? | ? | U |
| South America, Argentina, Chubut Province | Mesozoic; Jurassic | | Macrofossil | Low | Monotaxic | Alluvial | Overbank | Calcareous tuff | WTLND |
| South America, Argentina, Neuquen Province, Plaza de Huincul | Mesozoic, Cretaceous, Aptian-Cenomanian | Rio Limay Grp., Huincul Fm. | Macrofossil | Low | Monodominant | Alluvial | Channel | Sandstone | CHLSP |
| North America, Canada, Saskatchewan | Mesozoic; Cretaceous, Cenomanian | | Microfossil | High | ? | Marine | Marine | ? | MAR |
| North America, Canada, Alberta, Dinosaur Provincial Park | Mesozoic; Cretaceous | Dinosaur Park Fm. | Macrofossil | High, low | Monodominant, multidominant | Alluvial | Channel, splay, overbank | Sandstone, mudstone | CHLSP, PSOL, WTLND |
| North America, Canada, Alberta, Dry Island Buffalo Jump Provincial Park | Mesozoic; Cretaceous | Horseshoe Canyon Fm. | Macrofossil | Low | ? | Alluvial | Channel | Sandstone | CHLSP, WTLND |
| North America, Canada, Alberta, Dinosaur Provincial Park | Mesozoic; Cretaceous, Campanian | Judith River Fm. | Macrofossil | Low | Monodominant | Alluvial | Channel | Sandstone | CHLSP |
| Asia, China, Mongolia, Inner Mongolia, Iren Nur | Mesozoic; Cretaceous | Iren Dabasu Fm. | Macrofossil | High | Multidominant | Alluvial | Channel, splay, overbank | Conglomerate, sandstone, mudstone | CHLSP, PSOL, WTLND |

*(Continued)*

| Author | Date | Title | Citation | Taxa | Source | Bonebed Name | # of Sites |
|---|---|---|---|---|---|---|---|
| Currie, P.J., D. Trexler, E.B. Koppelhus, K. Wicks, and N. Murphy. | 2005 | An unusual multi-individual tyrannosaurid bonebed in the Two Medicine Formation (Late Cretaceous, Campanian) of Montana (USA). | Pp. 313–324 *in* The carnivorous dinosaurs. K. Carpenter, ed. Indiana University Press, Bloomington. | d | Ms | Daspletosaur Bonebed | 1 |
| Curtis, K.M. | 1989 | A taxonomic analysis of a microvertebrate fauna from the Kayenta Formation (Early Jurassic) of Arizona and its comparison to an Upper Triassic microvertebrate fauna from the Chinle Formation. | Master's thesis. University of California, Berkeley. | ndr | Th | ? | 2+ |
| Daeschler, E.B., and W. Cessler. | 1999 | Sampling an early continental ecosystem: Late Devonian bone beds at Red Hill, Clinton County, PA. | Journal of Vertebrate Paleontology 19(supplement to 3):41A. | f | Ab | ? | 2+ |
| Dalquest, W.W., and S.H. Mamay. | 1963 | A remarkable concentration of Permian amphibian remains in Haskall County, Texas. | Journal of Geology 71:641–644. | a | Ms | ? | 1 |
| Dashzveg, D., M.J. Novacek, M.A. Norel, J.M. Clark, L.M. Chiappe, A. Davidson, M.C. McKenna, L. Dingus, C. Swisher, and P. Altangerel. | 1995 | Extraordinary preservation in a new vertebrate assemblage of the Late Cretaceous of Mongolia. | Nature 374:446–449. | d | Ms | Ukhaa Tolgod | 2+ |
| Davies, K.L. | 1987 | Duck-bill dinosaurs (Hadrosauridae, Ornithischia) from the north slope of Alaska. | Journal of Paleontology 61:198–200. | d | Ms | Liscomb Bonebed | 1 |
| Deeming, D.C., L.B. Halstead, M. Manabe, and D.M. Unwin. | 1995 | An ichthyosaur embryo from the Lower Lias (Jurassic: Hettangian) of Somerset, England, with comments on the reproductive biology of ichthyosaurs. | Pp. 463–482 *in* Vertebrate fossils and the evolution of scientific concepts. W.A.S. Sarjeant, ed. Gordon and Breach Publishers, Amsterdam. | ndr | Ms | ? | 1 |
| Demere, T. | 1985 | The Mission Hills bone-bed: A palaeontological excavation. | Environment Southwest 506:9–12. | lm | Ms | Mission Hills Bone Bed | 1 |

| Location | Period / Epoch | Stratigraphy | Element Size | Diversity | Relative Abundance | Depositional Setting | Depositional Environment | Facies | Paleoenv Code |
|---|---|---|---|---|---|---|---|---|---|
| North America, USA, Montana, Teton County | Mesozoic; Cretaceous | Two Medicine Fm. | Macrofossil | Low | Monodominant | Alluvial | ? | Muddy siltstone | U |
| North America, USA, Arizona, Petrified Forest National Park | Mesozoic; Jurassic | Kayenta Fm. | Microfossil | High | ? | Alluvial | Diverse, ? | ? | CHLSP, PSOL, WTLND |
| North America, USA, Pennsylvania, Red Hill | Paleozoic; Devonian | Catskill Fm. | Microfossil | High | Multidominant | Alluvial, lacustrine | Strandline, channel, pond, overbank | Mudstone, siltstone, sandstone | CHLSP, WTLND, LAC |
| North America, USA, Texas, Jones County | Paleozoic; Permian | Vale Fm. | Macrofossil | Low | Monodominant | Alluvial | Channel, pond | Mudstone (siltstone) | WTLND |
| Asia, China, Mongolia, Gobi Desert | Mesozoic; Cretaceous | Djadokhta Fm. | Macrofossil | High | ? | Eolian | Eolian, interdune | Sandstone | EOL |
| North America, USA, Alaska | Mesozoic; Cretaceous | Prince Creek Fm., Colville Grp. | Macrofossil | ? | ? | Alluvial | Overbank, paleosol | Mudstone | WTLND |
| Europe, United Kingdom, England, Somerset, Blue Anchor Bay | Mesozoic; Jurassic, Hettangain | ? | Macrofossil | Low | Monotaxic | Marine | Marine shale | Shale | MAR |
| North America, USA, California, San Diego County | Cenozoic; Tertiary, Pliocene | San Diego Fm. | Macrofossil | High | ? | Paralic | Nearshore, lagoon | ? | PESH |

*(Continued)*

| Author | Date | Title | Citation | Taxa | Source | Bonebed Name | # of Sites |
|---|---|---|---|---|---|---|---|
| Denton, R.K. Jr., and R.C. O'Neill. | 1998 | *Parrisa neocesariensis*, a new batrachosauroidid salamander and other amphibians from the Campanian of eastern North America. | Journal of Vertebrate Paleontology 18:484–494. | a | Ms | Ellisdale Site | 1 |
| Derstler, K. | 1995 | The Dragons Grave: An *Edmontosaurus* bonebed containing theropod eggshells and juveniles, Lance Formation (Uppermost Cretaceous), Niobrara County, Wyoming. | Journal of Vertebrate Paleontology 15(supplement to 3):26A. | d | Ab | The Dragons Grave | 1 |
| W.A. DiMichele, and R.W. Hook. | 1992 | Paleozoic terrestrial ecosystems. | Pp. 205–325 *in* Terrestrial ecosystems through time. A.K. Behrensmeyer, J.D. Damuth, W.A. DiMichele, R. Potts, H.-D. Sues, and S.L. Wing, eds. University of Chicago Press, Chicago. | a, b, d, f, lm, ndr, sm | Ms | multiple sites | 1+ |
| Dingus, L., J. Clarke, G.R. Scott, C.C. Swisher III, L.M. Chiappe, and R.A. Coria. | 2000 | Stratigraphy and magnetostratigraphic/faunal constraints for the age of sauropod embryo-bearing rocks in the Neuquen Group (Late Cretaceous, Neuquen Province, Argentina). | Novitates 3290:1–11. | d | Ms | Auca Mahuevo | 2+ |
| Dodson, P. | 1971 | Sedimentology and taphonomy of the Oldman Formation (Campanian), Dinosaur Provincial Park, Alberta (Canada). | Palaeogeography, Palaeoclimatology, Palaeoecology 10:21–74. | d | Ms | multiple sites | 2+ |
| Dodson, P. | 1987 | Microfaunal studies of dinosaur paleoecology, Judith River Formation of southern Alberta. | Pp. 70–75 *in* Fourth symposium on Mesozoic terrestrial ecosystems, short papers. Occasional paper 3. P.J. Currie and E.H. Koster, eds. Royal Tyrrell Museum of Palaeontology, Drumheller, Alberta. | d | Ms | multiple sites | 2+ |
| Dodson, P., A.K. Behrensmeyer, R.T. Bakker, and J.S. Mcintosh. | 1980 | Taphonomy and paleoecology of the dinosaur beds of the Jurassic Morrison Formation. | Paleobiology 6:208–232. | d, ndr | Ms | multiple sites | 2+ |
| Dollo, M.L. | 1882 | Premier note sur les dinosauriens de Bernissart. | Bulletin du Musee Royal D'Histoire Naturelle de Belgique I:161-178. | d | Ms | multiple sites | 2+ |

| Location | Period / Epoch | Stratigraphy | Element Size | Diversity | Relative Abundance | Depositional Setting | Depositional Environment | Facies | Paleoenv Code |
|---|---|---|---|---|---|---|---|---|---|
| North America, USA, New Jersey, Monmouth County | Mesozoic; Cretaceous, Campanian | Marshalltown Fm. | Microfossil | ? | ? | Paralic | Nearshore marine, estuarine, paludal | ? | PESH |
| North America, USA, Wyoming, Niobrara County | Mesozoic; Cretaceous | Lance Fm. | Macrofossil | Low | Monodominant | Alluvial | Overbank | ? | PSOL |
| numerous | Paleozoic | numerous | Macrofossil, microfossil, mixed | ? | ? | ? | ? | ? | U |
| South America, Argentina, Patagonia, Neuquen Province | Mesozoic; Cretaceous | Anacleto Fm. | Macrofossil | Low | Monotaxic | Alluvial | Overbank | Mudstone | PSOL |
| North America, Canada, Alberta, Dinosaur Provincial Park | Mesozoic; Cretaceous | Oldman Fm. | Macrofossil, microfossil, mixed | High, low | Monodominant, multidominant | Alluvial | Channel, splay, overbank | Sandstone, mudstone | CHLSP, PSOL, WTLND |
| North America, Canada, Alberta | Mesozoic; Cretaceous | Judith River Fm. | Microfossil | High | Multidominant | Alluvial | Channel, splay | Sandstone | CHLSP |
| North America, USA | Mesozoic; Jurassic | Morrison Fm. | Macrofossil, microfossil, mixed | High, low | Monodominant, multidominant | Alluvial | Channel, splay, overbank, pond, paleosol | Sandstone, mudstone | CHLSP, PSOL, WTLND |
| Europe, Belgium, Bernissart | Mesozoic; Cretaceous | Wealden Grp. | Macrofossil | Low | Monodominant | ? | ? | ? | U |

*(Continued)*

| Author | Date | Title | Citation | Taxa | Source | Bonebed Name | # of Sites |
|---|---|---|---|---|---|---|---|
| Donelan, C., and G.D. Johnson. | 1997 | *Orthacanthus platypternus* (Chondrichthyes, Xenacanthida) occipital spines from the lower Permian Craddock bonebed, Baylor County, Texas. | Journal of Vertebrate Paleontology 17(supplement to 3):43A. | f | Ab | Craddock Bonebed | 1 |
| Donovan, S.K., D.P. Domning, F.A. Garcia, and H.L. Dixon. | 1990 | A bone bed in the Eocene of Jamaica. | Journal of Paleontology 64:660–662. | lm, ndr, sm | Ms | Dump Limestone Bone Bed | 1 |
| Downing, K.F., and E.H. Lindsay. | 1991 | Early Miocene vertebrate assemblages preserved in coastal paleoenvironments from Zinda Pir Dome, west-central Pakistan. | Journal of Vertebrate Paleontology 11(supplement to 3):27A. | f, lm, ndr, sm | Ab | ? | 2+ |
| Eberth, D.A. | 1985 | The skull of *Spenacodon ferocior* and comparison with other sphenacodontines (Reptilia: Pelycosauria). | New Mexico Bureau of Mines and Mineral Resources, Circular, 190:1–39. | ndr | Ms | Anderson Quarry Bone Bed | 1 |
| Eberth, D.A. | 1990 | Stratigraphy and sedimentology of vertebrate microfossil sites in the uppermost Judith River Formation (Campanian), Dinosaur Provincial Park, Alberta, Canada. | Palaeogeography, Palaeoclimatology, Palaeoecology 78:1–36. | d | Ms | ? | 24 |
| Eberth, D.A. | 1993 | Depositional environments and facies transitions of dinosaur-bearing Upper Cretaceous redbeds at Bayan Mandahu (Inner Mongolia, People's Republic of China). | Canadian Journal of Earth Sciences 30:2196–2213. | d | Ms | ? | 2+ |
| Eberth, D.A. | 1998 | Clustered ceratopsian bonebeds. Southern Alberta, Canada: Primary evidence for the size of ceratopsian herd death assemblages. | Abstract 13. The Dinofest Symposium. Academy of Natural Sciences, Philadelphia. | d | Ab | multiple sites | 14 |
| Eberth, D.A., A107 and D.S Berman. | 1993 | Stratigraphy, sedimentology and vertebrate paleoecology of the Cutler Formation redbeds (Pennsylvanian-Permian) of north-central New Mexico. | Vertebrate Paleontology in New Mexico. S.G. Lucas and J. Zidek, eds. New Mexico Museum of Natural History and Science Bulletin 2:33–48. | a, f, ndr | Ms | multiple sites | 2+ |

| Location | Period / Epoch | Stratigraphy | Element Size | Diversity | Relative Abundance | Depositional Setting | Depositional Environment | Facies | Paleoenv Code |
|---|---|---|---|---|---|---|---|---|---|
| North America, USA, Texas, Baylor County | Paleozoic; Permian | | Microfossil | Low | ? | Alluvial | Channel, pond | Mudstone | CHLSP, WTLND |
| North America, Caribbean, Jamaica, Parish of Manchester | Cenozoic; Tertiary, Eocene | Chaplton Fm., Guys Hill Mbr. | Microfossil | High | ? | Paralic | Estuarine channel | Mudstone | CHLSP |
| Asia, Pakistan, Dalana | Cenozoic; Tertiary, Miocene | Chitarwata Fm. | Microfossil | High | Multidominant | Paralic | Estuarine channel; tidal channel | Sandstone | CHLSP |
| North America, USA, New Mexico, Rio Arriba County | Paleozoic; Permian | Cutler Fm. | Macrofossil | High | Monodominant | Alluvial | Pond | Mudstone | WTLND |
| North America, Canada, Alberta, Dinosaur Provincial Park | Mesozoic; Cretaceous, Campanian | Judith River Fm. | Microfossil | ? | ? | Alluvial | Channel, splay | Sandstone, mudstone | CHLSP |
| Asia, China, Mongolia, Inner Mongolia, Bayan Mandahu | Mesozoic; Cretaceous | Djadohkta Fm. | Macrofossil | High, low | Monodominant, multidominant | Eolian, alluvial | Eolian, interdune, channel, sheet splay, braidplain | Sandstone | DBF, EOL |
| North America, Canada, Alberta, South Saskatchewan River | Mesozoic; Cretaceous | Dinosaur Park Fm. | Macrofossil | Low | Monodominant | Alluvial | Overbank, pond | Mudstone, organic rich | WTLND |
| North America, USA, New Mexico | Paleozoic; Pennsylvanian, Permian | Cutler Fm.; Abo Fm.; Sangre de Cristo Fm. | Macrofossil | High | Monodominant, multidominant | Alluvial | Pond, channel, splay | Mudstone, sandstone | CHLSP, WTLND |

*(Continued)*

| Author | Date | Title | Citation | Taxa | Source | Bonebed Name | # of Sites |
|---|---|---|---|---|---|---|---|
| Eberth, D.A., and D.B. Brinkman. | 1997 | Paleoecology of an estuarine, incised-valley fill in the Dinosaur Park Formation (Judith River Group, Upper Cretaceous) of southern Alberta, Canada. | Palaios 12:43–58. | d | Ms | multiple sites | 11 |
| Eberth, D.A., and M.A. Getty. | 2005 | Ceratopsian bonebeds at Dinosaur Provincial Park: Occurrence, origins, and significance. | Pp. 501–536 in Dinosaur Park: A spectacular ancient ecosystem revealed. P.J. Currie and E.B. Koppelhus, eds. Indiana University Press, Bloomington. | d | Ms | Bonebeds 30, 43, 41a, 91, 138 | 5 |
| Eberth, D.A., and R.T. McCrea. | 2002 | Were large theropods normally gregarious? | Alberta Paleontological Society, Abstracts 2002:13–15. | d | Ab | *Albertosaurus* Bonebed; Mapusaurus Bonebed | 2 |
| Eberth, D.A., D.S Berman, S.S. Sumida, and H. Hopf. | 2000 | Lower Permian terrestrial paleoenvironments and vertebrate paleoecology of the Tambach Basin (Thuringia, Central Germany): The upland holy grail. | Palaios 15:293–313. | a, ndr | Ms | Bromacker Locality | 1 |
| Eberth, D.A., D.R. Braman, and T.T. Tokaryk. | 1990 | Stratigraphy, sedimentology and vertebrate paleontology of the Judith River Formation (Campanian) near Muddy Lake, west-central Saskatchewan. | Bulletin of Canadian Petroleum Geology 38:387–406. | d | Ms | ? | 2+ |
| Eberth, D.A., B.B. Britt, R. Scheetz, K.L. Stadtman, and D.B. Brinkman. | 2006 | Dalton Wells: Geology and significance of debris-flow-hosted dinosaur bonebeds in the Cedar Mountain Formation (Lower Cretaceous) of eastern Utah, USA. | Palaeogeography, Palaeoclimatology, Palaeoecology 236:217–245 | d, ndr | Ms | Dalton Wells | 4 |
| Eberth, D.A., P.J. Currie, D.B. Brinkman, M.J. Ryan, D.R. Braman, J.D. Gardner, V.D. Lam, D.N. Spivak, and A.G. Neuman. | 2001 | Alberta's dinosaurs and other fossil vertebrates: Judith River and Edmonton Groups (Campanian-Maastrichtian). | Pp. 47–75 in Mesozoic and Cenozoic paleontology in the western Plains and Rocky Mountains. Occasional paper 3. C.L. Hill, ed. Museum of the Rockies, Bozeman, Montana. | d | Ms | multiple sites | 2+ |

| Location | Period / Epoch | Stratigraphy | Element Size | Diversity | Relative Abundance | Depositional Setting | Depositional Environment | Facies | Paleoenv Code |
|---|---|---|---|---|---|---|---|---|---|
| North America, Canada, Alberta, Onefour | Mesozoic; Cretaceous | Dinosaur Park Fm. | Microfossil | High | Multidominant | Paralic | Estuarine channel | Mudstone | CHLSP |
| North America, Canada, Alberta, Dinosaur Provincial Park | Mesozoic; Cretaceous | Oldman Fm.; Dinosaur Park Fm. | Macrofossil | High, low | Monodominant | Alluvial | Channel and overbank | Sandstone, mudstone | CHLSP, WTLND |
| North America, Canada, Alberta, Dry Island Buffalo Jump; Argentina | Mesozoic; Cretaceous, Maastrichtian, Cenomanian | Horseshoe Canyon Fm.; Rio Limay Grp. | Macrofossil | Low | Monodominant | Alluvial | Channel, pond | Sandstone, organic rich with coalified trees | CHLSP, WTLND |
| Europe, Germany, Gotha, Thurigian Forest | Paleozoic; Permian | Tambach Fm. | Macrofossil | Low | Multidominant | Alluvial, lacustrine | Overbank, debris flow, pond | Mudstone | DBF, WTLND |
| North America, Canada, Saskatchewan, Unity | Mesozoic; Cretaceous | Judith River Fm. | Macrofossil, microfossil, mixed | High | Multidominant | Alluvial | Channel, splay, pond | Sandstone, mudstone | CHLSP, WTLND |
| North America, USA, Utah, Moab | Mesozoic; Cretaceous | Cedar Mountain Fm. | Macrofossil | High | Multidominant | Alluvial | Debris flow | Sandy mudstone | DBF, WTLND, LAC |
| North America, Canada, Alberta, Southern | Mesozoic; Cretaceous, Campanian, Maastrichtian | Judith River Grp.; Edmonton Grp. | Macrofossil, microfossil, mixed | High, low | Monodominant, multidominant | Alluvial, paralic | Channel, overbank, pond, estuarine Channel | Sandstone, siltstone, mudstone | CHLSP, PSOL, WTLND |

*(Continued)*

| Author | Date | Title | Citation | Taxa | Source | Bonebed Name | # of Sites |
|---|---|---|---|---|---|---|---|
| Eberth, D.A., P.J. Currie, R.A. Coria, A.C. Garrido, and J.P. Zonneveld. | 2000 | Large-theropod bonebed, Neuquen, Argentina: Paleoecological importance. | Journal of Vertebrate Paleontology 20(supplement to 3):39A. | d | Ab | Mapusaurus Bonebed | 1 |
| Elliot, D.K., and R.G. Thomas. | 1991 | Problems in Siluro-Devonian fish taphonomy: Examples from the Canadian Arctic. | Journal of Vertebrate Paleontology 11(supplement to 3):27A. | f | Ab | ? | 2+ |
| Emry, R.J., S.G. Lucas, and B.U. Bayshashov. | 1997 | Brontothere bone bed in the Eocene of eastern Kazakstan. | Journal of Vertebrate Paleontology 17(supplement to 3):44A. | lm | Ab | Kyzal Murun | 1 |
| Emslie, S.D., and G.S. Morgan. | 1994 | A catastrophic death assemblage and paleoclimatic implications of Pliocene seabirds of Florida. | Science 264:684–685. | b | Ms | ? | 1 |
| Emslie, S.D., W.D. Allmon, F.J. Rich, J.H. Wrenn, and S.D. de France. | 1996 | Integrated taphonomy of an avian death assemblage in marine sediments from the late Pliocene of Florida. | Palaeogeography, Palaeoclimatology, Palaeoecology 124:107–136. | b | Ms | Richardson Road Shell Pits | 1 |
| Erickson, B.R. | 1991 | Flora and fauna of the Wannagan Creek quarry: Late Paleocene of North America. | Scientific Publications of the Science Museum of Minnesota, New Series 7(3):1–19. | a, f, ndr | Ms | Wannagan Creek Quarry | 1 |
| Estes, R., and P. Berberian. | 1970 | Paleoecology of a late Cretaceous vertebrate community from Montana. | Breviora 343:1–35. | a, b, d, f, ndr, sm | Ms | Bug Creek Anthills | 1 |
| Evanoff, E., and K. Carpenter. | 1998 | History, sedimentology, and taphonomy of Felch Quarry 1 and associated sandbodies, Morrison Formation, Garden Park, Colorado. | Modern Geology 22:145–169. | d | Ms | Flech Quarry 1 | 1 |
| Farlow, J.O., J.A. Sunnerman, J.J. Havens, A.L. Swinehart, J.A. Holman, R.L. Richards, N.G. Miller, R.A. Martin, R.M. Hunt, G.L. Storrs Jr., B.B. Curry, R.H. Fluegeman, M.R. Dawson, and M.E.T. Flint. | 2000 | The Pipe Creek Sinkhole biota, a diverse Late Tertiary continental fossil assemblage from Grant County, Indiana. | The American Midland Naturalist 145:367–378. | a, b, lm, ndr, sm | Ms | Pipe Creek Sinkhole | 1 |

| Location | Period / Epoch | Stratigraphy | Element Size | Diversity | Relative Abundance | Depositional Setting | Depositional Environment | Facies | Paleoenv Code |
|---|---|---|---|---|---|---|---|---|---|
| South America, Argentina | Mesozoic; Cretaceous, Cenomanian | Rio Limay Fm., Huincul Mbr. | Macrofossil | Low | Monotaxic | Alluvial | Channel | Sandstone | CHLSP |
| North America, Canada, Northwest Territories | Paleozoic; Silurian, Devonian | Cape Storm Fm.; Peel Sound Fm. | Microfossil | ? | ? | Paralic | Supratidal/ intertidal; shallow marine; fan-delta | Mudstone; carbonate; sandstone; conglomerate | PESH |
| Asia, Kazakstan | Cenozoic; Tertiary, Eocene | Kyzylbulak Fm. | Macrofossil | ? | ? | Alluvial | Pond | Mudstone | WTLND |
| North America, USA, Florida, Sarasota County | Cenozoic; Tertiary, Pliocene | Tamiami Fm. | Microfossil | High | Monodominant | Paralic | Barrier beach, back barrier | Sandstone, conglomerate (shell gravel) | PESH |
| North America, USA, Florida, Sarasota County | Cenozoic; Tertiary, Pliocene | Tamiami Fm. | Macrofossil | High | Monodominant | Paralic | Barrier beach, back barrier | Sandstone, conglomerate (shell gravel) | PESH |
| North America, USA, North Dakota, Billings County | Cenozoic; Tertiary, Paleocene, Tiffanian | Bullion Creek (Tongue River) Fm. | Macrofossil | High | Multidominant | Lacustrine | Paludal wetlands | Mudstone (lignitic, silty clay) | LAC |
| North America, USA, Montana, McCone County | Mesozoic; Cretaceous | Hell Creek Fm. | Microfossil | High | Multidominant | Alluvial | Channel | Sandstone | CHLSP |
| North America, USA, Colorado, Canyon City | Mesozoic; Jurassic | Morrison Fm. | Macrofossil | High | Multidominant | Alluvial | Channel | Coarse grained sandstone | CHLSP |
| North America, USA, Indiana, Grant County | Cenozoic; Tertiary, Pliocene, Hemphillian, Blancan | ? | Mixed | High | Multidominant | Sinkhole, pond | Pond, spring | Mud | LAC, O |

*(Continued)*

| Author | Date | Title | Citation | Taxa | Source | Bonebed Name | # of Sites |
|---|---|---|---|---|---|---|---|
| Fastovsky, D.E., J.M. Clark, N.H. Strater, M.R. Montellano, R. Hernandez, and J.A. Hopson. | 1995 | Depositional environments of a Middle Jurassic terrestrial vertebrate assemblage, Huizachal Canyon, Mexico. | Journal of Vertebrate Paleontology 15:561–575. | d | Ms | ? | 2+ |
| Fiorillo, A.R. | 1987 | Significance of juvenile dinosaurs from Careless Creek Quarry. | Pp. 88–95 in Fourth symposium on Mesozoic terrestrial ecosystems, short papers. Occasional paper 3. P.J. Currie and E.H. Koster, eds. Royal Tyrell Museum of Palaeontology, Drumheller, Alberta. | d | Ms | Careless Creek Quarry | 1 |
| Fiorillo, A.R. | 1988 | Taphonomy of Hazard Homestead Quarry (Ogallala Group), Hitchcock County, Nebraska. | Contributions to Geology, University of Wyoming 26:57–97. | lm | Ms | Hazard Homestead Quarry | 1 |
| Fiorillo, A.R. | 1989 | Taphonomy and paleoecology of the Judith River Formation (Late Cretaceous) of south-central Montana. | Ph.D. thesis, University of Pennsylvania, Philadelphia. | d | Th | Careless Creek Quarry; Hidden Valley Quarry; Karen's Quarry; Top Cat Quarry; Antelope Head Quarry; SPA Quarry | 5 |
| Fiorillo, A.R. | 1991 | Taphonomy and depositional setting of Careless Creek Quarry (Judith River Formation), Wheatland County, Montana, U.S.A. | Palaeogeography, Palaeoclimatology, Palaeoecology 81:281–311. | d | Ms | Careless Creek Quarry | 1 |
| Fiorillo, A.R. | 1998 | Bone modification features on sauropod remains (Dinosauria) from the Freezeout Hills Quarry N (Morrison formation) of southeastern Wyoming and their contribution to fine-scale paleoenviornmental interpretation. | Modern Geology 23:111–126. | d | Ms | Freezeout Hills Quarry N | 1 |
| Fiorillo, A.R. | 1998 | Measuring fossil reworking within a fluvial system: An example from the Hell Creek Formation (Upper Cretaceous) of Eastern Montana. | Pp. 243–251 in Advances in vertebrate paleontology and geochronology. Monograph 14. Y. Tomida, L.J. Flynn, and L.L. Jacobs, eds. National Science Museum, Tokyo. | d | Ms | multiple sites | 108 |

| Location | Period / Epoch | Stratigraphy | Element Size | Diversity | Relative Abundance | Depositional Setting | Depositional Environment | Facies | Paleoenv Code |
|---|---|---|---|---|---|---|---|---|---|
| North America, Mexico | Mesozoic; Jurassic | La Boca Fm. | Mixed | High | Multidominant | Alluvial fan, volcanic | Debris flow in a volcanic setting | Conglomerate | DBF |
| North America, USA, Montana, Wheatland County | Mesozoic; Cretaceous, Campanian | Judith River Fm. | Macrofossil | High | ? | Alluvial | Channel, log jam | Sandstone | CHLSP |
| North America, USA, Nebraska, Hitchcock County | Cenozoic; Tertiary, Miocene, Barstovian | Ogallala Grp. | Macrofossil | High | Multidominant | Alluvial | Overbank | ? | U |
| North America, USA, Montana | Mesozoic; Cretaceous | Judith River Fm. | Macrofossil, microfossil, mixed | High | Multidominant | Alluvial | Channel, splay, overbank, pond | ? | CHLSP, WTLND |
| North America, USA, Montana, Wheatland County | Mesozoic; Cretaceous Campanian | Judith River Fm. | Mixed | High | Multidominant | Alluvial | Channel, log jam | Sandstone | CHLSP |
| North America, USA, Wyoming | Mesozoic; Jurassic | Morrison Fm. | Macrofossil | High | Multidominant | Alluvial | Overbank, paleosol | Siltstone | PSOL |
| North America, USA, Montana | Mesozoic; Cretaceous | Hell Creek Fm. | Microfossil | High | Multidominant | Alluvial | Channel? | Sandstone | CHLSP |

*(Continued)*

| Author | Date | Title | Citation | Taxa | Source | Bonebed Name | # of Sites |
|---|---|---|---|---|---|---|---|
| Fiorillo, A.R. | 1998 | Preliminary report on a new sauropod locality in the Javelina Formation (Late Cretaceous), Big Bend National Park, Texas. | Pp. 29–31 *in* Research in the national parks. V. Santucci, ed. Geologic Resources Division Technical Report NPS/NRGRD/GRDTR-98/01:29–31. | d | Ms | *Alamosaurus* Bonebed | 1 |
| Fiorillo, A.R., and C.L. May. | 1996 | Preliminary report on the taphonomy and depositional setting of a new dinosaur locality in the Morrison Formation (Brushy Basin Member) of Curecanti National Recreational Area, Colorado. | The continental Jurassic. M. Morales, ed. Museum of Northern Arizona, Bulletin 60:555–561. | d | Ms | ? | 1 |
| Fiorillo, A.R., K. Padian, and C. Musikasinthorn. | 2000 | Taphonomy and depositional setting of the Placerias Quarry (Chinle Formation: Late Triassic, Arizona). | Palaios 15:373–386. | ndr | Ms | *Placerias* Quarry | 1 |
| Fix, M.F., and G.E. Darrough. | 2004 | Dinosauria and associated vertebrate fauna of the Late Cretaceous Chronister site of southeast Missouri. | Geological Society of America, North-Central Section, Abstracts with Programs 36:14. | a, d, f, ndr | Ab | Chronister Site | 1 |
| Foreman, B.C., and L.D. Martin. | 1988 | A review of Paleozoic tetrapod localities of Kansas and Nebraska. | Pp. 133–145 *in* Regional geology and paleontology of Upper Paleozoic Hamilton Quarry Area in southeastern Kansas. Guidebook series 6. G. Mapes and R. Mapes, eds. Kansas Geological Survey, Lawrence. | a, f, ndr | Ms | Garnett, Rasmussen, Hamilton Quarry, Du Bois, Topeka, Robinson, Peru, McDowell Creek Roadcut, Williston Tooth, Westmoreland, Keats, Eskridge, Junction City, Bushong, Winfield. | 10+ |
| Forster, C.A. | 1990 | Evidence for juvenile groups in the ornithopod dinosaur *Tenontosaurus tilletti ostrom*. | Journal of Paleontology 64:164–165. | d | Ms | ? | 2 |
| Franzen, J.L. | 1985 | Exceptional preservation of Eocene vertebrates in the lake deposit of Grube Messel (West Germany). | Philosophical Transactions of the Royal Society of London 331:181–186. | a, b, f, ndr, sm | Ms | Grube Messel | 1 |
| Fraser, N.C., and A.C. Dooley Jr. | 2000 | A diverse marine assemblage from a locality in the Calvert Formation of Virginia. | Journal of Vertebrate Paleontology 20(supplement to 3):42A. | f, lm, ndr, sm | Ab | ? | 1 |

| Location | Period / Epoch | Stratigraphy | Element Size | Diversity | Relative Abundance | Depositional Setting | Depositional Environment | Facies | Paleoenv Code |
|---|---|---|---|---|---|---|---|---|---|
| North America, USA, Texas, Big Bend National Park | Mesozoic; Cretaceous | Javelina Fm. | Macrofossil | Low | Monotaxic | Alluvial | Floodplain waterhole | Mudstone, siltstone, calcareous concretions | WTLND |
| North America, USA, Colordao, Curecanti National Recreational Area | Mesozoic; Jurassic | Morrison Fm., Brushy Basin Mbr. | Macrofossil | Low | ? | Alluvial | Pond | Mudstone | WTLND |
| North America, USA, Arizona, St. Johns | Mesozoic; Triassic | Chinle Fm., Petrified Forest Mbr. | Mixed | High | Multidominant | Alluvial | Pond, paleosol, water-hole? | Mudstone, calcareous | PSOL, WTLND |
| North America, USA, Missouri, Ozarks | Mesozoic; Cretaceous, Campanian | ? | Mixed | High | Multidominant | Alluvial | Wetland, oxbow, debris flows, | Claystone | WTLND, O |
| North America, USA, Kansas | Paleozoic; Carboniferous, Permian | numerous | Macrofossil, microfossil, mixed | High, low | Multidominant, monotaxic | Alluvial, paralic, marine | Open marine, estuarine channel, bay-fill, paleochannel, overbank, strand line | Mudstone, siltstone, limestone, sandstone | MAR, PESH |
| North America, USA, Montana | Mesozoic; Cretaceous | Cloverly Fm. | Macrofossil | Low | Monodominant | ? | ? | ? | U |
| Europe, Germany, Frankfurt am Main | Cenozoic; Tertiary, Eocene | Messel Fm. | Macrofossil | High | Multidominant | Lacustrine | Lacustrine | Mudstone (oil shale) | LAC |
| North America, USA, Virginia, Caroline County | Cenozoic; Tertiary, Miocene | Calvert Fm. | Microfossil | High | ? | Paralic | Channel lag | Sandstone | CHLSP |

*(Continued)*

| Author | Date | Title | Citation | Taxa | Source | Bonebed Name | # of Sites |
|--------|------|-------|----------|------|--------|--------------|------------|
| Frey, W.R., M.R. Voohries, and J.D. Howard. | 1975 | Estuaries of the Georgia coast U.S.A. sedimentology and biology. VIII. Fossil and recent skeletal remains in Georgia estuaries. | Zeitschrift für Meeresgeologie und Meeresbiologie der Senckenbergischen Naturforschenden Gesellschaft. G. Hertweck and S. Little-Gadow, eds. Senckenbergiana Maritima 7: 257–295. | lm, sm | Ms | multiple sites | 2+ |
| Gale, T.M. | 1988 | Comments on a "nest" of juvenile dicynodont reptiles. | Modern Geology 13:119–124. | ndr | Ms | ? | 1 |
| Garcia, W. | 2000 | Preliminary description of a Mississippian (Namurian) tetrapod locality from Hancock County, Kentucky. | Journal of Vertebrate Paleontology 20(supplement to 3):43A. | a, f | Ab | ? | 2+ |
| Gates, T. | 2002 | Murder in Jurassic Park: The Cleveland-Lloyd dinosaur quarry as a drought-induced assemblage. | Journal of Vertebrate Paleontology 22(supplement to 3):56A–57A. | d | Ab | Cleveland-Lloyd Quarry | 1 |
| Gates, T.A. | 2005 | The Late Jurassic Cleveland-Lloyd Dinosaur Quarry as a drought-induced assemblage. | Palaios 20:363–375. | d | Ms | Cleveland-Lloyd Quarry | 1 |
| Getty, M, . D.A. Eberth, D.B. Brinkman, and M.J. Ryan. | 1998 | Taphonomy of three Centrosaurus bonebeds in the Dinosaur Park Formation, Alberta, Canada. | Journal of Vertebrate Paleontology 18(supplement to 3):46A. | d | Ab | BB 91, BB 30, BB 41a | 3 |
| Getty, M., D.A. Eberth, D.B. Brinkman, D. Tanke, M. Ryan, and M. Vickaryous. | 1997 | Taphonomy of two Centrosaurus bonebeds in the Dinosaur Park Formation, Alberta, Canada. | Journal of Vertebrate Paleontology 17(supplement to 3):48A–49A. | d | Ab | BB 91, BB 30 | 2 |
| Gilbert, B.M., and L.D. Martin. | 1984 | Late Pleistocene fossils of Natural Trap Cave, Wyoming, and the climatic model of extinction. | Pp. 138–147 in Quaternary extinctions: A prehistoric revolution. P.S. Martin and R.G. Klein, eds. The University of Arizona Press, Tuscon. | lm | Ms | Natural Trap Cave | 1 |
| Gilmore, C.W. | 1936 | Osteology of Apatosaurus, with special reference to specimens in the Carnegie Museum. | Memoirs of the Carnegie Museum, 11(4):175–300. | d | Ms | Douglass Quarry | 1 |

| Location | Period / Epoch | Stratigraphy | Element Size | Diversity | Relative Abundance | Depositional Setting | Depositional Environment | Facies | Paleoenv Code |
|---|---|---|---|---|---|---|---|---|---|
| North America, USA, Georgia | Cenozoic; Tertiary, Miocene, Quater-nary, Pleistocene | | Microfossil | High | Multidominant | Paralic | Coastal plain, estuarine, proximal marine | Sandstone, mudstone | PESH |
| Africa, Zambia, Luangawa Valley | Paleozoic; Permian | Madumabisa Mudstones | Macrofossil | Low | Monotaxic | Alluvial | Burrow | Mudstone | O |
| North America, USA, Kentucky | Paleozoic; Carbonifer-ous, Namurian | | Microfossil | High | Multidominant | Paralic | Brackish lake, channel, paleosol | Mudstone, sandstone | CHLSP, PESH |
| North America, USA, Utah, Cleveland | Mesozoic; Jurassic | Morrison Fm. | Macrofossil | High | Monodominant | Alluvial | Overbank; pond margin | Mudstone, calcareous | WTLND |
| North America, USA, Utah, Cleveland-Lloyd | Mesozoic; Jurassic | Morrison Fm., Brushy Basin Mbr. | Macrofossil | High | Monodominant | Alluvial | Pond | Calcareous mudstone | WTLND |
| North America, Canada, Alberta, Dinosaur Provincial Park | Mesozoic; Cretaceous | Dinosaur Park Fm. | Macrofossil | Low | Monodominant | Alluvial | Overbank pond | Mudstone, organic-rich | WTLND |
| North America, Canada, Alberta, Dinosaur Provincial Park | Mesozoic; Cretaceous | Dinosaur Park Fm. | Macrofossil | Low | Monodominant | Alluvial | Overbank pond | ? | WTLND |
| North America, USA, Wyoming | Cenozoic; Quater-nary, Pleistocene | | Macrofossil | High | ? | Karst | Karst | ? | O |
| North America, USA, Utah, Jensen | Mesozoic; Jurassic | Morrison Fm. | Macrofossil | High | Multidominant | Alluvial | Paleochannel | Sandstone, conglomerate | CHLSP |

*(Continued)*

| Author | Date | Title | Citation | Taxa | Source | Bonebed Name | # of Sites |
|--------|------|-------|----------|------|--------|--------------|------------|
| Gingerich, P.D. | 1989 | New earliest Wasatchian mammalian fauna from the Eocene of Northwestern Wyoming: Composition and diversity in a rarely sampled high-floodplain assemblage. | University of Michigan, Papers on Paleontology 28:1–97. | b, f, lm, ndr, sm | Ms | multiple sites | 20 |
| Godefroit, P., Z.M. Dong, P. Bultynck, H. Li, and L. Feng. | 1998 | Sino-Belgian Cooperation Program "Cretaceous dinosaurs and mammals from Inner Mongolia." | Bulletin de L'Institut Royal Des Sciences Naturelles de Belgique. Sciences de la Terre 68(supplement):3–70. | d | Ms | Bactrosaurus Bonebed SBDE 95E5 | 1 |
| Godefroit, P., S. Zan, and L. Jin. | 2000 | *Charonosaurus jiayinensis* n.g., n.sp., a lambeosaurine dinosaur from the late Maastrichtian of northeastern China. | Compte Rendus Academie des Sciences du Paris, Sciences de la Terre et des Planetes 330:875–882. | d | Ms | ? | 1+ |
| González-Rodríguez, K.A. and S.P. Applegate. | 2000 | Muhi Quarry, a new Cretaceous fish locality in central Mexico. | Journal of Vertebrate Paleontology 20(supplement to 3):45A. | f | Ab | Muhi Quarry | 1 |
| Goodwin, M.B., W.A. Clemens, C.R. Schaff, and C.B. Wood. | 1996 | New occurrences of Mesozoic vertebrates from the Upper Blue Nile Gorge, Ethiopia. | Journal of vertebrate Paleontology 16(supplement to 3):38A. | d | Ab | ? | 12 |
| Graham, R., E. Evanoff, L. Juliusson, and B. Weiss. | 1999 | Stratigraphy, paleontology and taphonomy of Bones Galore, a Chadronian site in Northeastern Colorado. | Journal of Vertebrate Paleontology 19(supplement to 3):48A. | lm | Ab | Bones Galore | 1 |
| Graham, R.W., J.A. Holman, and P.W. Parmalee. | 1983 | Taphonomy and paleoecology of the Christensen Bog Mastadon bone bed, Hancock County, Indiana. | Illinois State Museum, Reports of Investigations 38:1–29. | lm | Ms | Christensen Bog Mastadon Site | 1 |
| Gunnell, G.F. | 1994 | Paleocene mammals and faunal analysis of the Chappo type locality (Tiffanian), Green River Basin, Wyoming. | Journal of Vertebrate Paleontology 14:81–104. | sm | Ms | Chappo | 1 |
| Hammer, W.R. | 1995 | New Therapsids from the upper Fremouw Formation (Triassic) of Antarctica. | Journal of Vertebrate Paleontology 15:105–112. | d | Ms | ? | 1 |

| Location | Period / Epoch | Stratigraphy | Element Size | Diversity | Relative Abundance | Depositional Setting | Depositional Environment | Facies | Paleoenv Code |
|---|---|---|---|---|---|---|---|---|---|
| North America, USA, Wyoming, Powell, Bighorn & Clarks Fork Basins | Cenozoic; Tertiary, Eocene, Wasatchian | Willwood Fm. | Microfossil | High | Multidominant | Alluvial | Channel and paloesol | Sandstone, mudstone | CHLSP, WTLND |
| Asia, China, Mongolia, Inner Mongolia, Erenhot | Mesozoic; Cretaceous | Iren Dabasu Fm. | Macrofossil | Low | Monodominant | Alluvial | Small channel | Sandstone | CHLSP |
| Asia, China, Heilongjiang | Mesozoic; Cretaceous, Maas-trichtian | Yuliangze Fm. | Macrofossil | Low | Monodominant | Alluvial | Channel | ? | CHLSP |
| North America, Mexico, Hidalgo, Zimapán | Mesozoic; Cretaceous | | Mixed | High | Multidominant | Marine | Marine | Limestone | MAR |
| Africa, Ethiopia, Upper Blue Nile Gorge | Mesozoic; Jurassic, Cretaceous | Mugher Mudstone | Macrofossil, microfossil, mixed | High | ? | Paralic, alluvial | ? | Variety | U |
| North America, USA, Colorado | Cenozoic; Tertiary, Oligocene, Chadro-nian | White River Fm. | Macrofossil | High | Multidominant | Alluvial | Pond (wa-terhole) | Mudstone | WTLND |
| North America, USA, Indiana, Hancock County | Cenozoic; Quater-nary, Pleistocene | | Macrofossil | Low | Monodominant | Bog | Bog | ? | WTLND, O |
| North America, USA, Wyoming, Lincoln County | Cenozoic; Tertiary, Paleocene, Tiffanian | Wasatch Fm., Chappo Mbr. | Microfossil | High | Multidominant | Alluvial | Overbank | Mudstone | PSOL |
| Antarctica, Beardmore Glacier | Mesozoic; Triassic | Fremouw Fm. | Macrofossil | High | Multidominant | Alluvial | Channel | Sandstone | CHLSP |

*(Continued)*

| Author | Date | Title | Citation | Taxa | Source | Bonebed Name | # of Sites |
|--------|------|-------|----------|------|--------|--------------|------------|
| Hammer, W.R. | 1997 | Jurassic dinosaurs from Antarctica. | Pp. 249–251 *in* Dinofest international: Proceedings of a symposium held at Arizona State University. D.L. Wolberg, E. Stump, and G.D. Rosenberg, eds. Academy of Natural Sciences, Philadelphia. | d | Ms | Mt. Kirkpatrick | 1 |
| Hanna, R.R., J.W. LaRock, and J.R. Horner. | 1999 | Pathological brachylophosaur bones form the Upper Cretaceous Judith River Formation, northeastern Montana. | Journal of Vertebrate Paleontology 19(supplement to 3):49A. | d | Ab | Malta Bonebed | 1 |
| Heaton, T.H., and F. Grady. | 1999 | Late Quaternary birds and fishes from On Your Knees Cave, Prince of Wales Island, southeast Alaska. | Journal of Vertebrate Paleontology 19(supplement to 3):50A. | b, f | Ab | On Your Knees Cave | 1 |
| Hebert, B.L., and J.H. Calder. | 2004 | On the discovery of a unique terrestrial faunal assemblage in the classic Pennsylvanian section at Joggins, Nova Scotia. | Canadian Journal of Earth Sciences 41:247–254. | a | Ms | Hebert Sandstone | 1 |
| Heckert, A.B. | 2004 | Late Triassic microvertebrates from the lower Chinle Group (Otischalkian-Adamanian: Carnian), southwestern U.S.A. | New Mexico Museum of Natural History and Science Bulletin 27:1–170. | a, d, f, ndr | Ms | multiple sites | 1+ |
| Heckert, A.B., S.G. Lucas, and R.M. Sullivan. | 2000 | Triassic dinosaurs in New Mexico. | New Mexico Museum of Natural History and Science Bulletin 17:17–26. | d | Ms | Fort Wingate (NMMNH locality 2739); Lamy (NMMNH locality 149; 1171); Synder Quarry; Ghost Ranch | 5+ |
| Heckert, A.B., S.G. Lucas, and L.F. Rinehart. | 1999 | From decapods to dinosaurs: A diverse new fauna from a bonebed in the Upper Triassic (Norian) Petrified Forest Formation. | Journal of Vertebrate Paleontology 19(supplement to 3):50A. | d | Ab | NMMNH Locality 3845 | 1 |

| Location | Period / Epoch | Stratigraphy | Element Size | Diversity | Relative Abundance | Depositional Setting | Depositional Environment | Facies | Paleoenv Code |
|---|---|---|---|---|---|---|---|---|---|
| Antarctica, Transantarctic Mountains | Mesozoic; Jurassic | Falla Fm. | Macrofossil | High | ? | ? | ? | Mudstone (tuffaceous siltstone) | U |
| North America, USA, Montana, Malta | Mesozoic; Cretaceous | Judith River Fm. | Macrofossil | Low | Monodominant | ? | ? | Sandstone | U |
| North America, USA, Alaska, Pirnce of Wales Island | Cenozoic; Quaternary, Pleistocene, Wisconsin, Holocene | | Microfossil | High | ? | Paralic | ? | ? | PESH |
| North America, Canada, Nova Scotia, Joggins | Paleozoic; Pennsylvanian | Cumberland Grp., Joggins Fm., Hebert Sandstone | Macrofossil, microfossil | High | Multidominant | Alluvial | Channel | Sandstone | CHLSP |
| North America, USA, New Mexico, southwestern | Mesozoic; Triassic, Carnian | | Microfossil | ? | ? | ? | ? | ? | U |
| North America, USA, New Mexico | Mesozoic; Triassic, Carian-Norian | Chinle Grp., Bluewater Creek Fm., Santa Rosa Fm., Los Esteros Mbr., Bull Canyon Fm., Rock Point Fm. | Macrofossil | ? | ? | Alluvial | Distal floodplain or paleosol | ? | PSOL, WTLND |
| North America, USA, New Mexico, Rio Arriba County | Mesozoic; Triassic, Norian | Chinle Grp., Petrified Forest Fm. | Macrofossil | High | Multidominant | Alluvial | Channel | Conglomerate | CHLSP |

*(Continued)*

| Author | Date | Title | Citation | Taxa | Source | Bonebed Name | # of Sites |
|---|---|---|---|---|---|---|---|
| Heckert, A.B., S.G. Lucas, K.E. Zeigler, R.E. Peterson, R.E. Peterson, and N.V. D'Andrea. | 2000 | Stratigraphy, taphonomy, and new discoveries from the Upper Jurassic (Morrison Formation: Brushy Basin Member) Peterson Quarry, Central New Mexico. | Dinosaurs of New Mexico. S.G. Lucas and A.B. Heckert, eds. New Mexico Museum of Natural History and Science Bulletin 17:51–60. | d | Ms | Peterson Quarry | 1 |
| Hendey, Q.B. | 1981 | Paleoecology of the Late Tertiary fossil occurrences in "E" quarry, Langebaanweg, South Africa, and a reinterpretation of their geological context. | Annals of the South African Museum 84:1–104. | a, b, f, lm, ndr, sm | Ms | "E" Quarry | 1 |
| Henrici, A.C., and A.M. Baez. | 2001 | First occurrence of *Xenopus* (Anura: Pipidae) on the Arabian Peninsula: A new species from the Upper Oligocene of Yemen. | Journal of Paleontology 75:870–882 | a | Ms | Freshwater Interbed | 1 |
| Henrici, A.C., and A.R. Fiorillo. | 1993 | Catastrophic death assemblage of *Chelomophrynus bayi* (Anura, Rhinophrynidae) from the Middle Eocene Wagon Bed Formation of Central Wyoming. | Journal of Paleontology 7:1016–1026. | a | Ms | Frog Quarry | 1 |
| Hill, R.V., J.G. Honey, and M.A. O'Leary. | 2000 | New fossils from the early Eocene Four Mile area and improved relative dating of vertebrate localities. | Journal of Vertebrate Paleontology 20(supplement to 3):48A. | ndr, sm | Ab | ? | 1 |
| Hogler, J.A. | 1992 | Taphonomy and paleoecology of *Shonisaurus popularis* (Reptilia: Ichthyosauria). | Palaios 7:108–117. | ndr | Ms | Fossil Horse Quarry | 1 |
| Holdaway, R.N., and T.H. Worthy. | 1997 | A reappraisal of the late Quaternary fossil vertebrates of Pyramid Valley Swamp, North Canterbury, New Zealand. | New Zealand Journal of Zoology 24:69–121. | b, ndr, sm | Ms | Pyramid Valley | 1 |
| Holman, J.A. | 1981 | A herpetofauna from an eastern extension of the Harrison Formation (Early Miocene: Arikareean), Cherry County, Nebraska. | Journal of Vertebrate Paleontology 1:49–56. | a, ndr | Ms | Mouth of McCann's Canyon Site | 1 |
| Holman, J.A., A.J. Stuart, and J.D. Clayden. | 1990 | A Middle Pleistocene herpetofauna from Cudmore Grove, Essex, England, and its paleogeographic, and paleoclimatic implications. | Journal of Vertebrate Paleontology 10:86–94. | a, ndr | Ms | Cudmore Grove | 1 |

| Location | Period / Epoch | Stratigraphy | Element Size | Diversity | Relative Abundance | Depositional Setting | Depositional Environment | Facies | Paleoenv Code |
|---|---|---|---|---|---|---|---|---|---|
| North America, USA, New Mexico | Mesozoic; Jurassic | Morrison Fm. | Macrofossil | Low | Multidominant | Alluvial | Channel | Sandstone, minor conglomerate | CHLSP |
| Africa, South Africa, Langebaanweg | Cenozoic; Tertiary, Miocene, Pliocene | Varswater Fm. | Macrofossil | High | Multidominant | Paralic | Estaurine channel | Sandstone (phosphatic sandstone) | CHLSP, PESH |
| Africa, Yemen, Rada | Cenozoic; Tertiary, Oligocene | Yemen Volcanic Grp. | Macrofossil | Low | Monodominant | Lacustrine | Lacustrine | Marl | LAC |
| North America, USA, Wyoming, Hot Springs County | Cenozoic; Tertiary, Eocene, Uintan | Wagon Bed Fm. | Microfossil | Low | Monotaxic | Lacustrine | Shoreline | Sandstone | LAC |
| North America, USA, Colorado | Cenozoic; Tertiary, Eocene | | Microfossil | High | Multidominant | ? | ? | ? | U |
| North America, USA, Nevada, Berlin-ichthyosaur State Park | Mesozoic; Triassic, Carnian | Luning Fm. | Macrofossil | Low | Monotaxic | Marine | Nearshore | Limestone (bioclastic wackestone) | MAR |
| Australia, New Zealand, South Island, North Canterbury, Waikari | Cenozoic; Quaternary, Pleistocene | ? | Macrofossil | High | Multidominant | Paludal | Swamp | Peat, silt | WTLND |
| North America, USA, Nebraska, Cherry County | Cenozoic; Tertiary, Miocene | Harrison Fm. | Microfossil | High | Multidominant | ? | ? | ? | U |
| Europe, United Kingdom, England, Essex | Cenozoic; Quaternary, Pleistocene | | Microfossil | High | Multidominant | Paralic | Estuarine | Mudstone (detritus mud) | PESH |

*(Continued)*

| Author | Date | Title | Citation | Taxa | Source | Bonebed Name | # of Sites |
|--------|------|-------|----------|------|--------|--------------|------------|
| Hook, R.W., and D. Baird. | 1986 | The Diamond Coal Mine of Linton, Ohio, and it's Pennsylvanian age vertebrates. | Journal of Vertebrate Paleontology 6:174–190. | a, f, ndr | Ms | Linton | 2+ |
| Hooker, J.S. | 1987 | Late Cretaceous ashfall and the demise of a hadrosaurian "herd." | Geological Society of America, Abstracts with Programs 19:284. | d | Ab | ? | 1 |
| Horner, J.R. | 1982 | Evidence for colonial nesting and "site fidelity" among ornithischian dinosaurs. | Nature 297:675–676. | d | Ms | ? | 1 |
| Horner, J.R., and R. Makela. | 1979 | Nest of juveniles provides evidence of family structure among dinosaurs. | Nature 282:296–298. | d | Ms | ? | 1 |
| Hubert, J.F., P.T. Panish, D.J. Chure, and K.S. Prostak. | 1996 | Chemistry, microstructure, petrology, and diagenetic model of Jurassic Dinosaur bones, Dinosaur National Monument, Utah. | Journal of Sedimentary Research 66:531–547. | d | Ms | Carnegie Quarry | 1 |
| Hungerbühler, A. | 1998 | Taphonomy of the prosauropod dinosaur *Sellosaurus*, and its implications for carnivore faunas and feeding habits in the Late Triassic. | Palaeogeography, Palaeoclimatology, Palaeoecology 143:1–29. | d | Ms | Weisser Steinbruch Quarry, Pfaffenhofen Quarry | 2 |
| Hunt, A.P. | 1991 | Integrated vertebrate, invertebrate and plant taphonomy of the Fossil Forest area (Fruitland and Kirtland formations: Late Cretaceous), San Juan County, New Mexico, USA. | Palaeogeography, Palaeoclimatology, Palaeoecology 88:85–107. | d | Ms | Dinosaur Graveyard; Big Bone Quarry; Dinosaur Quarry; Quarries 1–4 | 7 |
| Hunt, R., and R. Skolnick, R. | 1989 | Geometry and sedimentology of the Carnegie Hill bone bed, Agate Fossil Beds National Monument, Nebraska. | Proceedings of the Nebraska Academy of Sciences 99:51. | lm | Ab | Agate Spring Bone Bed at Carnegie Hill | 1 |
| Hunt, R.M. Jr. | 1972 | Miocene amphicyonids (Mammalia, Carnivora) from the Agate Spring Quarries, Sioux County, Nebraska. | American Museum Novitiates 2506:1–39. | lm | Ms | Agate Springs Quarries (Quarry No. 3) | 2+ |
| Hunt, R.M. Jr. | 1978 | Depositional setting of a Miocene mammal assemblage, Sioux County, Nebraska (U.S.A.). | Palaeogeography, Palaeoclimatology, Palaeoecology 24:1–52. | lm | Ms | Harper Quarry | 1 |

| Location | Period / Epoch | Stratigraphy | Element Size | Diversity | Relative Abundance | Depositional Setting | Depositional Environment | Facies | Paleoenv Code |
|---|---|---|---|---|---|---|---|---|---|
| North America, USA, Ohio, Linton | Paleozoic; Pennsylva-nian | Allegheny Grp. | Macrofossil | High | Multidominant | Paralic | Coal-swamp, distribu-tary channel | Channel + U71 | CHLSP |
| North America, USA, Montana | Mesozoic; Cretaceous | Two Medicine Fm. | Macrofossil | Low | Monodominant | Alluvial | Overbank, ashfall | Mudstone | DBF, PSOL |
| North America, USA, Montana, Willow Creek Anticline | Mesozoic; Cretaceous | Two Medicine Fm. | Macrofossil | Low | Monodominant | Alluvial | Overbank | ? | PSOL |
| North America, USA, Montana, Willow Creek Anticline | Mesozoic; Cretaceous | Two Medicine Fm. | Macrofossil | Low | Monodominant | Alluvial | Overbank | ? | PSOL |
| North America, USA, Utah, Dinosaur National Monument | Mesozoic; Jurassic | Morrison Fm., Brushy Basin Mbr. | Macrofossil | High | Multidominant | Alluvial | Channel | Sandstone | CHLSP |
| Europe, Germany, Trossingen | Mesozoic; Triassic, Norian | Keuper Succession, Middle Stubensand-stein | Macrofossil | Low | Monodominant | Alluvial | Playa | Sandstone, mudstone | CHLSP, WTLND |
| North America, USA, New Mexico, San Juan County | Mesozoic; Cretaceous | Fruitland Fm.; Kirtland Fm. | Macrofossil, microfossil, mixed | High, low | Monodominant, multidominant | Alluvial | Channel, splay, overbank | Conglomerate; sandstone; mudstone | CHLSP, PSOL, WTLND |
| North America, USA, Nebraska, Agate Fossil Beds National Monument | Cenozoic; Tertiary, Miocene | Harrison Fm., Upper Harrison Fm. | Macrofossil | High | Multidominant | Alluvial | Channel, pond, waterhole | Fine-grained, tuffaceous sandstone | CHLSP, WTLND |
| North America, USA, Nebraska, Sioux County | Cenozoic; Tertiary, Miocene | Harrison Fm. | Macrofossil | Low | Multidominant | ? | ? | ? | U |
| North America, USA, Nebraska, Sioux City | Cenozoic; Tertiary, Miocene | Marsland Fm. | Macrofossil | High | Multidominant | Alluvial | Channel | Sandstone (calcareous ashy Sandstone) | CHLSP |

*(Continued)*

| Author | Date | Title | Citation | Taxa | Source | Bonebed Name | # of Sites |
|---|---|---|---|---|---|---|---|
| Hunt, R.M. Jr. | 1990 | Taphonomy and sedimentology of Arikaree (Lower Miocene) fluvial, eolian, and lacustrine paleoenvironments, Nebraska and Wyoming: A paleobiota entombed in fine-grained volcaniclastic rocks. | Pp. 69–111 *in* Volcanism and fossil biotas. Special paper 244. M. Lockley and A. Rice, eds. Geological Society of America, Boulder, Colorado. | lm | Ms | ? | 1+ |
| Hunt, R.M. Jr. | 1999 | Depositional setting of the Miocene Diatomite Quarry Bonebed, Shanwang Locality, Shandong Province, China. | Journal of Vertebrate Paleontology 19(supplement to 3):53A. | a, b, f, lm, ndr, sm | Ab | Shanwang Locality (Miocene Diatomite Quarry Bonebed) | 1 |
| Hunt, R.M., Jr., X. Xiang-Xu, and J. Kaufman. | 1983 | Miocene burrows of extinct bear dogs: Indication of early denning behavior of large mammalian carnivores. | Science 221:364–366. | lm | Ms | ? | 1 |
| Ishii, K.I, M. Watabe, S. Suzuki, S. Ishigaki, R. Rasobold, and K. Tsogtbaatarm, eds. | 2000 | Results of Hayashibara Museum of Natural Sciences-Mongolian Academy of Sciences-Mongolian Paleontological Center-Joint Paleontological Expedition. | Hayashibara Museum of Natural Sciences Research Bulletin 1:1–137. | d | Bk | ? | 2+ |
| Jacobs, L.L., and P.A. Murry. | 1980 | The vertebrate community of the Triassic Chinle Formation near St. Johns, Arizona. | Pp. 55–71 *in* Aspects of vertebrate history. L.L. Jacobs, ed. Museum of Northern Arizona Press, Flagstaff. | ndr | Ms | *Placerias* Quarry; Downs Quarry | 2 |
| Jacobsen, A.R., and M.J. Ryan. | 1999 | Taphonomic aspects of theropod tooth-marked bones from an *Edmontosaurus* bonebed (Lower Maastrichtian), Alberta Canada. | Journal of Vertebrate Paleontology 19(supplement to 3):55A. | d | Ab | Day Digs Bonebed (1) | 1 |
| Jamniczky, H.A., D.B. Brinkman, and A.P. Russell. | 2001 | Vertebrate microsite analysis: Enough is enough. | Journal of Vertebrate Paleontology 21(supplement to 3):65A. | d | Ab | ? | 1 |
| Jerzykiewicz, T., P.J. Currie, D.A. Eberth, P.A. Johnston, E.H. Koster, and J.J. Zheng. | 1993 | Djadokhta Formation correlative strata in Chinese Inner Mongolia: An overview of the stratigraphy, sedimentary geology, and paleontology and comparisons with type locality in the pre-Altai Gobi. | Canadian Journal of Earth Sciences 30:2180–2195. | d | Ms | ? | 2+ |

| Location | Period / Epoch | Stratigraphy | Element Size | Diversity | Relative Abundance | Depositional Setting | Depositional Environment | Facies | Paleoenv Code |
|---|---|---|---|---|---|---|---|---|---|
| North America, USA, Nebraska & Wyoming | Cenozoic; Tertiary, Miocene | Arikaree | Macrofossil | ? | ? | Alluvial | ? | ? | U |
| Asia, China, Shandong Province | Cenozoic; Tertiary, Miocene | | Macrofossil | High | Multidominant | lacustrine | Lacustrine, associated anoxia and basalt flows, trap | Diatomite | LAC |
| North America, USA, Nebraska, Agate Fossil Beds National Monument | Cenozoic; Tertiary, Miocene | | Macrofossil | Low | Monodominant | Alluvial | Burrows | Sandstone | O |
| Asia, China, Mongolia, Gobi Desert | Mesozoic; Cretaceous | Multiple | Macrofossil | Low | monotaxic | Eolian, alluvial, lacus-trine | Alluvial overbank, inter-dunes, pond, lacustrine margin | Sandstone | EOL, CHLSP, PSOL, WTLND, LAC |
| North America, USA, Arizona, St. Johns | Mesozoic; Triassic | Chinle Fm. | Mixed | High | Multidominant | Alluvial | ? | Mudstone | CHLSP, WTLND |
| North America, Canada, Alberta, Bleriot Ferry | Mesozoic; Cretaceous | Horseshoe Canyon Fm. | Macrofossil | Low | Monodominant | Alluvial | Pond, paleosol | Mudstone, organic-rich | WTLND |
| North America, Canada, Alberta | Mesozoic; Cretaceous | Oldman Fm. | Microfossil | High | Multidominant | Alluvial | Channel, splay | Sandstone | CHLSP |
| Asia, China, Mongolia, Inner Mongolia, Bayan Mandahu | Mesozoic; Cretaceous | Djadokhta Fm. | Macrofossil | Low | Monotaxic | Eolian, alluvial | Interdune, sheet-flood, channel | Sandstone | DBF, EOL, CHLSP |

*(Continued)*

| Author | Date | Title | Citation | Taxa | Source | Bonebed Name | # of Sites |
|---|---|---|---|---|---|---|---|
| Johnson, G.D. | 1992 | Early Permian vertebrate microfossils from Archer City bonebed 3, Archer County, Texas. | Journal of Vertebrate Paleontology 12(supplement to 3):35A. | a, f, ndr | Ab | Archer City Bonebed 3 | 1 |
| Johnston, P.A. | 1980 | Late Cretaceous and Paleocene mammals from southwestern Saskatchewan. | Masters thesis, University of Alberta, Department of Geology, Edmonton, Alberta. | sm | Ms | Rav W-1; Long Fall; Croc Pot | 3 |
| Johnston, P.A., and R.C. Fox. | 1984 | Paleocene and Late Cretaceous mammals from Saskatchewan, Canada. | Palaeontographica A 186:166–222. | sm | Ms | Rav W-1; Long Fall | 2 |
| Jordan, D.S. | 1920 | A Miocene catastrophe. | Natural History 20:18–22. | f | Ms | ? | 1 |
| Kaye, F.T., and K. Padian. | 1994 | Microvertebrates from the Placerias Quarry: A window on Late Triassic vertebrate diversity in the American Southwest. | Pp. 171–196 *in* In the shadow of the dinosaurs: Early Mesozoic tetrapods. N.C. Fraser and H.-D. Sues, eds. Cambridge University Press, Cambridge. | ndr | Ms | *Placerias*/Downs Quarry | 1 |
| Khajuria, C.K., and G.V.R. Prasad. | 1998 | Taphonomy of a Late Cretaceous mammal-bearing microvertebrate assemblage from the Deccan inter-trappean beds of Naskal, peninsular India. | Palaeogeography, Palaeoclimatology, Palaeoecology 137:153–172. | sm | Ms | ? | 1 |
| Kimmel, P.G. | 2000 | Tilefish mass mortality in the late Miocene. | Geological Society of America, Southeastern and South-Central Sections, Abstracts with Programs 2000:249. | f | Ab | ? | 2 |
| Kirkland, J.I., and H.J. Armstrong. | 1992 | Taphonomy of the Mygatt-Moore (M & M) Quarry, Middle Brushy Basin Member, Morrison Formation (Upper Jurassic), western Colorado. | Journal of Vertebrate Paleontology 12(supplement to 3):37A. | d | Ab | M&M Quarry | 1 |
| Kiser, E.L. | 1978 | The re-examination of Pedro De Castaneda's bone bed by geological investigations. | Bulletin of the Texas Archeological Society 49:331–339. | lm | Ms | Pedro Casteneda Site/Silver Lake Bison Site | 1 |

| Location | Period / Epoch | Stratigraphy | Element Size | Diversity | Relative Abundance | Depositional Setting | Depositional Environment | Facies | Paleoenv Code |
|---|---|---|---|---|---|---|---|---|---|
| North America, USA, Texas, Archer County | Paleozoic; Permian | Archer City Fm. | Microfossil | ? | ? | Alluvial, lacustrine | Pond, lacustrine | Mudstone | WTLND, LAC |
| North America, Canada, Saskatchewan | Mesozoic; Cenozoic; Cretaceous, Tertiary, Paleocene | Ravenscrag Fm.; Frenchman Fm. | Microfossil | High | Multidominant | Alluvial | Channel | Sandstone | CHLSP |
| North America, Canada, Saskatchewan | Mesozoic; Cretaceous, Maastrichtian | Ravenscrag Fm. | Microfossil | High | Multidominant | Alluvial | Overbank | Sandstone | CHLSP, PSOL, WTLND |
| North America, USA, California, Santa Barbara County | Cenozoic; Tertiary, Miocene | | Macrofossil | Low | Monotaxic | Marine | Nearshore | Diatomaceous mudstone | MAR |
| North America, USA, Arizona, St. Johns | Mesozoic; Triassic | Chinle Fm. | Mixed | High | Multidominant | Alluvial | Pond, paleosol | Mudstone, limestone concretions | WTLND |
| Asia, India, Andhra Pradesh, Naskal | Mesozoic; Cretaceous | | Microfossil | High | Multidominant | Lacustrine | Shoreline | Mudstone | LAC |
| North America, USA, Virginia & Maryland | Cenozoic; Tertiary, Miocene | Calvert Fm. | Macrofossil | Low | Monotaxic | Marine | Offshore | Marl | MAR |
| North America, USA, Colorado | Mesozoic; Jurassic | Morrison Fm. | Macrofossil | High | Multidominant | Alluvial | Pond, waterhole | Mudstone (claystone) | WTLND |
| North America, USA, Texas, Silver Lake | Cenozoic; Quaternary, Holocene | | Macrofossil | Low | ? | Lacustrine | Pond | Mudstone | LAC |

*(Continued)*

| Author | Date | Title | Citation | Taxa | Source | Bonebed Name | # of Sites |
|--------|------|-------|----------|------|--------|--------------|-----------|
| Krause, D.W., and P.D. Gingerich. | 1983 | Mammalian fauna from Douglas Quarry, earliest Tiffanian (Late Paleocene) of the eastern Crazy Mountain Basin, Montana. | Contributions from the Museum of Paleontology, The University of Michigan 26:157–196. | sm | Ms | Douglass Quarry | 1 |
| Krissek, L.A., T.C. Horner, and J.W. Collinson. | 1991 | Sedimentology of a vertebrate bone-bearing bed in the mid Triassic Fremouw Formation, Beardmore Glacier region, Antarctica. | Geological Society of America, Abstracts with Programs 23:A290. Also: Antarctic Journal of the U.S. 26:17–19. | d | Ab, Ms | Gordon Valley Site | 1 |
| Lam, V.D., and M.J. Ryan. | 2001 | The Royal Tyrrell Museum Day Digs Program: Public supported dinosaur research in the Drumheller Valley. | Alberta Palaeontological Society, Fifth Annual Symposium, Abstracts 1953:37–38. | d | Ms | Bleriot Ferry; Fox Coulee | 1 |
| Langston, W. Jr. | 1953 | Permian amphibians from New Mexico. | University of California Publications in Geological Sciences 29:349–416. | a, f, ndr | Ms | Baldwin; Miller; Camp; Welles; Anderson; Spanish Queen Mine; Johnson; Arroyo de la Parida; Glorieta Pass; VanderHoof; Quarry Butte | 9 |
| Langston, W. Jr. | 1975 | The ceratopsian dinosaurs and associated lower vertebrates from the St. Mary River Formation (Maestrichtian) at Scabby Butte, southern Alberta. | Canadian Journal of Earth Sciences 12:1576–1608. | d | Ms | Scabby Butte Sites 1& 2 | 2 |
| Laury, R.L. | 1980 | Paleoenvironments of a Late Quaternary mammoth-bearing sinkhole deposit, Hot Springs, South Dakota. | Geological Society of America Bulletin 91:465–475. | lm | Ms | Hot Springs Mammoth Site | 1 |
| Lavocat, R. | 1953 | Sur la presence de quelques restes de Mammiferes dans le bone-bed Eocene de Tamaguilel (Soudan francais). | Societe Geologique de France. Compte Rendu Sommaire de Seances 7–8:109–110. | lm, sm | Ab | Tamaguilel Bone-Bed | 1 |

| Location | Period / Epoch | Stratigraphy | Element Size | Diversity | Relative Abundance | Depositional Setting | Depositional Environment | Facies | Paleoenv Code |
|---|---|---|---|---|---|---|---|---|---|
| North America, USA, Montana | Cenozoic; Tertiary, Paleocene, Tiffanian | Tongue River Fm. | Microfossil | High | Multidominant | ? | ? | ? | U |
| Antarctica, Gordon Valley | Mesozoic; Triassic | Fremouw Fm. | Macrofossil | High | Multidominant | Alluvial | Channel | Sandstone, conglomerate | CHLSP |
| North America, Canada, Alberta, Drumheller | Mesozoic; Cretaceous | Horseshoe Canyon Fm. | Macrofossil | Low | Monodominant | Paralic | Overbank, pond, paleosol | Mudstone | WTLND |
| North America, USA, New Mexico, Rio Arriba, Sandoval, Sororro, San Miruel & Santa Fe Counties | Paleozoic; Permian | Cutler Fm.; Abo Fm.; Sangre de Cristo Fm. | Macrofossil | High | Monodominant, multidominant | Alluvial | Channel, pond | ? | CHLSP, PSOL, WTLND |
| North America, Canada, Alberta, Scabby Butte | Mesozoic; Cretaceous | St Mary River Fm. | Macrofossil | Low | Monodominant | Alluvial | overbank | Mudstone | WTLND |
| North America, USA, South Dakota, Hot Springs | Cenozoic; Quaternary, Pleistocene | | Macrofossil | Low | ? | Sinkhole | Sinkhole | Sand, gravel, clay | O |
| Africa, Sudan | Cenozoic; Tertiary, Eocene | | Mixed | High | Multidominant | Alluvial | ? | ? | U |

*(Continued)*

| Author | Date | Title | Citation | Taxa | Source | Bonebed Name | # of Sites |
|---|---|---|---|---|---|---|---|
| Lawton, R. | 1977 | Taphonomy of the dinosaur quarry, Dinosaur National Monument, University of Wyoming. | University of Wyoming Contributions to Geology 9:119–126. | d | Ms | Dinosaur National Monument | 1 |
| Lederer, E.M., and B.J. Small. | 1999 | An upper Jurassic microvertebrate site, Moffat County, Colorado. | Journal of Vertebrate Paleontology 19(supplement to 3):58A. | ndr | Ab | Wolf Creek | 1 |
| Leggitt, V.L., and H.P. Buchheim. | 1997 | Bird bone taphonomic data from recent lake margin strandlines compared with an Eocene *Presbyornis* (Aves: Anseriformes) bone strandline. | Geological Society of America, Abstracts with Programs A-105. | b | Ab | ? | 1 |
| Lehman, T.M. | 1982 | A ceratopsian bone bed from the Aguja Formation (Upper Cretaceous), Big Bend National Park, Texas. | Masters thesis, University of Texas, Austin. | d | Th | WPA Quarries | 2 |
| Loope, D.B., L. Dingus, C.C. Swisher III, and C. Minjin. | 1998 | Life and death in a Late Cretaceous dune field, Nemegt basin, Mongolia. | Geology 26:27–30 | d | Ms | ? | 2+ |
| Loope, D.B., J.A. Mason, and L. Dingus. | 1999 | Lethal sandslides from eolian dunes. | Journal of Geology 107:707–713. | d | Ms | multiple sites | 1+ |
| Loyal, R.S. | 2004 | Basal Paleogene biostratigraphy and biochronology of the collision-induced paralic-fluvial package: Interpretatations for paleoenvironments and paleobiogeography of a narrowing Tethys, NW Himalaya, India. | Geological Society of America, Abstracts with Programs 36:93. | f, ndr | Ab | ? | 1 |
| Maas, M.C. | 1985 | Taphonomy of a late Eocene microvertebrate locality, Wind River Basin, Wyoming (U.S.A.). | Palaeogeography, Palaeoclimatology, Palaeoecology 52:123–142. | ndr, sm | Ms | Badwater Locality 20 | 1 |
| Maisch, M.W., A.T. Matzke, H.U. Pfretzschner, G. Sun, H. Stohr, and F. Grossmann, F. | 2003 | Fossil vertebrates from the Middle and Upper Jurassic of the Southern Junggar Basin (NW China): Results of the Sino-German Expeditions 1999–2000. | Neues Jahrbuch für Geologie und Paläontologie 2003:82–98. | a, d, f, ndr | Ms | Bonebed | 1 |

| Location | Period / Epoch | Stratigraphy | Element Size | Diversity | Relative Abundance | Depositional Setting | Depositional Environment | Facies | Paleoenv Code |
|---|---|---|---|---|---|---|---|---|---|
| North America, USA, Utah | Mesozoic; Jurassic | Morrison Fm. | Macrofossil | High | Multidominant | Alluvial | Channel | Sandstone | CHLSP |
| North America, USA, Colorado, Moffat County | Mesozoic; Jurassic | Morrison Fm., Brushy Basin Mbr. | Microfossil | High | Multidominant | Alluvial | Overbank | Mudstone | WTLND |
| North America, USA | Cenozoic; Tertiary, Eocene | | Macrofossil | Low | ? | Lacustrine | Strandline | ? | LAC |
| North America, USA, Texas, Big Bend National Park | Mesozoic; Cretaceous | Aguja Fm. | Mixed | High | Monodominant, multidominant | Alluvial | Overbank | Mudstone (claystone) | WTLND |
| Asia, China, Mongolia, Nemegt Basin | Mesozoic; Cretaceous | Djadokhta Fm. | Macrofossil | High, low | Monodominant, multidomi- nant, monotaxic | Eolian | Dune, interdune | Sandstone | EOL |
| Asia, China, Mongolia, Gobi Desert | Mesozoic; Cretaceous | Djadokhta Fm. (equivalent) | Macrofossil | Low | Monotaxic | Eolian | Dunes | Sandstone | EOL |
| Asia, India | Cenozoic; Tertiary, Paleocene, Eocene | Subathu Fm. | Macrofossil | High | ? | Paralic | Lagoonal muds | Limey mudstone | PESH |
| North America, USA, Wyoming, Wind River Basin | Cenozoic; Tertiary, Eocene | Wagon Bed Fm., Hendry Ranch Mbr. | Microfossil | High | ? | Alluvial | Overbank, paleosol | Mudstone (volcanigenic) | PSOL |
| Asia, China, Xinjiang, Junggar Basin, Liuhonggou | Mesozoic; Jurassic | Toutunhe Fm. | Mixed | High | Multidominant | Alluvial | ? | Sandstone | CHLSP |

*(Continued)*

| Author | Date | Title | Citation | Taxa | Source | Bonebed Name | # of Sites |
|---|---|---|---|---|---|---|---|
| Martin, J.E., and J.R. Foster. | 1998 | First Jurassic mammals from the Black Hills, northeastern Wyoming. | Modern Geology 23:381–392. | sm | Ms | Little Houston Quarry; Mammal Quarry | 2 |
| Martin, L.D., B.M. Gilbert, and D.B. Adams. | 1977 | A cheetah-like cat in the North American Pleistocene. | Science 195:981–982. | lm | Ms | Natural Trap Cave | 1 |
| Matthew, W.D. | 1923 | Fossil bones in the rock. | Natural History 23:359–369. | lm | Ms | Agate Spring Fossil Quarry | 1 |
| Matthew, W.D., and J.W. Gidley. | 1904 | New or little known mammals from the Miocene of South Dakota. American Museum Expedition of 1903. | American Museum of Natural History Bulletin 20:241–268. | lm, sm | Ms | ? | 2+ |
| Maxwell, W.D., and J.H. Ostrom. | 1995 | Taphonomy and paleobiological implications of *Tenontosaurus-Deinonychus* associations. | Journal of Vertebrate Paleontology 15:707–712. | d | Ms | Cloverly Sites | 2+ |
| May, K., and R.A. Gangloff. | 1999 | New dinosaur bonebed from the Prince Creek Formation, Colville River, National Petroleum Reserve, Alaska. | Journal of Vertebrate Paleontology 19(supplement to 3):62A. | d | Ab | ? | 1 |
| McCool, K.E. | 1988 | Taphonomy of the Valentine Railway Quarry "B" Bone Bed (Late Barstovian), north-central Nebraska. | Masters thesis, University of Nebraska, Lincoln. | lm | Ms | Valentine Railway Quarry "B" | 1 |
| McGee, E.M. | 1993 | The taphonomy of Roehler's *Coryphodon* catastrophe quarry (Lower Eocene, Wasatch Formation, Washakie Basin, Wyoming). | Journal of Vertebrate Paleontology 13(supplement to 3):49A. | lm | Ab | *Coryphodon* Castatrope Quarry | 1 |
| McGee, E.M. | 1999 | An analysis of bone orientation at the Cleveland-Lloyd dinosaur quarry using vector summation: A new spin on an old technique. | Journal of Vertebrate Paleontology 19(supplement to 3):62A. | d | Ab | Cleveland-Lloyd Dinosaur Quarry | 1 |
| Merriam, J.C. | 1906 | Recent discoveries of Quaternary mammals in southern California. | Science 24:248–250. | lm | Ms | La Brea Tar Pits | 1 |

| Location | Period / Epoch | Stratigraphy | Element Size | Diversity | Relative Abundance | Depositional Setting | Depositional Environment | Facies | Paleoenv Code |
|---|---|---|---|---|---|---|---|---|---|
| North America, USA, Wyoming, Black Hills | Mesozoic; Jurassic | Morrison Fm. | Mixed | High | Multidominant | Alluvial | Overbank | Siltstone | PSOL, WTLND |
| North America, USA, Wyoming, Big Horn Mountains | Cenozoic; Quaternary, Pleistocene | Pleistocene sinkhole deposit | Macrofossil | High | Multidominant | Sinkhole | Sinkhole | Not applicable | O |
| North America, USA, Nebraska, Agate Springs | Cenozoic; Tertiary, Miocene | Harrison Fm. | Macrofossil | Low | Monodominant | Alluvial | Pond | Sandstone | WTLND |
| North America, USA, South Dakota, Little White River | Cenozoic; Tertiary, Miocene | Loup Fork beds, Rosebud beds | Macrofossil | High | Multidominant | Alluvial | Floodplain, channel, mire | Sandstone, mudstone | CHLSP, WTLND |
| North America, USA, Montana & Wyoming | Mesozoic; Cretaceous, Aptian, Albian | Cloverly Fm. | Mixed | Low | Monodominant | Alluvial | Overbank | Mudstone | PSOL |
| North America, USA, Alaska, National Petroleum Reserve | Mesozoic; Cretaceous | Prince Creek Fm. | Macrofossil | Low | Monodominant | Alluvial | Overbank | Sandstone | WTLND |
| North America, USA, Nebraska | Cenozoic; Tertiary, Miocene, Barstovian | Valentine Fm., Crookston Bridge Mbr. | Macrofossil | High | Multidominant | Alluvial | Channel | Sandstone | CHLSP |
| North America, USA, Wyoming | Cenozoic; Tertiary, Eocene | Wasatch Fm. | Macrofossil | Low | Monodominant | Alluvial | ? | ? | U |
| North America, USA, Utah | Mesozoic; Jurassic | Morrison Fm. | Macrofossil | High | Monodominant | Alluvial | Overbank, bog, pond | Calcareous smectitic mudstone | WTLND |
| North America, USA, California, Los Angeles | Cenozoic; Quaternary, Pleistocene | La Brea | Macrofossil | High | Multidominant | Seep | Seep | Tar | O |

(Continued)

| Author | Date | Title | Citation | Taxa | Source | Bonebed Name | # of Sites |
|--------|------|-------|----------|------|--------|--------------|------------|
| Merriam, J.C. | 1909 | A death trap which antedates Adam and Eve: The discovery of a Californian tar swamp that holds the bones of extinct monsters. | Harper's Weekly 53:11–12. | lm | Ms | La Brea Tar Pits | 1 |
| Merriam, J.C. | 1911 | The fauna of Rancho La Brea. I. Occurrence. | Memoirs of the University of California 1:199–213. | lm | Ms | La Brea Tar Pits | 1 |
| Miller, W.E., R.D. Horrocks, and J.H. Madsen, Jr. | 1996 | The Cleveland-Lloyd dinosaur quarry, Emery County, Utah: A U.S. natural landmark (including history and quarry map). | Brigham Young University, Geology Studies 41:3–24. | d | Ms | Cleveland-Lloyd Quarry | 1 |
| Monaco, P.E. | 1998 | A short history of dinosaur collecting in the Garden Park Fossil area, Canon City, Colorado. | Modern Geology 23:465–480. | d | Ms | Felch Quarry 1; Cope's Nipple Quarries | 5+ |
| Morris, T.H., D.R. Richmond, and S.D. Grimshaw. | 1996 | Orientation of dinosaur bones in riverine environments: Insights into sedimentary dynamics and taphonomy. | The continental Jurassic. M.Morales, ed. Museum of Northern Arizona Bulletin 60:521–530. | d | Ms | Cleveland-Lloyd Quarry; Dry Mesa Quarry; Dinosaur National Monument | 3 |
| Murchison, R.I. | 1856 | On the Bone Beds of the Upper Ludlow Rock and the Base of the old Redstone. | Reports of the British Association for the Advancement of Science 70–71. | f | Ms | ? | 2+ |
| Murry, P.A. | 1989 | Microvertebrate fossils from the Petrified Forest and Owl Rock members (Chinle Formation) in Petrified Forest National Park and vicinity, Arizona. | Pp. 29–64 in Dawn of the age of dinosaurs in the American Southwest. S.G. Lucas and A.P. Hunt, eds. New Mexico Museum of Natural History, Albuquerque. | a, d, f, ndr | Ms | Placerias/Downs Quarry | 1 |
| Murry, P.A., and G.D. Johnson. | 1987 | Clear Fork vertebrates and environments for the Lower Permian of north-central Texas. | The Texas Journal of Science, 39:253–266. | a, f, ndr | Ms | multiple sites | 10+ |
| Myers, T. | 2004 | Evidence for age segragation in a herd of diplodocid sauropods. | Journal of Vertebrate Paleontology 24(supplement to 3):97A. | d, ndr | Ab | Mother's Day Quarry | 1 |
| Nelms, L.G. | 1989 | Late Cretaceous dinosaurs from the north slope of Alaska. | Journal of Vertebrate Paleontology 9(supplement to 3):34A. | d | Ab | ? | 1 |

| Location | Period / Epoch | Stratigraphy | Element Size | Diversity | Relative Abundance | Depositional Setting | Depositional Environment | Facies | Paleoenv Code |
|---|---|---|---|---|---|---|---|---|---|
| North America, USA, California, Los Angeles | Cenozoic; Quaternary, Pleistocene | La Brea | Macrofossil | High | Multidominant | Seep | Seep | Tar | O |
| North America, USA, California, Los Angeles | Cenozoic; Quaternary, Pleistocene | | Macrofossil | High | Multidominant | Seep | Seep | Tar | O |
| North America, USA, Utah, Emery County | Mesozoic; Jurassic | Morrison Fm. | Macrofossil | High | Monodominant | Alluvial | Pond | Mudstone (calcareous claystone) | WTLND |
| North America, USA, Colorado, Canyon City | Mesozoic; Jurassic | Morrison Fm. | Macrofossil | High, low | Monodominant, multidominant | Alluvial | ? | ? | U |
| North America, USA, Utah | Mesozoic; Jurassic | Morrison Fm. | Macrofossil | High | Monodominant, multidominant | Alluvial | Channel, pond | Sandstone; mudstone | CHLSP, WTLND |
| Europe, United Kingdom, Wales, Ludlow | Paleozoic; Silurian, Devonian | Upper Ludlow Rock; Old Red Sandstone | Microfossil | High | Multidominant | Marine | Nearshore | Limestone; sandstone | MAR |
| North America, USA, Arizona, Petrified Forest National Park | Mesozoic; Triassic | Chinle Fm., Petrified Forest Mbr.; Owl Rock Mbr. | Microfossil | High | Multidominant | Alluvial | Overbank, pond | Mudstone | WTLND |
| North America, USA, Texas | Paleozoic; Permian | Choza Fm.; Vale Fm.; Arroyo Fm. | Macrofossil, microfossil, mixed | High, low | Monodominant, multidominant, monotaxic | Alluvial, paralic, marine | ? | ? | PESH |
| North America, USA, Montana | Mesozoic; Jurassic | Morrison Fm., Salt Wash Mbr. | Macrofossil | Low | Monodominant | Alluvial | Mudflow | Mudstone | DBF, CHLSP |
| North America, USA, Alaska, Ocean Point | Mesozoic; Cretaceous | Prince Creek Fm., Kogosukruk tongue | Macrofossil | Low | ? | Alluvial | Overbank, channel | ? | CHLSP, WTLND |

*(Continued)*

| Author | Date | Title | Citation | Taxa | Source | Bonebed Name | # of Sites |
|---|---|---|---|---|---|---|---|
| Nessov, L.A., and K. Gao. | 1993 | Cretaceous lizards from the Kizylkum desert, Uzbekhistan. | Journal of Vertebrate Paleontology 13(supplement to 3):51A | ndr | Ab | ? | 2+ |
| Norell, M.A., J.M. Clark, L.M. Chiappe, and D. Dashzeveg. | 1995 | A nesting dinosaur. | Nature, 378:774–776. | d | Ms | ? | 1 |
| Norman, D.B. | 1987 | A mass-accumulation of vertebrates from the Lower Cretaceous of Nehden (Sauerland), West Germany. | Proceedings of the Royal Society of London 230:215–255. | d | Ms | ? | 1 |
| Olson, E.C. | 1967 | Early Permian vertebrates. | Oklahoma Geological Survey, Circular 74:1–111. | a, f, ndr | Ms | ? | 2+ |
| Olson, E.C. | 1977 | Permian lake faunas: A study in community evolution. | Journal of the Palaeontological Society of India 20:146–163. | a, f, ndr | Ms | Waurika and Orlando Sites | 2 |
| Olson, E.C. | 1989 | The Arroyo Formation (Leonardian: Lower Permian) and its vertebrate fossils. | Texas Memorial Museum bulletin 35. Texas Memorial Museum, Austin. | a, f, ndr | Ms | multiple sites | 2+ |
| Ostrom, J.H. | 1969 | Osteology of Deinonychus antirrhopus, an unusual theropod from the Lower Cretaceous of Montana. | Peabody Museum of Natural History, Bulletin 30:1–165. | d | Ms | Yale Quarry | 1 |
| Paik, I.S., H.J. Kim, K.H. Park, Y.S. Song, Y.I. Lee, J.Y. Hwang, and M. Huh. | 2001 | Paleoenvironments and taphonomic preservation of dinosaur bone-bearing deposits in the Lower Cretaceous Hasandong Formation, Korea. | Cretaceous Research 22:627–642. | d | Ms | multiple sites | 2+ |
| Palmqvist, P., and A. Arribas. | 2001 | Taphonomic decoding of the paleobiological information locked in a lower Pleistocene assemblage of large mammals. | Paleobiology 27:512–530. | lm | Ms | Venta Micena Quarry | 1 |
| Parris, D.C., and B.S. Grandstaff. | 1989 | Nonmarine microvertebrates of the Ellisdale local fauna: Campanian of New Jersey. | Journal of Vertebrate Paleontology 9(supplement to 3):35A | d | Ab | ? | 1 |
| Parrish, W.C. | 1978 | Paleoenviornmental analysis of a Lower Permian bone bed and adjacent sediments, Wichita County, Texas. | Palaeogeography, Palaeoclimatology, Palaeoecology 24:209–237. | a, f, ndr | Ms | Thrift Bone Bed | 1 |

| Location | Period / Epoch | Stratigraphy | Element Size | Diversity | Relative Abundance | Depositional Setting | Depositional Environment | Facies | Paleoenv Code |
|---|---|---|---|---|---|---|---|---|---|
| Asia, Uzbekhistan, Kizylkum Desert | Mesozoic; Cretaceous | Chodzhakul Fm.; Bissekty Fm. | Microfossil | High | Multidominant | Alluvial, paralic | Estuarine | ? | PESH |
| Asia, China, Mongolia, Gobi Desert, Ukhaa Tolgod | Mesozoic; Cretaceous | | Macrofossil | Low | Monotaxic | Eolian | Interdune | Sandstone | EOL |
| Europe, Germany, Sauerland, Nehden | Mesozoic; Cretaceous, Aptian | | Macrofossil | ? | ? | Alluvial | Pond | Mudstone (claystone) | WTLND |
| North America, USA, Texas | Paleozoic; Permian | | Macrofossil, microfossil, mixed | High, low | Monodominant, multidominant | Alluvial | Channel, pond, levee | Sandstone, mudstone | CHLSP, WTLND |
| North America, USA, Oklahoma | Paleozoic; Permian | Oscar Fm.; Welington Fm. | Mixed | High | Multidominant | Lacustrine | Lacustrine shale and sandstone | Mudstone, sandstone | LAC |
| North America, USA, Texas & Oklahoma | Paleozoic; Permian | Arroyo Fm. | Macrofossil, microfossil, mixed | High, low | Monodominant, multidominant | Alluvial, paralic | Shoreline, overbank, pond, channel | Conglomerate, sandstone, mudstone, limestone | CHLSP, WTLND, PESH |
| North America, USA, Montana | Mesozoic; Cretaceous | Cloverly Fm. | Macrofossil | Low | Monodominant | Alluvial | Overbank | ? | PSOL |
| Asia, Korea, Gyeongsang Basin | Mesozoic; Cretaceous | Hasandong Fm. | Macrofossil | ? | ? | Alluvial | Overbank, paleosol | Mudstone, sandstone | PSOL |
| Europe, Spain, Granada, Venta Micena | Cenozoic; Quater-nary, Pleistocene | Pleistocene | Macrofossil | High | Multidominant | Lacustrine | Lake margin | Micritic mudstone | WTLND, LAC |
| North America, USA, New Jersey | Mesozoic; Cretaceous, Campa-nian | Marshalltown Fm. | Microfossil | High | Multidominant | Paralic | Estuarine | ? | PESH |
| North America, USA, Texas, Wichita County | Paleozoic; Permian | Bead Mountain Fm. | Macrofossil | Low | Monodominant | Paralic | Beach, mudflat, pond | Conglomerate | WTLND, PESH |

*(Continued)*

| Author | Date | Title | Citation | Taxa | Source | Bonebed Name | # of Sites |
|---|---|---|---|---|---|---|---|
| Pearson, D.A., T. Schaefer, K.R. Johnson, D.J. Nichols, and J.P. Hunter. | 2002 | Vertebrate biostratigraphy of the Hell Creek Formation in southwestern North Dakota and northwestern South Dakota. | Pp. 145–167 *in* The Hell Creek Formation and the Cretaceous-Tertiary boundary in the northern Great Plains: An integrated continental record of the end of the Cretaceous. Special paper 361. J.H. Hartman, K.R. Johnson, and D.J. Nichols, eds. Geological Society of America, Boulder, Colorado. | d | Ms | multiple sites | 53 |
| Peng, J.H., A.P. Russell, and D.B. Brinkman. | 2001 | Vertebrate microsite assemblages (exclusive of mammals) from the Foremost and Oldman formations of the Judith River Group (Campanian) of southeastern Alberta: An illustrated guide. | Provincial Museum of Alberta, Natural History Occasional Papers 25:1–54. | d | Ms | multiple sites | 19 |
| Peterson, O.A. | 1905 | Preliminary note of a gigantic mammal from the Loup Fork beds of Nebraska. | Science 22:211–212. | lm | Ms | Agate Spring Fossil Quarry | 1 |
| Peterson, O.A. | 1906 | Preliminary description of two new species of the genus *Diceratherium* Marsh, from the Agate Spring Fossil Quarry. | Science 24:281–283. | lm | Ms | Agate Spring Fossil Quarry | 1 |
| Peterson, O.A. | 1906 | The Agate Spring fossil quarry. | Annals of the Carnegie Museum 3:487–494. | lm | Ms | Agate Spring Fossil Quarry | 1 |
| Peterson, R., R. Peterson, N.V. D'Andrea, S.G. Lucas, and A.B. Heckert. | 1999 | Geological context and preliminary taphonomic analysis of the Peterson site, a Late Jurassic dinosaur quarry in New Mexico. | Journal of Vertebrate Paleontology 19(supplement to 3):68A. | d | Ab | Peterson Site | 1 |
| Pinsof, J.D. | 1985 | The Pleistocene vertebrate localities of South Dakota. | Fossiliferous Cenozoic deposits of western South Dakota and northwestern Nebraska. J.E. Martin, eds. Dakoterra 2: 233–264. | lm | Ms | multiple sites | 2+ |
| Prasad, G.V.R., and J.C. Rage. | 1995 | Amphibians and squamates from the Maastrichtian of Naskal, India. | Cretaceous Research 16:95–107. | a, ndr | Ms | Naskal | 1 |
| Pratt, A.E. | 1989 | Fossil vertebrates from the Marks Head Formation (lower Miocene) of southeastern Georgia. | Journal of Vertebrate Paleontology 9(supplement to 3):35A | f, lm, ndr, sm | Ab | ? | 1 |

| Location | Period / Epoch | Stratigraphy | Element Size | Diversity | Relative Abundance | Depositional Setting | Depositional Environment | Facies | Paleoenv Code |
|---|---|---|---|---|---|---|---|---|---|
| North America, USA, North & South Dakota | Mesozoic; Cretaceous | Hell Creek Fm.; Tullock Fm.; Ludlow Fm. | Microfossil | High | Multidominant | Alluvial | Paleochannel, splay, overbank, paleosol, pond | Mudstone, sandstone | CHLSP, PSOL, WTLND |
| North America, Canada, Alberta, Southern | Mesozoic; Cretaceous | Judith River Grp.; Oldman Fm.; Foremost Fm. | Microfossil | High | Multidominant | Alluvial, paralic | Transgressive lag, channel, splay | Sandstone, mudstone | CHLSP, PESH |
| North America, USA, Nebraska, Sioux County | Cenozoic; Tertiary, Miocene | Harrison Fm. | Macrofossil | Low | Monodominant | Alluvial | Channel? | ? | CHLSP |
| North America, USA, Nebraska, Agate Springs | Cenozoic; Tertiary, Miocene | Harrison Fm. | Macrofossil | Low | Monodominant | Alluvial | Channel, pond | Light colored sediments | CHLSP, WTLND |
| North America, USA, Nebraska, Agate Springs | Cenozoic; Tertiary, Miocene | Harrison Fm. | Macrofossil | Low | Monodominant | Alluvial | Channel, pond | Light colored sediments | CHLSP, WTLND |
| North America, USA, New Mexico | Mesozoic; Jurassic | Morrison Fm. | Macrofossil | ? | ? | Alluvial | Channel | Sandstone | CHLSP |
| North America, USA, South Dakota | Cenozoic; Quaternary, Pleistocene | | Macrofossil, microfossil, mixed | High, low | Monodominant, multidominant | Alluvial | ? | Gravel, sand, silt | U |
| Asia, India, Andhra Pradesh, Naskal | Mesozoic; Cretaceous, Maastrichtian | ? | Microfossil | High | ? | Lacustrine | shoreline | ? | LAC |
| North America, USA, Georgia, Porter's Landing | Cenozoic; Tertiary, Miocene | Marks Head Fm. | Microfossil | High | Multidominant | Marine | nearshore | ? | MAR |

*(Continued)*

| Author | Date | Title | Citation | Taxa | Source | Bonebed Name | # of Sites |
|--------|------|-------|----------|------|--------|--------------|------------|
| Pratt, A.E. | 1989 | Taphonomy of the microvertebrate fauna from the early Miocene Thomas Farm locality, Florida (U.S.A.). | Palaeogeography, Palaeoclimatology, Palaeoecology 76:125–151. | sm | Ms | Thomas Farm | 1 |
| Prideaux, G.J., G.A. Gully, L.K. Ayliffe, M.I. Bird, and R.G. Roberts. | 2000 | Tight Entrance Cave, Southwestern Australia: A Late Pleistocene vertebrate deposit spanning more than 180 KA. | Journal of Vertebrate Paleontology 20(supplement to 3):62–63A. | lm | Ab | Tight Entrance Cave | 1 |
| Raath, M.A. | 1980 | The theropod dinosaur *Syntarsus* (Saurischia: Podokesauridae) discovered in South Africa. | South African Journal of Science 76:375–376 | d, ndr | Ms | Mequatling Locality | 1 |
| Raath, M.A. | 1990 | Morphological variation in small theropods and its meaning in systematics: Evidence from *Syntarsus rhodesiensis*. | Pp. 91–106 *in* Dinosaur systematics, approaches and perspective. K. Carpenter and P.J. Currie, eds. Cambridge University Press, Cambridge. | d | Ms | Chintake | 1 |
| Reif, W.E. | 1971 | On the Genesis of the bone bed at the Muschelkalk-Keuper-Boundary "Grenzbonebed" in S.W. Germany. | Neues Jahrbuch fuer Geologie und Palaeontologie 139:82–98. | f, ndr | Ms | Muschelkalk-Keuper-Boundary Bonebed | 1 |
| Richmond, D., and G. McDonald. | 1999 | Sedimentology and taphonomy of the Smithsonian horse quarry. | Journal of Vertebrate Paleontology 19(supplement to 3):70A. | lm | Ab | Horse Quarry | 1 |
| Richmond, D.R., and T.H. Morris. | 1996 | The dinosaur death-trap of the Cleveland-Lloyd Quarry, Emery County, Utah. | The continental Jurassic. M.J. Morales, ed. Museum of Northern Arizona Bulletin 60:533–545. | d | Ms | Cleveland-Lloyd Quarry | 1 |
| Richmond, D.R., and T.H. Morris. | 1998 | Stratigraphy and cataclysmic deposition of the Dry Mesa Dinosaur Quarry, Mesa County, Colorado. | Modern Geology 22:121–143. | d | Ms | Dry Mesa Dinosaur Quarry | 1 |
| Richter, A.E. | 1982 | Fossile Knochenlager. | Mineralien Magazine 6:63–66. | f, ndr | Ms | ? | 1 |
| Rodrigues-De La Rosa, R.A., and S.R.S. Cevallos-Ferriz. | 1998 | Vertebrates of the El Pelillal locality (Campanian, Cerro Del Pueblo Formation), southeastern Couahuila, Mexico. | Journal of Vertebrate Paleontology 18:751–764. | d | Ms | El Pelillal | 1 |

| Location | Period / Epoch | Stratigraphy | Element Size | Diversity | Relative Abundance | Depositional Setting | Depositional Environment | Facies | Paleoenv Code |
|---|---|---|---|---|---|---|---|---|---|
| North America, USA, Florida | Cenozoic; Tertiary, Miocene, Hemingfordian | | Microfossil | High | Multidominant | Sinkhole | Sinkhole | ? | O |
| Australia | Cenozoic; Quaternary, Pleistocene | | Macrofossil | ? | ? | Sinkhole | Sinkhole | ? | O |
| Africa, South Africa, Mequatling Farm | Mesozoic; Triassic, Jurassic | Elliot Fm., Tritylodon Band | Macrofossil | High | Monodominant | Alluvial | Floodplain | Iron-rich silty mudstone | WTLND |
| Africa | Mesozoic; Jurassic | Forest Sandstone Fm. | Macrofossil | Low | Monodominant | Alluvial | Overbank | Mudstone | U |
| Europe, Germany | Mesozoic; Triassic | Hauptmuschelkalk Fm. | Microfossil | High | Multidominant | Marine | Nearshore | ? | MAR |
| North America, USA, Idaho, Hagerman Fossil Beds National Monument | Cenozoic; Tertiary, Pliocene | | Macrofossil | ? | ? | Alluvial | Channel | Sandstone | CHLSP |
| North America, USA, Utah, Cleveland | Mesozoic; Jurassic | Morrison Fm. | Macrofossil | High | Monodominant | Alluvial | Overbank, pond, trap | Calcareous smectitic mudstone | PSOL, WTLND |
| North America, USA, Utah | Mesozoic; Jurassic | Morrison Fm. | Macrofossil, mixed | High | Multidominant | Alluvial, lacustrine | Waterhole drought, flash flood | Silty sandstone | CHLSP |
| Europe, Germany | Mesozoic; Triassic | Muschelkalk | Microfossil | High | Multidominant | Marine | Transgressive lag | Conglomeratic limestone | PESH |
| North America, Mexico, Coahuila | Mesozoic; Cretaceous, Campanian | Cerro Del Pueblo Fm. | Microfossil | High | ? | Paralic, alluvial | Overbank | ? | WTLND, PESH |

*(Continued)*

| Author | Date | Title | Citation | Taxa | Source | Bonebed Name | # of Sites |
|---|---|---|---|---|---|---|---|
| Rogers, R.R. | 1988 | Taphonomy of a hadrosaur bonebed, Two Medicine Formation, Northwestern Montana. | Journal of Vertebrate Paleontology 8(supplement to 3):24A. | d | Ab | Hadrosaur Bonebed | 1 |
| Rogers, R.R. | 1990 | Taphonomy of three dinosaur bone beds in the Upper Cretaceous Two Medicine Formation of Northwestern Montana: Evidence for drought related mortality. | Palaios 5:394–413. | d | Ms | Canyon Bone Bed; Dino Ridge Quarry; West Side Quarry | 3 |
| Rogers, R.R. | 2005 | Fine-grained debris flows and extraordinary vertebrate burials in the Late Cretaceous of Madagascar. | Geology 33:297–300. | a, d, f, ndr, sm | Ms | multiple sites | 15 |
| Rogers, R.R., and S.M. Kidwell. | 2000 | Associations of vertebrate skeletal concentrations and discontinuity surfaces in terrestrial and shallow marine records: A test in the Cretaceous of Montana. | Journal of Geology 108:131–154. | d | Ms | ? | 2+ |
| Rogers, R.R., A.B. Arcucci, F. Abdala, P. Sereno, C.A. Forster, and C.L. May. | 2001 | Paleoenvironments and taphonomy of the Chañares Formation tetrapod assemblage (Middle Triassic), Northwestern Argentina: Spectacular preservation in volcanogenic concretions. | Palaios 16:461–481. | d | Ms | Los Chañares | 2+ |
| Rogers, R.R., J.H. Hartman, and D.W. Krause. | 2000 | Stratigraphic analysis of Upper Cretaceous rocks in the Mahajanga Basin, northwestern Madagascar: Implications for ancient and modern faunas. | The Journal of Geology 108:275–301. | d | Ms | ? | 2+ |
| Rogers, R.R., D.W. Krause, and K.C. Rogers. | 2003 | Cannibalism in the Madagascan dinosaur *Majungatholus atopus*. | Nature 422:515–518. | a, d, f, ndr, sm | Ms | MAD96-01; MAD96-21; MAD93-18 | 3 |
| Rolfe, W.D.I., G.P. Durant, A.E. Fallick, A.J. Hall, D.J. Large, A.C. Scott, T.R. Smithson, and G.M. Walkden. | 1990 | An early terrestrial biota preserved by Visean vulcanicity in Scotland. | Pp. 13–24 *in* Volcanism and fossil biotas. Special paper 244. M.G. Lockley and A. Rice, eds. Geological Society of America, Boulder, Colorado. | a, f, ndr | Ms | East Kirkton Site | 1+ |

| Location | Period / Epoch | Stratigraphy | Element Size | Diversity | Relative Abundance | Depositional Setting | Depositional Environment | Facies | Paleoenv Code |
|---|---|---|---|---|---|---|---|---|---|
| North America, USA, Montana | Mesozoic; Cretaceous, Campanian | Two Medicine Fm. | Macrofossil | ? | ? | Alluvial | Overbank, pond | ? | PSOL, WTLND |
| North America, USA, Montana | Mesozoic; Cretaceous | Two Medicine Fm. | Macrofossil | Low | Monodominant | Alluvial | Overbank, pond | Sandstone, siltstone, mudstone | WTLND |
| Africa, Madagascar, | Mesozoic; Cretaceous | Maevarano Fm., Anembalemba Mbr. | Macrofossil | High | Multidominant | Alluvial | Channel belt debris flow | Silty, clayey sandstone | DBF, CHLSP |
| North America, USA, Montana | Mesozoic; Cretaceous | Two Medicine Fm. & Judith River Fm. | Microfossil | High | Multidominant | Alluvial, paralic | Transgressive lag, channel, overbank | Sandstone, mudstone | WTLND, PESH |
| South America, Argentina, Ladinian | Mesozoic; Triassic | Chañares Fm. (Los Rastros Fm.) | Macrofossil | High | ? | Alluvial | Alluvial, paludal, eolian, volcanic | Concretions | U |
| Africa, Madagascar | Mesozoic; Cretaceous | Maevarano Fm. | Macrofossil, microfossil, mixed | High, low | ? | Paralic, alluvial | Shoreline, estuarine | Sandstone, conglomertae, mudstone | PESH |
| Africa, Madagascar, | Mesozoic; Cretaceous | Maevarano Fm., Anembalemba Mbr. | Macrofossil | High | Multidominant | Alluvial | Channel belt | Sandstone | CHLSP |
| Europe, United Kingdom, Scotland, Midland Valley | Paleozoic; Carboniferous, Visean | Upper Oil Shale Group, East Kirkton Limestone | Macrofossil | High | Multidominant | Alluvial, lacustrine | Hot springs | Limestone | O |

*(Continued)*

| Author | Date | Title | Citation | Taxa | Source | Bonebed Name | # of Sites |
|--------|------|-------|----------|------|--------|--------------|------------|
| Rowe, T. | 1989 | A new species of the theropod dinosaur *Syntarsus* from the early Jurassic Kayenta Formation of Arizona. | Journal of Vertebrate Paleontology 9:125–136. | d | Ms | *Syntarsus kayentakatae* Type Quarry; Shake-N-Bake Locality | 2+ |
| Ryan, M.J., J.G. Bell, and D.A. Eberth. | 1995 | Taphonomy of a Hadrosaur (Ornithischia: Hadrosauridae) bonebed from the Horseshoe Canyon Formation (Early Maastrichtian), Alberta, Canada. | Journal of Vertebrate Paleontology 15(supplement to 3):51A. | d | Ab | Day Digs Bonebed (1) | 1 |
| Ryan, M.J., P.J. Currie, J.D. Gardner, M.K. Vickaryous, and J.M. Lavigne, J.M. | 1998 | Baby hadrosaurid material associated with an unusually high abundance of *Troodon* teeth from the Horseshoe Canyon Formation, Upper Cretaceous, Alberta, Canada. | Gaia 15:123–133. | d | Ms | L2000 | 1 |
| Ryan, M.J., A.P. Russell, D.A. Eberth, and P.J. Currie. | 2001 | The taphonomy of a Centrosaurus (Ornithischia: Certopsidae [sic]) bone bed from the Dinosaur Park Formation (Upper Campanian), Alberta, Canada, with comments on cranial ontogeny. | Palaios 16:482–506. | d | Ms | BB 43 | 1 |
| Sahni, A. | 1972 | The vertebrate fauna of the Judith River Formation, Montana. | Bulletin of the American Museum of Natural History 147:1–412. | d | Ms | multiple sites | 2 |
| Sampson, S.D. | 1995 | Horns, herds, and hierarchies. | Natural History 104:36–40. | d | Ms | ? | 2+ |
| Sampson, S.D., Ryan, M.J., and Tanke, D.H. | 1997 | Craniofacial ontogeny in centrosaurine dinosaurs (Ornithischia: Ceratopsidae): Taxonomic and behavioral implications. | Zoological Journal of the Linnean Society 121:293–337. | d | Ms | multiple sites | 2+ |
| Sander, P.M. | 1987 | Taphonomy of the Lower Permian Geraldine Bonebed in Archer County, Texas. | Palaeogeography, Palaeoclimatology, Palaeoecology 61:221–236. | a, f, ndr | Ms | Geraldine Bone Bed | 1 |

| Location | Period / Epoch | Stratigraphy | Element Size | Diversity | Relative Abundance | Depositional Setting | Depositional Environment | Facies | Paleoenv Code |
|---|---|---|---|---|---|---|---|---|---|
| North America, USA, Arizona, Navajo Nation | Mesozoic; Jurassic | Kayenta Fm. | Macrofossil | Low | Monodominant, monotaxic | Alluvial | ? | Sandstone | U |
| North America, Canada, Alberta, Bleriot Ferry | Mesozoic; Cretaceous, Maastrichtian | Horseshoe Canyon Fm. | Macrofossil | Low | Monodominant | Alluvial | Overbank, pond | Mudstone, organic-rich | WTLND |
| North America, Canada, Alberta, Drumheller | Mesozoic; Cretaceous | Horseshoe Canyon Fm. | Microfossil | High | Monodominant | Alluvial | Overbank | Mudstone | PSOL |
| North America, Canada, Alberta, Dinosaur Provincial Park | Mesozoic; Cretaceous | Dinosaur Park Fm. | Macrofossil | High | Monodominant | Alluvial | Channel | Sandstone | CHLSP |
| North America, USA, Montana, Judith Landing | Mesozoic; Cretaceous | Judith River Fm. | Microfossil | High | Multidominant | Alluvial | Channel, splay | Sandstone, mudstone | CHLSP |
| North America, USA, Montana | Mesozoic; Cretaceous | Two Medicine Fm.; Wapiti Fm.; Dinosaur Park Fm. | Macrofossil | High, low | Monodominant | Alluvial | Channel, overbank, pond | Sandstone, mudstone | CHLSP, PSOL, WTLND |
| North America, USA & Canada, Montana & Alberta | Mesozoic; Cretaceous | Dinosaur Park Fm.; Two Medicine Fm.; St. Mary River Fm.; Wapiti Fm. | Macrofossil | High, low | Monodominant | Alluvial | Channel, overbank, pond | Sandstone, mudstone | CHLSP, PSOL, WTLND |
| North America, USA, Texas, Archer County | Paleozoic; Permian | Nocona Fm. | Macrofossil | High | ? | Alluvial, lacustrine | Overbank, pond | Mudstone | WTLND |

*(Continued)*

| Author | Date | Title | Citation | Taxa | Source | Bonebed Name | # of Sites |
|---|---|---|---|---|---|---|---|
| Sander, P.M. | 1989 | Early Permian depositional environments and pond bonebeds in central Archer County, Texas. | Palaeogeography, Palaeoclimatology, Palaeoecology 69:1–21. | a, f, ndr | Ms | Archer City 1–4; Coprolite Bonebed; Loftin Bonebed; Rattlesnake Canyon Bonebed | 7 |
| Sander, P.M. | 1992 | The Norian *Plateosaurus* bonebeds of central Europe and their taphonomy. | Palaeogeography, Palaeoclimatology, Palaeoecology 93:255–299. | d | Ms | ? | 3 |
| Sankey, J.T. | 1995 | A Late Cretaceous small vertebrate fauna from the Upper Aguja Formation, Big Bend National Park, Texas. | Geological Society of America, Abstracts with Programs 27:387. | d | Ab | ? | 1 |
| Sankey, J.T. | 2001 | Late Campanian southern dinosaurs, Aguja Formation, Big Bend, Texas. | Journal of Paleontology 75:208–215. | d | Ms | ? | 5 |
| Sankey, J.T., R.C. Fox, D.B. Brinkman, and D.A. Eberth. | 1999 | Paleoecology of mammals from the Dinosaur Park Formation (Judith River Group, Late Campanian), southern Alberta. | Journal of Vertebrate Paleontology 19(supplement to 3):73A | sm | Ab | ? | 36 |
| Sanz, L.J., L.M. Chiappe, Y. Fernández-Jalvo, B. Sánchez-Chillón, F.J. Poyato-Ariza, and B.P. Pérez-Moreno. | 2001 | An Early Cretaceous pellet. | Nature 409:998–999. | ndr, sm | Ms | Las Hoyas | 1 |
| Schmitt, J.G., J.R. Horner, R.R. Laws, and F. Jackson. | 1998 | Debris-flow deposition of a hadrosaur-bearing bone bed, Upper Cretaceous Two Medicine Formation, Northwest Montana. | Journal of Vertebrate Paleontology 18(supplement to 3):76A | d | Ab | *Maiasaura* Bone Bed | 1 |
| Schmude, D.E., and C.J. Weege. | 1996 | Stratigraphic relationship, sedimentology, and taphonomy of Meilyn, a dinosaur quarry in the basal Morrison Formation of Wyoming. | The continental Jurassic. M. Morales, ed. Museum of Northern Arizona Bulletin 60:547–554. | d | Ms | Meilyn | 1 |

| Location | Period / Epoch | Stratigraphy | Element Size | Diversity | Relative Abundance | Depositional Setting | Depositional Environment | Facies | Paleoenv Code |
|---|---|---|---|---|---|---|---|---|---|
| North America, USA, Texas, Archer County | Paleozoic; Permian | Petrolia Fm., Nocona Fm., Archer City Fm. & Markley Fm. | Macrofossil | High | Monodominant, multidominant | Alluvial, lacus- trine | Overbank, pond | Mudstone, siltstone, sandstone | WTLND |
| Europe, Switzerland & Germay | Mesozoic; Triassic, Norian | Knollenmerge beds | Mixed | High, low | Monodominant | Alluvial | Overbank | Mudstone | WTLND |
| North America, USA, Texas, Big Bend National Park | Mesozoic; Cretaceous | Aguja Fm. | Microfossil | High | Multidominant | Alluvial | Channel lag | Conglomerate | CHLSP |
| North America, USA, Texas, Big Bend National Park | Mesozoic; Cretaceous | Aguja Fm. | Microfossil | High | Multidominant | Alluvial | Channel lag | Calcareous, nodular, sandy conglomerate | CHLSP |
| North America, Canada, Alberta | Mesozoic; Cretaceous | Dinosaur Park Fm. | Microfossil | High | Multidominant | Alluvial | Channel, splay, overbank | Various | CHLSP |
| Europe, Spain, Cuenca, Cuenca, Las Hoyas | Mesozoic; Cretaceous | | Microfossil | Low | Multidominant | Lacustrine? | | Limestone | LAC |
| North America, USA, Montana, Willow Creek Anticline | Mesozoic; Cretaceous, Campa- nian | Two Medicine Fm. | Macrofossil | Low | Monodominant | Alluvial | Debris flow, volcanic | Mudstone, sandy | DBF, WTLND |
| North America, USA, Wyoming, Medicine Bow | Mesozoic; Jurassic | Morrison Fm. | Macrofossil | Low? | ? | Alluvial | Channel | Sandstone | CHLSP |

*(Continued)*

| Author | Date | Title | Citation | Taxa | Source | Bonebed Name | # of Sites |
|---|---|---|---|---|---|---|---|
| Schroder-Adams, C.J., S.L. Cumbaa, J. Bloch, D.A. Leckie, J. Craig, S.A. Seif El-Dein, D-J.H.A.E. Simons, and F. Kenig. | 2001 | Late Cretaceous (Cenomanian to Campanian) paleoenviornmental history of the eastern Canadian margin of the western Interior Seaway: Bonebeds and anoxic events. | Palaeogeography, Palaeoclimatology, Palaeoecology 170:261–289. | f | Ms | ? | 2+ |
| Schülter, T., and W. Schwarzans. | 1978 | Eine Bonebed-Lagerstätte aus dem Wealden Süd-Tunesiens (Umgebung Ksar Krerachfa). | Berlinger Geowissenshftliche Abhandlungen 9:53–65. | d, f, ndr | Ms | Ksar Krerachfa Bone Bed | 1 |
| Schwartz, H.L., and D.D. Gillette. | 1994 | Geology and taphonomy of the Coelophysis quarry, Upper Triassic Chinle Formation, Ghost Ranch, New Mexico. | Journal of Paleontology 68:1118–1130. | d | Ms | *Coelophysis* Quarry | 1 |
| Schwimmer, D.R. | 1997 | Late Cretaceous dinosaurs in eastern USA: A taphonomic and biogeographic model of occurrences. | Pp. 203–211 *in* Dinofest international, proceedings of a symposium sponsored by Arizona State University. D.L. Wolberg, E. Stump, and G.D. Rosenberg, eds. Academy of Natural Sciences, Philadelphia. | d | Ms | multiple sites | 20+ |
| Semonche, P.D., J.B. Smith, and J.R. Smith. | 1999 | A diverse dinosaur bonebed near the K/T boundary in Wyoming. | Journal of Vertebrate Paleontology 19(supplement to 3):75A. | d | Ab | Baculum Draconis | 1 |
| Shotwell, J.A. | 1955 | An approach to the paleoecology of mammals. | Ecology 36:327–337. | sm | Ms | McKay Reservoir Quarry; Hemphill Fauna Quarry | 3 |
| Sloan, R.E., and L. Van Valen. | 1965 | Cretaceous mammals from Montana. | Science 148:220–227. | sm | Ms | Bug Creek Anthills | 1 |
| Small, B. | 1989 | Post Quarry. | Pp. 145–148 *in* Dawn of the age of dinosaurs in the American Southwest. S.G. Lucas and A.P. Hunt, eds. New Mexico Museum of Natural History, Albuquerque. | a, ndr | Ms | Post Quarry | 1 |

| Location | Period / Epoch | Stratigraphy | Element Size | Diversity | Relative Abundance | Depositional Setting | Depositional Environment | Facies | Paleoenv Code |
|---|---|---|---|---|---|---|---|---|---|
| North America, Canada, Saskatchewan, Pasquia Hills | Mesozoic; Cretaceous, Cenomanian-Campanian | Ashville Fm. | Microfossil | High | Multidominant | Marine | Open marine | Mudstone | MAR |
| Africa, Tunisia, Ksar Krerachfa | Mesozoic; Cretaceous, Wealden | | Microfossil | High | Multidominant | ? | ? | Conglomerate | U |
| North America, USA, New Mexico, Abiquiu, Ghost Ranch | Mesozoic; Triassic | Chinle Fm. | Macrofossil | Low | Monodominant | Alluvial | Channel, pond | Mudstone (siltstone) | CHLSP, WTLND |
| North America, USA, Eastern USA | Mesozoic; Cretaceous | Multiple | Macrofossil | High | Multidominant | Marine | Offshore, bay-fill | Mudstone | MAR, PESH |
| North America, USA, Wyoming | Mesozoic; Cretaceous | Lance Fm. | Macrofossil | High | Multidominant | Alluvial | Channel | Sandstone | CHLSP |
| North America, USA, Texas & Oregon | Cenozoic; Tertiary, Pliocene, Hemphillian | | Macrofossil | High | Multidominant | ? | ? | ? | U |
| North America, USA, Montana, Bug Creek | Mesozoic; Cretaceous | Hell Creek Fm. | Microfossil | High | Multidominant | Alluvial | Channel | ? | CHLSP |
| North America, USA, Texas, Post | Mesozoic; Triassic | Dockum Fm. | Macrofossil | High | Multidominant | Paralic | Overbank | Mudstone | WTLND |

*(Continued)*

| Author | Date | Title | Citation | Taxa | Source | Bonebed Name | # of Sites |
|---|---|---|---|---|---|---|---|
| Smith, R., and J. Kitching. | 1997 | Sedimentology and vertebrate taphonomy of the *Tritylodon* Acme zone: A reworked paleosol in the Lower Jurassic Elliot Formation, Karoo Supergroup, South Africa. | Palaeogeography, Palaeoclimatology, Palaeoecology 131:29–50. | ndr | Ms | *Tritylodon* Acme Zone | 2+ |
| Smith, R.M.H. | 1987 | Helical burrow casts of therapsid origin from the Beaufort Group (Permian) of South Africa. | Palaeogeography, Palaeoclimatology, Palaeoecology 60:155–170. | ndr | Ms | one site | 1 |
| Smith, R.M.H. | 1993 | Vertebrate taphonomy of Late Permian floodplain deposits in the Southwestern Karoo basin of South Africa. | Palaios 8:45–67. | a, f, ndr | Ms | ? | 2+ |
| Spencer, L. | 1999 | Taphonomy at Rancho La Brea: A work in progress. | Journal of Vertebrate Paleontology 19(supplement to 3):77A. | lm | Ab | Pit 91 | 1 |
| Spencer, L.M., B. Van Valkenburgh, and J.M. Harris. | 2003 | Taphonomic analysis of large mammals recovered from the Pleistocene Rancho La Brea tar seeps. | Paleobiology 29:561–575. | lm | Ms | Pit 91 | 1 |
| Staron, R.M., B.S. Grandstaff, W.B. Gallagher, and D.E. Grandstaff. | 2001 | REE signatures in vertebrate fossils from Sewell, NJ: Implications for location of the K-T boundary. | Palaios 16:255–265. | ndr | Ms | Main Fossiliferous Layer; Navesink Assemblage; Upper Hornerstown Assemblage | 3 |
| Staron, R.M., D.E. Grandstaff, B.S. Grandstaff, and W.B. Gallagher. | 1999 | Mosasaur taphonomy and geochemistry: Implications for a K-T bonebed in the New Jersey coastal plain. | Journal of Vertebrate Paleontology 19(supplement to 3):78A. | ndr | Ab | Main Fossiliferous Layer | 1 |
| Stewart, K.M. | 1994 | A Late Miocene fish fauna from Lothagam, Kenya. | Journal of Vertebrate Paleontology 14:592–594. | f | Ms | ? | 1 |
| Stokes, W.L. | 1985 | The Cleveland-Lloyd Dinosaur Quarry. | United States Government Printing Office, Washington, D.C. | d | Ms | Cleveland-Lloyd Quarry | 1 |
| Straight, W.H., and D.A. Eberth. | 2002 | Testing the utility of vertebrate remains in recognizing patterns in fluvial deposits: An example from the lower Horseshoe Canyon Formation, Alberta. | Palaios 17:472–490. | d | Ms | multiple sites | 10+ |

| Location | Period / Epoch | Stratigraphy | Element Size | Diversity | Relative Abundance | Depositional Setting | Depositional Environment | Facies | Paleoenv Code |
|---|---|---|---|---|---|---|---|---|---|
| Africa, South Africa, Eastern Orange Free State | Mesozoic; Jurassic | Elliot Fm. | Macrofossil | ? | ? | Alluvial | Overbank, paleosol | Conglomerate | PSOL |
| Africa, South Africa, Karoo Basin | Paleozoic; Permian | Beaufort Grp., Teekloof Fm. | Macrofossil | Low | Monotaxic | Alluvial | Burrow | Siltstone | O |
| Africa, South Africa, Karoo Basin | Paleozoic; Permian | Hoedemaker Fm. | Macrofossil | Low | Monodominant | Alluvial | Levee, waterhole concentrate; trampled | Sandstone; siltstone; calcareous | WTLND |
| North America, USA, California, Rancho La Brea | Cenozoic; Quaternary, Pleistocene | Rancho La Brea Tar Seeps | Macrofossil | High | Multidominant | Alluvial | Tar seep, channels | Tar, sand, silt | CHLSP, O |
| North America, USA, California, Rancho La Brea | Cenozoic; Quaternary, Pleistocene | Rancho La Brea Tar Seeps | Macrofossil | High | Multidominant | Alluvial | Tar seep, channels | Tar, sand, silt | CHLSP, O |
| North America, USA, New Jersey | Mesozoic; Cretaceous | Navesink Fm.; Hornerstown Fm. | Mixed | High | Multidominant | Marine | Offshore | Glauconitic sandstone and silty sandstone | MAR |
| North America, USA, New Jersey | Mesozoic; Cretaceous | Hornerstown Fm. | Macrofossil | ? | ? | Paralic | ? | ? | MAR, PESH |
| Africa, Kenya, Lothagam | Cenozoic; Tertiary, Miocene | Nawata Fm. | Mixed | High | Multidominant | Alluvial | Meandering channel | Shale | CHLSP |
| North America, USA, Utah, Cleveland | Mesozoic; Jurassic | Morrison Fm. | Macrofossil | High | Multidominant | Alluvial | Marsh | Siltstone, calcareous | WTLND |
| North America, Canada, Alberta | Mesozoic; Cretaceous | Horseshoe Canyon Fm. | Macrofossil, microfossil, mixed | High, low | Monodominant, multidominant | Alluvial, paralic | Channel sandstones, overbank mudstones | Sandstone, mudstone | CHLSP, WTLND |

*(Continued)*

| Author | Date | Title | Citation | Taxa | Source | Bonebed Name | # of Sites |
|---|---|---|---|---|---|---|---|
| Strait, S.G. | 1999 | A new Earliest Wasatchian locality from the Bighorn Basin, Wyoming. | Journal of Vertebrate Paleontology 19(supplement to 3):79A. | lm, sm | Ab | Castle Gardens | 1 |
| Stucky, R.K., L. Krishtalka, and A.D. Redline. | 1990 | Geology, vertebrate fauna, and paleoecology of the Buck Spring Quarries (early Eocene, Wind River Formation), Wyoming. | Pp. 169–186 *in* Dawn of the age of mammals in the northern part of the Rocky Mountain interior, North America. Special paper 243. T.M. Bown and K.D. Rose, eds. Geological Society of America, Boulder, Colorado. | a, b, f, lm, ndr, sm | Ms | Buck Spring Quarries | 2+ |
| Suarez, C.A. | 2004 | Taphonomy and rare earth element geochemistry of the *Stegosaurus* sp. at the Cleveland Lloyd dinosaur quarry. | Geological Society of America, Abstracts with Programs 36:97. | d | Ab | Cleveland-Lloyd Quarry | 1 |
| Sullivan, C., and R.R. Reisz. | 1999 | The age, depositional environment and tetrapod paleofauna of the Lower Permian fissure fills at Richards Spur, Oklahoma. | Journal of Vertebrate Paleontology 19(supplement to 3):79A. | a, f, ndr | Ab | Dolese Brothers Limestone Quarry; Fort Sill | 1 |
| Sullivan, C.S., and R.R. Reisz. | 2002 | Lower Permian fissure deposits in the Slick Hills, Oklahoma, the oldest known fossiliferous paleokarst. | Journal of vertebrate Paleontology 22(supplement to 3):112A. | a, f, ndr | Ab | Bally Mountain Site; Dolese Brothers Quarry | 2 |
| Sun, A.-L. | 1978 | On the occurrence of *Parakannemeyeria* in Sinkiang. | Memoirs of the IVPP 13:47–54. | ndr | Ms | ? | 1+ |
| Sundell, K.A. | 1999 | Taphonomy of a multiple *Poebrotherium* kill site: An *Archeotherium* meat cache. | Journal of Vertebrate Paleontology 19(supplement to 3):79A. | lm | Ab | ? | 1 |
| Sykes, J.H. | 1977 | British Rhaetic bone-beds. | Mercian Geologist 6:197–240. | f | Ms | Basal Bone-Bed; The Clough; The Bone-Bed | 2+ |
| Tapanila, L., E. Roberts, M. Bouare, F. Sissoko, and M. O'Leary. | 2004 | Bivalve borings in phosphatic coprolites and bone, Cretaceous-Paleogene, northwestern Mali. | Palaios 19:565–573. | f, lm, ndr | Ms | Mali 20; Mali 8; Mali 18; Mali 19 | 4 |
| Taylor, M.A. | 1995 | Fossil reptiles from the Jurassic of the Isle of Skye, Scotland. | Journal of Vertebrate Paleontology 15(supplement to 3):55A. | a, d, ndr, sm | Ab | ? | 2+ |

| Location | Period / Epoch | Stratigraphy | Element Size | Diversity | Relative Abundance | Depositional Setting | Depositional Environment | Facies | Paleoenv Code |
|---|---|---|---|---|---|---|---|---|---|
| North America, USA, Wyoming, Bighorn Basin | Cenozoic; Tertiary, Eocene, Wasatchian | | Microfossil | High | Multidominant | Alluvial | Overbank | ? | PSOL |
| North America, USA, Wyoming | Cenozoic; Tertiary, Eocene | Wind River Fm. | Macrofossil | High | Multidominant | Alluvial | Pond | ? | WTLND |
| North America, USA, Utah, Cleveland | Mesozoic; Jurassic | Morrison Fm. | Macrofossil | High | Multidominant | Alluvial | Waterhole, drought, flash flood | Siltstone, calcareous | PSOL, WTLND |
| North America, USA, Oklahoma, Fort Sill, Richards Spur | Paleozoic; Permian | | Microfossil | High | Multidominant | Karst | Fissure fill | Limestone, karst | O |
| North America, USA, Oklahoma | Paleozoic; Permian | ? | Microfossil | High | Multidominant | Karst | Fissure fill | Limestone, karst | O |
| Asia, China | Paleozoic; Permian | ? | Macrofossil | Low | Monodominant | ? | ? | ? | U |
| North America, USA, Wyoming | Cenozoic; Tertiary, Oligocene, Orellan | White River Fm. | Macrofossil | Low | ? | Alluvial | Overbank | ? | PSOL |
| Europe, United Kingdom, England, Somerset, Blue Anchor Bay | Mesozoic; Triassic, Rhaetian | Brassington Fm. ? | Microfossil | High | Multidominant | Paralic | ? | Sandstone, mudstone | MAR, PESH |
| Africa, Mali | Mesozoic, Cenozoic; Cretaceous, Tertiary, Paleocene, Eocene | ? | Mixed | High | Multidominant | Marine | Shallow marine lags | phosphatic conglomer- ates | MAR |
| Europe, United Kingdom, Scotland | Mesozoic; Jurassic | Bathonian | Microfossil | High | Multidominant | Paralic | ? | ? | PESH |

*(Continued)*

| Author | Date | Title | Citation | Taxa | Source | Bonebed Name | # of Sites |
|---|---|---|---|---|---|---|---|
| Terry, D.O. Jr. | 1996 | Stratigraphy, paleopedology and depositional environment of the Conata Picnic Ground bone bed (Orellan), Brule Formation, Badlands National Park, South Dakota. | Geological Society of America, Abstracts with Programs, Rocky Mountain Section 28:4. | lm | Ab | Conata Picnic Ground Bone Bed; Big Pig Dig Bone Bed | 2 |
| Theobald, N. | 1951 | Presence d'un bonebed a la base du calcaire a ceratites. | Compte Rendus Hebomadaires des Seances de L'Academie des Sciences 233:1377–1378. | f | Ab | ? | 1 |
| Therrien, F. | 2005 | Paleoenvironments of the latest Cretaceous (Maastrichtian) dinosaurs of Romania: Insights from fluvial deposits and paleosols of the Transylvanian and Haţeg Basins. | Palaeogeography, Palaeoclimatology, Palaeoecology 218:15–56. | d | Ms | ? | 4+ |
| Therrien, F., and D.E. Fastovsky. | 2000 | Paleoenvironments of early theropods, Chinle Formation (Late Triassic), Petrified Forest National Park, Arizona. | Palaios 15:194–211. | d | Ms | Dinosaur Wash; Dinosaur Hill; Dinosaur Hollow | 3 |
| Tidwell, W.D., B.B. Britt, and S.R. Ash. | 1998 | Preliminary floral analysis of the Mygatt-Moore Quarry in the Jurassic Morrison Formation, west-central Colorado. | Modern Geology 22:314–378. | d | Ms | Mygatt-Moore Quarry | 1 |
| Trexler, D., and F.G. Sweeney. | 1995 | Preliminary work on a recently discovered ceratopsian (Dinosauria: Ceratopsidae) bonebed from the Judith River Formation of Montana suggests the remains are of *Ceratops montanus* Marsh. | Journal of Vertebrate Paleontology 15(supplement to 3):57A. | d | Ab | Ceratopsian Bonebed | 1 |
| Triebold, M. | 1997 | The Sandy Site: Small dinosaurs from the Hell Creek Formation of South Dakota. | Pp. 245–248 *in* Dinofest international, proceedings of a symposium sponsored by Arizona State University. D.L. Wolberg, E. Stump, and G.D. Rosenberg, eds. Academy of Natural Sciences, Philadelphia. | d | Ms | Sandy Site | 1 |
| Tumarkin, A.R., D.H. Tanke, and L.L. Malinconico, Jr. | 1997 | Sedimentology, taphonomy, and faunal review of a multigeneric bonebed (bone bed 47) in the Dinosaur Park Formation Campanian of southern Alberta, Canada. | Geological Society of America, Abstracts with Programs 29:86. | d | Ab | Bone Bed 47 | 1 |

| Location | Period / Epoch | Stratigraphy | Element Size | Diversity | Relative Abundance | Depositional Setting | Depositional Environment | Facies | Paleoenv Code |
|---|---|---|---|---|---|---|---|---|---|
| North America, USA, South Dakota, Badlands National Park | Cenozoic; Tertiary, Oligocene | Brule Fm., Scenic Mbr. | Macrofossil | ? | ? | Alluvial | Overbank, pond, paleosol | Mudstone | PSOL, WTLND |
| Europe, France, Saar and Vosges massif regions | Mesozoic; Triassic | | Microfossil | High | Multidominant | Marine | Open marine | Limestone | MAR |
| Europe, Romania, Transylvanian and Hațeg Basins | Mesozoic; Cretaceous, Maastrichtian | Red Continental Strata, Densuș-Ciula Fm., Pui beds, Sânpetru Fm. | Macrofossil, microfossil | High | Multidominant | Alluvial | Channel, overbank | Sandstone, mudstone, caliche | CHLSP, PSOL, WTLND, |
| North America, USA, Arizona, Petrified Forest National Park | Mesozoic; Triassic | Chinle Fm. | Macrofossil | High | Multidominant | Alluvial | Overbank, paleosol | Mudstone | PSOL |
| North America, USA, Colorado | Mesozoic; Jurassic | Morrison Fm. | Mixed | High | Multidominant | Alluvial, lacustrine | Lacustrine | Mudstone, conglomerate | LAC |
| North America, USA, Montana | Mesozoic; Cretaceous | Judith River Fm. | Macrofossil | ? | ? | Alluvial | ? | ? | U |
| North America, USA, South Dakota | Mesozoic; Cretaceous | Hell Creek Fm. | Mixed | High | Multidominant | Alluvial | Channel or splay | Sandstone | CHLSP |
| North America, Canada, Alberta, Dinosaur Provincial Park | Mesozoic; Cretaceous, Campanian | Dinosaur Park Fm. | Macrofossil | High | ? | Alluvial | Channel lag | Sandstone | CHLSP |

*(Continued)*

| Author | Date | Title | Citation | Taxa | Source | Bonebed Name | # of Sites |
|---|---|---|---|---|---|---|---|
| Turnbull, W.D., and D.M. Martill. | 1988 | Taphonomy and preservation of a monospecific titanothere assemblage from the Washakie Formation (Late Eocene), southern Wyoming: An ecological accident in the fossil record. | Palaeogeography, Palaeoclimatology, Palaeoecology 63:91–108. | lm | Ms | ? | 1 |
| Van Huet, S. | 1999 | The taphonomy of the Lancefield swamp megafaunal accumulation, Lancefield, Victoria. | Records of the Western Australian Museum, 57(supplement):331–340. | lm | Ms | Mayne Site; Classic Site; South Site | 3 |
| Van Itterbeeck, J., P. Bultynck, G.W. Li, and N. Vandenberghe. | 2001 | Stratigraphy, sedimentology and paleoecology of the dinosaur-bearing Cretaceous strata at Dashuiguo (Inner Mongolia, People's Republic of China). | Bulletin de L'Institut Royal Des Sciences Naturelles de Belgique. Sciences de la Terre 71(supplement):51–70. | d | Ms | ? | 2+ |
| Van Valkenburgh, B., F. Grady, and B. Kurtén. | 1990 | The Plio-Pleistocene cheetah-like cat *Miracinonyx inexpectatus* of North America. | Journal of Vertebrate Paleontology 10:434–454. | lm | Ms | Hamilton Cave | 1 |
| Varricchio, D.J. | 1995 | Taphonomy of Jack's Birthday Site, a diverse dinosaur bonebed from the Upper Cretaceous Two Medicine Formation of Montana. | Palaeogeography, Palaeoclimatology, Palaeoecology 114:297–323. | d | Ms | Jack's Birthday Site | 1 |
| Varricchio, D.J. | 2001 | Gut contents from a Cretaceous tyrannosaurid: Implications for theropod dinosaur digestive tracts. | Journal of Paleontology 75:401–406. | d | Ms | Old Trail Museum Locality L-6; Bob Tuesday's Site | 1 |
| Varricchio, D.J., and J.R. Horner. | 1993 | Hadrosaurid and lambeosaurid bone beds from the Upper Cretaceous Two Medicine Formation of Montana: Taphonomic and biologic implications. | Canadian Journal of Earth Sciences 30:997–1006. | d | Ms | Camposaur; West Hadrosaur Bonebed; Westside Quarry; Blacktail Creek North; Lambeosite; Jack's Birthday Site | 3 |
| Varricchio, D.J., F. Jackson, J.J. Borkowski, and J.R. Horner. | 1997 | Nest and egg clutches of the dinosaur *Troodon formosus* and the evolution of avian reproductive traits. | Nature 385:247–250. | d | Ms | multiple sites | 8 |

| Location | Period / Epoch | Stratigraphy | Element Size | Diversity | Relative Abundance | Depositional Setting | Depositional Environment | Facies | Paleoenv Code |
|---|---|---|---|---|---|---|---|---|---|
| North America, USA, Wyoming | Cenozoic; Tertiary, Eocene | Washakie Fm. | Macrofossil | Low | Monodominant | Alluvial | Channel | Sandstone (chaotic) | CHLSP |
| Australia, Victoria, Lancefield | Cenozoic; Quaternary | Lancefield swamp deposits | Macrofossil | High | Multidominant | Alluvial | Channel | Sandstone | CHLSP |
| Asia, China, Mongolia, Inner Mongolia, Dashuiguo | Mesozoic; Cretaceous | Dashuiguo Fm. | Macrofossil | High, low | Monodominant, multidominant | Alluvial | Channel lag | Pebbly sandstone | CHLSP |
| North America, USA, West Virginia | Cenozoic; Tertiary, Pliocene, Quaternary, Pleistocene | | Macrofossil | ? | ? | Alluvial | Den | ? | O |
| North America, USA, Montana, Glacier County, Badger Creek | Mesozoic; Cretaceous | Two Medicine Fm. | Mixed | High | Multidominant | Alluvial | Overbank, pond | Mudstone | WTLND |
| North America, USA, Montana, Teton County, Seven Mile Hill Badlands | Mesozoic; Cretaceous, Campanian | Two Medicine Fm. | Macrofossil | Low | Multidominant | Alluvial | Floodplain pond | claystone, calcitic nodules | WTLND |
| North America, USA, Montana | Mesozoic; Cretaceous | Two Medicine Fm. | Macrofossil | Low | Monodominant | Alluvial | Splay, overbank, pond | Mudstone | CHLSP, WTLND |
| North America, USA, Montana | Mesozoic; Cretaceous | Two Medicine Fm. | Macrofossil | Low | Monotaxic | Alluvial | Overbank, paleosol | Mudstone | PSOL |

*(Continued)*

| Author | Date | Title | Citation | Taxa | Source | Bonebed Name | # of Sites |
|---|---|---|---|---|---|---|---|
| Vaughn, P.P. | 1966 | *Seymouria* from the Lower Permian of southeastern Utah, and possible sexual dimorphism in that genus. | Journal of Paleontology 40:603–612. | a | Ms | *Seymouria* Site | 1 |
| Vergoossen, J.M.J. | 1999 | Late Silurian fish microfossils from Helvetesgraven, Skåne (southern Sweden). | Geologie en Mijnbouw 78:267–280. | f | Ab | ? | 1 |
| Vickaryous, M.K., M.J. Ryan, J.G. Bell, and D.A. Eberth. | 2000 | An excavation of a hadrosaur (Ornithischia) bone bed from the Horseshoe Canyon Formation (Early Maastrichtian), Alberta, Canada, as a cost recovery, public participation program. | Canadian Paleontology Conference, Programs and Abstracts 5:34. | d | Ab | ? | 1 |
| Visser, J. | 1986 | Sedimentology and taphonomy of a Styracosaurus bonebed in the Late Cretaceous Judith River Formation, Dinosaur Provincial Park, Alberta. | Masters thesis, University of Calgary, Calgary. | d | Th | Bonebed 42 | 1 |
| von Huene, E. | 1933 | Zur kenntnis de Württembergischen Rätbonebeds mit Zahnfunden neuer Säuger und säugerähnlicher Reptilien. | Jahreshefte des Vereins fur Vaterldndischen Naturkunde in Wuerttemberg 1933:65–128. | ndr | Ms | ? | 2+ |
| Von Koenigswald, W., A. Braun, and T. Pfeiffer. | 2004 | Cyanobacteria and seasonal death: A new taphonomic model for the Eocene Messel Lake. | Paläontologische Zeitschrift 78:417–424. | f, lm, ndr, sm | Ms | Messel Lake | 1+ |
| Voorhies, M.R. | 1969 | Taphonomy and population dynamics of an early Pliocene vertebrate fauna, Knox County, Nebraska. | Contributions to Geology special paper 1. University of Wyoming, Laramie. | lm | Ms | Verdigre Quarry | 1 |
| Voorhies, M.R. | 1981 | Ancient ashfall creates Pompeii of prehistoric animals. | National Geographic Research 159:66–75. | lm | Ms | Ashfall Fossil Beds | 1 |
| Voorhies, M.R. | 1985 | A Miocene rhinoceros herd buried in a volcanic ash. | National Geographic Society, Research Reports 19:671–688. | lm | Ms | Poison Ivy Quarry | 1 |
| Voorhies, M.R. | 1992 | Ashfall: Life and death at a Nebraska waterhole ten million years ago. | Museum Notes, University of Nebraska State Museum 81:4 pp. | lm | Ms | Poison Ivy Quarry | 1 |

| Location | Period / Epoch | Stratigraphy | Element Size | Diversity | Relative Abundance | Depositional Setting | Depositional Environment | Facies | Paleoenv Code |
|---|---|---|---|---|---|---|---|---|---|
| North America, USA, Utah | Paleozoic; Permian | Organ Rock Shale | Macrofossil | Low | Monotaxic | Alluvial | Overbank mudstone | Mudstone | WTLND |
| Europe, Sweden, Övedskloster (Skane) | Paleozoic; Silurian | Snajdri Conodont Interval Zone | Microfossil | High | Multidominant | Marine | ? | ? | MAR |
| North America, Canada, Alberta, Bleriot Ferry | Mesozoic; Cretaceous, Maastrichtian | Horseshoe Canyon Fm. | Macrofossil | Low | Monodominant | Paralic | Swamp margin | Carbonaceous mudstone | WTLND |
| North America, Canada, Alberta, Dinosaur Provincial Park | Mesozoic; Cretaceous | Judith River Fm. | Macrofossil | High | Multidominant | Alluvial | Channel | Mixed sandstone and mudstone (IHS) | CHLSP |
| Europe, Germany | Mesozoic; Triassic | | Microfossil | ? | ? | ? | ? | ? | U |
| Europe, Germay, Messel near Darmstadt | Cenozoic; Tertiary, Eocene | Messel Fm. | Macrofossil | High | Multidominant | Lacustrine | Lacustrine | Oil shale | LAC |
| North America, USA, Nebraska, Knox County | Cenozoic; Tertiary, Pliocene | Valentine Fm. | Mixed | High | Multidominant | Alluvial | Channel | Gravelly sand | CHLSP |
| North America, USA, Nebraska, Orchard | Cenozoic; Tertiary, Miocene | | Macrofossil | High | Monodominant | Alluvial | Overbank, pond | Volcanic ash | WTLND |
| North America, USA, Nebraska | Cenozoic; Tertiary, Miocene | Ash Hollow Fm. | Macrofossil | High | Monodominant | Alluvial | Overbank, pond | Volcanic ash | WTLND |
| North America, USA, Nebraska | Cenozoic; Tertiary, Miocene | Ash Hollow Fm. | Macrofossil | High | Monodominant | Alluvial | Overbank, pond, ashfall | Volcanic ash | WTLND |

*(Continued)*

| Author | Date | Title | Citation | Taxa | Source | Bonebed Name | # of Sites |
|---|---|---|---|---|---|---|---|
| Walcott, C.D. | 1892 | Preliminary notes on the discovery of a vertebrate fauna in Silurian (Ordovician) strata. | Geological Society of America, Bulletin 3:153–172. | f | Ms | Harding Quarry; Harding Sandstone | 1 |
| Walsh, S.A. | 1999 | A new Mio-Pliocene marine bonebed from north-central Chile. | Journal of Vertebrate Paleontology 19(supplement to 3):82A. | lm | Ab | ? | 1 |
| Webb, D.S., B.J. MacFadden, and J.A. Baskin. | 1981 | Geology and paleontology of the Love Bone Bed from the Late Miocene of Florida. | American Journal of Science 281:513–544. | a, b, f, lm, ndr, sm | Ms | Love Bone Bed | 1 |
| Webb, M.W. | 1999 | Hewett's foresight one: A new and unusual Lancian mammal locality, southwestern Bighorn Basin, Wyoming. | Journal of Vertebrate Paleontology 19(supplement to 3):83A. | d | Ab | Hewitt's Foresight One | 1 |
| Wedel, M. | 2000 | New material of sauropod dinosaurs from the Cloverly Formation. | Journal of Vertebrate Paleontology 20(supplement to 3):77A. | d | Ab | Wolf Creek | 1 |
| Wegweiser, M.D. | 2000 | Barbecued bones in central Wyoming: Evidence of a wildfire near the end of the Cretaceous? | Journal of Vertebrate Paleontology 20(supplement to 3):77A. | d | Ab | ? | 1 |
| Weil, A. | 1991 | Biostratigraphy and correlation of a microvertebrate site, Brewster County, Texas. | Journal of Vertebrate Paleontology 11(supplement to 3):61A. | d | Ab | Terlingua Microvertebrate Site 1 | 1 |
| Wells, J.W. | 1944 | Middle Devonian bone beds of Ohio. | Geological Society of America, Bulletin 55:273–302. | f | Ms | ? | 7 |
| Westgate, J.W. | 1989 | Lower vertebrates from an estuarine facies of the Middle Eocene Laredo Formation (Clairborne Group) Webb County, Texas. | Journal of Vertebrate Paleontology 9:282–294. | a, f, ndr | Ms | ? | 2+ |
| Westgate, J.W. | 1990 | Uintan land mammals (excluding rodents) from an estuarine facies of the Laredo Formation (Middle Eocene, Clairborne Group) of Webb County, Texas. | Journal of Paleontology 64:454–468. | sm | Ms | ? | 2+ |
| Westgate, J.W., and C.T. Gee. | 1990 | Paleoecology of a middle Eocene mangrove biota (vertebrates, plants and invertebrates) from southwest Texas. | Palaeogeography, Palaeoclimatology, Palaeoecology 78:163–177. | a, f, ndr | Ms | TMM 42486 | 1 |

| Location | Period / Epoch | Stratigraphy | Element Size | Diversity | Relative Abundance | Depositional Setting | Depositional Environment | Facies | Paleoenv Code |
|---|---|---|---|---|---|---|---|---|---|
| North America, USA, Colorado, Canyon City | Paleozoic; Ordovician, Silurian | Harding Sandstone | Microfossil | High | Multidominant | Marine | Nearshore | Sandstone, shale | MAR |
| South America, Chile | Cenozoic; Tertiary, Miocene, Pliocene | | Macrofossil | High | ? | Marine | Nearshore | Sandstone, coquina | MAR |
| North America, USA, Florida | Cenozoic; Tertiary, Miocene | Alachua Fm. | Microfossil | High | Multidominant | Alluvial | Channel | Sandstone (phosphatic) | CHLSP |
| North America, USA, Wyoming, Bighorn Basin | Mesozoic; Cretaceous, Maastrichtian | Lance Fm. | Microfossil | High | ? | Alluvial | Channel | Sandstone | CHLSP |
| North America, USA, Montana | Mesozoic; Cretaceous, Aptian, Albian | Cloverly Fm. | Macrofossil | ? | ? | Alluvial | Overbank | Mudstone (claystone) | WTLND |
| North America, USA, Wyoming, Elkhorn Basin | Mesozoic; Cretaceous, Campanian | Meeteesee Fm. | Macrofossil | High | Multidominant | Paralic | ? | Mudstone | U |
| North America, USA, Texas, Big Bend National Park | Mesozoic; Cretaceous | Aguja Fm. | Microfossil | High | Multidominant | Paralic | Estuarine | ? | PESH |
| North America, USA, Ohio | Paleozoic; Devonian | Delaware Fm.; Columbus Fm. | Microfossil | High | ? | Marine | Nearshore | Mudstone; sandstone | MAR |
| North America, USA, Texas, Webb County | Cenozoic; Tertiary, Eocene | Laredo Fm. | Microfossil | High | Multidominant | Paralic | Estuarine channel | Sandstone; minor mudstone | CHLSP |
| North America, USA, Texas, Webb County | Cenozoic; Tertiary, Eocene | Laredo Fm. | Microfossil | High | Multidominant | Paralic | Estuarine | ? | CHLSP |
| North America, USA, Texas, Webb County | Cenozoic; Tertiary, Eocene | Laredo Fm. | Microfossil | High | Multidominant | Paralic | Estuarine channel | Sandstone | CHLSP |

*(Continued)*

| Author | Date | Title | Citation | Taxa | Source | Bonebed Name | # of Sites |
|---|---|---|---|---|---|---|---|
| Whistler, D.P., and S.D. Webb. | 2005 | New goatlike camelid from the Late Pliocene of Tecopa Lake Basin, California. | Natural History Museum of Los Angeles County, Contributions in Science 503:1–40. | b, lm, sm | Ms | Standing Camel Quarry | 1 |
| White, P.D., D.E. Fastovsky, and P.M. Sheehan. | 1998 | Taphonomy and suggested structure of the dinosaurian assemblage of the Hell Creek Formation (Maastrichtian), eastern Montana and western North Dakota. | Palaios 13:41–51. | d | Ms | multiple sites | 2+ |
| Whybrow, P.J., and H.A. McClure. | 1981 | Fossil mangrove roots and paleoenvironments of the Miocene of the eastern Arabian Peninsula. | Palaeogeography, Palaeoclimatology, Palaeoecology 32:213–225. | lm | Ms | ? | 1 |
| Wilson, M.V.H. | 1996 | Taphonomy of a mass death layer of fishes in the Paleocene Paskapoo Formation at Joffre Bridge, Alberta, Canada. | Canadian Journal of Earth Science 33:1487–1498. | f | Ms | Fish Layer | 1 |
| Wilson, M.V.H., and D.G. Barton. | 1996 | Seven centuries of taphonomic variation in Eocene freshwater fishes preserved in varves: Paleoenvironments and temporal averaging. | Paleobiology + D380 22:535–542. | f | Ms | ? | 2+ |
| Windom, K.E., and J.M. Poort. | 1972 | Environmental interpretation of a Pliocene bone bed in northwestern Kansas. | Bulletin of the Georgia Academy of Science 30:81–82. | lm | Ab | ? | 1 |
| Winkler, D.A. | 1989 | Reeves bonebed: Mammalian predator-prey interactions in the early Oligocene, Trans Pecos, Texas. | Geological Society of America, Abstracts with Programs 21:1. | lm | Ab | Reeves Bonebed | 1 |
| Winkler, D.A., L.L. Jacobs, and P.A. Murry. | 1997 | Jones Ranch: An Early Cretaceous sauropod bone bed in Texas. | Journal of Vertebrate Paleontology 17(supplement to 3):85A. | d | Ab | Jones Ranch | 1 |
| Winkler, D.A., P.A. Murry, L.L. Jacobs, W.R. Downs, J.R. Branch, and P. Trundel. | 1988 | The Proctor Lake dinosaur locality, Lower Cretaceous of Texas. | Hunteria 2:1–8. | d | Ms | Proctor Lake | 1 |
| Wojcik, Z. | 1961 | Bone-bed sedimentation in the Tatra Mountains. | Die Hoehle 2/3:103–104. | lm, sm | Ab | ? | 2+ |

| Location | Period / Epoch | Stratigraphy | Element Size | Diversity | Relative Abundance | Depositional Setting | Depositional Environment | Facies | Paleoenv Code |
|---|---|---|---|---|---|---|---|---|---|
| North America, USA, California, Tecopa Lake Basin | Cenozoic; Tertiary, Pliocene | Lake Tecopa Allogroup | Macrofossil | Low | Monodominant | Lacustrine | Playa, lake | Gypsiferous mudstone | LAC |
| North America, USA, Montana & North Dakota | Mesozoic; Cretaceous, Maastrichtian | Hell Creek Fm. | Macrofossil, microfossil, mixed | High, low | Multidominant | Alluvial | Channel, splay, overbank, pond, paleosol | ? | CHLSP, PSOL, WTLND |
| Africa, Saudia Arabia, Eastern Arabian Peninsula, Jabal Barakah | Cenozoic; Tertiary, Miocene | Dam Fm.; Hofuf Fm. | Macrofossil | Low | Multidominant | Paralic | Estuarine channel, sheet flood | conglomerate | CHLSP, PESH |
| North America, Canada, Alberta, Joffre Bridge | Cenozoic; Tertiary, Paleocene, Tiffanian | Paskapoo Fm. | Macrofossil | Low | Monodominant | Lacustrine | Lacustrine | Mudstone | LAC |
| North America, Canada, British Columbia, Horsefly | Cenozoic; Tertiary, Eocene | Horsefly beds | Macrofossil | Low | Multidominant | Lacustrine | Lacustrine | Varved mudstone | LAC |
| North America, USA, Kansas, Phillipsburg | Cenozoic; Tertiary, Pliocene | Ogallala Fm. | Macrofossil | Low | Monotaxic | Alluvial | Overbank | Sandstone | WTLND |
| North America, USA, Texas | Cenozoic; Tertiary, Oligocene | | Macrofossil | Low | Monodominant | Alluvial | Overbank, den | Mudstone | O |
| North America, USA, Texas, Hood County | Mesozoic; Cretaceous | Twin Mountains Fm. | Macrofossil | Low | Monodominant | Alluvial | Channel, splay | Sandstone, mudstone | CHLSP |
| North America, USA, Texas, Proctor Lake | Mesozoic; Cretaceous | Twin Mountains Fm. | Macrofossil | Low | Monodominant | Alluvial | Overbank | Mudstone, sandstone | WTLND |
| Europe, Poland, Tatra | Cenozoic; Quaternary, Pleistocene, Holocene | | Macrofossil | ? | ? | Alluvial | Channel | ? | CHLSP |

*(Continued)*

| Author | Date | Title | Citation | Taxa | Source | Bonebed Name | # of Sites |
|---|---|---|---|---|---|---|---|
| Wolfe, D., Beekman, S., McGuiness, D., Robira, T., and Denton, R. | 2004 | Taphonomic characterization of a *Zuniceratops* bone bed from the Middle Cretaceous (Turonian) Moreno Hill Formation. | Journal of Vertebrate Paleontology 24(supplement to 3):131A. | d | Ab | Haystack Butte Bone Bed | 1 |
| Wolff, R.G. | 1973 | Hydrodynamic sorting and ecology of a Pleistocene mammalian assemblage from California (U.S.A.). | Palaeogeography, Palaeoclimatology, Palaeoecology 13:91–101. | sm | Ms | ? | 7 |
| Wood, J.M., R.G. Thomas, and J. Visser. | 1988 | Fluvial processes and vertebrate taphonomy: The Upper Cretaceous Judith River Formation, south-central Dinosaur Provincial Park, Alberta, Canada. | Palaeogeography, Palaeoclimatology, Palaeoecology 66:127–143. | d | Ms | BB 42; BB 10 | 2+ |
| Worthy, T.H., and R.N. Holdaway. | 1996 | Quaternary fossil faunas, overlapping taphonomies, and palaeofaunal reconstruction in North Canterbury, South Island, New Zealand. | Journal of the Royal Society of New Zealand 26:275–361. | b, f, ndr, sm | Ms | Omihi Stream; Pyramid Valley; Glencrieff Swamp; North Dean; Waikari Cave | 6+ |
| Wroblewski, A., and B. Mclaurin. | 1999 | A tale of two bonebeds: Comparison of an Upper Triassic and Upper Cretaceous assemblage. | Journal of Vertebrate Paleontology 19(supplement to 3):86A. | d | Ab | ? | 2 |
| Xia, W.J., and X.H. Li. | 1988 | The burial environment of dinosaurs and characteristics of lithofacies and paleogeography. | *In* The middle Jurassic dinosaur fauna from Dashanpu, Zigong, Sichuan, volume 5. Sichuan Publishing House of Science and Technology, Chengdu, China (English summary). | d | Ms | Zigong Bonebed | 1 |
| Xu, X., J.M. Clarke, C.A. Forster, M.A. Norell, G.M. Erickson, D.A. Eberth, C. Jia, and Q. Zhoa. | 2006 | A basal tyrannosauroid dinosaur from the Late Jurassic of China. | Nature 439:715–718. | d | Ms | *Guanlong* BB | 1 |
| Zeigler, K.E., A.B. Heckert, and S.G. Lucas. | 2002 | A fire related Late Triassic vertebrate fossil assemblage from north central New Mexico. | Geological Society of America, Abstracts with Programs 34:535. | d, f, ndr | Ab | Snyder Quarry | 1 |

| Location | Period / Epoch | Stratigraphy | Element Size | Diversity | Relative Abundance | Depositional Setting | Depositional Environment | Facies | Paleoenv Code |
|---|---|---|---|---|---|---|---|---|---|
| North America, USA, New Mexico | Mesozoic; Cretaceous | Moreno Hill Fm. | Macrofossil | Low | Monodominant | Alluvial | Alluvial plain, channel | Sandstone | CHLSP, WTLND |
| North America, USA, California, Rodeo | Cenozoic; Quaternary, Pleistocene | Montezuma Fm. | Microfossil | High | Multidominant | Paralic | Channel, delta-distributary | Sand and silt, pebbly | CHLSP |
| North America, Canada, Alberta, Dinosaur Provincial Park | Mesozoic; Cretaceous | Judith River Fm. | Macrofossil, microfossil, mixed | High, low | Monodominant, multidominant | Alluvial | Channel, splay, overbank, pond, paleosol | No specific descriptions | CHLSP, PSOL, WTLND |
| Australia, New Zealand, South Island, North Canterbury, Waikari | Cenozoic; Quaternary, Pleistocene | ? | Macrofossil | High, low | Monodominant, multidominant, monotaxic | Swamps, uplands, caves, pits, ledges | Pitfs, ledges, alluvial, swamp, lake | Colluvium, sand, peat, silt, soil | WTLND, O |
| North America, USA, Utah & Wyoming, Book Cliffs | Mesozoic; Triassic, Cretaceous | Popo Agie Fm.; Castlegate Fm. | Macrofossil | ? | ? | Alluvial, paralic | ? | Mudstone | WTLND, PESH |
| Asia, China, Sichuan, Zigong | Mesozoic; Jurassic | Xiashaximiao Fm. | Macrofossil | High | Multidominant | Lacustrine alluvial | Lacustrine shoals | Sandstone | LAC |
| Asia, China, Xinjiang, Junggar Basin, Wucaiwan area | Mesozoic; Jurassic | Shishugou Fm. | Macrofossil | Low | Monodominant | Paludal | Wetland | Tuffaceous mudstone | WTLND |
| North America, USA, New Mexico, Abiquiu | Mesozoic; Triassic, Norian | Chinle Grp., Petrified Forest Fm. | Mixed | High | Multidominant | Alluvial | Floodplain wildfire debris | Wildfire debris deposit | DBF, WTLND |

*(Continued)*

| Author | Date | Title | Citation | Taxa | Source | Bonebed Name | # of Sites |
|--------|------|-------|----------|------|--------|--------------|------------|
| Zeigler, K.E., S.G. Lucas, A.B. Heckert, A.C. Henrici, and D.S Berman. | 2003 | A time-averaged tetrapod bonebed: Taphonomy of the Early Permian Cardillo Quarry, Chama Basin, north-central New Mexico. | Geological Society of America, Abstracts with Programs 35:10. | a, ndr | Ab | Cardillo Quarry | 1 |
| Zinsmeister, W.J. | 1998 | Discovery of fish mortality horizon at the K-T boundary on Seymour Island: Re-evaluation of events at the end of the Cretaceous. | Journal of Paleontology 72:556–571. | f | Ms | fish bone layer | 1 |

| Location | Period / Epoch | Stratigraphy | Element Size | Diversity | Relative Abundance | Depositional Setting | Depositional Environment | Facies | Paleoenv Code |
|---|---|---|---|---|---|---|---|---|---|
| North America, USA, New Mexico, Arroyo del Agua | Paleozoic; Permian | Cutler Fm. | Macrofossil, microfossil | High | Multidominant | Alluvial | Crevasse splay | Muddy conglomerate | CHLSP |
| Antarctica, Seymour Island | Cenozoic; Tertiary, Paleocene | Lopez de Bertodano Fm. | Mixed | ? | ? | Marine | Offshore, below storm wave base | Mudstone | MAR |

# CHAPTER 4

# From Bonebeds to Paleobiology: Applications of Bonebed Data

*Donald B. Brinkman, David A. Eberth, and Philip J. Currie*

## INTRODUCTION

Bonebeds are prized by paleobiologists because they yield large numbers of fossils and, thus, provide important morphologic and taxonomic data sets that are the primary basis for integrative studies of paleobehavior, paleoecology, and paleocommunity structure (Table 4.1). "What can these bones tell us about the lives and the world of extinct animals?" has been a common paleobiological query throughout the history of bonebed studies (e.g., Peterson, 1906a, 1906b; Merriam, 1909; Jordan, 1920; Cook, 1926; Wells, 1944; Shipman, 1979; Sander, 1987; Currie, 2000). However, the "leap of faith" from compilations of descriptive data to paleobiological inference can be great, and the accuracy of paleobiological and paleoecological reconstructions can vary dramatically.

Rogers and Kidwell (Chapter 1 in this volume) identify two broad categories of bonebed-generating mechanisms—biogenic and physical—and it can be reasonably argued that all bonebeds form through complex combinations of biotic and abiotic mechanisms. Accordingly, every bonebed provides an opportunity to glimpse some aspect of paleobiology. However, moving from the bonebed assemblage to a reasonable reconstruction of paleobiological phenomena involves a complex mixture of data acquisition and analysis (Eberth et al., Chapter 5 in this volume; Blob and Badgley, Chapter 6 in this volume). Even in the best of circumstances, paleobiological insights gained from the study of bonebeds in the geologic

Table 4.1. Relative usefulness of bonebeds in different kinds of paleobiological studies

| Paleobiological Application | Usefulness |
|---|---|
| Anatomy | Excellent |
| Paleospecies characterization/ taxonomy | Excellent |
| Intraspecies variation | Excellent |
| Ontogeny and growth | Excellent |
| Intraspecies behavior and physiology | Good (somewhat limited) |
| Evolution/extinction | Fair |
| Faunal studies | Excellent |
| Paleocommunity/paleoecology | Good (requires taphonomic and geologic data) |

record are of lower resolution than biological conclusions drawn from observations of death events in the modern. Quite simply, all bonebed assemblages have passed through taphonomic filters, and as a result, it is usually not possible to retrieve complete paleobiological information about the source biocoenoses from the preserved assemblages.

When compared in detail, bonebeds often have complex histories, and although there are some broad relationships between types of bonebeds and the kinds of paleobiological data that they can be expected to yield (Behrensmeyer, Chapter 2 in this volume; Eberth et al., Chapter 3 in this volume), each bonebed data set must be carefully compiled and examined to determine what paleobiological hypotheses can be tested and supported. For example, it is generally safe to infer that in the case of a strictly mono-taxic bonebed some aspect of the paleobiology of the contained taxon—such as breeding, nesting, and herding behaviors—has caused individuals to gather before death (see below). However, such is not necessarily the case for monodominant or multitaxic bonebeds (for definitions of these and other terms see Eberth et al., Chapter 3 in this volume), where paleoenvironmental influences may also concentrate or disperse individuals before and after death (e.g., compare "intrinsic" and "forced intrinsic" biogenic concentrations of Rogers and Kidwell [Chapter 1 in this volume], and Eberth and Getty [2005, their Fig. 15]).

This chapter reviews the variety of paleobiological inferences that can be drawn from bonebed data sets and explores the pertinent issues and biases that relate to framing testable paleobiological hypotheses (Fig. 4.1). The first section reviews the data gathered from bonebeds that are often used to characterize species (morphology and variation) and to describe patterns

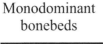

Monodominant
bonebeds

Multitaxic
bonebeds

Paleospecies
characterization
&
Paleobehavior

Paleofaunal composition
Paleocommunity structure
&
Species interactions

*Figure 4.1.* In general, monotaxic and monodominant bonebeds are more useful for species characterization, whereas multitaxic bonebeds are frequently a source of paleocommunity data.

of ontogenetic growth and sexual dimorphism. The second section explores
various characteristics of bonebed assemblages that provide information
leading to reconstructions of inter- and intraspecific paleobehaviors. The
third and final section examines the more synthetic paleobiological issues
that relate to faunal analyses and paleocommunity reconstructions. Throughout these sections, we consider the ways in which taphonomic and geological biases affect the paleobiological signals recorded in bonebeds, and how
understanding these biases allows for a reasonable reconstruction of the paleobiology of extinct organisms. Ultimately, the paleobiological inferences
drawn from bonebeds relate to fundamental questions in evolution and
ecology, including the stability of communities over time, the nature of com-
munity-level response to climate and other paleoenvironmental changes,
and the relationships between evolution and paleoenvironmental change.

In this chapter, the term "paleobiology" relates broadly to all aspects
of the biology of fossil species including (1) the biological characterization
of a paleospecies (taxonomy, morphology and morphological variation,
ontogeny, life history, behavioral patterns, intraspecies interactions);
(2) paleofaunal composition and paleocommunity organization (stratigraphically and paleogeographically); and (3) inter- and intraspecific interactions within a community. Associations of bonebed assemblages with
host sediments, fossilized microbes, plants, and invertebrates also provide
important paleobiological information, but they are beyond the scope of
this chapter. This chapter employs the bonebed terminology of Eberth et al.
(Chapter 3 in this volume), especially in the use of the terms "monotaxic,"
"multitaxic," "monodominant," and "multidominant." Lastly, we ac-
knowledge our scientific bias toward the literature on Late Paleozoic and
Mesozoic vertebrates, in particular, dinosaurs and dinosaur communities.
Although we present examples from different taxonomic, stratigraphic,

and geographic perspectives, many of the examples discussed are drawn from the literature with which we are most familiar.


## BONEBEDS AND THE CHARACTERIZATION OF SPECIES

Paleontologists are keenly aware that variation within ancient morphospecies must be taken into account if the species are to have biological validity. However, paleontological data are notoriously incomplete. Many fossil species are represented by only a few specimens and often only a single kind of skeletal element. Only in rare cases is it possible to fully address variation in an extinct species. Bonebeds are of great value for taxonomic work because they can yield multiple associated specimens of a given species. And with multiple specimens it is possible to assess patterns of morphological variation that relate to individual variation, sexual dimorphism, or ontogenetic development. However, the kind and quality of morphotaxic data varies significantly among bonebeds. For example, faunal assemblages formed during a sudden or catastrophic mortality event (e.g., flooding) may offer only a limited sample of morphological variation within a living population, whereas attritional assemblages formed over longer time spans may preserve a broader range of such variation.


### Intraspecific Variation

There are many kinds of variation within a species, including morphological variation in features that do not change with age, variation in structures during growth, and sexual variation. In any case, large numbers of specimens are usually required to quantify intraspecific variation. The Late Triassic theropod dinosaur *Coelophysis bauri* is an example of an extinct species in which morphological variation has been documented using specimens from the monodominant *Coelophysis* bonebed at Ghost Ranch, New Mexico. The *Coelophysis* assemblage at Ghost Ranch is thought to be a single population that succumbed to one or more mass mortality events linked in some way to drought (Schwartz and Gillette, 1994). Whereas fossil remains of small theropod dinosaurs are generally rare and fragmentary (e.g., Therrien and Fastovsky, 2000), the quality and abundance of material in this bonebed have made *Coelophysis* one of the best-understood dinosaurs and have been used to resolve taxonomic disagreements (ICZN, 1996). Consequently, the *Coelophysis* bonebed represents an exceptional opportunity to study individual, ontogenetic, and sexual variation within

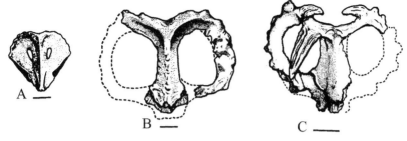

*Figure 4.2.* Ontogenetic series of parietals of the ceratopsian dinosaur *Centrosaurus* from Sampson et al. (1997). (A) juvenile; (B) subadult; (C) adult. Juvenile and subadult parietals without spikes (A and B) had been previously identified as *Monoclonius*. All specimens are from the Dinosaur Park Formation at Dinosaur Provincial Park, Alberta, Canada. Scale bar equals 10 cm. Used with permission.

a population (Colbert, 1989, 1990). Skeletal size series from the assemblage show a threefold increase in linear dimensions and demonstrate ontogenetic changes in the relative size of the orbit, length of the neck, and length of the hind limb. Large specimens of subequal size that vary in the degree of fusion of sacral vertebrate and the relative size of the skull have been interpreted as reflecting differences in gender. Variations in other features that are independent of size have been interpreted as reflecting natural morphological variation within the population.

Bonebeds have also played a critical role in establishing the range of variation in ceratopsian dinosaur species. Early studies of ceratopsians based on articulated individual specimens (e.g., Brown and Schlaikjer, 1937) underestimated the potential for variation within a given ceratopsian species. Instead, variation was misinterpreted as an expression of taxonomic diversity. However, by using data from a group of low-diversity monodominant bonebeds, Sampson et al. (1997) were able to describe ontogenetic changes in the skulls of a number of genera within the subfamily Centrosaurinae (Fig. 4.2). They showed that two of the previously recognized genera, *Brachyceratops* and *Monoclonius*, had been defined in large part on the basis of characters that are typical of juveniles. For example, *Monoclonius* had originally been distinguished largely on the basis of the lack of epoccipital spikes on the frill (Sternberg, 1940). Sampson et al. (1997) showed that spike and horn cores develop late in centrosaurine ontogeny and argued that specimens previously attributed to *Monoclonius* should be reinterpreted as juveniles of preexisting centrosaurine genera.

The skull morphology of adult ceratopsians is also variable, especially the shape of horn cores. Horn core shape was used to differentiate species in early taxonomic descriptions (e.g. Sternberg, 1940). For example, *Centrosaurus longirostris* was originally characterized in part by a nasal horn core that

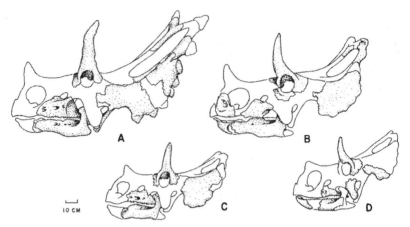

Figure 4.3. Variation in the skull of *Chasmosaurus mariscalensis* showing ontogentic and individual variation of horn core shape. (A) based on University of Texas, El Paso specimens, UTEP P.37.7.086, 046, 065, and 066; (B) based on UTEP P.37.7.062, 064, 067, 073, and 082; (C) based on UTEP P.37.7.045, 087, 088, and 091; (D) based on UTEP P.37.7.059, 072, 081, 085, and 095. From Lehman (1989), used with permission.

curves forward, whereas *C. nasicornus* was described as having a tall and straight horn core. Lehman (1989), using *Chasmosaurus* material from a low-diversity monodominant bonebed, demonstrated that horn core size and curvature is highly variable within that taxon (Fig. 4.3). Tall and straight versus curved nasal horn cores also co-occur in a *Centrosaurus apertus*–dominated bonebed in Dinosaur Provincial Park (BB 43) (Ryan et al., 2001), indicating that the shape of the nasal horn core is neither a defining feature for that species nor a reliable feature with which to diagnose other ceratopsian taxa.

Bonebed data sets have also been employed to document morphological variation in nondinosaurian vertebrate taxa. For example, Olson (1951) studied a series of skulls of the peculiar Permian amphibian *Diplocaulus* preserved in a bonebed (as defined in this volume) and documented changes in skull shape associated with size and, thus, presumably, growth. Small individuals have a narrow skull with no "horns" or short "horns" (extensions of the tabular) that extend posteriorly, whereas large individuals have a boomerang-shaped skull with "horns" that extend posterolaterally (Fig. 4.4). When large numbers of specimens from the bonebed are considered, these two morphologies are best interpreted as end points in an ontogenetic growth series. Olson recognized that five previously named species are included in this growth series, but that only a single taxon is actually present. Along these same lines, Eberth (1985) described variation in the skull of the Permian pelycosaur *Sphenacodon* from the Anderson Quarry bonebed in north-central New Mexico, and Donelan and Johnson (1997) quantified morphological variation and allometric changes in a population of selachian shark spines from the Lower Permian Craddock Bonebed of Texas.

Bonebed data sets also yield insight into sexual dimorphism in extinct species, and are therefore an important means of documenting and assessing aspects of intraspecific competition (Short and Balaban, 1994). In living vertebrates sexual dimorphism is commonly expressed in many different ways, including body size, tooth size, ornamentation, and coloration. However, documenting sexual dimorphism in extinct species often requires complex statistical and morphometric analyses of large data sets (e.g., Van Valkenburgh and Sacco, 2002). Dimorphism in the diameter of the lower tusk and other features of the limbs were interpreted as sexual dimorphic characteristics in Mead's (2000) study of 35 articulated adult skeletons of the Miocene North American rhinoceros *Teleoceras* from the Ashfall Fossil Beds of Nebraska. Sullivan et al. (2003) studied the Permian dicynodont *Diictodon* collected from numerous burrow fills and floodplain bonebeds (originally described by Smith [1987, 1993]). In their

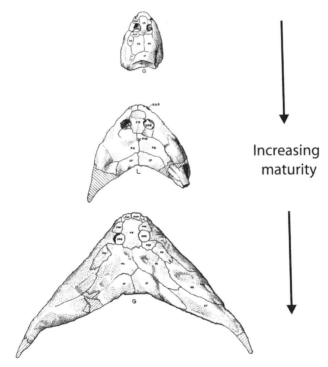

Increasing
maturity

*Figure 4.4.* Ontogenetic series (top to bottom) showing change in skull shape of the Permian amphibian *Diplocaulus*. Modified from Olson (1951). Used with permission.

interpretation of sexual dimorphism they concluded that the presumed males possessed formidable tusks and a cranial boss on the skull roof. Berman et al. (1987) described an aggregate of six articulated individuals of the Permian amphibian *Seymouria* and suggested sexual dimorphism in that genus based on varying morphological aspects of the cranium and vertebral column. In each of these studies, bonebed assemblages provided the large sample sizes required to effectively demonstrate that the dimorphs co-occurred in the source biocoenoses.

### Growth Rate

Bonebed assemblages also provide data about life-history traits for some species. For example, growth rate is a fundamental life-history trait that has implications for energy requirements and generational turnover rates. Where individuals of a species can be aged using criteria such as tooth wear or osteochronology, rates of growth can be inferred directly. For example, Voorhies (1969) was able to use tooth wear to age individuals in a population of ungulates recovered from a Pliocene bonebed. More recently, criteria have been developed using lines of arrested growth in dinosaurs to age individuals (e.g., Horner et al., 2000; Padian et al., 2001; Erickson et al., 2004; Xu et al., 2006). Where individuals can be aged directly, differences in size of successive age classes provide a measure of the rate of growth in that species.

This approach, however, cannot be applied to many extinct vertebrates because, in general, determining the age of individuals with a high degree of accuracy is not possible. In such cases indirect approaches may be employed. One approach used in the case of modern amphibians and reptiles is to identify age classes using size-frequency distributions. If taphonomic processes have not altered the size-frequency distribution of the fossil assemblage, and the mortality event was not biased in favor of a particular age of individual, any size classes that were present in the original population will arguably be preserved in the fossil assemblage. Thus, the number of size classes in a bonebed assemblage gives a measure of the number of age classes present in the living population.

Age classes may be represented by distinct size classes in extant vertebrates that have a seasonal breeding period and relatively rapid growth rates (e.g., Andrews, 1982; Halliday and Verrell, 1988). This approach is particularly applicable to data from bonebeds that result from mass mortality events, and was used by Varricchio and Horner (1993) to support their proposal that dinosaurs had very rapid growth rates. They recognized that a single subadult size class comprised of individuals 3.0 to 3.5 m long is present in many hadrosaur-dominated bonebeds from the Upper Cretaceous Two Medicine Formation of Montana. They interpreted this as a first-year age class and calculated a rate of growth for these hadrosaurs equal to that of modern ungulates. Subsequently, this interpretation of rapid growth has been supported by independent histological and osteological evidence (Horner et al., 2000).

Using computer simulations, Craig and Oertel (1966) demonstrated that age classes might also be reflected as size classes in attritional assemblages of fossil taxa. They showed that age classes are represented by distinct size classes in the death assemblage if (1) a taxon has a seasonal breeding period and/or a seasonal period of high mortality and (2) the size increase during a year's growth exceeds the range in variation in size of individuals within a single age class. The results of this simulation were validated by studies of the size-frequency distributions in the catastomid fish *Amyzon* from Eocene lake beds of British Columbia (Wilson, 1984). *Amyzon* specimens were recovered from varved lakebeds and are preferentially preserved in fine-grained layers. Thus, it appears that *Amyzon* experienced high mortality during the winter seasons when there was little or no growth. Repeated mortality of the fish during this season has resulted in a death assemblage in which size classes represent different ages.

In addition to determining growth rates of juveniles and subadults, the change in the rate of growth of adults can be interpreted from the size frequency distribution of a fossil population preserved in a bonebed assemblage. For example, it is widely recognized that in general, growth rates of large vertebrates decrease or cease once an individual becomes sexually mature. Thus, within populations of adults, size differences may reflect individual or gender variation, but not relative age. It is not surprising therefore that Varrichio and Horner (1993) were unable to recognize discrete size classes within adult hadrosaur populations.

### Stratigraphic and Paleogeographic Variation in Species

In theory, when a taxon's morphological traits are compared from multiple bonebeds, patterns of morphological variation can be documented stratigraphically and paleogeographically. Documenting such patterns has long been one of the goals for vertebrate paleontologists. For example, during the development of the punctuated equilibrium model of evolutionary change, studies of change in the patterns of variation through time in Tertiary mammals were used to test whether theoretical predictions of rates of evolutionary change have a basis in nature (Gingerich, 1977; Vrba, 1980). More recently, stratigraphically superposed bonebeds have been used to quantify patterns of change within single lineages in limited geographic areas. For example, in a study of small theropod teeth from a series of vertebrate microfossil localities in the Santonian through Maastrichtian of Alberta, Baszio (1997) observed changes in the pattern of serrations on teeth of *Richardoestesia gilmorei* and *Saurornitholestes langstoni*. He showed that the teeth of these two morphotypes are less distinct in the older Milk River Formation (Santonian) than they are in the succeeding Dinosaur Park Formation (Late Campanian). Along these same lines, Ryan (2003) studied monodominant centrosaurine bonebeds in Dinosaur Provincial Park in southern Alberta and proposed the possibility that stratigraphic changes in the taxa of ceratopsians present in the assemblages are due to anagenetic evolution of *Centrosaurus apertus* to *Styracosaurus albertensis*.

A comparison of data from geographically separated but coeval bonebeds can be used to demonstrate patterns of paleogeographic variation in a species. Among mammals, an increase in size at higher (and cooler) latitudes is sufficiently common that the pattern is known as Bergmann's rule (however see McNab [1971] and Ochocinska and Taylor [2003] for more complex descriptions of patterns and interpretations of change in size with latitude). In contrast, among ectothermic mesoreptiles, such as crocodiles and turtles, a decrease in size with increasing latitude is the expected pattern (Hutchison, 1982). Such observations from the modern have been used to assess geographically related morphological variation in the ceratopsian genus *Pachyrhinosaurus* (Langston, 1975; Tanke, 1988). Langston compared *Pachyrhinosaurus* material from two localities: the Scabby Butte bonebed in southern Alberta and an *Edmontosaurus*-dominated bonebed locality about 200 km to the north, near Drumheller. The discovery of a *Pachyrhinosaurus* bonebed 600 km farther to the northwest, near Grande Prairie, Alberta, allowed Tanke (1988) to note that these more northerly specimens are smaller than those from southern Alberta, and that they also differ in the development of a heavily constructed "unicorn" horn core on the anterior portion of the parietal bar. A *Pachyrhinosaurus* bonebed from the North Slope of Alaska (Fiorillo, 2004), when fully described, should eventually yield more data with which to test this size-paleolatitude hypothesis.

## BONEBEDS AND AGGREGATION PALEOBEHAVIOR

Aggregation paleobehavior emerged as a dominant theme in our review of paleobehavioral inferences drawn from bonebed assemblages. We regard an aggregation as a group of animals that live in close association during some portion of their lives, and respond similarly to stimuli. In this section, we explore different kinds of aggregation-related inferences based on bonebeds, discuss the methods used to develop them, and identify what we believe are some biases in these interpretations.

Commonly cited examples of aggregation paleobehavior inferred from bonebeds include

- intraspecific aggregations for purposes of reproduction and parenting (e.g., breeding, nesting, and juvenile care),
- intraspecific aggregations resulting from some degree of social interaction (e.g., herds, schools, flocks, and/or family groups), and
- intra- and interspecific aggregations that develop in response to resource availability (e.g., feeding grounds, water holes) or extreme/unusual environmental conditions (e.g., drought, flood, fire).

Aggregation paleobehavior is frequently cited as an influence in the formation of bonebeds (Rogers and Kidwell, Chapter 1 in this volume), especially in those cases where a site contains enormous numbers of individuals of one species (e.g., Jordan, 1920), spectacularly preserved specimens (e.g., Voorhies, 1985), or both (e.g., Merriam, 1909; Dalquest and Mamay, 1963). Accordingly, the existence of a monotaxic and monodominant bonebed, alone, is sometimes cited as the primary evidence for aggregation paleobehavior in a taxon that experienced a mass kill. However, because a monotaxic and monodominant bonebed can form as a result of repeated mortality over time among selected members of a community (attrition rather than mass kill [e.g., Laury, 1980; Sander, 1992; Rogers and Kidwell, Chapter 1 in this volume]), independent geological or taphonomic evidence is required before it can be accepted that mass mortality of an aggregation occurred. Furthermore, researchers should also be critical of using the presence of the death assemblage itself as the primary evidence for normal or day-to-day aggregation paleobehavior. Paleoenvironmental stresses such as floods, fire, drought, volcanic eruptions, and even disease often result in temporary and atypical aggregations and mass-death events among modern animals, and the same is likely to have occurred in the past. In short, the presence of a bonebed does not necessarily support the hypothesis that the individuals comprising it died together or lived together in day-to-day aggregates or complex social groups (Voorhies, 1985, p. 683). Such interpretations require a better understanding of the cause of mortality as well as the taphonomic and geological influences on the assemblage (e.g., Voorhies, 1985; Jerzykiewicz et al., 1993; Loope et al., 1998).

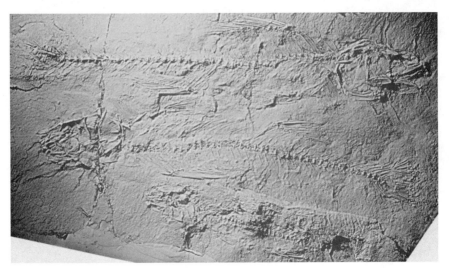

*Figure 4.5.* Specimens from a mass-mortality assemblage of Paleocene fish from the Joffre Bridge locality, southern Alberta. Site was described in detail and interpreted by Wilson (1996). Photo provided by M.V.H. Wilson.

### Aggregations during Breeding, Nesting, and Early Ontogeny

Many extant organisms aggregate for breeding or nesting, or for protection during their juvenile and subadult growth stages, and evidence for such aggregations is sometimes preserved in the fossil record. Above, we referred to a classic article by Jordan (1920), in which Miocene herring, believed to have been spawning, experienced a mass kill. Similarly, a fish assemblage from the Paleocene Joffre Bridge locality in central Alberta contains thousands of individuals preserved on a single bedding plane and is interpreted as a mass kill—likely resulting from a change of water chemistry or temperature—at the time of spawning (Wilson, 1996) (Fig. 4.5).

Among dinosaurs, communal nesting behavior has been documented in bonebeds containing monotaxic concentrations of egg and eggshell, perinatal elements (including hatchlings), and sometimes elements of adults (Horner and Makela, 1979; Horner, 1982; Varricchio et al., 1997; Chiappe et al., 1998, 2001; Huh and Zelenitsky, 2002). Horner and Makela (1979) documented parental care in hadrosaurs, citing occurrences of recently hatched hadrosaurs in nests. Because the individuals in the nests had more than doubled their hatchling size, some form of care during early growth was inferred.

Aggregations of juveniles and subadults have also been documented in bonebeds and have been interpreted as mirroring a behavioral pattern occasionally seen in extant vertebrates in which individuals in these age classes group together as a cohort. Gale (1988) described a death assemblage of 17 juvenile dicynodonts (?*Diictodon*) from the Permian of northern Zambia and hypothesized that the group had died during a flood event while occupying a nesting site. A similar explanation was proposed by Sun (1978) for an association of juvenile dicynodonts, *Parakannemeyeria*, found in China. Together, data from these bonebeds suggest that some form of aggregation behavior was normal for juvenile dicynodonts. Jerzykiewicz et al. (1993) described two bonebeds consisting of parallel-aligned juvenile skeletons of the ankylosaur *Pinacosaurus* (five and seven individuals each) from the Upper Cretaceous Djadokhta redbeds of Inner Mongolia and interpreted these occurrences as evidence of gregarious behavior in individuals that died during severe windstorms. The identical sizes of these juveniles suggest that they may have been siblings or a cohort of individuals from a seasonal hatch.

### Aggregations Due to Herding

Mass kills of herding species are common and well documented in the modern, and thus evidence for ancient mass kills of herding species should be expected in the fossil record. However, as discussed above, it can be very difficult to demonstrate with confidence that individuals were part of a socially interactive or structured herd. Herding paleobehavior is often inferred in the case of low-diversity monodominant bonebeds that contain the remains of large mobile herbivores. In some localities, such as the Hagerman Horse Quarry (Richmond et al., 1999), the interpretation of herding behavior is the most plausible interpretation when considered in the context of modern horses. However, for many extinct species, closely related living analogues are absent, and other evidence must be mustered to support the inference. For example, Turnbull and Martill (1988) interpreted the titanothere *Mesatirhinus,* which has no close living relatives, as a herding mammal based on the sedimentary and taphonomic features associated with a monotaxic assemblage. Exquisite preservation and articulated remains preserved in the deposits of a high-energy alluvial flow event suggested that the assemblage resulted from a flood-induced mass-kill event.

Herding in herbivorous dinosaurs was first proposed on the basis of the giant ceratopsian bonebeds from the Late Cretaceous of Alberta that contain abundant remains of *Centrosaurus apertus* (Currie, 1982; Currie and Dodson, 1984; Eberth and Getty, 2005) (Fig. 4.6). Herding was

*Figure 4.6. Centrosaurus* bonebeds. A. Bonebed 43 (described by Ryan et al., 2001). Note the complete disarticulation of the material. B. Quarry map from Bonebed 91 (illustration by Donna Sloan). These are both monodominant bonebeds (sensu Eberth et al., Chapter 3 in this volume) that contain the remains of herbivorous dinosaurs believed to have perished in a mass kill while herding (Currie and Dodson, 1984; Eberth and Getty, 2005).

proposed primarily on the basis of faunal and demographic information. In each bonebed multiple growth stages are present, and although *Centrosaurus* is relatively rare in the overall formational assemblage, the taxon is overwhelmingly abundant in the ceratopsian bonebeds at the Park, far outnumbering any other taxon. For example, an excavation of less than 15% of one bonebed yields a minimum number of individuals of 50 individuals, and it has been projected that the skeletal remains of many hundreds of individuals are present across the full extent of that bonebed. Excavated elements share similar taphonomic signatures, suggestive of a shared taphonomic history and further supporting the interpretation of mass kill. The occurrence of multiple bonebeds with similar compositions makes it very likely that aggregating with others of its species was normal for *Centrosaurus apertus*. Studies of other low-diversity monodominant bonebeds imply some degree of herding behavior in many species of ceratopsians, as well as hadrosaurs and sauropods (e.g., Hooker, 1987; Fiorillo, 1991; Coria, 1994; Ryan et al., 2001).

In cases where the occurrence of a nonselective mass-mortality event has been documented and a herding hypothesis proposed, the age structure of the herd can sometimes be inferred from the fossil assemblage. Voorhies (1981, 1985, 1992) (Fig. 4.7) described 35 articulated adult skeletons of the rhinoceros *Teleoceras* that were presumably killed during a volcanic eruption and ash fallout. Sexual and age distribution data drawn from this sample suggested that *Teleoceras* was likely a herding and polygynous species, different from modern rhinoceroses and more like extant hippopotamuses (Mead 2000).

Some low-diversity, monodominant bonebeds containing the remains of large-bodied carnivores can be linked to mass-kill events based on sedimentological context and taphonomic analyses, thus indicating that the biocoenose consisted of an aggregation prior to death. Currie (2000) proposed that the remains of nine individuals of *Albertosaurus* from a bonebed in the Lower Maastrichtian portion of the Horseshoe Canyon Formation could be the remains of a gregarious group that was engaged in cooperative hunting. Although the evidence for this specific interpretation was drawn largely from modern analogues, and the behavioral interpretation has been challenged (Eberth and McCrea, 2001), the facts that the assemblage includes a mixture of juveniles, subadults, and adults and that there is little variability of the taphonomic features of the bones clearly underscore the possibility that these animals died together near the site of burial. In this context, the aggregation may also be reasonably interpreted as having formed due to environmental stresses (see below).

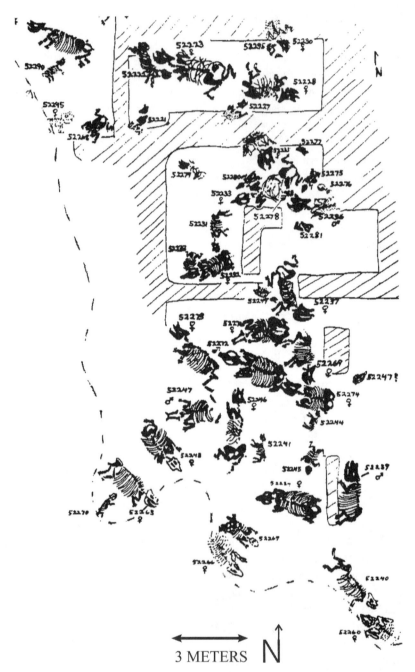

3 METERS N

*Figure 4.7.* Quarry map of a monodominant assemblage of *Teleoceras* from the Miocene of Nebraska (from Voorhies, 1985). Note the complete articulation of specimens (compared to Fig. 6). Data from this site provide compelling evidence for herding in *Teleoceras*. Illustration used with permission.

Modern populations of fish, amphibians, and many archosaurs and lepido-saurs often aggregate to feed or bask, or come together in response to environmental stress. Accordingly, their extinct relatives are inferred to have exhibited similar behaviors. Where fossil species are interpreted as having occupied geographically restricted habitats, such as a pond or small lake, unusually rich fossil assemblages of the species are often interpreted as resulting from paleoenvironmental stresses that reduced the physical extent of the habitat or otherwise degraded paleoenvironmental conditions. For example, Henrici and Fiorillo (1993) concluded that disease was the most likely agent of death for an assemblage of rhinophrynid frogs from the Eocene of Wyoming, based on an understanding of modern frog autecology. Similarly, monodominant bonebeds containing the obligatorily aquatic amphibian *Trimerorachis* from the Lower Permian of the southwestern United States have been interpreted as forced aggregations within desiccating and shrinking pools of water (Case, 1935; Langston, 1953; Berman and Reisz, 1980) (Fig. 4.8).

Olson (1971a) described Permian bonebeds from Texas that consist largely of the remains of the amphibian *Lysorophus*. Specimens are preserved as coiled skeletons with one individual per burrow, and groups of skeletons concentrated in areas of up to a few square meters. In a single concentration, individuals tend to be the same size. Death during aestivation was proposed as an explanation for these associations because of the postures and orientations of the skeletons, and geologic evidence for seasonal flooding and aridity. Similarly, Berman et al. (1988) reported an assemblage of more than eight *Stegotretus* microsaur amphibian skulls and tightly coiled skeletons from a caliche-rich locality in New Mexico in which death during aestivation was considered likely. Vaughn (1966) and Berman et al. (1987) (Fig. 4.8) described small but well-preserved assemblages of the Early Permian amphibian *Seymouria* preserved in sediments deposited in seasonally arid alluvial systems preserved in Utah and New Mexico, respectively. We consider it a good possibility that these individuals also died during aestivation, given the parallel alignment of skeletons and the absence of evidence for a hydraulic concentrating mechanism. Smith (1987, 1993) described helical burrow casts from the Permian Leeukloof Formation of South Africa and documented a burrow-fill with two intertwined individuals of the dicynodont *Diictodon*, inferred to have died at the same time, although, in this case, the cause of death was unknown.

Stress-induced aggregation has also been proposed to explain monotaxic and monodominant bonebeds that preserve large and very mobile

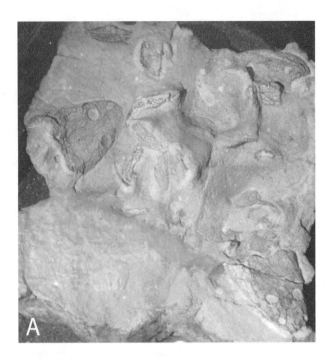

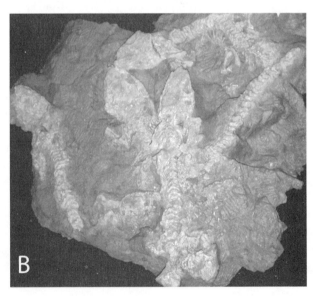

*Figure 4.8.* Permian bonebeds consisting of A, the obligate aquatic amphibian, *Trimerorhachis*; and B, the terrestrial amphibian *Seymouria*. At many localities, *Trimerorhachis* is believed to have perished in pond environments due to drought. The cause of mortality for *Seymouria* is not understood.

*Figure 4.9.* A nearly monotaxic assemblage of the small theropod dinosaur *Coelophysis* from Ghost Ranch, New Mexico. Note the articulated hindlimb at the center of the photograph, the small skull and neck segment in the lower left, and the large abundance of skeletal material.

terrestrial vertebrates. Rogers (1990) interpreted monodominant hadrosaur and ceratopsian bonebeds from the Two Medicine Formation of northwestern Montana as drought-related assemblages, based on the abundance of juveniles—a demographic profile characteristic of drought-induced mass kills (Shipman, 1975; Gates et al., 2003; Gates, 2005). The Late Triassic *Coelophysis* bonebed at Ghost Ranch, New Mexico, is one of the most famous carnivore bonebeds in the world and preserves articulated and semiarticulated skeletons estimated to number in the thousands (Schwartz and Gillette, 1994) (Fig. 4.9). Because of the large number of individuals present and evidence for seasonal aridity, this bonebed is also thought to represent an assemblage of several groups of individuals that died over time due to drought.

### Aggregations through Time

When considered in a stratigraphic framework, interpretations of aggregation paleobehavior may provide insight into the evolution of behaviors.

*Figure 4.10.* Two *Diictodon* specimens in intimate contact. Specimens were collected from the mouth of a burrow fill in the Upper Permian Beaufort Group of the South African Karoo Basin. Photo provided by R.M.H. Smith.

For example, an Oligocene assemblage of three virtually complete skeletons of erycinine boid snakes (*Ogmophis* and *Calamagras*) reported by Breithaupt and Duvall (1986) was interpreted as evidence that aggregation behavior in these taxa extends back in geologic time at least 32 million years. Hunt et al. (1983) and Hunt (1993) provided evidence of denning behavior in Miocene bear dogs, implying that it is a primitive behavior for members of Carnivora. Smith (1993) and Sullivan et al. (2003) described Permian burrow-fills containing multiple remains of the dicynodont *Diictodon* (Fig. 4.10), which suggests a long geologic history for communal denning behaviors.

It is also worth noting a potential bias in some kinds of behavioral reconstructions as a function of the age of the taxon. Assemblages of large terrestrial carnivorous vertebrates such as *Sphenacodon* and *Dimetrodon* are common in the Lower Permian fossil record but have never been interpreted in terms of gregarious "packs." Instead, these assemblages are almost always interpreted as reflecting forced assembly due to resource limitation or local resource abundance, and/or extreme environmental conditions. In contrast, and in spite of virtually identical taphonomic signatures and modes, vertebrate assemblages that preserve large dinosaurs or mammals are routinely interpreted in terms of modern mammalian social behaviors (e.g., pack hunting, herding, etc.). Given that the geologic and taphonomic data used to support these differing interpretations are essentially the same, it appears that there is a tendency to attribute socially based gregarious behaviors to extinct vertebrates that existed closer in time to the present.


## PALEOFAUNAS AND PALEOCOMMUNITIES

In a paleontologic context, the concept of "paleofauna" (also "fauna") is frequently employed in reference to the taxonomic composition of the fossil assemblage from a stratigraphic interval or locality (e.g., the fauna of a quarry, bed, member, formation, group [e.g., Estes, 1964; Estes and Berberian, 1970; Erickson, 1991; Chure et al., 1998; Eberth et al., 2001]) (Fig. 4.11). It also is used to refer to a taxonomic subset of an assemblage, such as "the mammalian fauna" (e.g., Krause and Gingerich, 1983). In contrast, we regard a "paleocommunity" as consisting of a group of fossil species that habitually occupied a physically discrete paleoenvironment, and among which there were transfers of energy and resources, or ongoing interactions that influenced energy transfers. Thus, taxa included in a "faunal list" are not necessarily part of a single paleocommunity. For example, marine and nonmarine taxa are often mixed together in bonebeds situated in coastal settings (e.g., Dinosaur Park Formation [Eberth et al., 2001]).

Bonebed-based studies of paleofaunas and paleocommunities arguably form a threefold hierarchy characterized by differences in both the amount of information incorporated and the complexity of the inferred conclusions (Fig. 4.12). At the lowest or simplest level of the hierarchy are studies that focus strictly on paleofaunal composition: which taxa are present or absent in a fossil assemblage, formation, and/or geographic area. Such studies may have a broad stratigraphic and/or paleogeographic focus (e.g., Olson, 1952; Chure et al., 1998; Peng et al., 2001) or may be limited to a single bed or bonebed (e.g., Fastovsky et al., 1995).

At an intermediate level of complexity are studies that document changes in paleofaunal composition in a paleogeographic or stratigraphic context (e.g., Brinkman, 1990; Lillegraven and Eberle, 1999). In some of these studies, information about the relative abundance of species within fossil assemblages or inferred paleocommunities is included. Because relative abundances can be strongly influenced by taphonomic processes, such studies tend to include taphonomic and/or statistical analyses of data, and comparisons with modern analogues (e.g., Brinkman, 1990). They may also include information about the organismal biology of the

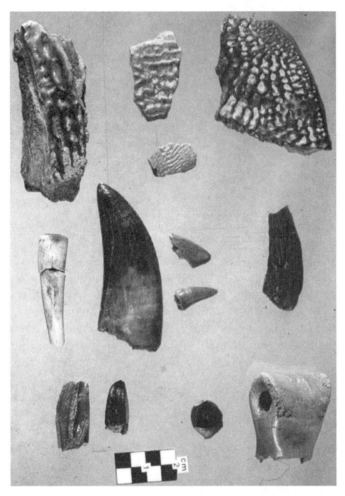

Figure 4.11. An example of a Late Cretaceous faunal assemblage from a vertebrate microfossil locality at Dinosaur Provincial Park. Specimens include aquatic and terrestrial turtles, fish, large and small theropods, hadrosaur, and crocodile. Note that the assemblage includes both obligate aquatic and terrestrial vertebrates and, thus, includes representatives from different paleocommunities.

animals including aggregation paleobehaviors (e.g., Gates, 2002, 2005) (see previous section).

At the most complex level are studies that seek to resolve a given fossil assemblage into discrete and/or multiple paleocommunities, or identify habitats and/or interspecies interactions (including energy transfer) within one or more paleocommunities (e.g., Behrensmeyer, 1975; Olson, 1977). In the following sections we explore the application of bonebed data to these different levels of study, and examine examples.

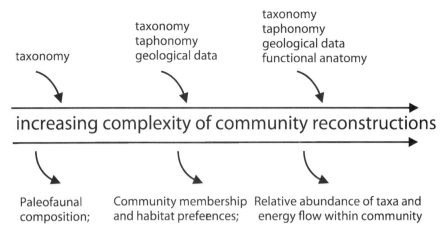

*Figure 4.12.* General relationship showing data required to build an increasingly deeper and more robust understanding of paleocommunities.

### Faunas

Taxonomic data are central in any faunal study, and multitaxic bone-beds—especially multitaxic microfossil assemblages—are often the main source for documenting the faunal composition of a stratigraphic unit. Bonebeds often yield large sample sizes, which, when sampled repeatedly lead to better representation of rare taxa and a full documentation of the faunal composition of a given stratigraphic unit (Jamniczky et al., 2003; Blob and Badgley, Chapter 6 in this volume).

In general, the goal of a faunal study is to be as taxonomically exhaustive as possible. Bonebed assemblages from a variety of facies in a given stratigraphic unit provide an excellent opportunity to document taxonomic composition in a variety of habitats (e.g., Gingerich, 1989; White et al., 1998; Straight and Eberth, 2002). Similarly, bonebeds that contain time averaged or paleoenvironmentally mixed faunal assemblages are especially significant because they likely contain a broad spectrum of the ancient vertebrate biodiversity and are likely to include representatives that were rare in the original paleocommunities from which the bonebeds were derived. For example, bonebeds preserved in paleochannel fills often preserve the most complete record of taxonomic diversity from the ancient floodplain due to reworking and concentration (Behrensmeyer, 1988; Badgley et al., 1996). Such bonebeds also often preserve our only record of rare species not commonly represented by isolated skeletons or elements outside the channel (e.g., Clemens, 1963; Brinkman, 1990).

## Faunas through Time

Faunas are frequently compared within or between stratigraphic units in order to (1) identify and assess evolutionary patterns, (2) reveal biotic responses to paleoenvironmental change, and (3) define biogeographic and biostratigraphic patterns. For example, Lillegraven and Eberle (1999) documented faunal changes up to and across the Lancian-Puercan (Cretaceous-Tertiary) boundary in Wyoming by examining a group of 76 fossil-bearing localities, most of which are high-diversity microfossil bonebeds preserved in alluvial deposits. Using samples from these sites they documented the timing and duration of extinction and diversification events among non-avian dinosaurs and mammals during a time of profound reorganization in the regional terrestrial paleoecosystem. Similar studies of faunal composition and patterns of extinction up to and across the Cretaceous-Tertiary boundary have been founded on comparisons of taxonomic composition among bonebeds (and other types of sites) that underlie or span the boundary (e.g., Sheehan et al., 1991; White et al., 1998).

## Paleocommunities

For a number of reasons, bonebeds in general, and multitaxic bonebeds in particular, play a key role in reconstructing paleocommunities. First, because bonebeds document a discrete subset of the time and geographic area in the host stratigraphic unit, they resolve patterns of association and distribution at a finer level than is possible with an assemblage of fossils amassed from the entire stratigraphic unit. Second, multitaxic bonebeds often yield numerous individuals of a taxon, which allows some level of statistical analysis with which to quantify patterns of diversity, association, and relative abundance of taxa (see Blob and Badgley, Chapter 6 in this volume). Finally, where there are multiple geographic and stratigraphic occurrences of bonebeds, all of the aforementioned patterns can be assessed through time and across paleogeography.

## Paleocommunity Membership and Reconstruction

Three broad approaches are used to initially assign a taxon membership in a paleocommunity: (1) fossil-facies associations, and the stratigraphic and paleogeographic distribution patterns of taxa (Brinkman, 1990; Lillegraven and Eberle, 1999); (2) consideration of functional morphology in analogous living species; and (3) biological aspects of a taxon inferred from phylogenetic analysis and other paleontologic associations (Sahni 1972). In the literature, vertebrate paleocommunities tend to be described in a more generalized fashion than their modern counterparts and are often categorized using terminology that relates to broad depositional settings such as upland, lowland, alluvial, fluvial, overbank, lacustrine, estuarine, eolian, and marine, etc. (Behrensmeyer, 1975; Eberth et al., 2000; Eberth et al., Chapter 3 in this volume).

Sedimentological setting and fossil-facies associations are frequently used as a first step to constrain the paleoenvironmental setting of a paleocommunity. Whitaker and Antia (1978) and Antia (1981) demonstrated that the historically significant Temeside bonebed (Silurian-Devonian) formed as a channel lag deposit in back beach lagoonal muds during a marine regression. Therrien and Fastovsky (2000) showed that in the Upper Triassic Chinle Formation of the southwestern United States, floodplain-water-hole settings were common sites for multitaxic bonebed formation. Olson (1971b) identified the elongate and short-limbed Permian amphibian *Trimerorachis* as a member of a "pond" paleocommunity because individuals are often found packed together with fish in lacustrine deposits. In contrast, Wu et al. (1993) identified the lepidosaurian *Sineoamphisbaenia* as the member of an eolian upland paleocommunity because it is found exclusively in eolian deposits along the margins of an alluvial fan, and it exhibits a morphology that indicates fossorial capabilities. Fossil-facies relationships combined with relative abundance data derived from vertebrate microfossil sites were used by Brinkman et al. (2005) to interpret the composition of a late Campanian nearshore marine community and to distinguish between nonmarine and euryhaline taxa.

Reconstructing a paleocommunity requires not only the development of a membership list, but also evidence that taxa co-occurred and inter acted. As discussed in the section on aggregation paleobehavior, independent geologic and taphonomic analyses often provide compelling evidence that an assemblage consists of individuals that died together or at nearly the same time (Voorhies, 1985) and, thus, must have been living together at the end of their lives (perimortem associations). A second approach involves documenting that members of an assemblage were physically restricted to the depositional setting in which they were buried and preserved, and that the assemblage was not transported prior to burial. For example, aquatic taxa found together in sediments deposited in calm water are often interpreted as autochthonous and contemporaneous paleocommunity members (Dalquest and Mamay, 1963; Olson, 1977; Erickson, 1991). A third approach is to document repeated taxonomic associations or repeated facies-taxa associations within a stratigraphic unit (taphonomic modes) (Olson, 1952; Clark et al., 1967; Behrensmeyer, 1975; Badgley, 1986; Gingerich, 1989; White et al., 1998; Straight and Eberth, 2002; Eberth and Currie, 2005). In this approach, repeated associations are assumed to reflect habitat preferences of the taxa. For example, Olson (1971b, 1989) documented frequent occurrences of the Permian vertebrates *Eryops* and *Seymouria* in pond and pond margin facies, and the rarity of these same taxa in channel-fill facies. He interpreted this pattern as reflecting a paleoenvironmental preference among these taxa for quiet water settings. Conversely, he noted a greater relative abundance of *Diplocaulus* and *Dimetrodon* in facies associated with flowing water (conglomeratic channel fills) and interpreted that pattern as reflecting a preference among these taxa for more riparian settings. These repeated associations were central in the development of the "chronofauna" concept (Olson, 1952, 1971b, 1989).

Brinkman (1990) used a similar approach in his study of paleocommunities in the Upper Cretaceous Judith River Group of southern Alberta. In that study the occurrence of taxa in 24 localities was considered with respect to stratigraphy and facies associations (Eberth, 1990). Taxa showing distinctly different stratigraphic distributions and facies associations were interpreted as members of separate paleocommunities. Blob and Fiorillo (1996) and Brinkman et al. (2004) further refined this approach by evaluating taphonomic influences on the taxonomic-abundance data as reflected in variations in element size and shape. For example, turtles, which are typically represented in the Belly River Group by elements in a 1 to 2 cm size range, might be under- or overrepresented at a given site due to taphonomic factors.

### Relative Abundance and Taphonomic Biases

By considering the relative abundance of taxa, paleocommunity reconstructions move beyond membership to considerations of how paleocommunities are structured and how energy flows through them (e.g., Gingerich, 1989). Arguably, reliable estimates of relative taxonomic abundance at the time of death are possible in multitaxic assemblages that accumulated in response to nonselective catastrophic mortality events and conditions of minimal transport and reworking. Such assemblages provide a snapshot of paleocommunity structure. Examples of bonebeds that form under such conditions are rare but include the Geraldine Bonebed from the Lower Permian of Texas. The Geraldine fossil assemblage is interpreted as resulting from a fire-related mass death of a group of vertebrates that habitually inhabited a marshy area at the margin of a lake (Sander, 1987). Compared to the overall composition of the host formation, the Geraldine Bonebed is unusual in yielding large numbers of the terrestrial herbivore *Edaphosaurus*. Accordingly, this taxon was reasonably interpreted as the most important primary plant consumer in this lacustrine-margin paleocommunity. Outside of this paleoenvironmental setting, however, terrestrial herbivores are rare, and the primary plant consumers appear to have been invertebrates (Olson, 1971b).

Paleosol horizons sometimes provide opportunities to assess relative abundance data in a paleocommunity. Bown and Kraus (1981) analyzed mammalian bone concentrations in paleosols from the Willwood Formation of Wyoming and proposed that the preserved fossil assemblages from these deposits more closely reflect the composition of the biocoenose than do the assemblages concentrated by predators or hydraulic transport and reworking.

Bonebeds consisting of time-averaged elements that have been transported and reworked are far more common than bonebeds in which composition reflects an unaltered biocoenose (Eberth et al., Chapter 3 in this volume). Thus, it has long been recognized that, in the vast majority of bonebeds, the relative abundance of taxa does not necessarily reflect their relative abundance in the original paleocommunities from which they were derived. Accordingly, approaches have been developed to derive accurate paleocommunity information from reworked assemblages.

In Shotwell's (1955) early but influential study, the abundance of a taxon was assumed to result from two factors: (1) original abundance and (2) distance from the paleocommunity site to the site of fossil preservation. In subsequent studies that considered this approach, attempts were made to assign bonebed taxa to "proximal" versus "distal" paleocommunities (Estes, 1964; Vaughn, 1969). However, as taphonomic influences on bonebed formation have become better understood, it has been recognized that taphonomic filters act differentially on taxa and skeletal components, and that transport and reworking do not act equally on different skeletal components. Along these lines, Korth (1979) argued that only limited paleoecological interpretations are possible using hydraulically sorted vertebrate microfossil assemblages. In contrast, Behrensmeyer (1975, 1991) argued that significant paleoecological information incorporating relative abundance data can be derived from fluvial assemblages once the taphonomic histories are established. In addition, Behrensmeyer (1975) argued that the validity of paleocommunity reconstruction can be tested by repeated sampling and comparison of localities preserved in the same or different taphonomic modes. This is now a standard approach among bonebed workers seeking to identify paleocommunity structure (Pearson et al., 2002; Straight and Eberth, 2002; Brinkman et al., 2004).

Dodson et al. (1980) were among the first to apply the comparative approach of Behrensmeyer (1975) to a dinosaurian paleocommunity. That study was based on data from 12 sites associated with paleochannels, oxbow lakes or ponds, swamps, and overbank settings in the Jurassic Morrison Formation. They found that the remains of dinosaurs such as *Camarasaurus, Apatosaurus, Diplodocus,* and *Allosaurus* are common and not confined to a single depositional setting. Thus, they argued that this sauropod assemblage most likely represents a true dinosaur community with overlapping habitats. In contrast, *Stegosaurus* was thought to be ecologically separate from the sauropod paleocommunity because of its notably greater association with fluvial facies.

Comparison of relative abundance data from assemblages exhibiting different taphonomic modes is also a powerful tool for assessing paleocommunity structure. Dodson (1983, 1987) recognized that Upper Cretaceous vertebrate microfossil assemblages (his taphonomic mode 1) from the Oldman and Dinosaur Park formations at Dinosaur Provincial Park in southern Alberta were formed largely as a result of ancient fluvial activity. Accordingly, he hypothesized that hydraulic sorting should have directly influenced and modified the relative abundance of taxa in the vertebrate microfossil assemblages compared to the historic collections of articulated and associated specimens (his taphonomic mode 2) from the host formations. However, when he tested the hypothesis with field data, he found the same relative abundance of dinosaur taxa in both the vertebrate microfossil and articulated specimen assemblages. In both modes, hadrosaurs are overwhelmingly abundant (47% of the microfossils and 41% of the articulated specimens), ceratopsians are second in abundance (21% of the microfossils and 22% of the articulated specimens), and ankylosaurs are third (13% of the microfossils and 12% of the articulated specimens). Concordance in the two data sets was subsequently interpreted as strong evidence that both sets correctly estimate the relative abundance of living dinosaur taxa in this ancient ecosystem.

### Assessing Paleoecosystem Complexity

Another aspect of reconstructing paleocommunities involves identifying the different paleohabitats or paleoenvironmental subsets that were present in ancient ecosystems. For localities of relatively young age in which the taxa are members of living groups with known ecological constraints, the habitat preferences of extant relatives can be used to interpret those of the fossil taxa. For example, Campbell (1979) identified several taxa of Pleistocene nonpasserines preserved in Peruvian tar seeps. One group of taxa, including the grebes, a number of ardeids, and *Porzana carolina*, is related to modern species that prefer to live next to large bodies of water, whereas another group has extant relatives that require forested or heavy-scrub habitats. The modern habitat preferences of the living relatives demonstrate that the Pleistocene paleoenvironments probably included savanna woodland rather than just the desert that is present in the region today.

However, the practice of using fossil taxa to reconstruct ancient habitats diminishes in effectiveness as one delves deeper in the fossil record, and links to modern analogues become more speculative. In these cases, an understanding of morphology (body shape, body size, jaw and tooth structure, etc.) and related biomechanical function and constraints become essential in order to decipher modes of life. For example, Milner (1980) combined knowledge of the skeletal anatomy in some Carboniferous tetrapods from Nyrany with their frequency of occurrence in different facies in order to document habitat preferences within an overall lacustrine-paludal depositional setting. Ultimately, he recognized open water, shallow water, lowland swamp, and lake margin habitats (Fig. 4.13). He interpreted open water habitats as occupied by a small number of specialized lineages, shallow water habitats as occupied by both permanently aquatic tetrapods and seasonally aquatic forms, and the lake margin and lowland swamp habitats as occupied by insectivores and carnivorous forms.

Where fossils exhibit varied taphonomic signatures in a single bonebed, additional insights into the complexity of ancient paleocommunities can be revealed. For example, Pratt (1989) identified groups of fossils with different taphonomic signatures in a study of the mammalian assemblage from the Miocene Thomas Farm locality of Florida, a rich multitaxic bonebed preserved in a sinkhole. The overall assemblage was interpreted as containing (1) a coprocoenosis in which the diversity of rodents reflects predator activity, (2) taxa that frequented cave habitats, and (3) a mixture of taxa whose remains were reworked by moving water. Similarly, Varricchio (1995) documented two taphonomic modes at Jack's Birthday Site, a multitaxic dinosaur bonebed in the Upper Cretaceous Two Medicine Formation of Montana. One mode consisted of predominately unassociated and isolated elements that were widely distributed vertically and horizontally. These elements were interpreted as an attritional accumulation of skeletal debris from a widespread alluvial-plain paleocommunity. The second mode, consisting of the associated remains of hadrosaurs and the small theropod *Troodon*, was interpreted as reflecting members of a

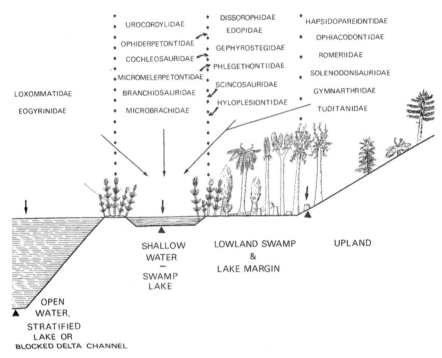

Figure 4.13. Reconstruction of paleocommunities of the Carboniferous Nyrany locality. From Milner (1980), used with permission. Assignment of taxa to habitats is based on data from multiple localities. Note the absence of vertebrate herbivores in these paleocommunities.

local paleocommunity that may have succumbed en masse to one or more drought or disease events.

### Predator-Prey Relationships

A significant emphasis in paleocommunity studies is the documentation of interactions among different species, particularly predator and prey. Some bonebeds provide spectacular opportunities to document predator-prey associations and evidence of scavenging. For example, many fish from the Green River Shale possess their last meals in their gut cavities. Evidence for agonistic predator-prey behavior in the terrestrial vertebrate fossil record has also been preserved on rare occasions. One of the most spectacular interspecies dinosaur associations known is referred to as the "fighting dinosaurs"—articulated remains of *Velociraptor* (carnivore) and *Protoceratops* (herbivore) that were collected in 1971 by members of the Polish-Mongolian Expeditions in the Upper Cretaceous beds of Mongolia.

Skeletal poses and facies relationships suggest that a *Protoceratops* was partially buried in sediment when it was attacked and possibly killed by a *Velociraptor*. From all indications, the *Protoceratops* locked its jaws on the arm of the *Velociraptor* before the latter could escape (Jerzykiewicz et al., 1993; Currie 1997). Both died during or shortly after the altercation and were buried by eolian sands (Loope et al., 1999).

Bonebeds such as the "fighting dinosaurs" that preserve such spectacularly direct evidence for predation are strikingly rare. More frequently, the evidence is indirect. For example, Ryan et al. (2000) documented the association of abundant embryonic and hatchling remains of an undetermined hadrosaur and teeth from the small theropod *Troodon* in the Upper Cretaceous Horseshoe Canyon Formation of Alberta. He cited the association as potential evidence for a predator-prey relationship between the taxa. Ostrom (1969) and Maxwell and Ostrom (1995) used the closely associated remains of the dinosaurs *Deinonychus* and *Tenontosaurus* from the Lower Cretaceous Cloverly Formation of Montana to suggest a predator-prey relationship between those taxa. Norell et al. (1994) interpreted the partial skull of a juvenile troodontid preserved within a nest containing embryonic oviraptorid skeletons as either evidence for nest parasitism, or as part of a meal brought to the nest by oviraptorid parents in order to feed their hatchings. In a study focusing on an Eocene multitaxic microfossil bonebed, Maas (1985) identified fossil material resulting from the direct activity of predators. In that study the occurrence of disarticulated and fragmentary microfossils as well as the relatively low abundance of teeth and jaws were interpreted as the result of selective ingestion and digestion by predators. She argued that even though the identity of the predators is unknown, prey assemblages such as these are important for paleocommunity reconstruction because they indicate prey-species richness.

Scavenging behaviors in the fossil record are supported by both direct and indirect evidence from bonebed studies. Shed (rootless) theropod teeth in bonebeds that are otherwise dominated by the skeletal remains of plant-eating dinosaurs are common and have been used to support the hypothesis of scavenging of carcasses by theropods (e.g., Rogers, 1990; Ryan et al., 2001). Rogers et al. (2003) proposed an even more unusual example of scavenging by providing evidence for late-stage cannibalism in the large theropod *Majungatholus*.

Predator trapping during scavenging has been proposed as an explanation for the origins of some bonebeds that contain notably large numbers of terrestrial carnivores, such as the La Brea Tar Pits and the Cleveland-Lloyd Quarry (Stock, 1972; Bilbey 1999). Whereas there appears to be broad support for a predator trap scenario in the case of the La Brea Tar

Pits, this kind of interpretation is not unanimously accepted in the case of the Cleveland Lloyd Quarry. Bilbey (1999) interpreted Cleveland Lloyd as a predator trap in part because of the abundance of the carnivorous dinosaur *Allosaurus* (MNI 44) and the rarity of herbivores (MNI 22). However, an alternative view proposes that the Cleveland-Lloyd Quarry formed as a drought assemblage (Gates, 2002, 2005; Suarez and Suarez, 2004).

### Paleocommunities in Time and Space

When constrained by a high-resolution stratigraphic framework, paleo-communities can be traced through time and paleogeographically in order to document paleocommunity responses to paleoenvironmental changes or simply to document long-term paleocommunity evolution. Whereas these studies are superficially similar to stratigraphically focused faunal studies, they differ in that the emphasis is on evaluating species interactions through time. Recently, the term "evolutionary paleoecology" has been applied to the study of changing paleoecological organization over long intervals of geologic time (Wing et al., 1992).

Olson (1952) analyzed data from Permian terrestrial vertebrate assemblages and bonebeds and proposed the concept of a chronofauna—a discrete paleocommunity with a stable taxonomic composition that ranges through measurable geologic time. He proposed that the long-term compositional stability of a chronofauna allows for local to regional paleoenvironmental and climatic influences on composition to be identified (e.g., Olson, 1977, 1989).

Integrative studies have also identified the changing role of predators in terrestrial paleocommunities through the Phanerzoic. Olson (1952) noted that, as a group, carnivores dominate Carboniferous and Early Permian assemblages and, presumably, paleocommunities. In contrast, vertebrate herbivores are rare in these paleocommunities. Accordingly, the primary consumers in these ancient ecosystems must have been invertebrates or fish. More recent studies using bonebed assemblages have supported this interpretation and have also shown that, by the Late Permian, vertebrate herbivores had replaced vertebrate carnivores as the dominant faunal element in terrestrial vertebrate paleocommunities (Eberth et al., 2000).

Bakker (1975) compared predator-prey ratios in terrestrial paleocommunities from the Late Paleozoic, Mesozoic, and Cenozoic by assessing a combination of bonebed data and the faunal compositions from a variety of stratigraphic units. He argued that the predator-prey ratios in a paleocommunity reflect the energetic requirements of the predators, so that paleocommunities dominated by ectothermic vertebrates (e.g., all Paleozoic paleocommunities) will have higher predator-prey ratios than communities of endothermic vertebrates (e.g., all Cenozoic mammalian paleocommunities). Because the predator-prey ratios in Mesozoic dinosaur assemblages are more similar to those in Cenozoic mammalian paleocommunities, he argued that predatory dinosaurs were endotherms.

Lastly, bonebed data sets have also been combined to pose questions about the geographic distribution of paleocommunities on the local, regional, and continental scale. For example, Holroyd and Hutchison (2002) identified patterns of geographic variation in the relative abundance of turtles from bonebeds in the Hell Creek Formation of North Dakota. On the larger scale, Lehman (1987, 1997) documented the distribution of some terrestrial vertebrates in western North America during successive stages of the Upper Cretaceous. Using these distribution patterns he inferred the structure and distribution of some Late Cretaceous dinosaur paleocommunities.

## CONCLUSIONS

Three distinct areas of paleobiological research can be addressed with bonebed data. The first relates to the characterization of extinct species. Bonebeds undeniably offer unique opportunities to characterize morphological variation within populations from both an ontogenetic and a sexual-dimorphic perspective, and some bonebed assemblages have allowed paleontologists to approximate the biological species concept in a paleospecies. Through comparisons of morphological data from different bonebeds, species variation can be explored through time and across geographic regions.

The second area of paleobiological research hinging upon bonebeds focuses on the reconstruction of group behavior. The existence of rich monotaxic and monodominant bonebeds tends to raise questions about their formation and, accordingly, encourages speculation about group behavior in the ancient record, including herding in large-bodied herbivores, breeding or nesting, response to environmental stress, and cooperative predation. Although it may not be possible to conclusively identify the cause of death for a given bonebed assemblage, in general, sedimentologic and taphonomic studies can provide important insights. Where bonebeds are shown to be the result of mass kills they can provide evidence for population age structures and growth rates.

A third major direction of paleobiological research that depends to a considerable extent on bonebed data sets involves the more synthetic issues of documenting and defining paleofaunas and paleocommunities. Bonebeds are often primary sources of faunal data in a stratigraphic unit and, thus, are also a primary source of data for reconstructing paleocommunities. Bonebed-based paleocommunity studies cover a range of topics including paleocommunity membership, relative abundance of taxa, relationships between taxa, relationships between taxa and the paleoenvironment, energy flow through a paleocommunity, and interactions between individual members of the community.

In conclusion, bonebeds are an invaluable source of paleobiological information, and many of the insights provided through the study of bonebeds are simply irretrievable from isolated bones or skeletons. Historically, the challenge for bonebed-based paleobiological studies has been to move from "storytelling" to the articulation of clear and testable hypotheses. When this challenge is met, bonebed-related research contributes significantly to the understanding of both local (e.g., predator-prey interactions) and large-scale (e.g., paleocommunity evolution) paleobiological and paleoecological phenomena.

## REFERENCES

Andrews, R.M. 1982. Patterns of growth in reptiles. Pp. 273–320 *in* Biology of the Reptilia, vol. 13. C. Gans and F.H. Pough, eds. Academic Press, London.

Antia, D.D.J. 1981. The Temeside bone bed and associated sediments from Wales and the Welsh borderland. Mercian Geologist 8:163–215.

Badgley, C. 1986. Taphonomy of mammalian fossil remains from Siwalik rocks of Pakistan. Paleobiology 12:119–142.

Bakker, R.T. 1975. Dinosaur renaissance. Scientific American 232:58–78.

Baszio, S. 1997. Systematic palaeontology of isolated dinosaur teeth from the latest Cretaceous of south Alberta, Canada. Courier Forschungsinstitut Senckenberg 196:33–77.

Behrensmeyer, A.K. 1975. The taphonomy and paleoecology of Plio-Pleistocene vertebrate assemblages east of Lake Rudolf, Kenya. Bulletin of the Museum of Comparative Zoology 146:474–578.

Behrensmeyer, A.K. 1988. Vertebrate preservation in fluvial channels. Palaeogeography, Palaeoclimatology, Palaeoecology 63:183–199.

Behrensmeyer, A.K. 1991. Terrestrial vertebrate accumulations. Pp. 291–335 *in* Taphonomy: Releasing the data locked in the fossil record. P.A. Allison and D.E.G. Briggs, eds. Plenum, New York.

Behrensmeyer, A.K. This volume. Bonebeds through time. Chapter 2 *in* Bonebeds: Genesis, analysis, and paleobiological significance. R.R. Rogers, D.A. Eberth, and A.R. Fiorillo, eds. University of Chicago Press, Chicago.

Berman, D.S, and R.R. Reisz. 1980. A new species of *Trimerorhachis* (Amphibia, Temnospondyli) from the Lower Permian of New Mexico, with discussions of Permian faunal distributions in that state. Bulletin of Carnegie Museum of Natural History 49:455–485.

Berman, D.S, R.R. Reisz, and D.A. Eberth. 1987. *Seymouria sanjuanensis* (Amphibia, Batrachosauria) from the Lower Permian Formation of north-central New Mexico and the occurrence of sexual dimorphism in that genus questioned. Canadian Journal of Earth Sciences 24:1769–1784.

Berman, D.S, D.A. Eberth, and D.B. Brinkman. 1988. *Stegotretus agyrus* a new genus and species of microsaur (amphibian) from the Permo-Pennsylvania of New Mexico. Annals of the Carnegie Museum 57:293–323.

Bilbey, S.A. 1999. Taphonomy of the Cleveland-Lloyd dinosaur quarry in the Morrison Formation, Central Utah: A lethal spring-fed pond. Miscellaneous publication 99-1. Pp. 121–133 *in* Vertebrate paleontology in Utah. D.D. Gillette, ed. Utah Geological Survey, Salt Lake City.

Blob, R.W., and Badgley, C. This volume. Numerical methods for bonebed analysis. Chapter 6 *in* Bonebeds: Genesis, analysis, and paleobiological significance. R.R. Rogers, D.A. Eberth, and A.R. Fiorillo, eds. University of Chicago Press, Chicago.

Blob, R.W., and A.R. Fiorillo. 1996. The significance of vertebrate microfossil size and shape distributions for faunal abundance reconstructions: A Late Cretaceous example. Paleobiology 22:422–435.

Bown, T.M., and M.J. Kraus. 1981. Vertebrate fossil-bearing paleosol units (Willwood Formation, Lower Eocene, Northwest Wyoming, U.S.A.): Applications for taphonomy, biostratigraphy, and assemblages analysis. Palaeogeography, Palaeoclimatology, Palaeoecology 34:31–56.

Breithhaupt, B.H., and D. Duvall. 1986. The oldest record of serpent aggregation. Lethaia 19:181–185.

Brinkman, D.B. 1990. Paleoecology of the Judith River Formation (Campanian) of Dinosaur Provincial Park, Alberta, Canada: Evidence from vertebrate microfossil localities. Palaeogeography, Palaeoclimatology, Palaeoecology 78:37–54.

Brinkman, D.B., A.P. Russell, D.A. Eberth, and J. Peng. 2004. Vertebrate palaeocommunities of the lower Judith River Group (Campanian) of southeastern Alberta, Canada, as interpreted from vertebrate microfossil assemblages. Palaeogeography, Palaeoclimatology, Palaeoecology 213:295–313.

Brinkman, D.B., D.R. Braman, A.G. Neuman, P.E. Ralrick, and T. Sato. 2005. A vertebrate assemblage from marine shales of the Lethbridge Coal Zone. Pp. 486–500 *in* Dinosaur Provincial Park: A spectacular ancient ecosystem revealed. P.J. Currie and E.B. Kopplelhus, eds. Indiana University Press, Bloomington.

Brown, B., and E.M. Schlaikjer. 1937. The skeleton of *Styracosaurus* with the description of a new species. American Museum Novitiates 955:1–12.

Campbell, K.E. Jr. 1979. The non-passerine Pleistocene avifauna of the Talara Tar Seeps northwestern Peru. Life science contribution 118. Royal Ontario Museum, Toronto.

Case, E.C. 1935. Description of a collection of associated skeletons of *Trimerorhachis*. Contributions of the Museum of Paleontology, University of Michigan 4:227–274.

Chiappe, L.M., R.A. Coria, L. Dingus, F. Jackson, A. Chinsamy, and M. Fox. 1998. Sauropod dinosaur embryos from the Late Cretaceous of Patagonia. Nature 396:258–261.

Chiappe, L.M., L. Dingus, and N. Frankfurt. 2001. Walking on eggs: The astonishing discovery of thousands of dinosaur eggs in the badlands of Patagonia. Scribner, New York.

Chure, D.J., K. Carpenter, R. Litwin, S. Hasiotis, and E. Evanoff. 1998. The fauna and flora of the Morrison Formation. Modern Geology 23:507–537.

Clark, J., J.R. Beerbower, and K.K. Kietzke. 1967. Oligocene sedimentation, stratigraphy, paleoecology, and paleoclimatology in the Big Badlands of South Dakota. Fieldiana Geology Memoirs 5:1–158.

Clemens, W.A. 1963. Fossil mammals of the type Lance Formation, Wyoming. I. Introduction and Multituberculata. University of California publications in the geological sciences 48. University of California Press, Berkeley.

Colbert, E.H. 1989. The Triassic dinosaur *Coelophysis*. Bulletin 57. Museum of Northern Arizona, Flagstaff.

Colbert, E.H. 1990. Variation in *Coelophysis bauri*. Pp. 81–90 *in* Dinosaur systematics: Approaches and perspectives. K. Carpenter and P J. Currie, eds. Cambridge University Press, New York.

Cook-, L. M. 1926. A rhinoceros bone bed. Black Hills Engineer 14:95–99.

Coria, R.A. 1994. On monospecific assemblage of sauropod dinosaurs from Patagonia: Implications for gregarious behavior. Gaia 10:209–213.

Craig, G.Y., and G. Oertel. 1966. Deterministic models of living and fossil populations of animals. Quarterly Journal of the Geological Society of London 122:315–355.

Currie, P.J. 1982. Long distance dinosaurs. Natural History 6:60–65.

Currie, P.J. 1997. Theropods. Pp. 216–233 *in* The complete dinosaur. J. O. Farlow and M.K. Brett-Surman, eds. Indiana University Press, Bloomington.

Currie, P.J. 2000. Possible evidence of gregarious behaviour in tyrannosaurids. Gaia 15:271–277.

Currie, P.J., and P. Dodson. 1984. Mass death of a herd of ceratopsian dinosaurs. Pp. 61–66 *in* Third symposium on Mesozoic terrestrial ecosystems, short papers. W.E. Reif and F. Westphal, eds. Attempto Verlag, Tübingen.

Dalquest, W.W., and S.H. Mamay. 1963. A remarkable concentration of Permian amphibian remains in Haskall County, Texas. Journal of Geology 71:641–644.

Dodson, P. 1983. A faunal review of the Judith River (Oldman) Formation, Dinosaur Provincial Park, Alberta. The Mosasaur 1:89–118.

Dodson, P. 1987. Microfossil studies of dinosaur palaeoecology, Judith River Formation of southern Alberta. In: Currie, P.J., Koster, E.H. (Eds.), Fourth Symp. Mesozoic Terrestrial Ecosystems, Short papers. Occasional Paper of the Tyrrell Museum 3:70–75.

Dodson, P., A.K. Behrensmeyer, R.T. Bakker, and J.S. McIntosh. 1980. Taphonomy and paleoecology of the dinosaur beds of the Jurassic Morrison Formation. Paleobiology 6:208–232.

Donelan, C., and G.D. Johnson. 1997. *Orthacanthus platypternus* (Chondrichthyes, Xenacanthida) occipital spines from the Lower Permian Craddock bonebed, Baylor County, Texas. Journal of Vertebrate Paleontology 17(supplement to 3): 43A.

Eberth, D.A. 1985. The skull of *Sphenacodon ferocior* and comparison with other sphenacodontines (Reptilia: Pelycosauria). Circular 190. New Mexico Bureau of Mines and Mineral Resources, Socorro.

Eberth, D.A. 1990. Stratigraphy and sedimentology of vertebrate microfossil sites in the uppermost Judith River Formation (Campanian), Dinosaur Provincial Park, Alberta, Canada. Palaeogeography, Palaeoclimatology, Palaeoecology 78:1–36.

Eberth, D.A., and P.J. Currie. 2005. Vertebrate taphonomy and taphonomic modes. Pp. 453–477 in Dinosaur Provincial Park: A spectacular ancient ecosystem revealed. P.J. Currie, and E.B. Koppelhus, eds. Indiana University Press, Bloomington.

Eberth, D.A., and M.A. Getty. 2005. Ceratopsian bonebeds at Dinosaur Provincial Park: Occurrence, origin, and significance. Pp. 501–536 in Dinosaur Provincial Park: A spectacular ancient ecosystem revealed. P.J. Currie and E.B. Koppelhus, eds. Indiana University Press, Bloomington.

Eberth, D.A., and R.T. McCrea. 2001. Were large theropods gregarious? Journal of Vertebrate Paleontology 21(supplement to 3):46A–47A.

Eberth, D.A., D.S Berman, S.S. Sumida, and H. Hopf. 2000. Lower Permian terrestrial paleoenvironments and vertebrate paleoecology of the Tambach Basin (Thuringia, Central Germany): The upland holy grail. Palaios 15:293–313.

Eberth, D.A., P.J. Currie, D.B. Brinkman, M.J. Ryan, D.R. Braman, J.D. Gardner, V.D. Lam, D.N. Spivak, and A.G. Neuman. 2001. Alberta's dinosaurs and other fossil vertebrates: Judith River and Edmonton Groups (Campanian-Maastrichtian). Pp. 47–75 in Mesozoic and Cenozoic paleontology in the Western plains and Rocky Mountains. Occasional paper 3. C.L. Hill, ed. Museum of the Rockies, Bozeman, Montana.

Eberth, D.A., R.R. Rogers, and A.R. Fiorillo. This volume. A practical approach to the study of bonebeds. Chapter 5 in Bonebeds: Genesis, analysis, and paleobiological significance. R.R. Rogers, D.A. Eberth, and A.R. Fiorillo, eds. University of Chicago Press, Chicago.

Eberth, D.A., M. Shannon, and B.G. Noland. This volume. A bonebeds database: Classification, biases, and patterns of occurrence. Chapter 3 in Bonebeds: Genesis, analysis, and paleobiological significance. R.R. Rogers, D.A. Eberth, and A.R. Fiorillo, eds. University of Chicago Press, Chicago.

Erickson, B.R. 1991. Flora and fauna of the Wannagan Creek quarry: Late Paleocene of North America. Scientific Publications of the Science Museum of Minnesota New Series 7(3):1–19.

Erickson, G.M., P.J. Makovicky, P.J. Currie, M.A. Norell, S.A. Yerby, and C.A. Brochu. 2004. Gigantism and comparative life-history parameters of tyrannosaurid dinosaurs. Nature 430:772–774.

Estes, R. 1964. Fossil vertebrates from the Late Cretaceous Lance Formation, eastern Wyoming. University of California Publications in Geological Sciences 49. University of California Press, Berkeley.

Estes, R., and P. Berberian. 1970. Paleoecology of a late Cretaceous vertebrate community from Montana. Breviora 343:1–35.

Fastovsky, D.E., J.M. Clark, N.H. Strater, M. Montellano, R. Hernandez, and J.A. Hopson, J.A, 1995. Depositional environments of a Middle Jurassic terrestrial vertebrate assemblage, Huizachal Canyon, Mexico. Journal of Vertebrate Paleontology 15:561–575.

Fiorillo, A.R. 1991. Taphonomy and depositional setting of Careless Creek Quarry (Judith River Formation), Wheatland County, Montana, U.S.A. Palaeogeography, Palaeoclimatology, Palaeoecology 81:281–311.

Fiorillo, A.R. 2004. The dinosaurs of arctic Alaska. Scientific American (December):84–91.

Gale, T.M. 1988. Comments on a "nest" of juvenile dicynodont reptiles. Modern Geology 13:119–124.

Gates, T. 2002. Murder in Jurassic Park: The Cleveland-Lloyd dinosaur quarry as a drought-induced assemblage. Journal of Vertebrate Paleontology 22(supplement to 3):56A–57A.

Gates, T. 2005. The Late Jurassic Cleveland-Lloyd dinosaur quarry as a drought-induced assemblage. Palaios 20:363–375.

Gates, T., E. Roberts, and R. Rogers. 2003. Drought in the vertebrate fossil record: A review of fossil and modern drought-related assemblages. Journal of Vertebrate Paleontology 23(supplement to 3):53A–54A.

Gingerich, P.D. 1977. Paleontology and phylogeny: Patterns of evolution at the species level in early Tertiary mammals. American Journal of Science 276:1–28.

Gingerich, P.D. 1989. New earliest Wasatchian mammalian fauna from the Eocene of northwestern Wyoming: Composition and diversity in a rarely sampled high-floodplain assemblage. University of Michigan Papers on Paleontology 28:1–97.

Halliday, T.R., and P.A. Verrell. 1988. Body size and age in amphibians and reptiles. Journal of Herpetology 22:253–265.

Henrici, A.C., and A.R. Fiorillo. 1993. Catastrophic death assemblage of *Chelomophrynus bayi* (Anura, Rhinophrynidae) from the Middle Eocene Wagon Bed Formation of central Wyoming. Journal of Paleontology 67:1016–1026.

Holroyd, P.A., and J.H. Hutchison. 2002. Patterns of geographic variation in latest Cretaceous vertebrates: Evidence from the turtle component. Pp. 177–190 *in* The Hell Creek Formation and the Cretaceous-Tertiary boundary in the Northern Great Plains: An integrated continental record of the end of the Cretaceous. Special paper 361. J.H. Hartman, K.R. Johnson, and D.J. Nichols, eds. Geological Society of America, Boulder, Colorado.

Hooker, J.S. 1987. Late Cretaceous ashfall and the demise of a hadrosaurian "herd." Geological Society of America Abstracts with Programs 19:284.

Horner, J.R. 1982. Evidence for colonial nesting and "site fidelity" among ornithischian dinosaurs. Nature 297:675–676.

Horner, J.R., and R. Makela. 1979. Nest of juveniles provides evidence of family structure among dinosaurs. Nature 282:296–298.

Horner, J.R., A.J. de Ricqlès, and K. Padian. 2000. Long bone histology of the hadrosaurid dinosaur *Maiasaura peeblesorum*: Growth dynamics and physiology based on an ontogenetic series of skeletal elements. Journal of Vertebrate Paleontology 20:115–129.

Huh, M., and D.K. Zelenitsky. 2002. Rich dinosaur nesting site from the Cretaceous of Bosung County, Chullanam-Do Province, South Korea. Journal of Vertebrate Paleontology 22:716–718.

Hunt, R.M. Jr. 1993. The Miocene carnivore dens of Agate Fossil Beds National Monument, Nebraska: Oldest known denning behavior of large mammalian carnivores. Technical report NPS/NRPO/NRTR-93/11. Paleontological research, vol. 2. V.L. Santucci, ed. National Park Service, Department of the Interior, Washington, D.C.

Hunt, R.M. Jr., X. Xiang-Xu, and J. Kaufman. 1983. Miocene burrows of extinct bear dogs: Indication of early denning behavior of large mammalian carnivores. Science 221:364–366.

Hutchison, J.H. 1982. Turtle, crocodilian, and champsosaur diversity changes in the Cenozoic of the north-central region of the western United States. Palaeogeography, Palaeoecology, Palaeoclimatology 37:149 –164.

ICZN (International Commission on Zoological Nomenclature). 1996. Opinion 1842. *Coelurus bauri* Cope, 1887 (currently *Coelophysis bauri*; Reptilia, Saurischia): Lectotype replaced by a neotype. Bulletin of Zoological Nomenclature 53:142–144.

Jamniczky, H.A., D.B. Brinkman, and A.P. Russell. 2003. Vertebrate microsite sampling: How much is enough? Journal of Vertebrate Paleontology 23:725–734.

Jerzykiewicz, T., P.J. Currie, D.A. Eberth, P.A. Johnston, E.H. Koster, and J.-J. Zheng. 1993. Djadokhta Formation correlative strata in Chinese Inner Mongolia: An overview of the stratigraphy, sedimentary geology, and paleontology and comparisons with type locality in the pre-Altai Gobi. Canadian Journal of Earth Sciences 30:2180–2195.

Jordan, D.S. 1920. A Miocene catastrophe. Natural History 20:18–22.

Korth, W.W. 1979. Taphonomy of microvertebrate fossil assemblages. Annals of the Carnegie Museum 48:235–284.

Krause, D.W., and P.D. Gingerich. 1983. Mammalian fauna from Douglas Quarry, earliest Tiffanian (Late Paleocene) of the eastern Crazy Mountain Basin, Montana. Contributions from the Museum of Paleontology, University of Michigan 26:157–196.

Langston, W. Jr. 1953. Permian amphibians from New Mexico. University of California Publications in Geological Sciences 29:349–416.

Langston, W. Jr. 1975. The ceratopsian dinosaurs and associated lower vertebrates from the St. Mary River Formation (Maastrichtian) at Scabby Butte, southern Alberta. Canadian Journal of Earth Sciences 12:1576–1608.

Laury, R.L. 1980. Paleoenvironment of a Late Quaternary mammoth-bearing sinkhole deposit, Hot Springs, South Dakota. Geological Society of America Bulletin 91:465–475.

Lehman, T.M. 1987. Late Maastrichtian paleoenvironments and dinosaur biogeography in the western interior of North America. Palaeogeography, Palaeoclimatology, Palaeoecology 60:189–217.

Lehman, T.M. 1989. *Chasmosaurus mariscalensis* sp. nov., a new ceratopsian dinosaur from Texas. Journal of Vertebrate Paleontology 9:137–162.

Lehman, T.M. 1997. Late Campanian dinosaur biogeography in the western interior of North America. Pp. 223–240 *in* Dinofest International: Proceedings of a symposium held at Arizona State University. D.L. Wolberg, E. Stump, and G.D. Rosenberg, eds. Academy of Natural Sciences, Philadelphia.

Lillegraven, J.A., and J.J. Eberle. 1999. Vertebrate faunal changes through Lancian and Puercan time in southern Wyoming. Journal of Paleontology 73:691–710.

Loope, D.B., L. Dingus, C.C. Swisher III, and C. Minjin. 1998. Life and death in a Late Cretaceous dune field, Nemegt Basin, Mongolia. Geology 26:27–30.

Loope, D.B., J.A. Mason, and L. Dingus. 1999. Lethal sandslides from eolian dunes. Journal of Geology 107:707–713.

Maas, M.C. 1985. Taphonomy of a Late Eocene microvertebrate locality, Wind River Basin, Wyoming (U.S.A.). Palaeogeography, Palaeoclimatology, Palaeoecology 52:123–142.

Maxwell, W.D., and J.H. Ostrom. 1995. Taphonomy and paleobiological implications of *Tenontosaurus-Deinonychus* associations. Journal of Vertebrate Paleontology 15:707–712.

McNab, B.K. 1971. On the ecological significance of Bergmann's rule. Ecology 52:845–854.

Mead, A.J. 2000. Sexual dimorphism and paleoecology in *Teleoceras*, a North American Miocene rhinoceros. Paleobiology 26:689–706.

Merriam, J.C. 1909. A death trap which antedates Adam and Eve; the discovery of a Californian tar swamp that holds the bones of extinct monsters. Harper's Weekly 53:11–12.

Milner, A.R. 1980. The tetrapod assemblage from Nyrany, Czechoslovakia. Pp. 439–496 *in* The terrestrial environment and the origin of land vertebrates. Systematics Association special vol. 15. A.L. Panchen, ed. Academic Press, London.

Norell, M.A., J.M. Clark, D. Dashzeveg, R. Barsbold, L.M. Chiappe, A.R. Davidson, M.C. McKenna, A. Perle, and M.J. Novacek. 1994. A theropod dinosaur embryo and the affinities of the Flaming Cliffs dinosaur eggs. Science 256:779–782.

Ochocinska, D., and J.R.E. Taylor. 2003. Bergmann's rule in shrews: Geographical variation of body size in Palearctic *Sorex* species. The Linnean Society of London, Biological Journal of the Linnean Society 78:365–381.

Olson, E.C. 1951. *Diplocaulus*, a study in growth and variation. Fieldiana Geology 11:57–154.

Olson, E.C. 1952. The evolution of a Permian vertebrate chronofauna. Evolution 6:181–196.

Olson, E.C. 1971a. A skeleton of *Lysorophus tricarinatus* (Amphibia: Lepospondyli) from the Hennessey Formation (Permian) of Oklahoma. Journal of Paleontology 45:443–449.

Olson, E.C. 1971b. Vertebrate Paleozoology. Wiley Interscience, New York.

Olson, E.C. 1977. Permian lake faunas: A study in community evolution. Journal of the Palaeontological Society of India 20:146–163.

Olson, E.C. 1989. The Arroyo Formation (Leonardian: Lower Permian) and its vertebrate fossils. Bulletin 35. Texas Memorial Museum, Austin.

Ostrom, J.H. 1969. Osteology of *Deinonychus antirrhopus*, an unusual theropod from the Lower Cretaceous of Montana. Bulletin 30. Peabody Museum of Natural History, New Haven, Connecticut.

Padian, K., A.J. de Ricqlès, and J.R. Horner, J.R. 2001. Dinosaurian growth rates and bird origins. Nature 412:405–408.

Pearson, D.A., T. Schaefer, K.R. Johnson, D.J. Nichols, and J.P. Hunter. 2002. Vertebrate biostratigraphy of the Hell Creek Formation in southwestern North Dakota and northwestern South Dakota. Pp. 145–167 *in* The Hell Creek Formation and the Cretaceous-Tertiary boundary in the Northern Great Plains: An integrated continental record of the end of the Cretaceous. Special paper 361. J.H. Hartman, K.R. Johnson, and D.J. Nichols, eds. Geological Society of America, Boulder, Colorado.

Peng, J.-H., A.P. Russell, and D.B. Brinkman. 2001. Vertebrate microsite assemblages (exclusive of mammals) from the Foremost and Oldman formations of the Judith River Group (Campanian) of southeastern Alberta: An illustrated guide. Natural history occasional paper 25. Provincial Museum of Alberta, Edmonton.

Peterson, O.A. 1906a. The Agate Spring fossil quarry. Annals of the Carnegie Museum 3:487–494.

Peterson, O.A. 1906b. Preliminary description of two new species of the genus *Diceratherium* Marsh, from the Agate Spring Fossil Quarry. Science 24:281–283.

Pratt, A.E. 1989. Taphonomy of the microvertebrate fauna from the early Miocene Thomas Farm locality, Florida (U.S.A.). Palaeogeography, Palaeoclimatology, Palaeoecology 76:125–151.

Richmond, D., G. McDonald, and J. Bertog. 1999. Sedimentology and taphonomy of the Smithsonian horse quarry. Journal of Vertebrate Paleontology 19(supplement to 3):70A.

Rogers, R.R. 1990. Taphonomy of three dinosaur bone beds in the Upper Cretaceous Two Medicine Formation of northwestern Montana: Evidence for drought related mortality. Palaios 5:394–413.

Rogers, R.R., and S.M. Kidwell. This volume. A conceptual framework for the genesis and analysis of vertebrate skeletal concentrations. Chapter 1 *in* Bonebeds: Genesis, analysis, and paleobiological significance. R.R. Rogers, D.A. Eberth, and A.R. Fiorillo, eds. University of Chicago Press, Chicago.

Rogers, R.R., D.W. Krause, and K. Curry Rogers. 2003. Cannibalism in the Madagascan dinosaur *Majungatholus atopus*. Nature 22:515–518.

Ryan, M.J. 2003. Taxonomy, systematics and evolution of centrosaurine ceratopsids of the Campanian Western Interior Basin of North America. Unpublished Ph.D. thesis, University of Calgary, Calgary, Alberta.

Ryan, M.J., P.J. Currie, J.D. Gardner, M.K. Vickaryous, and J.M. Lavigne. 2000. Baby hadrosaurid material associated with unusually high abundance of *Troodon* teeth from the Horseshoe Canyon Formation, Upper Cretaceous, Alberta, Canada. Gaia 15:123–133.

Ryan, M.J., A.P. Russell, D.A. Eberth, and P.J. Currie. 2001. The taphonomy of a *Centrosaurus* (Ornithischia: Certopsidae [sic]) bone bed from the Dinosaur Park Formation (Upper Campanian), Alberta, Canada, with comments on cranial ontogeny. Palaios 16:482–506.

Sahni, A. 1972. The vertebrate fauna of the Judith River Formation, Montana. Bulletin of the American Museum of Natural History 147:321–412.

Sampson, S.D., M.J. Ryan, and D.H. Tanke. 1997. Craniofacial ontogeny in centrosaurine dinosaurs (Ornithischia: Ceratopsidae): Taxonomic and behavioral implications. Zoological Journal of the Linnean Society 121:293–337.

Sander, P.M. 1987. Taphonomy of the Lower Permian Geraldine Bonebed in Archer County, Texas. Palaeogeography, Palaeoclimatology, Palaeoecology 61:221–236.

Sander, P.M. 1992. The Norian *Plateosaurus* bonebeds of central Europe and their taphonomy. Palaeogeography, Palaeoclimatology, Palaeoecology 93.255–299.

Schwartz, H.L., and D.D. Gillette. 1994. Geology and taphonomy of the *Coelophysis* quarry, Upper Triassic Chinle Formation, Ghost Ranch, New Mexico. Journal of Paleontology 68:1118–1130.

Sheehan, P.M., D.E. Fastovsky, R.G. Hoffmann, C.B. Berghaus, and D.L. Gabriel. 1991. Sudden extinction of the dinosaurs: Latest Cretaceous, upper Great Plains, U.S.A. Science 254:835–839.

Shipman, P. 1975. Implications of drought for vertebrate fossil assemblages. Nature 257:667–668.

Shipman, P. 1979. What are all these bones doing here? Confessions of a taphonomist. Harvard Magazine 6:42–46.

Short, R.V., and E. Balaba. 1994. The differences between the sexes. University Press, Cambridge.

Shotwell, J.A. 1955. An approach to the paleoecology of mammals. Ecology 36:327–337.

Smith, R.M.H. 1987. Helical burrow casts of therapsid origin from the Beaufort Group (Permian) of South Africa. Palaeogeography, Palaeoclimatology, Palaeoecology 60:155–170.

Smith, R.M.H. 1993. Vertebrate taphonomy of Late Permian floodplain deposits in the southwestern Karoo Basin of South Africa. Palaios 8:45–67.

Sternberg, C.M. 1940. Ceratopsidae from Alberta. Journal of Paleontology 14:284–286.

Stock, C. 1972. Rancho La Brea, a record of Pleistocene life in California. Science series 20. Los Angeles County Museum of Natural History, Los Angeles.

Straight, W.H., and D.A. Eberth. 2002. Testing the utility of vertebrate remains in recognizing patterns in fluvial deposits: An example from the lower Horseshoe Canyon Formation, Alberta. Palaios 17:472–490.

Suarez, C., and M. Suarez. 2004. Use of facies, taphonomy, and rare earth element geochemistry analyses at the Cleveland-Lloyd dinosaur quarry: Tools for bone bed interpretation. Journal of Vertebrate Paleontology 24(supplement to 3): 119A.

Sullivan, C., R.R. Reisz, and R.M.H. Smith. 2003. The Permian mammal-like herbivore Diictodon, the oldest known example of sexually dimorphic armament. Proceedings of the Royal Society of London B 270:173–178.

Sun, A.L. 1978. On the occurrence of Parakannemeyeria in Sinkiang. Memoirs of the Institute of Vertebrate Paleontology and Paleoanthropology 13:47–54.

Tanke, D.H. 1988. Ontogeny and dimorphism in Pachyrhinosaurus (Reptilia, Ceratopsidae), Pipestone Creek, N.W. Alberta, Canada. Journal of Vertebrate Paleontology 8(supplement to 3):27A.

Therrien, F., and D.E. Fastovsky. 2000. Paleoenvironments of early theropods, Chinle Formation (Late Triassic), Petrified Forest National Park, Arizona. Palaios 15:194–211.

Turnbull, W.D., and D.M. Martill. 1988. Taphonomy and preservation of a monospecific titanothere assemblage from the Washakie Formation (Late Eocene), southern Wyoming: An ecological accident in the fossil record. Palaeogeography, Palaeoclimatology, Palaeoecology 63:91–108.

Van Valkenburgh, B., and T. Sacco. 2002. Sexual dimorphism, social behavior, and intrasexual competition in large Pleistocene carnivorans. Journal of Vertebrate Paleontology 22:164–169.

Varricchio, D.J. 1995. Taphonomy of Jack's Birthday Site, a diverse dinosaur bonebed from the Upper Cretaceous Two Medicine Formation of Montana. Palaeogeography, Palaeoclimatology, Palaeoecology 114:297–323.

Varricchio, D.J., and J.R. Horner. 1993. Hadrosaurid and lambeosaurid bone beds from the Upper Cretaceous Two Medicine Formation of Montana: Taphonomic and biologic implications. Canadian Journal of Earth Sciences 30:997–1006.

Varricchio, D.J., F. Jackson, J. Borlowski, and J.R. Horner. 1997. Nest and egg clutches for the theropod dinosaur *Troodon formosus* and the evolution of avian reproductive traits. Nature 385:247–250.

Vaughn, P.P. 1966. *Seymouria* from the Lower Permian of southeastern Utah, and possible sexual dimorphism in that genus. Journal of Paleontology 40:603–612.

Vaughn, P.P. 1969. Early Permian vertebrates from southern New Mexico and their paleozoogeographic significance. Contributions in science 166. Los Angeles County Museum of Natural History, Los Angeles.

Voorhies, M.R. 1969. Taphonomy and population dynamics of an early Pliocene vertebrate fauna, Knox County, Nebraska. Contributions to geology special paper 1. University of Wyoming, Laramie.

Voorhies, M.R. 1981. Ancient ashfall creates Pompeii of prehistoric animals. National Geographic Research 159:66–75.

Voorhies, M.R. 1985. A Miocene rhinoceros herd buried in a volcanic ash. National Geographic Society Research Reports 19:671–688.

Voorhies, M.R. 1992. Ashfall: Life and death at a Nebraska waterhole ten million years ago. Museum notes 81. University of Nebraska State Museum, Lincoln.

Vrba, E.S. 1980. Evolution, species and fossils: How does life evolve? South African Journal of Science 76:61–84.

Whitaker, J.H. McD., and D.D.J. Antia. 1978. A scanning electron microscope study of the genesis of the Upper Silurian Ludlow Bone Bed. Pp. 119–136 *in* Scanning electron microscopy in the study of sediments. W.B. Whalley, ed. Geoabstracts, Norwich.

White, P.D., D.E. Fastovsky, and P.M. Sheehan. 1998. Taphonomy and suggested structure of the dinosaurian assemblage of the Hell Creek Formation (Maastrichtian), eastern Montana and western North Dakota. Palaios 13:41–51.

Wilson, M.V.H. 1984. Year classes and sexual dimorphism in the Eocene catostomid fish *Amyzon aggregatum*. Journal of Vertebrate Paleontology 3:137–142.

Wilson, M.V.H. 1996. Taphonomy of a mass death layer of fishes in the Paleocene Paskapoo Formation at Joffre Bridge, Alberta, Canada. Canadian Journal of Earth Science 33:1487–1498.

Wing, S.L., H.D. Sues, R. Potts, W.A. DiMichele, and A.K. Behrensmeyer. 1992. Evolutionary Paleoecology. Pp. 1–13 *in* Terrestrial ecosystems through time, evolutionary paleoecology of terrestrial plants and animals. A.K. Behrensemeyer, J.D. Damuth, W.A. Dimichele, R. Potts, H.D. Sues, and S. Wing, eds. University of Chicago Press, Chicago.

Wu, X.-C, D.B. Brinkman A.P. Russell, Z.-M. Dong, P.J. Currie, L.-H. Hou, and G.-H. Cui. 1993. Oldest known amphisbaenian from the Upper Cretaceous of Chinese Inner Mongolia. Nature 366:57–59.

Xu, X., J.M. Clark, C.A. Forster, M. Norell, G.M. Erickson, D.A. Eberth, C.-K. Jia, and Q. Zhao. 2006. A basal tyrannosauroid dinosaur from the Late Jurassic of China. Nature 439:715–718.

# CHAPTER 5

# A Practical Approach to the Study of Bonebeds

*David A. Eberth, Raymond R. Rogers, and Anthony R. Fiorillo*

## INTRODUCTION

Every bonebed preserved in the stratigraphic record reflects the interplay of complex physical, chemical, and biological phenomena, and given the myriad events and interactions that follow death, deciphering the formative pathway(s) that led to skeletal accumulation is no easy task. This is especially true given the highly degraded and potentially biased subsamples of original death assemblages that many bonebeds in fact represent. Clearly, the best hope for a successful study hinges upon the careful collection of data, and fortunately the methods of data collection and analysis in relation to bonebeds have evolved and improved over time. While this is certainly true, even a brief survey of the literature demonstrates that approaches to the assembly of bonebed data sets still vary significantly among workers. This at least in part reflects differences in geological and geographic context, quality of preservation, and taxonomic composition. Some of the methodological disparity also reflects differences in scientific design and research goals, and the variable backgrounds of bonebed workers. Despite the wide latitude of approach, we maintain that many bonebed studies pose the same general questions (e.g., How did this spectacular deposit form, and what can be reliably inferred from it?), and we contend that a standardized approach to data collection can be employed in most circumstances. Employing a suite of standard techniques will increase the efficiency of data collection and analysis and provide unique opportunities to compare data sets among bonebeds (Behrens-meyer, 1991).

In this chapter we consider bonebeds from a practical standpoint and address four main concerns that pertain to effective site management and data collection. The first relates to preliminary site assessment, which is the critical first step in any comprehensive analysis of a bonebed. Decisions made at this point can promote or thwart success throughout the remainder of the study. The second relates to considerations that arise when the real work begins, such as the removal of overburden, the establishment of reference systems for mapping, and the extraction of elements. The third focus of this chapter pertains to the collection of geological data. All too often bonebed studies are undertaken with too little emphasis placed on the collection of related geological information. This is truly unfortunate because rocks provide the overall context for bonebeds preserved in the stratigraphic record, and a thorough understanding of sedimentology and stratigraphy is needed before the formative history of a bonebed can be accurately reconstructed. Moreover, assessing a bonebed as a facies (rather than simply a unique or unusual part of a facies) provides important insights into the origin of the bonebed and its host stratigraphic succession. The fourth and final major goal of this chapter is to review the types of taphonomic information that should be retrieved from a bonebed assemblage. We examine the taphonomic methodologies commonly employed by bonebed workers and review the spectrum of data typically collected. Examples are provided from both classic studies and more recent bonebed investigations. This chapter concludes with some collective wisdom (and hopefully some good ideas) that relates to the interpretive end game: the reconstruction of the taphonomic history of a bonebed assemblage. Regardless of the final scenario(s) promoted, it is crucial to keep in mind that the true value of any bonebed study rests not only in the eventual story told (be it dramatic or mundane), but also in the breadth and quality of data that lie behind the narrative.

## STEP 1: PRELIMINARY SITE ASSESSMENT

Before fieldwork begins in earnest on any bonebed project, a researcher (or team of researchers), preferably in conjunction with an experienced preparator and/or field crew manager, should carefully consider a variety of preliminary concerns. By reviewing the various issues related to site logistics and data retrieval, a bonebed crew could minimize potential frustrations and maximize scientific outcomes. A preliminary site assessment will clarify goals and hopefully elucidate any potential complications, whether they relate to the retrieval of data or nonscientific issues, such as infrastructure support, site access, worker morale, and environmental mitigation.

Because most bonebed studies require a long-term commitment of manpower and resources it is essential that a carefully considered plan is in place before excavation and data collection begins. This increases efficiency and keeps information loss to a minimum. A preliminary site assessment serves to (1) frame the entire study and keep it on track from start to finish; (2) identify the feasibility of collecting different kinds of data; (3) measure how "typical" the bonebed is in the context of other sites, regions, or stratigraphic intervals; and (4) identify future potential. In the following sections we outline a few of the more important things to consider when planning a field-based bonebed study.

### Collect and Permanently Record Accurate Locality Data

Tanke (1999, 2001) reported that literally hundreds of dinosaur quarries excavated between 1898 and the early 1970s in Dinosaur Provincial Park (southern Alberta) were "lost" due to a combination of incomplete data collection and/or poor archival practices. Loss of locality data is truly unfortunate, because a bonebed with no context is largely useless in studies that relate to paleoenvironment, paleoecology, or biostratigraphy. A suitably archived or published photograph of a bonebed is an effective practice that usually permits future workers to relocate the site. In Dinosaur Provincial Park today, bonebeds are typically marked with a quarry stake, but this practice may be inappropriate in areas where vandalism or illegal collecting occur. Staking a quarry may also prove ineffective in regions where surface erosion rates are high, or where researchers are in remote areas without the necessary materials.

Satellite-based triangulation technology such as the Global Positioning System (GPS) provides an easy means of determining and recording a site's location using a variety of reference systems (e.g., UTM or latitude/longitude). GPS is an especially powerful tool in areas where relief is poor and/or maps are of low resolution or nonexistent. GPS locality data should always include the datum employed by the receiver (e.g., WGS '84; NAD 27) in order to calibrate the reference systems with a map datum so that future workers can effectively relocate sites. Where datums differ between GPS units or surveyed maps, sites can be mislocated by as much as 250 m. Current practice is to utilize the WGS '84/NAD '83 datum (1984 World Geodetic Survey based on the North American Datum, 1983) when collecting locality data in areas where no maps (or only low-resolution maps) are available. With the removal of "selective availability" by the United States government in May of 2000, low-cost GPS units began to provide data with a resolution of 10–20 m in the x-y plane, and 33 m in the z plane (±1 standard deviation [Shaw et al., 2000]). Although this level of resolution is generally good for site location purposes, it is inappropriate for detailed mapping or stratigraphic analyses within bonebeds. Differential GPS technology that employs two or more units calibrated by a precisely surveyed three-dimensional benchmark can provide resolution in the range of a few centimeters (Pryor et al., 2001), and this technology arguably can be used to map within bonebeds (Chadwick et al., 2005).

### Assess Scale, Geometry, and Richness

A variety of methods can be used to estimate the dimensions of a bonebed, including traditional pace and compass techniques, two or more GPS units in differential mode, or potentially ground penetrating radar (Main et al., 2002; Brown et al., 2003), and the data can be recorded on air photos, topographic maps, or a self-made map that includes local landmarks. The scale and geometry of a bonebed figure in the drafting of excavation plans and also provide a crude estimate of the amount of fossil material that may be ultimately recovered (Eberth and Getty, 2005). In turn, these data can help define how much of a site need be excavated and collected to yield a sample amenable to meaningful statistical analyses (Badgley, 1986a; Jamniczky et al., 2003; Blob and Badgley, Chapter 6 in this volume).

   A preliminary assessment of the aerial extent of a bonebed also provides an opportunity to gauge variations in bone abundance, packing, and preservation throughout the targeted interval. Most bonebeds show some degree of spatial patchiness and stratigraphic heterogeneity in relation to the concentration and size distribution of elements (e.g., Gilmore, 1936; Camp and Welles, 1956; Voorhies, 1985; Eberth et al., 2000; Rogers et al., 2001). Correctly identifying and interpreting this kind of heterogeneity in a preliminary study can help the researcher correctly predict where the best prospects are for further excavation, and better understand the origins of the bonebed. For example, a map of the exposed and inferred extent of a *Centrosaurus* bonebed (BB 30) (Fig. 5.1) at Dinosaur Provincial Park that was made before the site was excavated served to effectively identify the highest concentration of elements (darkest area) and, thus, the best area for excavation. This preliminary map and associated preliminary cross sections indicated that the bone-bearing facies at BB 30 has a

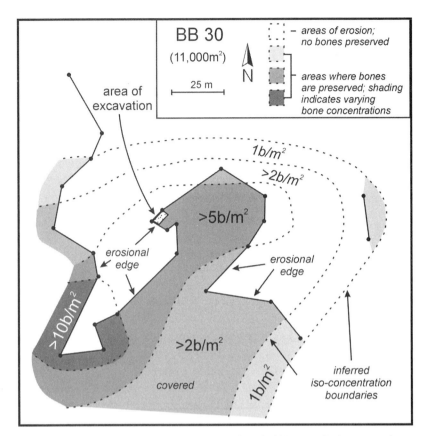

BB 30
(11,000m²)

area of
excavation

25 m

N

- areas of erosion;
no bones preserved

areas where bones
are preserved; shading
indicates varying
bone concentrations

1b/m²

>2b/m²

>5b/m²

erosional
edge

erosional
edge

>10b/m²

>2b/m²

covered

1b/m²

inferred
iso-concentration
boundaries

*Figure 5.1.* Plan view map of the exposed (erosional) edge of Bonebed 30, a monodominant ceratopsian bonebed (Upper Cretaceous) at Dinosaur Provincial Park, Alberta. The concentration of bones along the exposed edge was measured in bones per linear meter. This linear metric was then used to infer bone concentration in the deposit. The overall lenticular shape of the bonebed and the bull's-eye distribution of bone concentration suggest that bones were amassed in a landscape depression such as a pond or lake.

roughly oval shape (in plan view) and lenticular cross-sectional geometry with the highest bone concentration in the center of the deposit.

Careful examination of surface concentrates of fossils exposed by erosion can yield important specimens and can also provide insights into the potential abundance and variety of fossils still entombed within the bed (Leiggi et al., 1994). Although data derived from surface-picking a site can be used to help frame hypotheses about the abundance of material in the site, they must be supplemented by data collected from excavated areas. In most bonebed studies a preliminary excavation is undertaken on a small-scale to document in situ abundance relative to surface concentrates. This is particularly important at localities that are well known to the public. Sites that are regularly "high graded" or "picked over" can be misjudged as exhausted or unfavorable for further excavation. For example, the Red Deer River Valley at Drumheller is host to at least 425,000 tourists each year, and in this area the depletion of surface bone fragments and other fossils by the public is particularly problematic for scientists conducting fossil bone surveys (e.g., Straight and Eberth, 2002; Behrensmeyer and Barry, 2005).

Sites that preserve large, in situ elements may also be misjudged as particularly rich where intense weathering and geologic processes (e.g., the action of swelling clays) result in a surface concentration of small fragments. Conversely, sites that have excellent surface concentrates and that look like good prospects may prove difficult to excavate if the unweathered matrix is heavily cemented. Likewise, surface assemblages that consist of deflation lags (e.g., Berman et al., 1988; Rodrigues-De La Rosa and Cevallos-Ferriz, 1998) can suggest rich in situ assemblages when in fact fossils are actually quite sparse in the host bed.

For vertebrate microfossil localities, a preliminary assessment of surface versus in situ content and concentration can be critical for determining appropriate methods for specimen collection (e.g., surface picking, wet vs. dry screening). A preliminary assessment of the bulk matrix can aid in tailoring the separation processes after bulk collection (e.g., Cifelli et al., 1996). In this context, parameters such as the relative abundance of fossils to matrix, fossil size and robustness, and the lithology and durability of the matrix should be examined and considered carefully. For example, a site that yields only sparse fossils may require that large amounts of matrix are collected for processing, increasing both field and laboratory costs of the project. Likewise, because the durability and response of the matrix to processing procedures can vary—even within a site—an initial laboratory assessment may be required to judge the relative effectiveness of different separation methods and to assess the relative damage to the fossils. A multistep process may ultimately be employed using both physical and chemical means of separation.

### Determine Specimen Quality, Matrix Characteristics, and Mapping Protocols

One or more test pits excavated during the preliminary phase of site development can provide a realistic view of what's ahead, especially in cases when fragile in situ specimens occur as casehardened surface clasts. It is certainly better to know at the onset whether fossils are resilient or particularly susceptible to breakage. Specimens that are highly worn, weathered, fragmentary, or etched can be difficult to excavate without inflicting further damage.

In such cases, block collecting may be more suitable, with fine preparation and mapping undertaken in the laboratory. Likewise block collecting may be a preferable alternative to individual specimen collecting in bonebeds characterized by densely packed, overlapping elements (e.g., the *Coelophysis* quarry at Ghost Ranch, New Mexico [Colbert, 1989]; Agate Fossil Quarry [Peterson, 1906]; Lamy metoposaur quarry [Romer, 1939]). A preliminary examination of bonebed matrix can also help the investigator decide what tools will be required during the excavation phase (e.g., hand tools vs. jack hammers) and/or where the excavation should be focused. For example, variably developed diagenetic cements and localized concretions can impede progress in a bonebed.

With regard to mapping, it may be unrealistic to plan on mapping all elements preserved at the site when (1) a bonebed is densely packed (Colbert, 1989), (2) a bonebed includes a wide variety of bioclast sizes (e.g., a mixture of macro- and microfossil elements) (Fiorillo, 1991a; Fiorillo et al., 2000), (3) fossils in the bonebed are inherently unstable or the boundaries between fossil and matrix are unclear, and/or (4) fossils are widely distributed in three dimensions (e.g., Laury, 1980; Fastovsky et al., 1995). Although the absence of maps can seriously compromise the quality of a taphonomic study, overly complicated mapping can slow the overall collection of data. In such circumstances, it may be best to designate only a portion of the quarry for mapping, or it may be preferable to remove large oriented blocks that can be mapped in the laboratory. If mapping is required, there are several options available to the investigator (see below). Careful consideration of a site's "mapability" can significantly reduce frustrations in later stages of the excavation.

### Plan for Environmental Mitigation

As more rigorous environmental and reclamation regulations come into play, especially on government lands, it is becoming more important to consider postexcavation procedure. Before excavation begins, a bonebed researcher should consider (1) the impact that access to the site and excavation at the site will have, (2) where overburden and/or spoil piles will be stored or dumped, and (3) how a site will be reclaimed after the excavation is completed (seasonally or permanently). Often these issues must be dealt with during the permitting process (e.g., Barnes, 2000). Even when regulations do not require reclamation, some sites may be vulnerable to weathering, runoff, or even theft and vandalism between collecting seasons. Thus, whether mandated or not, a reclamation plan may provide the guidelines and impetus needed to protect and preserve the resource for future generations of researchers.

## STEP 2: WORKING THE SITE

Macrofossil bonebeds are typically documented in detail in the field prior to excavation in an effort to precisely ascertain and preserve spatial associations. Microfossil assemblages are generally collected in bulk with their associated matrix and then separated from the matrix with little or no documentation of original spatial relationships. In this overview of field-based methods, we treat macrofossil and microfossil bonebeds in turn.

### Macrofossil Bonebeds

Most macrofossil sites (single individuals and bonebeds) necessitate that the investigator(s) (1) remove overburden, (2) place reference points and/ or baselines for two- or three-dimensional mapping, and (3) stabilize and extract specimens. However, because each macrofossil site is unique, standardized methods are usually modified on-site to accommodate site-specific problems or complexities.

#### Excavation

Rock that is deeply weathered and appears at first easy to remove may actually be indurated and quite obstinate a few centimeters below the surface. In some instances heavy machinery may be required to prepare a site (e.g., bulldozers [Voorhies, 1985]). It is also often worth the effort to haul a jackhammer to a site (note that a jackhammer can also be used to trench fresh stratigraphic and sedimentologic sections within and outside the bonebed). Workers should carefully monitor the progress of the excavation, checking to see that the bonebed is not compromised and to spot check for fossils in the overburden. Workers should also be on the lookout for any lithologic clues that might indicate the dividing line between overburden and an underlying fossiliferous stratum.

A crew excavating a bonebed should also carefully consider the work to follow after overburden is removed. If overburden is cleared from an area that is greater than what can be excavated during a single season, subsequent weathering and seepage will almost certainly negatively impact portions of the exposed bonebed. Workers should also be careful to place

*Figure 5.2.* Single-element extraction in a macrofossil bonebed (MAD96-01) intercalated in the Anembalemba Member of the Upper Cretaceous Maevarano Formation, northwestern Madagascar. Arrows indicate elements that have been "pedestaled" and are ready for recovery. Several specimens jacketed in plaster are visible in the left portion of the photo.

spoil piles in appropriate locations that do not interfere with drainage or future excavations. Because a team may be required to backfill a site after the study is complete, the spoil pile should also be reasonably accessible at the end of the excavation season. Finally, it is worth mentioning that even spoil piles are worthy of scrutiny. Over time, spoil piles may yield unexpected specimens that were inadvertently exhumed with the overburden. For example, many small specimens collected by researchers with the Canada-China Dinosaur Project from the Iren Dabasu Formation at Iren Nor, Inner Mongolia, were retrieved from bulldozer-generated spoil piles of the Sino-Soviet expeditions in 1959 (Currie and Eberth, 1993).

Excavation methods are extremely varied and are influenced by the nature of the matrix, the durability of the fossils, the tools on hand, previous experience, and even the ambient weather (e.g., Greenwald, 1989; Leiggi et al., 1994). In general, the tendency in bonebed excavations is to expose the limits of individual elements and collect them one at a time, as opposed to collecting large blocks (Fig. 5.2). This approach is encouraged by the fact that smaller blocks are easier (and safer) to move and store than are large multiton packages. However, the process of excavating individual bones can be complex and time-consuming, especially when specimens overlap (e.g., Voorhies, 1985), and this type of careful practice often requires a multiseason commitment (e.g., Rogers, 1990; Fiorillo, 1991a; Ryan et al., 2001).

More rarely, bone density is so high and/or the matrix so hard that removal of large blocks for laboratory preparation is preferable to single-element extraction (Fig. 5.3). The 1981–1982 excavation team at the *Coelophysis* Quarry at Ghost Ranch, New Mexico, excised 15 multiton blocks that were manipulated ultimately by bulldozer, crane, and flatbed truck. Because the bonebed was densely packed with hundreds of overlapping bones, collection of the specimens in large blocks, with subsequent laboratory-based preparation, ultimately inflicted less damage (Colbert, 1989). Likewise, Peterson (1905) and Matthew (1923) describe the complex "boxing" technique used by the American Museum of Natural History to excavate the large blocks containing skeletons of *Diceratherium*, *Moropus*, and *Dinohyus* at the classic Agate Fossil Quarry. More recently, this boxing technique has been used successfully by Sino-American teams to collect bonebed assemblages in the Upper Jurassic Shishugou Formation of China.

Many bonebeds preserve concentrations of fossil bones that range from small unidentifiable fragments to pristine elements. In these instances it may be impractical to collect every exposed specimen. Indeed, the collection of bone pebbles and other taxonomically indistinct bioclasts may in fact be discouraged as a team works against time and weather to collect a "museum-quality" quarry sample. However, robust taphonomic data sets include all bioclasts, whether a pristine skull or a weathered bone pebble (see below). With taphonomic goals in mind, it is often necessary to establish a designated subarea of a bonebed where all fossil material is collected in a systematic fashion regardless of presumed significance (e.g., Rogers, 1990; Fiorillo, 1991a; Getty et al., 1997).

*Baseline*

A permanent physical baseline ensures that the positions of elements in a bonebed are consistently recorded with respect to stable reference points, and in relation to one another. The baseline often consists of a durable string (masonry or kite) that extends between two points (e.g., deeply or permanently anchored poles) in the bedrock. With reference to local magnetic declination, the baseline is generally oriented to form an east-west or north-south horizontal line (e.g., Hunt, 1978). However, the position, orientation, and material nature of the baseline are more often than not determined by convenience. For example, some workers place the baseline well out of the way of the excavators (Organ et al., 2003), while others will place it

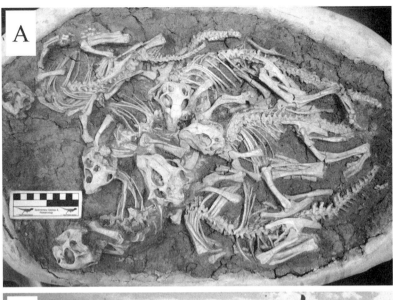

*Figure 5.3.* A. Bonebed block containing six juvenile psittacosaurs from the Lower Cretaceous of Liaoning, China. Block is 70 cm long. Collection of material from bonebeds having densities as seen here requires care in order to minimize specimen damage. Often, the best solution is to collect as large a block as resources allow. B. Three multiton blocks of fossils and matrix from the *Coelophysis* Quarry at Ghost Ranch, New Mexico, collected in 1982 and loaded by crane on flatbed truck for transport to the Carnegie Museum (courtesy of David S Berman, Carnegie Museum of Natural History).

through the middle of the excavation site. A bonebed excavation is often a multiyear undertaking, and thus it is important that the baseline be established in such a way as to be repositioned accurately year after year.

A baseline is typically marked off in meter and/or decimeter increments, and pieces of flagging tape are often added to make the line readily visible to workers. A baseline of string may be kept taut throughout the excavation process or may be extended only when mapping is in progress. In some unusual cases where the bonebed is quite large or there are multiple teams at work, two or more baselines may be utilized (e.g., east-west and north-south baselines of Eberth et al. [2000, their Fig. 12]). Over time a baseline may need to be shifted as excavations (and excavators) migrate across a bone-bearing stratum.

*Mapping*

Mapping specimens can be one of the most important and time-consuming activities in a macrofossil bonebed. A well-constructed map provides data that can be used to interpret the depositional conditions at the time the assemblage was concentrated and/or buried. Map associations and orientations also provide important clues about the taphonomic history of the assemblage. Maps are also essential tools for planning future excavations at a site (Fig. 5.4).

Mapping is most often accomplished by establishing a grid system over the bonebed and aligning it with the baseline or external reference points (e.g., Abler, 1984; Rogers, 1994; Organ et al., 2003). A portable 1 m$^2$ frame that is subdivided into smaller grids (10 cm$^2$) is often employed to improve accuracy. The portable mapping frame is often leveled using adjustable vertical rods that can be temporarily clamped into position after the frame has been positioned using simple bubble-levels attached to the $x$ and $y$ sides of the frame. A plumb bob can be used to ensure that frame is accurately positioned. Once in place above the area to be mapped, the positions of in situ elements relative to the frame can be sketched in plan view (planimetric mapping) on a map sheet. If Geographic Information System (GIS) data are being compiled, data may be directly imported into a computer. General practice is to assign each 1 m$^2$ grid a simple alphanumeric code, with each letter and number corresponding to the meter-wide $x$ and $y$ coordinate rows and columns (e.g., Miller et al., 1996, their Fig. 25). Once the baseline posts are established and a permanent baseline is in place, this alphanumeric grid system can be used year after year.

In the field, each 1 m$^2$ grid can be represented by a single sheet of paper attached to a clipboard or map case. This practice allows for the quarry

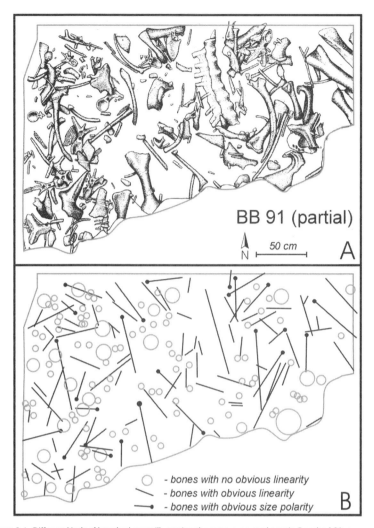

*Figure 5.4.* Different kinds of bonebed maps illustrating the same excavated area in Bonebed 91, a monodominant ceratopsian bonebed (Upper Cretaceous) in Dinosaur Provincial Park, Alberta. A. The stipple illustration provides a clear sense of depth and stacking. B. The use of simple lines and circles helps to document the (1) presence and concentrations of elements, (2) long bone trends, and (3) size polarity of elongate elements.

maps to be handled easily on site. Back at the lab, individual sheets can be digitally scanned and compiled and/or physically assembled to reconstruct the entire bonebed map for further analyses. Additional data, such as trend and plunge measurements and element-specific modifications (weathering, tooth marks, borings, etc.), can also be easily recorded in the field using predesigned hard copy or digital data sheets. Digital data collection also allows for subsets of data in the map (taxa, size ranges, elements) to be isolated and analyzed during later study (e.g., Lien et al., 2002). With regard to GIS applications, Breithaupt et al. (2000) demonstrated the utility of GIS and photogramatic mapping technology at a Morrison bonebed near the classic Howe Quarry, Wyoming. At this locality an on-site three-dimensional imaging system was used to collate data from stations on the ground and from an anchored helium balloon hovering above the site during excavation. Commercial GIS-based mapping and data compilation systems are also finding favor with bonebed workers (e.g., ArcView [Lien et al., 2002]).

### Microfossil Bonebeds

Methods for collecting specimens and data from vertebrate microfossil sites were introduced by Hibbard (1949) and McKenna (1962, 1965) and refined by numerous subsequent workers. Standard modern procedures are relatively straightforward and have been described by Ward (1984), Harris and Sweet (1989), Miller (1989), McKenna et al., (1994), and Cifelli et al. (1996). In short, the goal is to collect bulk matrix that contains fossils, and then to separate the fossils from the bulk matrix via physical (e.g., wet screening through a set of appropriately sized sieves) or chemical (acid preparation) techniques (Fig. 5.5). It is essential to record the volume, weight, and/or number of units of the matrix collected. This information provides a measure of relative fossil richness at the site and can also help guide future collecting.

The process of collecting and identifying fossils within concentrated residue is best managed by spreading a thin layer of the dried residue in a shallow tray marked by grid lines and picking the fossils using the unaided eye and/or a binocular microscope. Generally specimens are sorted on the basis of the research design and placed in labeled vials or trays. An automated point-counter can be used to tally the numbers of common specimens during the picking and identification phase, or specimens can be counted after they have been grouped. The advantage of point-counting during the picking phase is that an unbiased count can be established for a predetermined volume of residue or distribution area within a picking pan.

The picking phase also offers the researcher an important opportunity to assess taphonomic features such as size sorting, rounding, etching, polish, fractures, and weathering (see below). It is important to record these data during the picking phase because many microfossil residues will contain bone fragments whose only value is taphonomic in nature. Some

*Figure 5.5.* Microfossil collecting methods. A. Don Brinkman using a llama to pack raw, microfossil-rich matrix out of the field. B. Crew members of the Mahajanga Basin Project screening microfossil-bearing matrix from site MAD93-35 in Madagascar. Concentrate is drying on burlap sacks in the foreground. C. Advanced microfossil washing/screening system employed to concentrate fossils from the Pliocene Pipe Creek Jr. sinkhole assemblage located near Swayzee, Indiana (photo courtesy J. Farlow).

workers will retain all fossil material from this phase regardless of its apparent immediate value, whereas others may discard unidentifiable fragments (we recommend that you do not do this). The picking phase may also reveal the presence of other associated fossils such as invertebrate and plant remains and minerals that may be of value with regard to interpreting the paleoenvironment and/or paleoecology of the site.

Many of the published microfossil bonebed studies focus on the nature of the included fossil material and provide only cursory descriptions of the bonebed matrix and associated facies. However, detailed sedimentological descriptions in combination with fossil assemblage data can lead to informative taphonomic and paleoenvironmental reconstructions of microfossil bonebeds (e.g., Maas, 1985; Pratt, 1989; Brinkman, 1990; Eberth, 1990; Fastovsky et al., 1995; Rogers and Eberth, 1996; Eberth and Brinkman, 1997; Rogers and Kidwell, 2000).

## STEP 3: COLLECT GEOLOGICAL DATA

The collection of geological data should be a primary goal in every bonebed study, because knowledge that relates to the associated rocks can be used to constrain and/or further refine taphonomic, paleoenvironmental, and paleobiological interpretations, and can also provide significant insight into large-scale climatic and tectonic processes that influenced fossil preservation. Indeed, some bonebed studies focus almost exclusively on the relationship between the geology of a site (or suite of sites) and the associated fossils (e.g., Parrish, 1978; Badgley, 1986b; Hunt, 1987; Sander, 1987; Eberth, 1990; Smith, 1993; Schwartz and Gillette, 1994; Fastovsky et al., 1995; White et al., 1998; Bilbey, 1999; Rogers and Kidwell, 2000; Therrien and Fastovsky, 2000; Paik et al., 2001; Rogers et al., 2001; Rogers, 2005). In these geofocused taphonomic investigations, geological data are combined with fossil-derived information to address questions that relate to paleobiogeography, stratigraphic patterning in the fossil record, and unusual facies associations, among others. Here, we outline a basic approach to targeting and collecting geological data during the course of a bonebed study. Table 5.1 summarizes the types of geological data that generally serve to characterize a bonebed site. Some of this information can be recovered from amassed samples in the laboratory, whereas other information pertinent to geological goals must be collected in the field from sediments in direct association with the bones.

First, literature that pertains to the local stratigraphy and sedimentology should be compiled and reviewed so that a bonebed can be considered from an informed perspective. Stratigraphic reports will yield critical insights into the age of a bonebed assemblage and will also serve to place the site and its host facies in a framework that relates to the local and/or regional geological history. Sedimentological data in these same reports, or in studies specifically devoted to facies description and interpretation, will provide the researcher with invaluable insights into the nature of the rock under scrutiny, and the associated ancient environments that are yielding the desired fossils. When traveling to the field in pursuit of bonebeds we generally bring a small library of books and/or reprints that relate to the site(s) or region under investigation. We also bring a variety of tools, including a hand lens, hammer/hoe-pick, Jacob's staff, Brunton compass, GPS, hand level, hydrochloric acid, rock color assessment chart, sample bags, field notebook, camera, and scale bar, among other odds and ends (picks, brushes, permanent markers, etc.).

With regard to site-specific recovery of geological data, probably the most fundamental source of information is the measured section (Fig. 5.6). A measured section is basically a description of the local rock record from bottom to top (or vice versa, if conditions warrant). It provides a framework for assessing relative age relations among paleontological localities and significant stratigraphic horizons (e.g., ash beds, bonebeds, shell beds) or surfaces (e.g., flooding surfaces, sequence boundaries). It also serves as a framework for the collection and analysis of geological and paleontological samples. If multiple measured sections are described, they can be used to generate cross sections through a bonebed and thereby document the geometry of a productive unit and its facies relationships (Fig. 5.6). Detailed accounts of how to measure a stratigraphic section are provided by Compton (1962), Kottlowski (1965), and Rogers (1994). Differential GPS and GIS technology also holds promise for those studying the stratigraphic relationships of bonebeds (e.g., Lien et al., 2002).

In addition to measuring carefully positioned stratigraphic sections in and around a bonebed, geologic samples should also be collected from the bone-bearing stratum and associated units for use in a variety of laboratory analyses, including (1) petrographic description, (2) clay analysis (specifically X-ray diffraction), (3) geochemical characterization (major and trace element concentrations, isotopic signatures, etc.), and (5) microfossil identification (vertebrate, invertebrate, plant) (e.g., Hunt, 1978; Whitaker and Antia, 1978; Antia and Sykes, 1979; Voorhies, 1985; Thomas et al., 1990; Eberth, 1993; Varricchio, 1995; Coombs and Coombs, 1997; Bilbey, 1999; Eberth et al., 1990, 2000). Paleocurrent indicators both within the bonebed (e.g., sedimentary structures, elongate bones and logs) and in associated units should be measured and recorded as well. This

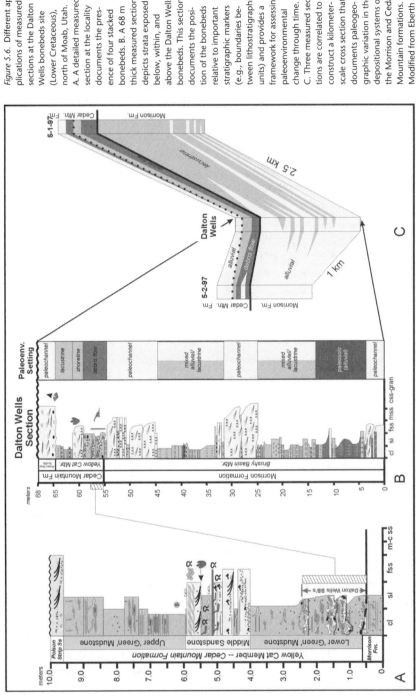

*Figure 5.6.* Different applications of measured sections at the Dalton Wells bonebeds site (Lower Cretaceous), north of Moab, Utah. A. A detailed measured section at the locality documents the presence of four stacked bonebeds. B. A 68 m thick measured section depicts strata exposed below, within, and above the Dalton Wells bonebeds. This section documents the position of the bonebeds relative to important stratigraphic markers (e.g., boundaries between lithostratigraphic units) and provides a framework for assessing paleoenvironmental change through time. C. Three measured sections are correlated to construct a kilometer-scale cross section that documents paleogeographic variation in the depositional systems of the Morrison and Cedar Mountain formations. Modified from Eberth et al. (2006).

type of detailed work within and around a bonebed can yield important insights into formative processes and can provide critical clues into taphonomic history (Trueman and Benton, 1997; Bilbey, 1999; Trueman, 1999, Chapter 7 in this volume; Staron et al., 1999; Rogers et al., 2001; Schröder-Adams et al., 2001). Detailed analyses of geochemistry, clay mineralogy, and microfossil content can also yield important insights into regional correlations, which in turn help to place a bonebed into a meaningful stratigraphic framework.

Finally, the description of a rock body that preserves a bonebed assemblage should always include color (fresh and weathered), texture (grain size, shape, and sorting), lithologic composition (mineralogy), sedimentary structures, degree of induration and type of cement, fossil content (above and beyond the bones), and pedogenic features (Table 5.1). Any type of structural deformation, such as folding, faulting, brecciation, or jointing, should also be noted. The nature of the contacts that bound a bone-bearing stratum is critical to the reconstruction of geological history and therefore deserves serious scrutiny. Contacts are most often characterized as erosional (underlying beds or structures truncated), sharp (no truncation, but knife-edge change in lithology), or gradational. The geometry of a bonebed-bearing stratum (e.g., lenticular, tabular, wedge-shaped) and lateral and vertical facies relationships should be determined whenever possible. Perhaps our best advice would be to always include a well-trained sedimentary geologist as part of the research team, whose main focus will be the careful description and analysis of sedimentary facies.

## STEP 4: COLLECT TAPHONOMIC DATA

Efremov (1940) originally defined taphonomy as the study of the transition of animal remains from the biosphere to the lithosphere. In the case of bonebeds, this already remarkable transition is by definition accomplished by multiple individuals. With this in mind, deciphering the timing of mortality is of particular significance. A given bonebed might record the catastrophic mortality of a gregarious group, at which time researchers may have the opportunity to explore group behavior and ecology (Brinkman et al., Chapter 4 in this volume; Rogers and Kidwell, Chapter 1 in this volume). Alternatively, a bonebed may reflect the long-term accumulation of vertebrate remains in a setting particularly amenable to bone preservation. These markedly disparate histories must be deciphered if accurate paleobiological reconstructions are to be achieved.

Table 5.1. Types of geologic and taphonomic information that should be collected and/or compiled from existing literature during a bonebed study. Based on taphonomic data sheets of Munthe and McLeod (1975) and Behrensmeyer (1991).

| Geologic Data | Taphonomic Data |
|---|---|
| Stratigraphy | Assemblage Data |
| Lithostratigraphy | Sample size (N) |
|   Stratigraphic unit: group, formation, member, bed | Number of individuals (NISP, MNE, MNI) |
|   Position within stratigraphic unit | Taxonomic representation |
| Age Assessment | Relative taxonomic abundance |
|   Radiometrics, magnetostratigraphy, biostratigraphy | Age and sex profiles |
|   Local and regional correlations | Body size |
|   Lithologic and temporal | Skeletal articulation and association |
|   Sequence stratigraphic context | Skeletal completeness/sorting |
|   Significant surfaces, cycles, etc. | *Careful sampling protocol is critical to the above metrics* |
| Sedimentology | Spatial Data |
|   Host-facies specific (focus on the bed that preserves the bones) | Scale and geometry of assemblage ($m^2$ or $m^3$) |
|   Nature of bounding contacts (erosional, sharp, gradational) | Density of accumulation (bones/$m^2$, bones/$m^3$) |
|   Thickness (document lateral variation) | Orientation of bones and carcasses |
|   Geometry (think three-dimensionally, map the bonebed facies) | *Be sure to distinguish structurally induced orientations* |
| | *from primary configurations; be wary of modern deflation lags* |

Grain size, shape, and sorting (consider the bones too)

Color (fresh and weathered, use a color chart)

Sedimentary structures (physical and biogenic)

Mineralogy (include petrographic study)

Cement type (calcite, silica, iron oxides, clay, etc.)

Pedogenic features (rooting, nodules, slickensides, etc.)

Halos, mineral crusts, microstructures around bioclasts

Other fossils (plants, invertebrates, microfossils, ichnofossils)

Local facies associations (focus on the depositional system)

Measured sections, facies maps, cross sections, etc.

Bone Modification Data

Breakage/fragmentation

Abrasion/rounding/corrosion/corrasion

Weathering

Trample/sediment scratch marks

Tooth marks (drags, punctures, gnaw marks)

Bioerosion features (root marks, insect borings, etc.)

Distortion/compaction

Fossilization, geochemical signatures (authigenic minerals, rare earth elements, isotopes, etc.)

*Be sure to distinguish ancient from modern modifications; also be careful to track side-specific features; finally, be aware that remnant soft tissues (e.g., keratin) might be associated with the bones*

Several workers have summarized the types of data that should be collected in a taphonomic investigation of a vertebrate fossil locality (e.g., Efremov, 1940; Munthe and McLeod, 1975; Shipman, 1981; Behrensmeyer, 1991; Lyman, 1994; Rogers, 1994). These previous treatments generally group data into categories that reflect natural subdivisions of inquiry (e.g., geology vs. biology). This very logical approach is also followed here in our consideration of bonebeds. The collection of geological data pertinent to the analysis of a bonebed was addressed in the previous section, and thus the following discussion focuses on three sets of data derived from the bones themselves. These three categories, which are distilled from the "taphonomic evidence" and "variables" of Behrensmeyer (1991, her Tables 1 and 3), include (1) assemblage data, (2) spatial data, and (3) bone modification data (Table 5.1).

After data are collected, the bonebed can be compared with other well-documented sites in an effort to explore taphonomic trends and large-scale paleobiological phenomena (e.g., Dodson et al., 1980; Brett and Baird, 1986; Behrensmeyer et al, 1992; Rogers, 1993; Smith, 1993; Fiorillo, 1999; Trueman, 1999; Behrensmeyer et al., 2000; Englemann and Fiorillo, 2000; Rogers and Kidwell, 2000; Therrien and Fastovsky, 2000; Rogers et al., 2001; Straight and Eberth, 2002). Behrensmeyer (1991) pioneered a graphic presentation method (taphograms) that can be used to compare multivariate data sets derived from several bonebeds (Lyman, 1994; Fiorillo, 1997; Colombi and Alcober, 2004).

**Assemblage Data**

Assemblage-derived data sets characterize the bonebed assemblage or its subsamples from a predominantly faunal perspective (Behrensmeyer, 1991). They serve to describe population statistics (number of individuals, number of taxa, relative abundance of taxa, age spectra, etc.), and also provide information on skeletal articulation and element sorting. Basic approaches to collecting these types of data are reviewed by Munthe and McLeod (1975), Shipman (1981), Behrensmeyer (1991), Rogers (1994), and Lyman (1994).

*Sample Size*

The sample size (N) of a bonebed is the total number of fossils recovered or encountered. It is an overall measure of the fossiliferous nature of a site and is critical to analytical aspects of taphonomic study (Blob and Badgley, Chapter 6 in this volume). Sample size also plays a vital role in subsequent comparisons to other sites (Behrensmeyer, 1991). Unfortunately, most bonebeds preserve a spectrum of elements that range from complete bones to unidentifiable fragments, and it can often be difficult to determine exactly what to recover and count and what to set aside. This problem is often exacerbated by the breakage of fossil material during excavation and/or screen-washing.

In microfossil bonebeds, N is often simply tallied as the total number of specimens that result from matrix removal procedures. This is presumably a scientifically sound approach because it is often virtually impossible to definitively associate fossils collected in these types of sites. In macrofossil bonebeds that preserve a mix of intact elements and fragments, counting specimens requires considerably more thought. Should every identifiable fragment be counted, or should there perhaps be an arbitrary cutoff based on considerations of element completeness? What should be done with fragments that cannot be readily identified? And given the all too often great costs of excavation in macrofossil sites, how much effort should be expended upon the collection and tabulation of unidentifiable fragments? There are not simple answers to these questions, and opinions will likely vary in relation to whether a study is primarily faunal or taphonomic in nature. Regardless, there should always be a final tally that reports either (1) the total number of fossils encountered (all specimens, including pristine elements and bone pebbles), or (2) the total number of identifiable specimens (NISP) (e.g., Voorhies, 1969; Fiorillo, 1991a; Coombs and Coombs, 1997).

*Counting Individuals*

Various tactics can be employed to establish the number of individuals preserved in a bonebed assemblage, and a few of the more common approaches are briefly reviewed here (Chaplin, 1971; Grayson, 1984; Klein and Cruz-Uribe, 1984; Badgley, 1986a). The reader is referred to the Chapter 6 by Blob and Badgley (this volume) for a detailed discussion of the protocols that relate to determining the number of identifiable specimens (NISP), minimum number of elements (MNE), and minimum number of individuals (MNI) in a bonebed.

NISP counts tally the number of identifiable specimens per taxon. Each bone, tooth, or bone/tooth fragment assignable to a given taxon is counted as an individual. This method is presumably most suitable for those sites that have experienced extensive postmortem disarticulation and mixing, either in a physical or temporal sense. With NISP there is an assumption of zero probability of association, and this assumption, while potentially valid for some sites, could easily lead to overcounting and the inflation of taxonomic representation. This is especially true at localities where fragmentation, whether occurring preburial or postexhumation, is pervasive (Badgley, 1986a). Arguably, many of the vertebrate microfossil bonebeds from the Cretaceous of Montana and Alberta (Eberth, 1990; Rogers and Kidwell, 2000) are amenable to this method of counting individuals.

A modification of the NISP method termed MNE strives to accommodate the count-inflating effects of breakage. The goal with MNE is to associate all fragments that could potentially comprise a single skeletal "element" (be it a single bone, such as the humerus, or a complex of bones, such as a hind limb or caudal series). The protocol is fairly lenient in that matching patterns of breakage are not generally required. Associations among MNE-assembled elements are not necessarily factored into the final counts, and thus each element "assembled" through consideration of its potential fragmentary components is typically counted as one individual (Badgley, 1986a).

MNI counts measure the minimum number of individuals in a sample based on a tally of unique skeletal elements. For example, a collection from a bonebed might include 223 identifiable elements, and eight of these bones might be left femora. This observation obviously translates to a minimum number of eight individuals. Additional individuals could be added to this total using arguments that relate to the age, size, and/or preservational state of specimens in a collection. For example, MNI estimates could be based on the frequencies of more than one unique element where size differences between elements are regarded as ontogenetically significant (e.g., four large left humeri and three small right humeri results in an MNI of 7) (e.g., Currie, 2000). The MNI count is arguably well suited for macrofossil bonebeds where there is clear indication of association among the majority of elements (e.g., Hunt, 1987; Rogers, 1990; Fiorillo, 1991a; Coombs and Coombs, 1997; Ryan et al., 2001).

*Taxonomic Representation and Relative Abundance*

Taxonomic representation in a bonebed is measured by compiling a list of taxa confidently identified in an assemblage. Thus, it is a crude measure of taxonomic diversity in a particular site. However, because taxonomic representation in a bonebed is often presented as a mixture of species, genera, and higher-level designations (e.g., Fiorillo, 1991a; Ryan et al., 2001), it can be underestimated. The relative abundance of taxa in a bonebed is determined by counting the estimated number of individuals for each taxon and expressing the number as a percentage of all the known individuals in the assemblage (e.g., Voorhies, 1969; Brinkman, 1990). Like taxonomic representation, the accuracy of this measure is dependent on the taphonomic nature of the site and the appropriateness of the method(s) used to determine the number of individuals. Jamniczky et al. (2003) proposed a methodology based on rarefaction analysis and empirical observations to test the degree of sampling required for NISP counts in microfossil bonebeds to accurately reflect the diversity and relative abundance of taxa. Such tests are potentially applicable to a wide spectrum of bonebed samples, and their use is encouraged to ensure that comparisons of taxonomic content and diversity between sites are valid. Pie charts are often used to illustrate and compare the relative abundance of taxa among bonebeds (e.g., Lehman, 1982; Brinkman et al., 1998).

The assessment of taxonomic diversity and relative abundance in a bonebed yields significant insights on the interpretive front. For example, low-diversity to monospecific assemblages are often interpreted to reflect stressed paleoenvironments or unusual preservational circumstances, or both (Behrensmeyer, 1991). In contrast, diverse assemblages that include a wide array of taxa that vary in body size are often considered to represent less-biased samples of standing vertebrate populations. In assemblages where the relative abundance of taxa is positively correlated with body size (i.e., elements of large taxa are more common than those of small taxa), it is often postulated that taphonomic processes such as hydraulic sorting and/or predation/scavenging have selectively removed more of the remains of smaller animals (e.g., Coombs and Coombs, 1997).

*Age Profiles*

Overall age spectra for an assemblage are obtained by assigning individuals to age categories (e.g., juveniles, subadults, adults) based on dental features, osteological traits, or some combination of both (Voorhies, 1969; Currie and Dodson, 1984; Henrici and Fiorillo, 1993; Ryan et al., 2000). The numbers of individuals in each category are tallied and usually expressed in frequency histograms (e.g., Voorhies, 1969). Ideally, sample size should be sufficiently large to provide statistical support for the interpretation of the resultant distributions. It is generally assumed that an assemblage that preserves a "catastrophic age profile" will include fewer individuals from each successive age class and will better reflect the standing vertebrate population. In contrast, an assemblage characterized by an "attritional age profile" will be biased toward those animals most susceptible to background mortality and will likely preserve more juveniles and

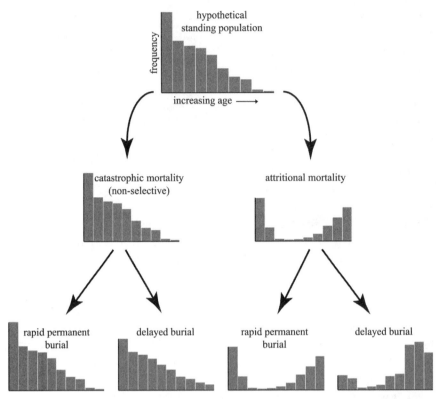

*Figure 5.7.* Theoretical size-frequency histograms for fossil assemblages that accumulate as a result of either a nonselective catastrophic kill event or long-term attrition of predominantly young and old members of a population. Postmortem burial histories further modify preserved age structure via preferential culling of juvenile remains by destructive taphonomic processes. Size-frequency diagrams such as these are typically employed to interpret ontogenetic patterns in fossil vertebrate assemblages (modified from Voorhies, 1969; Shipman, 1981).

old individuals (Fig. 5.7) (Kurten, 1953; Voorhies, 1969; Shipman, 1981; Korth and Evander, 1986; Behrensmeyer, 1991).

Ontogenetic data derived from bonebed assemblages can help a researcher determine whether a given accumulation formed as the result of mass mortality, selective culling, or long-term attrition (e.g., Lyman, 1994; Brinkman et al., Chapter 4 in this volume) and can also inform hypotheses related to modes of preservation given the generally lower resilience of juvenile bones (Behrensmeyer, 1978). For vertebrates in general, age classes are typically assigned based on the degree of skeletal fusion, the size distribution of specific skeletal elements, the texture of bone surfaces, and/or bone histology (e.g. Currie and Dodson, 1984; Henrici and Fiorillo, 1993; Varricchio, 1995; Fiorillo et al., 2000; Erickson et al., 2001; Ryan et al.,

2001). For mammals, eruption patterns in tooth-bearing elements and wear histories of individual teeth also provide telling clues into the age structure of an assemblage (Voorhies, 1969; Webb et al., 1981). This type of information is most readily available after fossils are prepared.

*Body Size*

Determining body size (typically expressed as mass estimates in kilograms) from a sample of fossilized bones or bone fragments is not necessarily a straightforward endeavor. When whole carcasses or sizeable fractions of carcasses are preserved (e.g., Gilmore, 1925; Voorhies, 1985), the estimation of body mass is relatively uncomplicated. However, it is more often the case that only partial remains of individuals are recovered, and when the remains are fragmentary, estimating body size can be problematic, especially when focused at the individual level (as opposed to general body size estimates for a taxon). There are several approaches to the estimation of body size among vertebrates, and it is almost always possible to relegate individuals to at least minimum size categories (Anderson et al., 1985; Behrensmeyer, 1991).

Body size variation among individuals is one attribute used to estimate the age structure of a vertebrate assemblage (see above). Body size also plays a significant role in preservation potential, with smaller individuals generally more susceptible to both transport and destructive taphonomic processes, such as scavenging and weathering (Behrensmeyer and Dechant Boaz, 1980). This bias works both ways, however, in that smaller individuals (and their smaller body parts) are also more readily buried. Given these considerations, determining body size trends among individuals represented in a bonebed assemblage can provide important insights that relate to potential mortality scenarios and postmortem taphonomic pathways.

*Skeletal Articulation and Association*

Bonebeds can preserve collections of articulated skeletons (Voorhies, 1985), assemblages of disarticulated and thoroughly dissociated skeletal elements (Smith and Kitching, 1997), or complex mixtures of skeletal remains in various states of articulation and association (Colbert, 1989; Fiorillo, 1991a; Schwartz and Gillette, 1994; Rogers, 2005). This varied signature of articulation and association reflects the diverse array of taphonomic histories that vertebrate carcasses experience postmortem, and also reflects to some degree ontogenetic and taxonomic controls on disarticulation

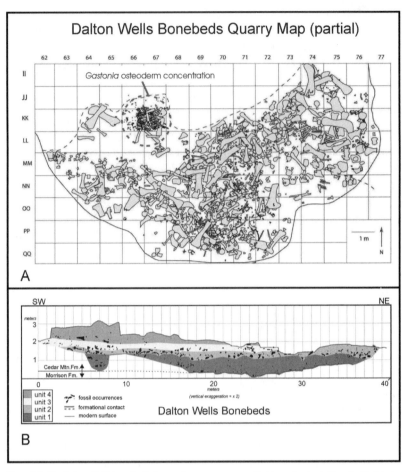

Figure 5.8. The Dalton Wells bonebeds (Lower Cretaceous, eastern Utah) are dominated by disarticulated and unassociated skeletal elements from a variety of dinosaur taxa. However, in plan view (A), a local concentration of osteoderms and limb bones from the ankylosaur *Gastonia* (circled area) indicates that associated partial skeletons with skin patches also occur in the assemblage (courtesy of B. B. Britt). B. A cross section through the site reveals the presence of at least four stratigraphically distinct bone concentrations (from Eberth et al. [2006], used with permission).

patterns (Fig. 5.8) (Gilmore, 1936; Fiorillo, 1991a; Varricchio, 1995; Fiorillo et al., 2000).

The degree of articulation in a bonebed assemblage is generally estimated in a qualitative sense (e.g., "partially articulated" vs. "fully articulated") from observations of maps and/or photographs. Alternatively, a researcher can calculate the number of expected articulations in an intact skeleton and express the degree of articulation at the individual level in a more exact fashion (e.g., 75% of potential articulations remain intact in specimen A, 5% in specimen B, etc.).

The degree of element association in a bonebed is generally gauged again in a qualitative fashion by evaluating the spatial proximity of bones (from maps and/or photographs) in relation to original anatomical associations (e.g., Is the femur close to pelvic elements?). Association is also sometimes considered from the perspective of individual carcasses: Do the disarticulated bones of carcass X remain in relative proximity, or are they dispersed widely among the remains of other animals?

Because the degrees of disarticulation and dissociation in a bonebed assemblage are at least partly a function of time (if burial is rapid or immediate, there is little to no time to disarticulate and scatter remains), their relative extent can be used to estimate the duration of exposure to surface processes (e.g., Dodson, 1971; Behrensmeyer, 1991; Currie, 2000). Numerous actualistic studies have focused on the timing and pattern of passive disarticulation in medium- to large-bodied ungulates (Toots, 1965a; Hill, 1979; Todd, 1983; Hill and Behrensmeyer, 1984; Micozzi, 1991). Lyman (1994) combined the results of studies by Hill (1979) and Hill and Behrensmeyer (1984) to show that during decay major body parts that are connected by synovial joints to the main body (e.g., forelimb, lower jaw, cranium, hindlimb) tend to separate consistently early in the disarticulation sequence (weeks to months) unless unusual environmental conditions (anoxia, rapid burial) come into play.

Living animals can also have an impact on the timing and extent of skeletal disarticulation and dissociation. Numerous studies describe the influence of modern predators/large scavengers on disarticulation (Weigelt, 1927, 1989; Haynes, 1980; Blumenschine, 1986; Domínguez-Rodrigo, 1999; see reviews in Behrensmeyer, 1991 and Lyman, 1994). Predation and scavenging of medium-to-large modern ungulates by mammalian carnivores appears to result in a relatively predictable sequence of disarticulation that correlates with the amounts of the soft tissue surrounding the targeted parts (Blumenschine, 1986). Fortunately, indication of predation and/or scavenging in a bonebed can often be inferred from the presence of shed teeth (e.g., Buffetaut and Suteethorn, 1989; Ryan et al., 2001) or tooth marks on elements (e.g., Voorhies, 1969; Fiorillo, 1991a, 1991b; Ryan et al., 2001; Rogers et al., 2003). Trampling in and around a bonebed may also accelerate disarticulation rates. Evidence for trampling is often preserved in the form of scratch marks, abrasion/polish, breakage, and bone orientation patterns (Hill and Walker, 1972; Fiorillo, 1984, 1987, 1988a, 1989; Lyman, 1994; and others).

Finally, although actualistic studies provide a wealth of taphonomic information about disarticulation patterns in modern large-mammal communities, it is critical to remember that some anatomical features in extinct vertebrates will likely result in disarticulation patterns that do not lend themselves to direct comparisons with modern ungulates. For example, it is very likely that dermal armor and ossified tendons inhibited or at least influenced disarticulation patterns in some groups within the Dinosauria. The bonebed worker should always keep in mind both the usefulness and potential limitations of modern analogs and actualistic observations when attempting to reconstruct phenomena in the fossil record.

## Skeletal Completeness

Skeletal completeness compares actual skeletal element representation in a bonebed with expected skeletal element representation calculated using MNI or, in some cases, MNE. For any given taxon, this completeness estimate provides a measure of the degree of preservational biases (Fig. 5.9). This measure of skeletal representation can help the researcher sort through the various impacts of physical and biological sorting agents and ultimately frame meaningful hypotheses related to overall taphonomic history. Skeletal completeness calculations can also provide a measure of preservational biases by taxon or age class and are a first step in evaluating relative degrees of time averaging and authochthony versus allochthony in a bonebed assemblage (Behrensmeyer, 1975).

Skeletal completeness is calculated by dividing the number of occurrences of a given element by the expected frequency of occurrence for this element in a complete skeleton multiplied by the MNI (e.g., Hunt, 1990; Fiorillo et al., 2000). Given the dependence on decisions related to counting protocols, skeletal completeness estimates are always subject to some degree of controversy and/or error (Blob and Badgley, Chapter 6 in this volume). Perhaps the best advice is to be entirely transparent in relation to the protocols used to determine completeness estimates, so that related findings can be critically evaluated by future workers.

With completeness estimates in hand, the researcher can entertain a variety of hypotheses that relate to the sorting and selective removal of skeletal elements. For example, if small elements such as phalanges, ribs, and caudal vertebrae are underrepresented it might reflect preferential culling via ingestion by carnivores feeding from the death assemblage. Alternatively, small and/or fragile elements may have been selectively winnowed from a site by vigorous fluvial currents, or perhaps obliterated by trampling or chemical dissolution. Fortunately, a wealth of information based on actualistic experimentation and observation can be considered in conjunction with bone census data to help diagnose the signatures of potential sorting agents in a bonebed assemblage (e.g., Voorhies, 1969;

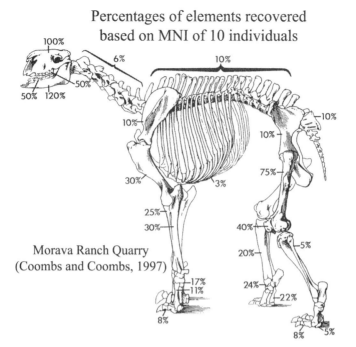

Percentages of elements recovered
based on MNI of 10 individuals

100%

6%        10%

50%

50%   120%

10%        10%

10%

30%        3%

75%

25%

30%        40%

Morava Ranch Quarry        20%
(Coombs and Coombs, 1997)

5%

17%        24%
11%

22%

8%

8%        5%

Figure 5.9. Diagram modified from work by Coombs and Coombs (1997) showing skeletal representation in the taxon *Moropus elatus*. Skeletal remains of at least 10 *M. elatus* individuals were recovered in
the Morava Ranch Quarry. Cranial components and large limb elements are in general better represented
than ribs, vertebrae, or components of the manus and pes.

Chaplin, 1971; Dodson, 1973; Frey et al., 1975; Hill, 1979, 1980; Korth,
1979; Haynes, 1981; Behrensmeyer, 1982, 1988; Grayson, 1984; Klein
and Cruz-Uribe, 1984; Lyman, 1984, 1994; Boaz and Behrensmeyer, 1986;
Coard and Dennell, 1995). It should also be stressed that accurate geological information can play a critical role in distinguishing among potential
sorting agents.

With regard to expressing "completeness" estimates and the extent of
sorting in a bonebed assemblage, most researchers utilize "Voorhies groups."
Voorhies (1969) conducted flume experiments using disarticulated coyote and sheep skeletons and recognized three hydraulic-transport groups.
Each successive group showed an increased resistance to movement and
dispersal, starting with (1) ribs, vertebrae, sacrum and sternum, some
phalanges, then (2) limb bones, some phalanges, and finally (3) skull
and jaws. Behrensmeyer (1975) formalized reference to these transport
categories as Voorhies groups I, II, and III, respectively. Determining the
relative abundance of skeletal elements in these categories can serve

to characterize the degree to which an assemblage is winnowed (leaving a parauthochthonous residual lag dominated by group III) or perhaps transported (yielding a more-or-less hydraulically compatible collection of allochthonous elements dominated by groups I and II) (e.g., Behrensmeyer, 1975; Coombs and Coombs, 1997). In an effort to further address the question of authochthony versus allochthony, Behrensmeyer and Dechant-Boaz (1980) developed the teeth/vertebrae ratio (T/V), which amplifies the significance of the presence/absence of Voorhies group I and Voorhies group III elements. Where the ratio is >1 in mammalian assemblages, some degree of hydraulic sorting is usually assumed (e.g., Fiorillo, 1988a, his Table 5). Skeletal data assigned to Voorhies groups are traditionally presented in tables (Voorhies, 1969; Ryan et al., 2001), ternary diagrams (Behrensmeyer, 1975; Fiorillo, 1988a; Fiorillo et al., 2000), and histograms (Coombs and Coombs, 1997).

Over the years numerous studies have relegated collections of fossil bones derived from a wide variety of taxa (e.g., dinosaurs, mammals, turtles, crocodiles) to the three Voorhies groups to reconstruct sorting histories (e.g., Dodson, 1973; Behrensmeyer, 1975; Hunt, 1978; Shipman, 1981; Lehman, 1982; Fiorillo, 1988a, 1991a; Rogers, 1990; Blob and Fiorillo, 1996; Blob, 1997; Coombs and Coombs, 1997; Ryan et al., 2001). However, bonebed workers are herein strongly cautioned to consider whether the Voorhies group concept is applicable to animal groups (living or extinct) that deviate in size and osteology from the original target group (sheep and coyotes).

### Spatial Data

Spatial data sets derived from direct measurements of bones within bonebed assemblages serve to clarify the geometry (e.g., lens, ribbon, sheet) and extent (generally expressed in square meters), or at least the minimum extent, of a bone concentration, and shed light on the two- and three-dimensional distribution of skeletal remains (packing, patchiness, trends in preservational quality, etc.). Orientation data collected from individual elements provide important insights into transport histories, sorting phenomena, and bioturbation. Data that serve to characterize a bonebed assemblage from a spatial or geometric perspective are derived from in situ measurements of fossils in their host facies prior to element extraction and are generally compiled in the form of two- or three-dimensional maps (Fig. 5.8), photographs, and tallies of directional measurements (trend, plunge). Directional measurements in turn are regularly displayed using rose diagrams and/or stereonets (see below).

Assessing the lateral extent and geometry of a bone accumulation is of primary importance because the overall scale of a deposit can help constrain potential formative scenarios. For example, a localized occurrence of skeletal remains that spans only a few square meters and consists of relatively few carcasses might reflect background mortality at a watering hole, den, or other frequently visited locale. In contrast, widespread concentrations of vertebrate remains that include many tens, hundreds, or even thousands of individuals may represent some type of event mortality (Behrensmeyer, 1991) or the accumulation of biological debris during periods of sediment starvation or erosion (Smith and Kitching, 1997; Schröder-Adams et al., 2001). Regardless of potential formative scenarios, tracking bonebeds to their termination is of utmost importance but is not always plausible. In cases where erosion or limited exposures inhibit the tracing of a bone-bearing stratum, minimum dimensions should be determined. In addition, laterally adjacent exposures should be thoroughly explored to determine if a bone-bearing stratum extends beyond the original locality. Ideally, a bonebed should be mapped throughout its entire extent. Aerial photographs and satellite imaging may prove particularly useful for tracking extensive bonebed horizons.

The geometry of a bone accumulation is also worthy of scrutiny, as telltale configurations do exist. For example, a linear assemblage of vertebrate remains could reflect a wave-generated concentration of bones and/or carcasses along a strandline (e.g., Sander, 1989; Rogers et al., 2001), or perhaps an accumulation of skeletal remains in a linear channel or trough. Similarly, a roughly circular accumulation could record death in and around a small pond or waterhole environment (Saunders, 1977).

Patterns in the distribution of skeletal remains within a bonebed should also be carefully documented with mapping and photography. Bones can be relatively evenly dispersed across a surface or within a stratum, or they can be highly patchy in their distribution (Fig. 5.10). A bone assemblage can be densely packed with skeletal material (e.g., Peterson, 1905; Hill and Walker, 1972; Lawton, 1977; Norman, 1987), or relatively sparse and patchy in its preservation of skeletal remains. Bone-on-bone contacts between anatomically unrelated elements can reflect a variety of physical and/ or biological processes, such as hydraulic transport, winnowing, erosion, biological gathering, trampling, and so forth. In contrast, assemblages showing few if any bone-on-bone contacts in graded to massive deposits may reflect mass flow or bioturbation, among other phenomena (Fastovsky et al., 1995; Eberth et al., 2006). The degree of bone concentration

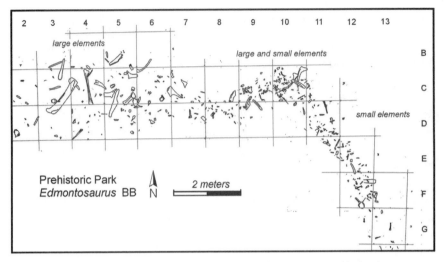

*Figure 5.10.* Patchiness in bone concentration, bone size, and taxonomic composition is a characteristic feature of the Prehistoric Park bonebed (Horseshoe Canyon Formation, Upper Cretaceous) near Drumheller, Alberta. Large limb elements of an *Edmontosaurus* are abundant in the western portion of the site. Small fragmentary ornithischian bones and scattered teeth of large and small theropods are predominant in the eastern sector of the site. Courtesy of the Royal Tyrrell Museum, Day Digs Program.

in macrovertebrate assemblages is generally reported as bones per unit area (e.g., bones/m$^2$). In microvertebrate localities, bone density is generally reported as bones per unit weight (e.g., bones/kg).

Size-sorting trends may also be evident within a bonebed (Fig. 5.10), with perhaps large individuals or large skeletal elements segregated to a particular sector of an accumulation. A bonebed assemblage may also show internal patterning in degrees of fragmentation or weathering. A two- or three-dimensional depiction of the spatial distribution of bones can also clarify whether there are potential associations among elements, especially when multiple animals are present and represented by broken and disarticulated remains (e.g., Britt et al., 2004). Such analyses can help to identify the presence, distribution, and taphonomic significance of subpopulations of bones in a bonebed (e.g., Voorhies, 1985; Britt et al., 2004). Accurate measurement of in situ bone density and distribution requires well-planned excavation and consistent mapping.

The thickness of a bonebed deposit should also be carefully ascertained and documented with cross sections and three-dimensional maps. Trends in thickness can reveal the topography of the surface upon which accumulation occurred, which in turn can help an investigator narrow potential depositional settings (Fig. 5.11). Freshly trenched vertical sections through a bonebed provide perspective on how specimens relate to beds and bounding surfaces and thus may yield insights into processes that

David A. Eberth, Raymond R. Rogers, and Anthony R. Fiorillo

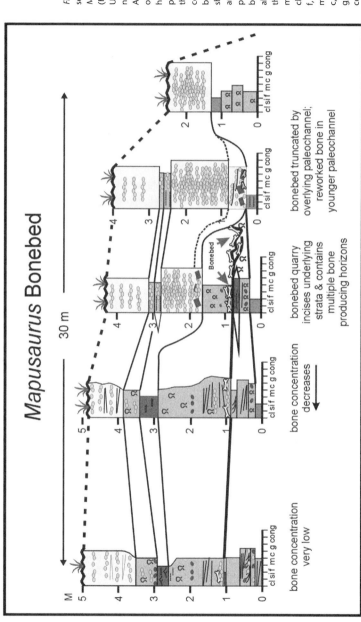

*Figure 5.11.* Cross section through the *Mapusaurus* bonebed (Huincal Formation, Upper Cretaceous) near Plaza Huincal, Argentina. The base of the bonebed exhibits a concave-up profile and indicates that the bones were concentrated near the base of a paleochannel shortly after avulsion and incision. Multiple concentrations of bone suggest recurrent alluvial reworking in the paleochannel. M, meters; m, meters; cl, claystone; si, siltstone; f, fine sandstone; m, medium sandstone, c, coarse sandstone; g, granules; cong, conglomerate.

*Mapusaurus* Bonebed

30 m

bone concentration very low

bone concentration decreases

bonebed quarry incises underlying strata & contains multiple bone producing horizons

bonebed truncated by overlying paleochannel; reworked bone in younger paleochannel

Bonebed

relate to original concentration and burial (Fig. 5.8). For example, where vertical profiles and maps show the majority of bones concentrated in thin pavements along bounding surfaces, hypotheses of formation should focus on scenarios that include active turbulent flows, or perhaps conditions of sediment starvation that encouraged enrichment in bioclastic debris prior to final burial. Further resolution can be ascertained by grain size analyses. Elements associated with coarse-grained facies may prove to be lag deposits modified and concentrated by low-viscosity, high-energy flows (Lawton, 1977; Ryan et al., 2001). Conversely, elements concentrated at the base of fine-grained facies (mudstones) may be lag deposits or an assemblage that was ultimately buried in a low-energy setting by suspension deposits (be it aqueous or subaerial). Finally, where bones occur dispersed throughout a massive sedimentary deposit, potential formative scenarios might include (1) incremental addition of bones over time to a slowly accumulating deposit (e.g., a lake slowly filling with sediment), (2) bioturbation and mixing of bones in a setting prone to biological activity (e.g., waterhole), (3) transport and deposition by high-viscosity flows (e.g., cohesive debris flows), or (4) some combination of the above (Andrews and Ersoy, 1990; Sander, 1992; Rogers, 2005).

### Orientation of Individual Bones and Carcasses

Studies of element orientation focus on the three-dimensional in situ arrangement of populations of fossils in the bonebed host facies. Its measures derive from an assessment of element orientations within a three-dimensional coordinate system (compass direction and deviation from the horizontal). With bone orientation data in hand, a researcher can evaluate a wealth of hypotheses that relate to the overall history of a bonebed assemblage.

Most of the emphasis on bone orientation has focused on deciphering the nature and influence of surface flows on bone assemblages. For example, experiments and field observations indicate that turbulent surface flows ("normal" stream flow) can orient the long axes of bones (and other elongate objects such as trees) either parallel or perpendicular to the trend of flow, depending on the energy of the flow, shape of the element, water depth, and the geometry of associated bedforms (ripples, dunes) and macroforms (in-channel bars) (Voorhies, 1969; MacDonald and Jefferson, 1985; Behrensmeyer, 1990; Morris et al., 1996). In addition to diagnosing the general sense of ancient currents, upstream and downstream directions can often be determined by examining the polarity of elongate elements. For example, Voorhies (1969, appendix) noted that larger ends of fresh limb bones from coyotes, sheep, badger, and rabbit were preferentially oriented upstream. Moreover, bones imbricate with an upstream dip.

Data derived from bone orientation can also be used to identify other causes of bone dispersion, such as trampling (Fiorillo, 1988a, 1988b, 1989) and convection (Van Valkenburgh et al., 2004). Similarly, high-viscosity flows can be diagnosed by the presence of a chaotic three-dimensional pattern in association with poorly sorted massive sediments (Eberth et al., 2006).

Orientations of elongate skeletal elements (those that have a maximum axis that is at least twice as long as the intermediate and minimum axes, such as limb bones and ribs) in a bonebed assemblage should be measured using a Brunton compass (see Compton [1962, p. 21] for a description of a Brunton) or comparable device (less expensive compasses will also do the job). Trend, plunge, and polarity data should be collected from all suitable elements. Be sure to note whether bones are in contact or close proximity, because such association may indicate bone interference or soft-tissue connections, both of which can influence ultimate orientations.

The trend of a bone is the compass direction of the vertical plane that includes the long axis of the bone. To measure the trend, orient the axial line of a horizontally leveled Brunton compass parallel to the vertical plane that includes the long axis of the bone (Rogers, 1994, Fig. 3.3A). If the elongate axis of a bone is horizontal (or very nearly so), trend is a bidirectional property (e.g., the bone trends N45E and S45W), and there is no plunge. The plunge of a bone is the vertical angle between the long axis of the bone and a horizontal plane (Rogers, 1994, Fig. 3.3B). To measure plunge, orient the edge of a Brunton compass parallel to the long axis of the bone and rotate the clinometer level until the bubble is centered. Remember to correct plunge measurement if bones occur in structurally dipping strata. If a bone is not horizontal, it by convention trends in the direction of plunge (e.g., N45E, 25°, where 25° indicates the angle of plunge). Alternatively, trend data can be presented as an azimuth, which in the previous example would be presented as 045, 25°. Bone polarity simply refers to whether one end of a bone is significantly larger than the other (e.g., fragmentary limb bones with one intact end and vertebral elements with elongate neural spines and centra [Hunt, 1978]). Data relating to bone trend, plunge, and polarity (if present) should be recorded on the field map and in a field notebook. As is the case when recording similar data from cross-stratified sedimentary rocks, it is always good to note the relative sizes of the bones from which the data are retrieved. For example, a given surface flow may act preferentially on smaller elements and leave larger bones unaffected.

Where bones show only minor divergence (<5°) from a horizontal datum, two-dimensional data sets that record the trends of long bones provide sufficient data for statistical analysis and are commonly displayed in rose diagrams (Shipman, 1981; Kreutzer, 1988; Hunt, 1990; Henrici and Fiorillo, 1993). Where bone orientations show a significant divergence from the horizontal due to complex biostratinomic or geologic (structural) histories (e.g., Hunt, 1978; Fiorillo, 1988a; Varricchio, 1995), three-dimensional stereographic presentations are much more informative (Fig. 5.12) (Toots, 1965b; Voorhies, 1969; Fiorillo, 1988b). In instances where original bone orientations have been shifted significantly by tectonics (e.g., Dinosaur National Monument), a stereographic projection can be used to rotate bones to their original predeformation orientations (Ragan, 1973; Knox-Robinson and Gardoll, 1998).

A variety of statistical tests can be applied to the spatial data set to determine whether the element orientations represent a significant departure from a random distribution. However, statistics applied to radial (and three-dimensional) data sets are notoriously complex (Curray, 1956; Davis, 1986). For example, simple analysis of variance (Kreutzer, 1988) may generate spurious results, depending on the size of the petals designated for a given rose diagram. Therefore, the reader is strongly encouraged to review the literature on this topic devoted to bones (Morris et al., 1996; Richmond and Morris, 1996). There are a variety of statistical packages available that provide a range of tests based on different conditions, any one of which may or may not be appropriate depending on site-specific circumstances (e.g., Oriana, www.kovcomp.co.uk/oriana/).

### Bone Modification Data

Bones and teeth are exposed to a wide variety of processes in the postmortem environment that can alter macroscopic and microscopic morphology. These processes include but are certainly not limited to trampling, weathering, scavenging, rooting, burrowing, and chemical/biologic etching and corrosion (e.g., Clark et al., 1967). Given their links to particular processes and phenomena, bone modification features provide some of the most telling insights into the origins of bonebed assemblages, and they provide pivotal clues into the duration of exposure and residence time in the "active zone" prior to final burial (discussions in Behrensmeyer, 1991; Lyman, 1994). We regard the geochemical alteration and diagenesis of bones as a separate topic (Trueman, Chapter 7 in this volume; Fricke, Chapter 8 in this volume).

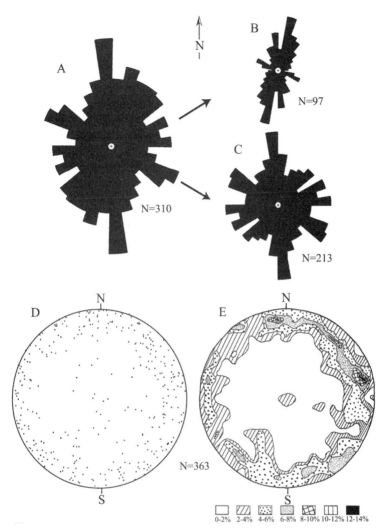

*Figure 5.12.* Rose diagrams depicting bone orientations in the lower Agate bonebed (modified from Hunt, 1990). A. In this plot azimuth readings derived from two superposed horizons are combined. Measurements are clustered in 10° classes. The overall pattern is consistent with a relatively random distribution of elements. When the measurements are segregated by horizon, a preferred alignment is evident in a fine-grained sandstone bed (B), whereas a more random distribution characterizes bones preserved in an associated calcareous tuff (C). D. Stereographic projection of bone orientation data collected in Hazard Homestead Quarry (modified from Fiorillo, 1988a). In this mode of graphic display bones plunging steeply are readily apparent, and they plot well inside the perimeter. E. Contour diagram of data presented in the stereoplot. No preferred orientation is apparent in the data.

Bone modification features can certainly be identified in situ as a bonebed is excavated, but they are more frequently discovered after preparation of specimens (e.g., Fiorillo 1988a, 1991a; Jacobsen, 1998; Rogers et al., 2003). However, many clues that pertain to the generation of modification features can be lost or obscured if the sample is excavated before thorough taphonomic characterization. Subtle features of the sedimentary matrix encasing bones, such as alteration halos, mineral crusts, fine root traces, and localized indication of invertebrate bioturbation (small burrows or borings), may be obliterated or otherwise obfuscated during collection and preparation (Rogers, 1994). Side-specific features (e.g., Rogers, 1990) may also be overlooked. Thus, it is always necessary to closely examine bone-sediment relationships prior to element extraction. Moreover, determining the relative abundance of bone modification features in a bonebed based on their occurrence in a prepared and curated collection raises concerns about sampling and collecting methods. For example, tooth marks may occur more frequently on fragmentary limb bones and ribs and less so on intact vertebrae and pristine cranial elements due to carnivore/scavenger preferences (e.g., Haynes, 1980). If certain elements are over- or underrepresented in a sample due to collecting bias (leave no fragments behind!), data that relate to the relative abundance and distribution patterns of tooth marks in the bonebed will be inaccurate.

With regard to the collection of bone modification data, we recommend an organized two-phase approach that incorporates both field and laboratory/collections-based observations. A spreadsheet that includes all potential modification features should be generated and distributed to those working a bonebed (Fiorillo, 1991a), and this checklist can be efficiently reviewed and completed as bones are exposed and extracted. A few moments spent educating the excavation team on the many manifestations of bone modification can yield copious amounts of critical data that might otherwise be lost. Once compiled, these field-based spreadsheets can be compared with a second set of observations derived from the examination of prepared specimens. We also strongly recommend thorough photo-documentation of all modification features both in the field and in the laboratory. Finally, collect the data at face value, and be aware that diagenesis and other postburial effects may mimic or obliterate original bone modification features.

Lyman (1994) explored the origin and significance of a wide variety of bone modification features, and the reader is referred to his text for a comprehensive consideration of the topic. Here we review a selection of some of the more common and telling bone modification features found in bonebed assemblages.

*Breakage*

Bone breakage refers to the occurrence of breaks (completely through-going), fractures, and punctures. Bones break when applied forces reach or exceed the work necessary to fracture. Fresh bones are particularly resilient, and it takes considerable force to break them. However, post-mortem degradation of bone tissue can greatly diminish bone strength, and weathered bone is more readily fractured and pulverized. The agents of bone breakage in natural systems are primarily biological in origin (e.g., carnivore mastication, trampling), and according to Behrensmeyer (1982, 1991), it is unlikely that many bones are broken by impact during phys-ical transport (e.g., fluvial currents). Thus, many of the broken bones in ancient fluvial deposits probably entered the system in a broken state (but see Ryan et al. [2001] and Britt et al. [2004] for an alternative view).

Morphological patterns of bone breakage have been intensely studied by the archaeological community (e.g., Biddick and Tomenchuk, 1975; Shipman, 1981; Davis, 1985; Johnson, 1985; Gifford-Gonzalez, 1989; Mar-shall, 1989; Lyman, 1994), and these workers have erected detailed classifi-cation systems that incorporate both macroscopic and microscopic aspects of osseous fracture. Interestingly, a review of the paleontological literature indicates that bonebed workers have consistently employed a rather sim-plified version of these classifications (e.g., Myers et al. 1980; Fiorillo, 1988a, 1991a; Rogers, 1990; Behrensmeyer, 1991; Coombs and Coombs, 1997; Fiorillo et al., 2000; Ryan et al., 2001) that serves essentially to iden-tify the relative proportions of spiral and oblique breaks versus stepped and transverse breaks. This basic approach reflects the primary goal of identifying the breakage of green versus weathered/mineralized bone tis-sue (Behrensmeyer, 1991; Coombs and Coombs, 1997; Ryan et al., 2001). This distinction arguably suffices for the purposes of most bonebed inves-tigations focused in deposits older than the Plio-Pleistocene (prehominid). Additional criteria useful for the identification of prefossilization breakage within a bonebed include the presence of matrix on broken surfaces and the isolation of broken elements.

Most macrofossil bonebeds contain a significant number of broken elements (e.g., Fiorillo, 1988a; Rogers, 1990; Varricchio, 1995; Coombs and Coombs, 1997), and breakage frequency may range from ubiquitous (in the case of bone-fragment lags [Smith and Kitching, 1997]) to nearly nonexistent (as in sites dominated by articulated skeletons [Voorhies, 1985]). Patterns of breakage frequency can usually be correlated with el-ement size and shape. When breakage is due to dynamic loading (e.g., high-force impacts resulting from trampling by megaherbivores), breakage

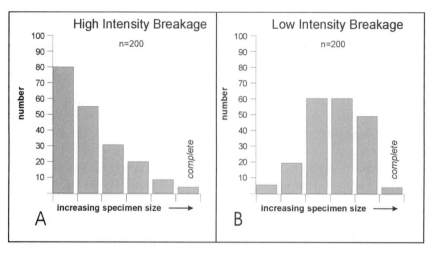

*Figure 5.13.* Graphs showing high (A) and low (B) breakage-intensity patterns in two hypothetical assemblages of fossil limb bones from the same taxon. Although both assemblages have the same number of complete elements, bones are more thoroughly broken in A than in B.

tends to occur preferentially in large and less-compact elements (e.g., Coombs and Coombs, 1997; Ryan et al., 2001). In contrast, breakage due to selective predation/scavenging should show little if any correlation to the size and shape of the original elements but should instead reflect anatomical preferences of the predator/scavenger (e.g., Voorhies, 1969). Breakage frequency can be measured as a function of total N, NISP, or both (Coombs and Coombs, 1997). A relative measure of breakage intensity in an assemblage can be approximated by examining the shape of size-frequency curves for given elements (e.g., limb bones and ribs, Fig. 5.13).

Microfossil bonebeds also tend to preserve abundant broken elements, and while a sizeable fraction of these elements may record prefossilization modification, many others probably reflect the impact of collection histories that include long-distance transport in sacks and subsequent screen-washing. Thus, it is very important to document the nature of breakage in microfossil bonebed collections so that pre- and postfossilization breaks can be distinguished. Studies by Mellet (1974) and Dodson and Wexlar (1979) have examined the nature of predator-induced breakage in vertebrate microfossil assemblages, but much remains to be learned. It is likely that overall patterns of breakage in microfossil sites will differ substantially from those typical of macrofossil sites given the dramatic differences in element size and taphonomic history.

David A. Eberth, Raymond R. Rogers, and Anthony R. Fiorillo

Finally, it is worth noting that a review of the bonebed literature (Behrensmeyer, Chapter 2 in this volume; Eberth et al., Chapter 3 in this volume) indicates that bone breakage in macrofossil bonebed sites may be less common in the Upper Paleozoic than it is in the Mesozoic and Cenozoic. This perhaps reflects (1) damped breakage potential in fine-grained aqueous depositional settings (much of the Paleozoic record is preserved as pond/lake assemblages) or (2) fewer large vertebrate tramplers, predators, and scavengers in the more ancient Paleozoic communities.

## Abrasion and Corrosion

Abrasion refers to the physical removal of bone/tooth surfaces (Bromage, 1984; Shipman and Rose, 1988; Behrensmeyer, 1991). It is often linked to an increase in the rounding of an element, and it may serve to obscure or remove evidence of other surface modification features, such as tooth marks and scratch marks (see below). Polish on bone surfaces is typically considered a form of abrasion (Shipman et al., 1984; Behrensmeyer, 1991; Rogers and Kidwell, 2000, see their Fig. 7).

Factors presumed to control the nature and extent of abrasion on bones include grain size of abrading particles, the hardness of those particles, distance of transport, duration of exposure, relative amount of soft tissue present, overall condition of the bone prior to abrasion, and the nature of the abrading agent (e.g., sand-blasting vs. bioturbation). Comparable forms and degrees of abrasion can potentially be generated by different agents (such as trampling, wind, and water flows), and skeletal elements can respond variably to abrasion depending on their condition (e.g., fresh vs. weathered). Thus, any interpretation of abrasion should be framed in the context of multiple potential causative scenarios, weighed against additional taphonomic indicators, and considered in an appropriate geological context.

Relative states of abrasion within a bonebed assemblage can be used to identify mixed assemblages and assess the likelihood of time averaging. Abrasion can also serve as a useful relative measure of distance of transport and/or duration of exposure to various hydraulic, aerodynamic, chemical, or biological processes (e.g., Shipman, 1981; Shipman and Rose, 1984, 1988; Lyman, 1994). Hunt (1978) proposed a simple subjective scheme for assessing the level of abrasion in a bonebed assemblage that ranked bones from 0 (least worn) to 3 (most worn). Shipman (1981) advocated a comparable scheme that employed three basic states of abrasion: (1) little or no abrasion, (2) moderate abrasion, and (3) heavy abrasion. Subsequent workers (e.g., Fiorillo, 1988a; Ryan et al., 2001) embellished these basic yet ultimately effective approaches.

Corrosion is a somewhat more enigmatic modification feature that generally refers to the loss of bone surface through chemical or biochemical means. The macroscopic and microscopic modifications that reflect corrosion are not always easy to identify or differentiate from abrasion and weathering in general (see below), and accordingly laboratory-based techniques that measure chemical changes in the composition of bones and teeth (such as energy dispersive spectrometry [Denys et al., 1992]) are sometimes used to document corrosion. The study of corrosion in modern bone assemblages has historically focused on vertebrate microfossil assemblages that have passed through the digestive tracts of avian or mammalian predators (Mayhew, 1977; Richardson, 1980; Kusmer, 1990; Denys et al., 1992). Some work has also focused on the nature of corrosion in and upon modern soils in relation to pH (e.g., Andrews, 1990), and this work has been extrapolated to the fossil record (see work by Fiorillo [1998] and Fiorillo et al. [2000] for examples from the Triassic and Jurassic). The potential for corrosion resulting from contact with organisms such as algae (e.g., Behrensmeyer, 1991) and bacteria and fungi (Hassan and Ortner, 1977) has also been addressed (and here this is considered bioerosion, see below).

Like abrasion, the extent of corrosion in a bonebed assemblage is probably best assessed with a subjective measure that pragmatically characterizes the extent of modification (e.g., none, minimal, moderate, maximal). Given the potential links to ingestion (e.g., Denys et al., 1992), it might also be quite informative to compare the degrees of corrosion with the desirability (from the carnivore perspective) of particular skeletal elements.

The combined effects of abrasion and corrosion are occasionally lumped together as "corrasion" (Brett and Baird, 1986; Brett, 1990). While this term is more commonly applied in studies of shelly macroinvertebrates, it works equally well as a broad-brush descriptor of bone surface deterioration in studies of bonebeds.

*Weathering*

The weathering of vertebrate skeletal remains occurs as bones are defleshed and subjected to physical and chemical degradation both at the ground surface and in soils (Behrensmeyer, 1978). Here weathering is distinguished from abrasion and biochemical corrosion in the sense that weathering transpires when bones and teeth are more or less at rest, and there is no implied contact with physical or biological agents such as mobile abrasive particles or digestive acids. Chemical corrosion related to sediment, soil, or surface water pH is arguably a component of weathering.

A variety of factors may act to degrade organic and inorganic components of bones and teeth as part of the weathering equation, including diurnal and seasonal temperature changes, wetting and drying, freezing and thawing, and UV exposure (Behrensmeyer, 1978; Shipman, 1981; Johnson, 1985; Lyman and Fox, 1989; Andrews, 1990; Tappen, 1994; Fiorillo, 1995). Many of the more vigorous agents of weathering operate in subaerial settings, and the actualistic studies that address the topic focus on a relatively small selection of terrestrial habitats (e.g., Behrensmeyer, 1978; Andrews, 1990; Tappen, 1994). However, there is no reason to exclude subaqueous settings in considerations of bone and/or tooth weathering.

Early actualistic studies of bone weathering sought to identify modification sequences or stages. A classic study by Behrensmeyer (1978) initially outlined six stages of bone weathering for mammals >5 kg in body weight exposed within the semiarid confines of Amboseli National Park, Kenya. Stages 0 (unweathered, still greasy) through 5 (bone disintegrating, detached splinters) were based on readily observable features such as grease content, bone cracking patterns and flaking, depth of cracking, and degree of splintering (see Behrensmeyer, 1978, for graphic examples). Several subsequent workers built upon the weathering stage categories of Behrensmeyer (1978) and applied their own slightly modified schemes to bones of different taxa and bones in different climatic and environmental settings in the modern, and to bones in the fossil record (e.g., Fiorillo, 1988a, 1989; Andrews, 1990; Smith, 1993; Tappen, 1994; Coombs and Coombs, 1997; Ryan et al., 2001).

Based on actualistic observations, it is often possible to use weathering stages to estimate the duration of exposure of bones and teeth (e.g., Potts, 1986; Lyman and Fox, 1989; Behrensmeyer, 1991; Tappen, 1994; Fiorillo, 1995). However, the complex interplay of factors that can influence the rate of weathering and the progression of weathering stages in modern systems beg caution when exploring this phenomenon in the fossil record. General approximations of the duration of surface exposure (e.g., buried quickly after death vs. long period of exposure) are probably best in most circumstances.

The relative frequency of bones in a bonebed assemblage that exhibit different weathering stages can be presented in a table or, more effectively, as a group of simple histograms (Behrensmeyer, 1978; Gifford, 1984; Potts, 1986; Fiorillo, 1989, 1995; Lyman and Fox, 1989; Smith, 1993; Cook, 1995; Varricchio, 1995; Coombs and Coombs, 1997). When data are presented in histogram format they define a weathering profile for the assemblage. These profiles can be generated for an entire assemblage (Smith, 1993, Fig. 17), or cast relative to a variety of covariables such as skeletal element (e.g., dense podials and compact vertebrae vs. scapulae and pelvic elements), taxon (mammal vs. fish), and ontogeny (adult vs. juvenile). This type of approach may be particularly insightful where a mixture of taphonomic modes is suspected (e.g., Bower et al., 1985; Smith, 1993, his Fig. 18).

If bone-weathering profiles are dominated by a high percentage of unweathered elements (stages 0–1 of Behrensmeyer [1978]) it is reasonable to conclude that bones in an assemblage were buried relatively quickly or accumulated in a setting prone to only minor weathering (e.g., Rogers, 1990; Cook, 1995; Varricchio, 1995; Fiorillo et al., 2000; Ryan et al., 2001). If multimodal, subequal, or normal distributions of weathering stages result (Fiorillo, 1988a; Smith, 1993; Cook, 1995; Coombs and Coombs, 1997; Britt et al., 2004), a given bonebed assemblage may (1) contain elements with variable taphonomic histories (e.g., accumulation via long-term attrition or concentration after weathering in different settings), (2) preserve elements that were exposed for a relatively long time (multiple years), or (3) include elements that reflect variable weathering due to micropaleoenvironmental variation (Behrensmeyer, 1991; Lyman, 1994). Profiles that are dominated by a high percentage of deeply weathered elements (stages 4–5 of Behrensmeyer [1978]) are generally rare and point to a lengthy or intense weathering history, and possibly multiyear exposure in regions that experienced significant hiatuses in sedimentation (e.g., distal floodplain of Smith, 1993, his Fig. 18).

### Trample/Sediment Scratch Marks

Surface scoring and scratching is a common modification feature in many bonebed assemblages, and the origin of this type of damage has received considerable attention because of its potential link to tool-generated cut marks. Numerous workers, including Brain (1967), Fiorillo (1984, 1989), Andrews and Cook (1985), Behrensmeyer et al. (1986), and Olsen and Shipman (1988) have examined the morphology and origins of discrete surface striations in a variety of modern substrates. These experimental studies have shown that the trampling of bones in clastic sediments generally results in surface scoring characterized by fields of subparallel scratches. Accordingly, bonebed workers regularly interpret the presence of patches of subparallel scratch marks as evidence for trampling (e.g., Fiorillo, 1987, 1988a, 1991a; Rogers, 1990; Varricchio, 1995; Fiorillo et al., 2000; Ryan et al., 2001). It should be noted, however, that the absence of these features should not be used to rule out the potential for trampling (e.g., Coombs and Coombs, 1997).

In the trampling scenario, scratch marks are generated on bones by the grinding of sediment grains against bone surfaces. The abundance and intensity (depth and width) of scratch marks depends on substrate conditions, the weathering state of the bones, the intensity of loading (mass of trampler), and the duration of trampling. To our knowledge, no experimental studies or empirical observations have convincingly shown that comparable traces can be produced by hydraulic flows that entrain coarse-grained sedimentary particles (Lyman, 1994). However, it is theoretically possible that bones could exhibit flow-induced tool scratches in settings where bone is transported across a coarse bed of sand or gravel or interacts with conglomeratic debris flows.

In bonebed studies, the presence or absence of scratch marks is often simply noted in a qualitative fashion and illustrated with photographs. However, in large bonebeds that exhibit evidence of preservational complexity, it may be very profitable to document the relative abundance and intensity of scratch marks in samples derived from different portions of a locality. Data related to the frequency of scratch marks can be presented as a percentage of N or NISP and also tracked in relation to element type or weathering stage. When combined with information pertaining to breakage, skeletal completeness, and bone orientation, such observations may provide important insights into overall taphonomic history and paleoecology. For example, given substrate characteristics and the nature of the modified bones, it may be possible to estimate the size and identity of the bioturbators (e.g., medium-sized ungulates vs. elephants) and the frequency of disturbance in a given setting (e.g., chance encounters along the banks of a stream vs. heavy utilization of a waterhole). Moreover, it may be possible to link the morphology of trample marks (e.g., shallow, closely spaced grooves vs. deep, more widely spaced gouges) to specific sedimentary textures, and this in turn may yield insights into transport histories and the sourcing and processing of bone material.

As is the case with many surface modification features, scratch marks are probably best studied after bones are extracted and prepared. However, they are usually readily apparent in the field if present, and so their abundance and distribution should be tracked as excavation proceeds.

*Tooth Marks*

The recognition and analysis of tooth marks in a bonebed assemblage can help the researcher identify potential agents of death, disarticulation, sorting, and breakage and can also shed significant light on feeding ecology and interactions among extinct species. For these reasons, tooth-marked bone has been well studied on both modern landscapes (Miller, 1969; Dodson and Wexlar, 1979; Korth, 1979; Haynes, 1980, 1981, 1982, 1983; Binford, 1981) and in the fossil record (Fiorillo, 1988a; Hunt, 1990; Rogers, 1990; Fiorillo, 1991a, 1991b; Farlow and Holtz, 1992; Hunt et al., 1994; Varricchio, 1995; Jacobsen, 1997, 1998; Carpenter, 1998; Chure et al., 2000; Jacobsen and Ryan, 1999; Fiorillo et al., 2000; Ryan et al., 2000; Rogers et al., 2003).

Tooth-marked bone is common in Mesozoic bonebeds, but the frequency of tooth marks at any particular site is usually low (commonly <4% of NISP) (Carpenter, 1998; Chure et al., 2000; Jacobsen and Ryan, 1999; Fiorillo et al., 2000; Ryan et al., 2000; but see Rogers et al., 2003). Tooth marks are seemingly more common in deposits of Cenozoic age (Fiorillo, 1991b; Chure et al., 2000), and this presumably reflects the evolution of focused bone-utilization as an ecological strategy (Hunt, 1987).

Several morphotypes of carnivore-generated tooth marks are recognized, including pits, punctures, scores, and furrows (Maguire et al., 1980; Binford, 1981; Fiorillo, 1988a; Hunt et al., 1994; Lyman, 1994). Pitting develops when bones are chewed and the cortical bone is resilient enough to inhibit through-going punctures. Pits can be discrete but are often found clustered on a given element. Tooth puncture marks (or perforations) are characterized by discrete localized depressions that typically develop when a single cusp or tooth crown impacts a bone surface at a high angle and penetrates or fractures cortical bone (Shipman, 1981). Scores and scratches are elongate traces that may be deep and V-shaped or U-shaped in profile. They can occur in isolation but tend to regularly occur in parallel sets and apparently reflect the dragging of tooth rows across bone surfaces (see Rogers et al. [2003] for several examples). Furrows are linear traces with more irregular orientations that tend to be focused more on the spongy ends of limb bones. Carnivores processing bone may also generate ragged or scalloped edges that indicate focused gnawing. Ungulates and rodents also gnaw and chew bone (e.g., Sutcliffe, 1973, 1977; Gauthier-Pilters and Dagg, 1981; Greenfield, 1988) as they strive to alleviate nutrient deficiencies (calcium and phosphorous) and, in the case of rodents, keep rapidly growing incisors in check. The potential for non-carnivore gnawing should always be considered when evaluating tooth traces, especially in post-Mesozoic localities.

Studies of tooth marks in bonebed assemblages should include macro- and microscopic examinations of all material both in the field and in the laboratory after preparation, with the goal of characterizing general morphology (e.g., puncture vs. pit), size, and spacing. Careful observation may also yield evidence of denticle or serration patterns, and these in turn may provide important clues as to the identity of the trace makers (Fiorillo, 1988a; Erickson and Olson, 1996; Jacobsen, 1998, 2001; Rogers et al., 2003). Any trends in utilization should be documented in relation to both taxon and element. For example, in a museum-based study of bonebed collections from the Late Cretaceous of Madagascar, Rogers et al. (2003) noted that the vast majority of tooth-marked elements belonged to the theropod dinosaur *Majungatholus atopus*, and that most of bones exhibiting tooth marks were ribs, vertebrae, and vertebral processes.

Examples of tooth marks preserved in a bonebed assemblage should be described and figured (e.g., Fiorillo, 1988a; Hunt, 1990; Rogers, 1990), and whenever possible the relative abundance of tooth-marked bone and tooth mark types in an assemblage should be reported (Fiorillo, 1988a, 1991a, 1991b; Fiorillo et al., 2000). These data can be expressed as a percentage of N or NISP and presented in tables or, more rarely, histograms (Varricchio, 1995; Fiorillo et al., 2000). Data relating to the nature and distribution of tooth marks are frequently evaluated in conjunction with breakage and skeletal representation data in order to assess the relative influences of scavenging and bone utilization in an assemblage (e.g., Fiorillo, 1991b; Coombs and Coombs, 1997).

*Bioerosion*

Bioerosion includes all biologically produced bone modifications that *do not* reflect the activity of vertebrates. Commonly cited examples include root etching (Behrensmeyer, 1978, 1991) and insect-generated borings (e.g., Kitching 1980; Rogers, 1992; Martin and West, 1995; Hasiotis et al., 1999; Fejfar and Kaiser, 2005; Oldrich and Kaiser, 2005; Roberts et al., 2007). Other potential bioeroders include mollusks (Frey et al., 1975) and various types of microorganisms including fungi and bacteria (Marchiafava et al., 1974; Hackett, 1981; Piepenbrink, 1986; Hanson and Buikstra, 1987; Child, 1995; Jans et al., 2004). Careful documentation of bioerosional features (or the lack thereof) in bonebed taphonomy can yield insights into the physical conditions of the syndepositional environment during the late stages of taphonomic history, the degree to which an assemblage may have a mixed taphonomic origin (Grayson, 1988), and the degree to which environmental conditions may have varied across a site of burial. Consideration of bioerosion processes can also provide alternative explanations for the degradation of bone and biological materials at a site (e.g., Behrensmeyer, 1978).

Recording the orientation and distribution of bioerosion features both on the bones and relative to the host sediment are important because bioerosion most often takes place at the sediment-bone interface (e.g., beetle larvae boring) or within specific portions of soil horizons (in the case of root/fungal etching). Comparing the distribution and orientation of bioerosion features on elements throughout a bonebed can reveal whether bones have been reworked or have remained in situ since the bioerosion event (e.g., Britt et al., 2004; Nolte et al., 2004). As with other types of bone damage, the researcher should strive to identify bioerosion features both in the field and in the laboratory. In some cases (e.g., fungal boring), thin sections or scanning electron microscope images may be needed to identify bioerosion features. The type and frequency of bioerosion in a bonebed assemblage should be tabulated and reported in tandem with other taphonomic variables.

*Exploring the History of a Bonebed*

All bonebed studies seek at some level to answer Pat Shipman's (1981) fundamental question, "What are these bones doing here?" This deceptively simple question is all too often difficult to answer. This is because it may be entirely unrealistic to ask the "big question" until the smaller pieces are in place. In this chapter we have described the methods needed to assemble the various pieces of the bonebed puzzle. Once these data are collected, it is then time to contemplate the history of the assemblage. This is undeniably the most exciting and integrative phase of any bonebed study, and also potentially the most frustrating (cherished hypotheses may not stand the test). Here we offer a few general guidelines to help get things started on the right path.

Framing testable hypotheses is the critical first step. How will the available evidence serve to distinguish among a variety of alternative hypotheses that relate to complex and often intertwined physical, chemical, and biological processes? Can the data be employed to distinguish assemblages linked to mass-death events from attritional time-averaged concentrations? Can the interpreted age profiles and weathering patterns be used to differentiate between attritional and catastrophic death scenarios? Can chemical or biological signals preserved in the fossil bones themselves be used to demonstrate a link between mass mortality and disease or predation? Taking the time to carefully consider the quality of the data and their overall suitability in relation to questions such as these can help the researcher focus energy on productive pursuits and avoid potentially embarrassing retractions and revisions.

David A. Eberth, Raymond R. Rogers, and Anthony R. Fiorillo

Hypotheses should be framed at a variety of stages during the course of a bonebed study. Posing questions at multiple stages in the analysis is a natural approach and builds toward answering the "big question." For example, early on a researcher may infer a scenario of mass mortality based on the examination of surface concentrations alone. This hypothesis based on a single data set may be modified or rejected as more data are brought to bear. After testing multiple hypotheses, a more robust interpretation of bonebed history will result. However, in the end, the "big question" may have more than one reasonable answer.

Insights from modern analogs should always be considered when attempting to interpret the taphonomic history of an ancient bonebed assemblage. In fact, many published studies that focus on the origins of bonebed concentrations hinge their major conclusions on parallels drawn from well-documented modern examples of death, decay, and disarticulation. The logic of this approach is obvious, but the reader is cautioned to remember that some extinct animal groups, such as the Mesozoic non-avian dinosaurs and armored fishes of the Early Paleozoic, do not have readily identifiable modern counterparts.

Researchers should also always consider the potential relationship between mortality and preservation. It is hard enough to become a fossil, and it is probably quite reasonable to assume that it is harder still to become a concentration of fossils representing the localized mortality of multiple individuals. With this basic premise in mind, we encourage the bonebed researcher to first consider scenarios that can accommodate both the localized demise of numerous individuals and their subsequent (or even concomitant) permanent burial. An example is drought, which often serves to concentrate animals near refuges such as waterholes and stream courses on modern landscapes. These same refuges also tend to be low-lying areas that receive abundant sediment when the drought breaks (Shipman, 1975; Hendy, 1981; Rogers, 1990). That said, it is equally important to keep in mind that death and eventual burial may be entirely unrelated.

Lastly, be sure to consider all previous studies that may offer insights, be they geologic or taphonomic in nature. Look for patterns, both in the literature and in the stratigraphic interval under scrutiny. If multiple bonebeds occur in a single formation or area, perhaps look first to a single overarching theme, and then diversify the potential formative scenarios as the evidence warrants (see taphonomic modes and taphofacies concepts of Behrensmeyer et al., [1992], and Behrensmeyer et al. [2000]). Remember that relatively few processes that serve to concentrate vertebrate remains operate with regularity in modern settings, and the same should hold true in the fossil record. There is no need to discover a novel scenario with every new bonebed prospect.

## CONCLUSION

In this chapter we outline many of the basic considerations that relate to the study of bonebeds. We have approached the problem from a practical perspective, and we have tried to provide at least a few good ideas that should serve to guide those excavating and/or planning to study a bonebed for the first time. However, we are well aware that no single suite of methods will serve under all circumstances, and we fully realize that every investigator must assess conditions and formulate plans and research expectations on a site-specific basis. We encourage the reader to go beyond this broad-ranging treatment and to explore the many detailed case studies referenced in this chapter and in other chapters in this book.

On first approach to a bonebed prospect, be sure to accurately determine the coordinates of the site and permanently archive the locality data. Once you know exactly where the site is, make a preliminary map supplemented with cross sections so that you can begin to visualize the geometry of the bone deposit. Carefully scour the bonebed exposures to get a feel for the abundance and quality of bone in the deposit, and perhaps check your premonitions by digging a test pit or two. As you excavate the pits, gauge the friability of the encasing matrix, and assess the resilience and spatial distribution of the exhumed fossil material. When you feel reasonably familiar with the site, begin removing overburden in earnest. Be sure to remove enough rock to allow effective mapping and to accommodate the crew, but do not overdo it. With regard to mapping, be sure to pin your efforts on a suitably permanent baseline. Use whatever mapping technique works best for the local conditions, and strive for consistency and accuracy.

Carefully collect taphonomic data as elements are exposed and extracted from the quarry. Be sure to encourage the entire crew to check for sediment-bone associations and side-specific taphonomic characteristics that might be hard or impossible to reconstruct once the bone is out of context and back in the laboratory. Perhaps designate a segment of the active quarry as the "taphonomic laboratory," and work here in a slow methodical fashion to document every potential taphonomic attribute of the site under study. Systematically photo-document the bone assemblage and do be compulsive (spare no film or disk space), because things change rapidly and critical associations are irreversibly lost as excavation proceeds. Be sure to supplement your field-based taphonomic observations with lab-based study once the collection is prepared.

A thorough geological study of the site, surrounding facies, and host stratigraphic interval should be conducted concurrently with excavation and the collection of taphonomic data. Strive to place the site in both a local and regional stratigraphic context, and characterize fully the sedimentology of the bone-bearing matrix. Carefully compare the sedimentological features of the bone-bearing horizon with surrounding beds in order to determine if unusual geological circumstances might have played a role in mortality and/or preservation. Is the bonebed a unique occurrence, or are there other bonebeds in the stratigraphic interval or geographic area? Gather as much geological data as feasible, because a bonebed data set coupled with accurate geological information is much more likely to yield credible insights. Whenever possible, include a trained geologist on the project who can focus on deciphering the rocks.

Finally, weigh all the evidence against a variety of potential formative scenarios, and work to produce a reasonable and scientifically robust reconstruction of taphonomic history. Try to stay grounded in the modern, but do not hesitate to apply an evolutionary view to at least some taphonomic processes and ecological strategies. Keep in mind that it is only in rare cases that definitive killing agents and/or accumulation scenarios can be diagnosed in the fossil record, and that most bonebed studies lack the clear signature of a proverbial "smoking gun." Be prepared to explore complex and at times downright confusing fossil deposits that demand the evaluation of multiple working hypotheses. Be sure that the hypotheses entertained are testable, and hold your ultimate reconstructions accountable to the available data, building and modifying your hypotheses as you go. In the end, a definitive conclusion related to the formative history of a bonebed may not even be reached, but rest assured that every well-documented locality adds to our collective understanding of the vertebrate fossil record.

## REFERENCES

Abler, W.L. 1984. A three-dimensional map of a paleontological quarry. Contributions to Geology, University of Wyoming 23:9–14.

Anderson, J.F., A. Hall-Martin, and D.A. Russell. 1985. Long-bone circumference and weight in mammals, birds, and dinosaurs. Journal of Zoology A207: 53–61.

Andrews, P. 1990. Owls, caves and fossils. The Natural History Museum, London.

Andrews, P., and J. Cook. 1985. Natural modification to bones in a temperate setting. Man 20:675–691.

Andrews, P., and A. Ersoy. 1990. Taphonomy of the Miocene bone accumulations at Pasalar, Turkey. Journal of Human Evolution 19:379–396.

Antia, D.D.J., and J.H. Sykes. 1979. The surface textures of quartz grains from a Rhaetian bone-bed, Blue Anchor Bay, Somerset. Mercian Geologist 7:205–210.

Badgley, C. 1986a. Counting individuals in mammalian fossil assemblages from fluvial environments. Palaios 1:328–338.

Badgley, C. 1986b. Taphonomy of mammalian fossil remains from Siwalik rocks of Pakistan. Paleobiology 12:119–142.

Barnes, C.V. 2000. Paleontological excavations in designated wilderness: Theory and practice. Pp. 155–159 in Proceedings: Wilderness science in a time of change. Proc. RMRS-P-000. D.N. Cole and S.F. McCool, eds. U.S. Department of Agriculture, Forest Service, Rocky Mountain Research Station 3, Ogden, Utah.

Behrensmeyer, A.K. 1975. The taphonomy and paleoecology of Plio-Pleistocene vertebrate assemblages east of Lake Rudolf, Kenya. Bulletin of the Museum of Comparative Zoology 146:474–578.

Behrensmeyer, A.K. 1978. Taphonomic and ecologic information from bone weathering. Paleobiology 4:150–162.

Behrensmeyer, A.K. 1982. Time resolution in fluvial vertebrate assemblages. Paleobiology 8:211–228.

Behrensmeyer, A.K. 1988. Vertebrate preservation in fluvial channels. Palaeogeography, Palaeoclimatology, Palaeoecology 63:183–199.

Behrensmeyer, A.K. 1990. Bones. Pp. 232–235 in Paleobiology: A synthesis. D.E.G. Briggs and P.R. Crowther, eds. Blackwell Scientific Publications, Oxford.

Behrensmeyer, A.K. 1991. Terrestrial vertebrate accumulations. Pp. 291–335 in Taphonomy: Releasing the data locked in the fossil record. P.A. Allison and D.E.G. Briggs, eds. Plenum, New York.

Behrensmeyer, A.K. This volume. Bonebeds through time. Chapter 2 in Bonebeds: Genesis, analysis, and paleobiological significance. R.R. Rogers, D.A. Eberth, and A.R. Fiorillo, eds. University of Chicago Press, Chicago.

Behrensmeyer, A.K., and J.C. Barry. 2005. Biostratigraphic surveys in the Siwaliks of Pakistan: A method for standardized surface sampling of the vertebrate fossil record. Palaeontologia Electronica 8(1):8.1.14A.

Behrensmeyer, A.K., and D.E. Dechant Boaz. 1980. The Recent bones of Amboseli National Park, Kenya, in relation to East African paleoecology. Pp. 72–92 in Fossils in the making. A.K. Behrensmeyer and A.P. Hill, eds. University of Chicago Press, Chicago.

Behrensmeyer, A.K., K.D. Gordon, and G.T. Yanagi. 1986. Trampling as a cause of bone surface damage and pseudo-cutmarks. Nature 319:768–771.

Behrensmeyer, A.K., R.S. Hook, C.E. Badgley, J.A. Boy, R.E. Chapman, P. Dodson, R.A. Gastaldo, R.W. Graham, L.D. Martin, P.E. Olsen, R.A. Spicer, R.E. Taggart, and M.V.H. Wilson. 1992. Paleoenvironments and taphonomy, Pp. 15–136 in Terrestrial ecosystems through time: The evolutionary paleoecology of terrestrial plants and animals. A.K. Behrensmeyer, J.D. Damuth, W.A.

DiMichelle, R. Potts, H.-D. Sues, and S.L. Wing, eds. University of Chicago Press, Chicago.

Behrensmeyer, A.K., S.M. Kidwell, and R.A. Gastaldo. 2000. Taphonomy and paleobiology. Paleobiology 26(supplement to 4):103–147.

Berman, D.S, D.A. Eberth, and D.B. Brinkman. 1988. *Stegotretus agyrus* a new genus and species of microsaur (amphibian) from the Permo-Pennsylvania of New Mexico. Annals of the Carnegie Museum 57:293–323.

Biddick, K.A., and J. Tomenchuck. 1975. Quantifying continuous lesions and fractures on long bones. Journal of Field Archaeology 2:239–249.

Bilbey, S.A. 1999. Taphonomy of the Cleveland-Lloyd dinosaur quarry in the Morrison Formation, Central Utah: A lethal spring-fed pond. Pp. 121–133 *in* Vertebrate paleontology in Utah. Miscellaneous Publication 99-1. D.D. Gillette, ed. Utah Geological Survey, Salt Lake City.

Binford, L.R. 1981. Bones, ancient men and modern myths. Academic Press, New York.

Blob, R. 1997. Relative hydrodynamic dispersal potentials of soft-shelled turtle elements: Implications for interpreting skeletal sorting in assemblages of non-mammalian terrestrial vertebrates. Palaios 12:151–164.

Blob, R.W., and A.R. Fiorillo. 1996. The significance of vertebrate microfossil size and shape distributions for faunal abundance reconstructions: A Late Cretaceous example. Paleobiology 22:422–435.

Blob, R.W., and C. Badgley. This volume. Numerical methods for bonebed analysis. Chapter 6 *in* Bonebeds: Genesis, analysis, and paleobiological significance. R.R. Rogers, D.A. Eberth, and A.R. Fiorillo, eds. University of Chicago Press, Chicago.

Blumenschine, R.J. 1986. Carcass consumption sequences and the archaeological distinction of scavenging and hunting. Journal of Human Evolution 15:639–659.

Boaz, N.T., and A.K. Behrensmeyer. 1986. Hominid taphonomy: Transport of human skeletal parts in an artificial fluviatile environment. American Journal of Physical Anthropology 45:53–60.

Bower, J.R.F., D.P. Gifford, and D. Livingston. 1985. Excavations at Loiyangalani Site, Serengeti National Park, Tanzania. National Geographic Society Research Reports 20:41–56.

Brain, C.K. 1967. Bone weathering and the problem of bone pseudo-tools. South African Journal of Science 63:97–99.

Breithaupt, B.H., M. Fox, N.C. Fraser, N.A. Matthews, and B. Wilborn. 2000. A new dinosaur bonebed in the Morrison Formation of Bighorn County, Wyoming. Journal of Vertebrate Paleontology 20(supplement to 3):31A.

Brett, C.E. 1990. Destructive taphonomic processes and skeletal durability. Pp. 223–236 *in* Paleobiology: A synthesis. D.E.G. Briggs and P.R. Crowther, eds. Blackwell Scientific Publications, Oxford.

Brett, C.E., and G.C. Baird. 1986. Comparative taphonomy: A key to paleoenvironmental interpretation based on fossil preservation. Palaios 1:207–227.

Brinkman, D.B. 1990. Paleoecology of the Judith River Formation (Campanian) of Dinosaur Provincial Park, Alberta, Canada: Evidence from microfossils localities. Palaeogeography, Palaeoclimatology, Palaeoecology 78:37–54.

Brinkman, D.B., M.J. Ryan, and D.A. Eberth. 1998. The paleogeographic and strati-graphic distribution of ceratopsids (Ornithischia) in the Upper Judith River Group of Western Canada. Palaios 13:160–169.

Brinkman, D.B., D.A. Eberth, and P.J. Currie. This volume. From bonebeds to paleobi-ology: Applications of bonebed data. Chapter 4 *in* Bonebeds: Genesis, analysis, and paleobiological significance. R.R. Rogers, D.A. Eberth, and A.R. Fiorillo, eds. Univer-sity of Chicago Press, Chicago.

Britt, B.B., D.A. Eberth, R. Scheetz, and B. Greenhalgh. 2004. Taphonomy of the Dal-ton Wells dinosaur quarry (Cedar Mountain Formation, Lower Cretaceous, Utah). Journal of Vertebrate Paleontology 24(supplement to 3):41A.

Bromage, T.G. 1984. Interpretation of scanning electron microscopic images of abraded forming bone surfaces. American Journal of Physical Anthropology 64:161–178.

Brown, L., J. Chiment, W. Perks, J. Haenlein, and J. Neville. 2003. Ground penetrating radar in support of Mastodon studies in New York. Geological Society of America, Abstracts with Programs 35(3):75.

Buffetaut, E., and V. Suteethorn. 1989. A sauropod skeleton associated with theropod teeth in the Upper Jurassic of Thailand: Remarks on the taphonomic and paleoe-cological significance of such associations. Palaeogeography, Palaeoclimatology, Palaeoecology 73:77–83.

Camp, C.L., and S.P. Welles. 1956. Triassic dicynodont reptiles. I. The North American genus *Placerias*. Memoirs of the University of California 13:255–304.

Carpenter, K. 1998. Evidence of predatory behavior by carnivorous dinosaurs. Gaia 15:135–144.

Chadwick, A.V., L.A. Spencer, and L.E. Turner. 2005. Taphonomic windows into an upper Cretaceous *Edmontosaurus* bonebed. Geological Society of America, Abstracts with Programs 37:159.

Chaplin, R.E. 1971. The study of animal bones from archaeological sites. Seminar Press, London.

Child, A.M. 1995. Towards an understanding of the microbial decomposition of ar-chaeological bone in the burial environment. Journal of Archaeological Science 22:165–174.

Chure, D.J., A.R. Fiorillo, and A. Jacobsen. 2000. Prey bone utilization by predatory dinosaurs in the Late Jurassic of North America with comments on prey bone use by dinosaurs throughout the Mesozoic. Gaia 15:227–232.

Cifelli, R.L., S.K. Madsen, and E.M. Larson. 1996. Screenwashing and associated tech-niques for the recovery of microvertebrate fossils. Pp. 1–24 *in* Techniques for recov-ery and preparation of microvertebrate fossils. Special publication 96-4. R.L. Cifelli, ed. Oklahoma Geological Survey, Norman.

Clark, J., J.R. Beerbower, and K.K. Kietzke. 1967. Oligocene sedimentation, stratigraphy, paleoecology, and paleoclimatology in the Big Badlands of South Dakota. Fieldiana Geology Memoirs 5:1–158.

Coard, R., and R.W. Dennell. 1995. Taphonomy of some articulated skeletal remains: Transport potential in an artificial environment. Journal of Archaeological Science 22:441–448.

Colbert, E.H. 1989. The Triassic dinosaur *Coelophysis*. Museum of Northern Arizona Bulletin 57:1–160.

Colombi, C., and O. Alcober. 2004. Taphofacies characterization of the Ischigualasto Formation (Upper Triassic). Journal of Vertebrate Paleontology 24(supplement to 3):47A.

Compton, R.R. 1962. Manual of field geology. Wiley and Sons, New York.

Cook, E. 1995. Taphonomy of two non-marine Lower Cretaceous bone accumulations from southeastern England. Palaeogeography, Palaeoclimatology, Palaeoecology 116:263–270.

Coombs, M.C., and W.P. Coombs Jr. 1997. Analysis of the geology, fauna, and taphonomy of Morava Ranch Quarry, Early Miocene of northwest Nebraska. Palaios 12:165–187.

Curray, J.R. 1956. The analysis of two-dimensional orientation data. Journal of Geology 64:117–131.

Currie, P.J. 2000. Possible evidence of gregarious behavior in tyrannosaurids. Gaia 15:123–133.

Currie, P.J., and P. Dodson. 1984. Mass death of a herd of ceratopsian dinosaurs, Pp. 61–66 in Third symposium on Mesozoic terrestrial ecosystems, short papers. W.E. Reif and F. Westphal, eds. Attempto Verlag, Tübingen.

Currie, P.J., and D.A. Eberth. 1993. Paleontology, sedimentology and paleoecology of the Iren Dabasu Formation (Upper Cretaceous), Inner Mongolia, People's Republic of China. Cretaceous Research 14:127–144.

Davis, J. 1986. Statistics and data analysis in geology. John Wiley, Toronto.

Davis, K.L. 1985. A taphonomic approach to experimental bone fracturing and applications to several South African Pleistocene sites. Unpublished Ph.D. dissertation. State University of New York, Binghamton.

Denys, C., K. Kowalski, and Y. Dauphin. 1992. Mechanical and chemical alterations of skeletal tissues in a Recent Saharian accumulation of faeces from Vulpes ruepelli (Carnivora, Mammalia). Acta Zoologica Cracoviensia 35:265–283.

Dodson, P. 1971. Sedimentology and taphonomy of the Oldman Formation (Campanian), Dinosaur Provincial Park, Alberta (Canada). Palaeogeography, Palaeoclimatology, Palaeoecology 10:21–74.

Dodson, P. 1973. The significance of small bones in paleoecological interpretation. Contributions to Geology, University of Wyoming 12:15–19.

Dodson, P., and D. Wexlar. 1979. Taphonomic investigations of owl pellets. Paleobiology 5:279–284.

Dodson, P., A.K. Behrensmeyer, R.T. Bakker, and J.S. McIntosh. 1980. Taphonomy of the dinosaur beds of the Jurassic Morrison Formation. Paleobiology 6:208–232.

Domínguez-Rodrigo, M. 1999. Flesh availability and bone modification in carcasses consumed by lions: Paleoecological relevance in hominid foraging patterns. Palaeogeography, Palaeoclimatology, Palaeoecology 149:373–388.

Eberth, D.A. 1990. Stratigraphy and sedimentology of vertebrate microfossil sites in the uppermost Judith River Formation (Campanian), Dinosaur Provincial Park, Alberta, Canada. Palaeogeography, Palaeoclimatology, Palaeoecology 78:1–36.

Eberth, D.A. 1993. Depositional environments and facies transitions of dinosaur-bearing Upper Cretaceous redbeds at Bayan Mandahu (Inner Mongolia, People's Republic of China). Canadian Journal of Earth Sciences 30:2196–2213.

Eberth, D.A., and D.B. Brinkman. 1997. Paleoecology of an estuarine, incised-valley fill in the Dinosaur Park Formation (Judith River Group, Upper Cretaceous) of southern Alberta, Canada. Palaios 12:43–58.

Eberth, D.A., and M.A. Getty. 2005. Ceratopsian bonebeds at Dinosaur Provincial Park: Occurrence, origin, and significance. Pp. 501–536 *in* Dinosaur Provincial Park. P.J. Currie and E.B. Koppelhus, eds. Indiana University Press, Bloomington, Indiana.

Eberth, D.A., D.R. Braman, and T.T. Tokaryk. 1990. Stratigraphy, sedimentology and vertebrate paleontology of the Judith River Formation (Campanian) near Muddy Lake, west-central Saskatchewan. Bulletin of Canadian Petroleum Geology 38:387–406.

Eberth, D.A., D.S Berman, S.S. Sumida, and H. Hopf. 2000. Lower Permian terrestrial paleoenvironments and vertebrate paleoecology of the Tambach Basin (Thuringia, Central Germany): The upland holy grail. Palaios 15:293–313.

Eberth, D.A., B.B. Britt, R. Scheetz, K.L. Stadtman, and D.B. Brinkman. 2006. Dalton Wells: Geology and significance of debris-flow-hosted dinosaur bonebeds in the Cedar Mountain Formation (Lower Cretaceous) of eastern Utah, USA. Palaeogeography, Palaeoclimatology, Palaeoecology 236:217–245.

Eberth, D.A., M. Shannon, and B.G. Noland. This volume. A bonebeds database: Classification, biases, and patterns of occurrence. Chapter 3 *in* Bonebeds: Genesis, analysis, and paleobiological significance. R.R. Rogers, D.A. Eberth, and A.R. Fiorillo, eds. University of Chicago Press, Chicago.

Efremov, I.A. 1940. Taphonomy: New branch of paleontology. Pan-American Geologist 74:81–93.

Englemann, G., and A.R. Fiorillo. 2000. The taphonomy and paleoecology of the Morrison Formation determined from a field survey of fossil localities. GeoResearch Forum 6:533–540.

Erickson, G.M., and K.H. Olson. 1996. Bite marks attributable to *Tyrannosaurus rex*: Preliminary description and implications. Journal of Vertebrate Paleontology 16:175–178.

Erickson, G.M., K.A. Curry Rogers, and S. Yerby. 2001. Dinosaur growth patterns and rapid avian growth rates. Nature 412:429–433.

Farlow, J.O., and T.R. Holtz, Jr. 2002. The fossil record of predation in dinosaurs. Paleontological Society Papers 8:251–265.

Fastovsky, D.E., J.M. Clark, N.H. Strater, M.R. Montellano, R. Hernandez, and J.A. Hopson. 1995. Depositional environments of a Middle Jurassic terrestrial vertebrate assemblage, Huizachal Canyon, Mexico. Journal of Vertebrate Paleontology 15:561–575.

Fejfar, O., and T.M. Kaiser. 2005. Insect bone-modification and paleoecology of Oligocene mammal-bearing sites in the Doupov Mountains, northwestern Bohemia. Palaeontologia Electronica 8.1.8A:1-11.

Fiorillo, A.R. 1984. An introduction to the identification of trample marks. Current Research in the Pleistocene 1:47–48.

Fiorillo, A.R. 1987. Trample marks: Caution from the Cretaceous. Current Research in the Pleistocene 4:73–75.

Fiorillo, A.R. 1988a. Taphonomy of Hazard Homestead Quarry (Ogallala Group), Hitchcock County, Nebraska. Contributions to Geology, University of Wyoming 26:57–97.

Fiorillo, A.R. 1988b. A proposal for graphic presentation of orientation data from fossils. Contributions to Geology, University of Wyoming 26:1–4.

Fiorillo, A.R. 1989. An experimental study of trampling: Implications for the fossil record. Pp. 61–72 *in* Bone modification. R. Bonnichsen and M.H. Sorg, eds. Center for the Study of the First Americans, University of Maine, Orono.

Fiorillo, A.R. 1991a. Taphonomy and depositional setting of Careless Creek Quarry (Judith River Formation), Wheatland County, Montana, U.S.A. Palaeogeography, Palaeoclimatology, Palaeoecology 81:281–311.

Fiorillo, A.R. 1991b. Prey bone utilization by predatory dinosaurs. Palaeogeography, Palaeoclimatology, Palaeoecology 88:157–166.

Fiorillo, A.R. 1995. The possible influence of cold temperatures on bone weathering in Curecanti National Recreation Area, southwest Colorado. Current Research in the Pleistocene 12:69–71.

Fiorillo, A.R. 1997. Taphonomy. Pp. 713–716 *in* The encyclopedia of dinosaurs. P.J. Currie and K. Padian, eds. Academic Press, San Diego.

Fiorillo, A.R. 1998. Bone modification features on sauropod remains (Dinosauria) from the Freezeout Hills Quarry N (Morrison Formation) of southeastern Wyoming and their contribution to fine-scale paleoenvironmental interpretation. Modern Geology 23:111–126.

Fiorillo, A.R. 1999. Determining the relative roles of climate and tectonics in the formation of the vertebrate fossil record: A perspective from the Late Cretaceous of western North America. Records of the Western Australian Museum Supplement 57:219–228.

Fiorillo, A.R., K. Padian, and C. Musikasinthorn. 2000. Taphonomy and depositional setting of the *Placerias* Quarry (Chinle Formation: Late Triassic, Arizona). Palaios 15:373–386.

Frey, W.R., M.R. Voorhies, J.D. Howard. 1975. Estuaries of the Georgia coast U.S.A. sedimentology and biology. VIII. Fossil and recent skeletal remains in Georgia estuaries. Senckenbergiana Maritima 7:257–295.

Fricke, H. This volume. Stable Isotope Geochemistry of Bonebed Fossils: Reconstructing Paleoenvironments, Paleoecology, and Paleobiology. Chapter 8 *in* Bonebeds: Genesis, analysis, and paleobiological significance. R.R. Rogers, D.A. Eberth, and A.R. Fiorillo, eds. University of Chicago Press, Chicago.

Gauthier-Pilters, H., and A.I. Dagg. 1981. The camel, its evolution, ecology, behavior, and relationship to man. University of Chicago Press, Chicago.

Getty, M., D.A. Eberth, D.B. Brinkman, D. Tanke, M. Ryan, and M. Vickaryous. 1997. Taphonomy of two *Centrosaurus* bonebeds in the Dinosaur Park Formation, Alberta, Canada. Journal of Vertebrate Paleontology 17(supplement to 3):48A–49A.

Gifford, D. 1984. Taphonomic specimens, Lake Turkana. National Geographic Research Reports 17:419–428.

Gifford-Gonzalez, D. 1989. Ethnographic analogues for interpreting modified bones: Some cases from East Africa. Pp. 179–246 *in* Bone modification. R. Bonnichsen and M.H. Sorg, eds. University of Maine Center for the Study of the First Americans, Orono.

Gilmore, C.W. 1925. A nearly complete articulated skeleton of *Camarasaurus*, a saurischian dinosaur from the Dinosaur National Monument, Utah. Memoirs of the Carnegie Museum 10:347–384.

Gilmore, C.W. 1936. Osteology of *Apatosaurus*, with special reference to specimens in the Carnegie Museum. Memoirs of the Carnegie Museum 11(4): 175–300.

Grayson, D.K. 1984. Quantitative zooarchaeology. Academic Press, New York.

Grayson, D.K. 1988. Danger Cave, Last Supper Cave, and Hanging Rock Shelter: The faunas. American Museum of Natural History Anthropological Papers 66:1–130.

Greenfield, H.J. 1988. Bone consumption by pigs in a contemporary Serbian village: Implications for the interpretation of prehistoric faunal assemblages. Journal of Field Archaeology 15:473–479.

Greenwald, M.T. 1989. Techniques for collecting large vertebrate fossils. Pp. 264–274 *in* Paleotechniques. Special publication 4. R.M. Feldman, R.E. Chapman, and J.T. Hannibal, eds. The Paleontological Society, Knoxville, Tennessee.

Hackett, C.J. 1981. Microscopical focal destruction (tunnels) in exhumed human bones. Medicine, Science, and the Law 21:243–265.

Hanson, D.B., and J.E. Buikstra. 1987. Histomorphological alteration in buried human bone from the Lower Illinois Valley: Implications for paleodietary research. Journal of Archaeological Science 14:549–563.

Harris, A.G., and W.C. Sweet. 1989. Mechanical and chemical techniques for separating microfossils from rock, sediment, and residue matrix. Pp. 70–86 *in* Paleotechniques. Special publication 4. R.M. Feldman, R.E. Chapman, and J.T. Hannibal, eds. The Paleontological Society, Knoxville, Tennessee.

Hasiotis, S.T., A.R. Fiorillo, and G.R. Laws. 1999. Borings in Jurassic dinosaur bones: Trace fossil evidence of beetle interactions with vertebrates. Pp. 193–200 *in* Vertebrate paleontology in Utah. Miscellaneous publication 99-1. D. Gillette, ed. Utah Geological Survey, Salt Lake City.

Hassan, A.A., and D.J. Ortner. 1977. Inclusions in bone material as a source of error in radiocarbon dating. Archaeometry 19:131–135.

Haynes, G. 1980. Evidence of carnivore gnawing on Pleistocene and Recent mammalian bones. Paleobiology 6:341–351.

Haynes, G. 1981. Prey bones and predators: Potential ecologic information from analysis of bone sites. Ossa 7:75–97.

Haynes, G. 1982. Utilization and skeletal disturbances of North American prey carcasses. Arctic 35:266–281.

Haynes, G. 1983. A guide for differentiating mammalian carnivore taxa responsible for gnaw damage to herbivore limb bones. Paleobiology 9:164–172.

Hendey, Q.B. 1981. Palaeoecology of the Late Tertiary fossil occurrences in "E" quarry, Langebaanweg, South Africa, and a reinterpretation of their geological context. Annals of the South African Museum 84:1–104.

Henrici, A.C., and A.R. Fiorillo. 1993. Catastrophic death assemblage of *Chelomophrynus bayi* (Anura, Rhinophrynidae) from the Middle Eocene Wagon Bed Formation of central Wyoming. Journal of Paleontology 67:1016–1026.

Hibbard, C.W. 1949. Techniques of collecting microvertebrate fossils. Contributions of the Museum of Paleontology, University of Michigan 3(2):7–19.

Hill, A.P. 1979. Disarticulation and scattering of mammal skeletons. Paleobiology 5:261–274.

Hill, A.P. 1980. Early postmortem damage to the remains of some contemporary east African mammals. Pp. 131–152 *in* Fossils in the making. A.K. Behrensmeyer and A.P. Hill, eds. University of Chicago Press, Chicago.

Hill, A.P., and A.K. Behrensmeyer. 1984. Disarticulation patterns of some modern East African mammals. Paleobiology 10:366–376.

Hill, A.P., and A. Walker. 1972. Procedures in vertebrate taphonomy. Journal of the Geological Society of London 128:399–406.

Hunt, A.P. 1987. Phanerozoic trends in nonmarine taphonomy: Implications for Mesozoic vertebrate taphonomy and paleoecology. Geological Society of America, Abstracts with Program 19:171.

Hunt, A.P., C.A. Meyer, M.G. Lockley, and S.G. Lucas. 1994. Archaeology, toothmarks and sauropod dinosaur taphonomy. Gaia 10:225–231.

Hunt, R.M. Jr. 1978. Depositional setting of a Miocene mammal assemblage, Sioux County, Nebraska (U.S.A.). Palaeogeography, Palaeoclimatology, Palaeoecology 24:1–52.

Hunt, R.M. Jr. 1990. Taphonomy and sedimentology of Arikaree (lower Miocene) fluvial, eolian, and lacustrine paleoenvironments, Nebraska and Wyoming: A paleobiota entombed in fine-grained volcaniclastic rocks. Pp. 69–111 *in* Volcanism and fossil biotas. Special paper 244. M.G. Lockley and A. Rice, eds. Geological Society of America, Boulder, Colorado.

Jacobsen, A.R. 1997. Tooth marks. Pp. 738–739 *in* The encyclopedia of dinosaurs. P.J. Currie and K. Padian, eds. Academic Press, San Diego.

Jacobsen, A.R. 1998. Feeding behaviour of carnivorous dinosaurs as determined by tooth marks on dinosaur bones. Historical Biology 13:17–26.

Jacobsen, A.R. 2001. Tooth-marked small theropod bone: An extremely rare trace. Pp. 58–63 *in* Mesozoic vertebrate life. D. Tanke and K. Carpenter, eds. Indiana University Press, Bloomington.

Jacobsen, A.R., and M.J. Ryan. 1999. Taphonomic aspects of theropod tooth-marked bones from an *Edmontosaurus* bonebed (Lower Maastrichtian), Alberta Canada. Journal of Vertebrate Paleontology 19(supplement to 3):55A.

Jamniczky, H.A., D.B. Brinkman, and A.P. Russell. 2003. Vertebrate microsite sampling: How much is enough? Journal of Vertebrate Paleontology 23:725–734.

Jans, M.M.E., C.M. Nielsen-Marsh, C.I. Smith, M.J. Collins, and H. Kars. 2004. Characterisation of microbial attack on archaeological bone. Journal of Archaeological Science 31:87–95.

Johnson, E. 1985. Current developments in bone technology. Pp. 157–235 *in* Advances in archaeological method and theory. M.B. Schiffer, ed. Academic Press, New York.

Kitching, J.M. 1980. On some fossil Arthropoda from the Limeworks, Makapansgat, Potgietersrus. Palaeontologia Africana 23:63–68.

Klein, R.G., and K. Cruz-Uribe. 1984. The analysis of animal bones from archeological sites. University of Chicago Press, Chicago.

Knox-Robinson, C.M., and S.J. Gardoll. 1998. GIS-stereoplot: an interactive stereonet plotting module for ArcView 3.0 geographic information system. Computers and Geosciences 24:243–250.

Korth, W.W. 1979. Taphonomy of microvertebrate fossil assemblages. Annals of the Carnegie Museum 48:235–284.

Korth, W.W., and R.L. Evander. 1986. The use of age-frequency distributions of micromammals in the determination of attritional and catastrophic mortality of fossil assemblages. Palaeogeography, Palaeoclimatology, Palaeoecology 52:227–236.

Kottlowski, F.E. 1965. Measuring stratigraphic sections. Holt, Rinehart and Winston, New York.

Kreutzer, L.A. 1988. Megafaunal butchering at Lubbock Lake, Texas: A taphonomic reanalysis. Quaternary Research 30:221–231.

Kurten, B. 1953. On the variation and population dynamics of fossil and recent mammal populations. Acta Zoologica Fennica 76:1–122.

Kusmer, K.D. 1990. Taphonomy of owl pellet deposition. Journal of Paleontology 64:629–637.

Laury, R.L. 1980. Paleoenvironment of a Late Quaternary mammoth-bearing sinkhole deposit, Hot Springs, South Dakota. Geological Society of America Bulletin 91:465–475.

Lawton, R. 1977. Taphonomy of the dinosaur quarry, Dinosaur National Monument. Contributions to Geology, University of Wyoming 15:119–126.

Lehman, T.M. 1982. A ceratopsian bone bed from the Aguja Formation (Upper Cretaceous) Big Bend National Park, Texas. Unpublished masters thesis, University of Texas, Austin.

Leiggi, P., C.R. Schaff, and P. May. 1994. Field organization and specimen collecting. Pp. 59–76 in Vertebrate paletonological techniques. P. Leiggi and P. May, eds. Cambridge University Press.

Lien, D., J. Cavin, S. Johnson, and C. Herbel. 2002. Interactive field mapping: Using the Pentax total Station and Arcview for the analysis of fossil bed accumulations. Journal of Vertebrate Paleontology 22(supplement to 3):79A.

Lyman, R.L. 1984. Bone density and differential survivorship of fossil classes. Journal of Anthropological Archaeology 3:259–299.

Lyman, R.L. 1994. Vertebrate taphonomy. Cambridge manuals in archaeology. Cambridge University Press, Cambridge.

Lyman, R.L., and G.L. Fox. 1989. A critical evaluation of bone weathering as an indication of bone assemblage formation. Journal of Archaeological Science 16:293–317.

Maas, M.C. 1985. Taphonomy of a Late Eocene microvertebrate locality, Wind River Basin, Wyoming (U.S.A.). Palaeogeography, Palaeoclimatology, Palaeoecology 52:123–142.

MacDonald, D.I.M., and T.H. Jefferson. 1985. Orientation studies of waterlogged wood: A paleocurrent indicator? Journal of Sedimentary Petrology 55:235–239.

Maguire, J.M., D. Pemberton, and M.H. Collett. 1980. The Makapansgat Limeworks Grey Breccia: Hominids, hyaenas, hystricids or hillwash? Paleontologia Africana 23:75–98.

Main, D., A.R. Fiorillo, and H. Montgomery. 2002. New methods in paleontology: GPR mapping and excavation of a titanosaurid (Dinosauria, Sauropoda) bonebed in Big Bend National Park. Geological Society of America, Abstracts with Programs 34(3):10.

Marchiafava, V., E. Bonucci, and A. Ascenzi. 1974. Fungal osteoclasia: A model of dead bone resorption. Calcified Tissue Research 14:195–210.

Marshall, L.G. 1989. Bone modification and "the laws of burial." Pp. 7–24 *in* Bone modification. R. Bonnichsen and M.H. Sorg, eds. University of Maine Center for the Study of the First Americans, Orono.

Martin, L.D., and D.L. West. 1995. The recognition and use of dermestid (Insecta, Coleoptera) pupation chambers in paleoecology. Palaeogeography, Palaeoclimatology, Palaeoecology 113:303–310.

Matthew, W.D. 1923. Fossil bones in the rock. Natural History 23:359–369.

Mayhew, D.F. 1977. Avian predators as accumulators of fossil material. Boreas 6:25–31.

McKenna, M.C. 1962. Collecting small fossils by washing and screening. Curator 5:221–235.

McKenna, M.C. 1965. Collecting microvertebrate fossils by washing and screening. Pp. 193–203 *in* Handbook of paleontological techniques. B. Kummel and D. Raup, eds. W.H. Freeman, San Francisco.

McKenna, M.C., A.R. Bleefield, and J.S. Mellet. 1994. Microvertebrate collecting: Large-scale wet sieving for fossil microvertebrates in the field. Pp. 93–111 *in* Vertebrate paletonological techniques. P. Leiggi and P. May, eds. Cambridge University Press, Cambridge.

Mellet, J.S. 1974. Scatological origin of microvertebrate fossil accumulations. Science 185:349–350.

Micozzi, M.S. 1991. Postmortem change in human and animal remains: A systematic approach. C.C. Thomas, Springfield, Illinois.

Miller, B.B. 1989. Screen-washing unconsolidated sediments for small macrofossils. Pp. 260–263 *in* Paleotechniques. Special publication 4. R.M. Feldman, R.E. Chapman, and J.T. Hannibal, eds. The Paleontological Society, Knoxville, Tennessee.

Miller, G.J. 1969. A study of cuts, grooves and other marks on recent and fossil bone. 1. Animal tooth marks. Tebiwa 12:20–26.

Miller, W.E., R.D. Horrocks, and J.H. Madsen Jr. 1996. The Cleveland-Lloyd dinosaur quarry, Emery County, Utah: A U.S. natural landmark (including history and quarry map). Brigham Young University Geology Studies 41:3–24.

Morris, T.H., D.R. Richmond, and S.D. Grimshaw. 1996. Orientation of dinosaur bones in riverine environments: Insights into sedimentary dynamics and taphonomy. Museum of Northern Arizona Bulletin 60:521–530.

Munthe, K., and S.A. McLeod. 1975. Collection of taphonomic information from fossil and recent vertebrate specimens with a selected bibliography. Paleobios 19:1–12.

Myers, T., M.R. Voorhies, and R.G. Corner. 1980. Spiral fractures and bone pseudotools at paleontological sites. American Antiquity 45:483–489.

Nolte, M.J., B.W. Greenhalgh, A. Dangerfield, R.D. Scheetz, and B.B. Britt. 2004. Invertebrate burrows on dinosaur bones from the Lower Cretaceous Cedar Mountain Formation near Moab, Utah, U.S.A. Geological Society of America, Abstracts with Programs 36:379.

Norman, D.B. 1987. A mass-accumulation of vertebrates from the Lower Cretaceous of Nehden (Sauerland), West Germany. Proceedings of the Royal Society of London 230:215–255.

Oldrich, F., and T. Kaiser. 2005. Insect bone-modification and paleoecology of Oligocene mammal-bearing sites in the Doupov Mountains, northwestern Bohemia. Palaeontologia Electronica 8(1):8.1.8A.

Olsen, S.L., and P. Shipman. 1988. Surface modification of bone: Trampling versus butchery. Journal of Archaeological Science 15:535–553.

Organ, C., J.B. Cooley, and T.L. Hieronymus. 2003. A non-invasive quarry mapping system. Palaios 18:74–77.

Paik, I.S., H.J. Kim, K.H. Park, Y.S. Song, Y.I. Lee, J.Y. Hwang, and M. Huh. 2001. Palaeonvironments and taphonomic preservation of dinosaur bone-bearing deposits in the Lower Cretaceous Hasandong Formation, Korea. Cretaceous Research 22:627–642.

Parrish, W.C. 1978. Paleoenvironmental analysis of a Lower Permian bone bed and adjacent sediments, Wichita County, Texas. Palaeogeography, Palaeoclimatology, Palaeoecology 24:209–237.

Peterson, O.A. 1905. Preliminary note of a gigantic mammal from the Loup Fork beds of Nebraska. Science 22:211–212.

Peterson, O.A. 1906. The Agate Spring Fossil Quarry. Annals of the Carnegie Museum 3:487–494.

Piepenbrink, H. 1986. Two examples of biogenous dead bone decomposition and their consequences for taphonomic interpretation. Journal of Archaeological Science 13:417–430.

Potts, R. 1986. Temporal span of bone accumulations at Olduvai Gorge and implications for early hominid foraging behavior. Paleobiology 12:25–31.

Pratt, A.E. 1989. Taphonomy of the microvertebrate fauna from the early Miocene Thomas Farm locality, Florida (U.S.A.). Palaeogeography, Palaeoclimatology, Palaeoecology 76:125–151.

Pryor, R., D.H. Tanke, and P.J. Currie. 2001. Precise mapping of fossil localities in Dinosaur Provincial park (Alberta, Canada) using advanced GPS technology. Pp. 58–60 in Alberta Palaeontological Society fifth annual symposium abstracts. University of Calgary Press, Calgary.

Ragan, D.M. 1973. Structural geology: An introduction to geometrical techniques, 2nd ed. John Wiley and Sons, New York.

Richardson, P.R.K. 1980. Carnivore damage to antelope bones and its archaeological implications. Paleontologia Africana 23:109–125.

Richmond, D.R., and T.H. Morris. 1996. The dinosaur death-trap of the Cleveland-Lloyd Quarry, Emery County, Utah. Museum of Northern Arizona Bulletin 60:533–545.

Roberts, E.M., R.R. Rogers, and B.Z. Foreman. 2007. Continental insect borings in dinosaur bone: Examples from the Late Cretaceous of Madagascar and Utah. Journal of Paleontology 81: 201–208.

Rodrigues-De La Rosa, R.A., and S.R.S. Cevallos-Ferriz. 1998. Vertebrates of the El Pelillal locality (Campanian, Cerro Del Pueblo Formation), southeastern Coahuila, Mexico. Journal of Vertebrate Paleontology 18:751–764.

Rogers, R.R. 1990. Taphonomy of three dinosaur bone beds in the Upper Cretaceous Two Medicine Formation of northwestern Montana: Evidence for drought-related mortality. Palaios 5:394–413.

Rogers, R.R. 1992. Non-marine borings in dinosaur bones from the Upper Cretaceous Two Medicine Formation, northwestern Montana. Journal of Vertebrate Paleontology 12:528–531.

Rogers, R.R. 1993. Systematic patterns of time-averaging in the terrestrial vertebrate record: A Cretaceous case study. Pp. 228–249 in Taphonomic approaches to time resolution in fossil assemblages. Short courses in paleontology 6. S.M. Kidwell and A.K. Behrensmeyer, eds. The Paleontological Society, Knoxville, Tennessee.

Rogers, R.R. 1994. Collecting taphonomic data from vertebrate localities. Pp. 47–58 *in* Vertebrate paleontological techniques. P. Leiggi and P. May, eds. Cambridge University Press, Cambridge.

Rogers, R.R. 2005. Fine-grained debris flows and extraordinary vertebrate burials in the Late Cretaceous of Madagascar. Geology 33:297–300.

Rogers, R.R., and D.A. Eberth. 1996. Stratigraphic utility of vertebrate microfossil assemblages in the Campanian of Montana and Alberta. Journal of Vertebrate Paleontology 16(supplement to 3):61A.

Rogers, R.R., and S.M. Kidwell. 2000. Associations of vertebrate skeletal concentrations and discontinuity surfaces in terrestrial and shallow marine records: a test in the Cretaceous of Montana. Journal of Geology 108:131–154.

Rogers, R.R., and S.M. Kidwell. This volume. A conceptual framework for the genesis and analysis of vertebrate skeletal concentrations. Chapter 1 *in* Bonebeds: Genesis, analysis, and paleobiological significance. R.R. Rogers, D.A. Eberth, and A.R. Fiorillo, eds. University of Chicago Press, Chicago.

Rogers, R.R., A.B. Arucci, F. Abdala, P.C. Sereno, C.A. Forster, and C.L. May. 2001. Paleoenvironment and taphonomy of the Chañares Formation tetrapod assemblage (Middle Triassic), northwestern Argentina: Spectacular preservation in volcanogenic concretions. Palaios 16:461–481.

Rogers, R.R., D.W. Krause, and K. Curry Rogers. 2003. Cannibalism in the Madagascan dinosaur *Majungatholus atopus*. Nature 422:515–518.

Romer, A.S. 1939. An amphibian graveyard. Scientific Monthly 49:337–339.

Ryan, M.J., P.J. Currie, J.D. Gardner, M.K. Vickaryous, and J.M. Lavigne. 2000. Baby hadrosaurid material associated with an unusually high abundance of *Troodon* teeth from the Horseshoe Canyon Formation, Upper Cretaceous, Alberta, Canada. Gaia 15:123–133.

Ryan, M.J., A.P. Russell, D.A. Eberth, and P.J. Currie. 2001. The taphonomy of a *Centrosaurus* (Ornithischia: Certopsidae [sic]) bone bed from the Dinosaur Park Formation (Upper Campanian), Alberta, Canada, with comments on cranial ontogeny. Palaios 16:482–506.

Sander, P.M. 1987. Taphonomy of the Lower Permian Geraldine Bonebed in Archer County, Texas. Palaeogeography, Palaeoclimatology, Palaeoecology 61:221–236.

Sander, P.M. 1989. Early Permian depositional environments and pond bonebeds in central Archer County, Texas. Palaeogeography, Palaeoclimatology, Palaeoecology 69:1–21.

Sander, P.M. 1992. The Norian *Plateosaurus* bonebeds of central Europe and their taphonomy. Palaeogeography, Palaeoclimatology, Palaeoecology 93:255–299.

Saunders, J.J. 1977. Late Pleistocene vertebrates of the western Ozark highland, Missouri. Reports of investigations 33. Illinois State Museum, Springfield.

Schröder-Adams, C.J., S.L. Cumbaa, J. Bloch, D.A. Leckie, J. Craig, S.A. Seif El-Dein, D.-J.H.A.E. Simons, and F. Kenig. 2001. Late Cretaceous (Cenomanian to Campanian) paleoenvironmental history of the Eastern Canadian margin of the Western Interior Seaway: Bonebeds and anoxic events. Palaeogeography, Palaeoclimatology, Palaeoecology 170:261–289.

Schwartz, H.L., and D.D. Gillette. 1994. Geology and taphonomy of *Coelophysis* Quarry, Upper Triassic Chinle Formation, Ghost Ranch New Mexico. Journal of Paleontology 68:1118–1130.

Shaw, M., K. Sandhoo, and D. Turner. 2000. Modernization of the global positioning system. GPS World Online, www.gpsworld.com/1000/1000shaw.html.

Shipman, P. 1975. Implications of drought for vertebrate fossil assemblages. Nature 257:667–668.

Shipman, P. 1981. Life history of a fossil: An introduction to taphonomy and paleoecology. Harvard University Press, Cambridge.

Shipman, P., and J. Rose. 1984. Cutmark mimics on modern and fossil bovid bones. Current Anthropology 25:116–117.

Shipman, P., and J. Rose. 1988. Bone tools: An experimental approach. Pp. 303–335 *in* Scanning electron microscopy in archaeology. British archaeological reports international series 452. S.L. Olson, ed. British Archaeological Reports, Oxford.

Shipman, P., D.C. Fisher, and J. Rose. 1984. Mastodon butchery: Microscopic evidence of carcass processing and bone tool use. Paleobiology 10:358–365.

Smith, R.M.H. 1993. Vertebrate taphonomy of Late Permian floodplain deposits in the southwestern Karoo Basin of South Africa. Palaios 8:45–67.

Smith, R.M.H., and J.M. Kitching. 1997. Sedimentology and vertebrate taphonomy of the *Tritylodon* Acme Zone; a reworked palaeosol in the Lower Jurassic Elliot Formation, Karoo Supergroup, South Africa. Palaeogeography, Palaeoclimatology, Palaeoecology 131:29–50.

Staron, R.M., D.E. Grandstaff, B.S. Grandstaff, and W.B. Gallagher. 1999. Mosasaur taphonomy and geochemistry: Implications for a K-T bonebed in the New Jersey coastal plain. Journal of Vertebrate Paleontology 19(supplement to 3):78A.

Straight, W.H., and D.A. Eberth. 2002. Testing the utility of vertebrate remains in recognizing patterns in fluvial deposits: An example from the lower Horseshoe Canyon Formation, Alberta. Palaios 17:472–490.

Sutcliffe, A.J. 1973. Similarity of bones and antlers gnawed by deer to human artifacts. Nature 246:428–430.

Sutcliffe, A.J. 1977. Further notes on bones and antlers chewed by deer and other ungulates. Deer 4:73–82.

Tanke, D.H. 1999. Relocating the lost dinosaur quarries of Dinosaur Provincial Park, Alberta, Canada. Journal of Vertebrate Paleontology 19(supplement to 3):80A.

Tanke, D.H. 2001. Historical archaeology: Solving the mystery quarries of Drumheller and Dinosaur Provincial Park, Alberta, Canada. Pp. 70–72 *in* Alberta Palaeontological Society, fifth annual symposium Abstracts. University of Calgary Press, Calgary.

Tappen, M. 1994. Bone weathering in the tropical rain forest. Journal of Archaeological Science 21:667–673.

Therrien, F., and D.E. Fastovsky. 2000. Paleoenvironments of early theropods, Chinle Formation (Late Triassic), Petrified Forest National Park, Arizona. Palaios 15:194–211.

Thomas, R.G., D.A. Eberth, A.L. Deino, and D. Robinson. 1990. Composition, radioisotopic ages, and potential significance of an altered volcanic ash (bentonite) from the Upper Cretaceous Judith River Formation, Dinosaur Provincial Park, southern Alberta, Canada. Cretaceous Research 11:125–162.

Todd, L.C. 1983. Taphonomy: Fleshing out the dry bones of Plains prehistory. The Wyoming Archaeologist 26:36–46.

Toots, H. 1965a. Sequence of disarticulation in mammalian skeletons. Contributions to Geology, University of Wyoming 4:37–39.

Toots, H. 1965b. Orientation and distribution of fossils as environmental indicators. Pp. 219–229 in Sedimentation of Late Cretaceous and Tertiary outcrops, Rock Springs Uplift. 19th annual field conference guidebook. R.H. DeVoto and R.K. Bitter, eds. Wyoming Geological Association, Casper.

Trueman, C.N. 1999. Rare Earth element geochemistry and taphonomy of terrestrial vertebrate assemblages. Palaios 14:555–568.

Trueman, C.N. This volume. Trace element geochemistry of bonebeds. Chapter 7 in Bonebeds: Genesis, analysis, and paleobiological significance. R.R. Rogers, D.A. Eberth, and A.R. Fiorillo, eds. University of Chicago Press, Chicago.

Trueman, C.N., and M.J. Benton. 1997. A geochemical method to trace the taphonomic history of reworked bones in sedimentary settings. Geology 25:263–266.

Van Valkenburgh, B., L. Spencer, J. Harris, J. Samuels, A. Friscia, and J. Meachen. 2004. Chronology and spatial distribution of large mammal bones in Pit 91, Rancho La Brea. Journal of Vertebrate Paleontology 24(supplement to 3):124A.

Varricchio, D.J. 1995. Taphonomy of Jack's Birthday Site, a diverse dinosaur bonebed from the Upper Cretaceous Two Medicine Formation of Montana. Palaeogeography, Palaeoclimatology, Palaeoecology 114:297–323.

Voorhies, M.R. 1969. Taphonomy and population dynamics of the early Pliocene vertebrate fauna, Knox County, Nebraska. Contributions to geology, special paper 1. University of Wyoming, Laramie.

Voorhies, M.R. 1985. A Miocene rhinoceros herd buried in volcanic ash. National Geographic Research Reports, Projects 1978:671–688.

Ward, D.J. 1984. Collecting isolated microvertebrate fossils. Zoological Journal of the Linnaean Society 82:245–259.

Webb, D.S., B.J. MacFadden, and J.A. Baskin. 1981. Geology and paleontology of the Love Bone Bed from the late Miocene of Florida. American Journal of Science 281:513–544.

Weigelt, J. 1927. Rezente wirbeltierleichen und ihre paläobiologische bedeutung. Verlag von Max Weg, Leipzig.

Weigelt, J. 1989. Recent vertebrate carcasses and their paleobiological implications. J. Schaefer, transl. University of Chicago Press, Chicago.

Whitaker, J.H.McD., and D.D.J. Antia. 1978. A scanning electron microscope study of the genesis of the Upper Silurian Ludlow Bone Bed. Pp. 119–136 in Scanning electron microscopy in the study of sediments. W.B. Whalley, ed. Geoabstracts, Norwich.

White, P.D., D.E. Fastovsky, and P.M. Sheehan. 1998. Taphonomy and suggested structure of the dinosaurian assemblage of the Hell Creek Formation (Maastrichtian), eastern Montana and western North Dakota. Palaios 13:41–51.

# CHAPTER 6

# Numerical Methods for Bonebed Analysis

*Richard W. Blob and Catherine Badgley*

## INTRODUCTION

Studies of fossil bonebeds range widely in scope, depending largely upon the mode of preservation at a site, the taxa preserved, and the research goals of the investigator. In monotaxic or paucitaxic assemblages, for example, fossilized skeletons can provide detailed information about populations of animals, including data on age structure, the proportions of males and females, and even some aspects of behavior, such as the tendency for adults to aggregate, the prevalence of parental care, or preferred habitat (e.g., Voorhies, 1969; Horner and Makela, 1979; Horner, 1982; Currie and Dodson, 1984). Other types of multispecies assemblages, such as microvertebrate concentrations, may not provide such detailed information about the habits of single species, but comparisons of multispecies assemblages provide opportunities to examine changes in faunal composition through time and across sedimentary basins or larger geographic regions. Such analyses facilitate studies of community evolution (e.g., Krause, 1986; Maas et al., 1988; Brinkman, 1990; Badgley and Behrensmeyer, 1995; Barry et al., 1995, 2002; Brinkman et al., 1998; Lillegraven and Eberle, 1999). Brinkman et al. (Chapter 4 in this volume) discuss the different kinds of information that can be derived from mono-, pauci-, and multitaxic bonebeds.

The primary data for many of the questions that bonebed deposits are used to address are counts of skeletal elements and taxa and the associated

Table 6.1. Research questions and analytical methods

1. How many individuals of each taxon occur at a bonebed/site?
   Protocols for counting specimens
   Taxon-specific modifications of counting protocols
2. Are two (or more) sites taphonomically equivalent? In what ways can faunal composition be legitimately compared among sites?
   Assessment of taphonomic equivalence: facies, environments, mode of accumulation, postdepositional processes, etc.
   Comparison of skeletal-element distributions, size distributions of taxa, or frequencies of particular skeletal elements
   Assessment of hydraulic equivalence
3. How can species richness be compared among sites?
   Assessment of sample-size, sampling intensity
   Simulation
   Rarefaction
4. How can original abundances of species in ancient ecosystems be estimated from taxonomic abundances in fossil assemblages?
   Inferential model of original community composition, accounting for preservation bias and variation in turnover rates among taxa.
5. Do vertebrate faunas change significantly along spatial or temporal environmental gradients?
   Analysis of variance
   Regression
   Cluster analysis
   Principal components analysis
   Correspondence analysis

contextual information preserved at fossil localities. The most appropriate methods of counting, as well as the most reasonable comparisons to make, depend not only upon the questions of the investigator, but also upon the depositional context of the bonebeds under study.

In this chapter, we discuss several kinds of quantitative analyses of bonebed assemblages in relation to particular research questions (Table 6.1). First, we review protocols for counting specimens recovered from bonebeds, a crucial initial step that lays the foundation for further analyses. Second, we discuss the evaluation of taphonomic equivalence among bonebeds, a consideration that is crucial for comparisons of faunal counts among fossil localities. Third, we discuss comparisons of species richness among localities, particularly with respect to sampling intensity (i.e., evaluating whether faunal differences among sites could be due to differences in the intensity of fossil collecting). Fourth, we consider approaches for estimating taxonomic abundances in ancient ecosystems based on species abundances in fossil assemblages. Finally, we discuss several techniques for the quantitative analysis of changes in lineages and faunas among fossil bonebeds. We illustrate all of the quantitative methods presented with case studies in which those methods have been applied. Although we refer to vertebrate faunas as our units of analysis in this review, the principles of these methods are suitable for analyses of a wide variety of fossil assemblages.

## PROTOCOLS FOR SPECIMEN COUNTING: THE FOUNDATION OF QUANTITATIVE FAUNAL ANALYSES

### Methods of Quantification

At the start of any quantitative analysis of a fossil bonebed, a question arises: How many individuals of each taxon are represented? This question is important, because its answer will form the basis for almost every subsequent analysis of the fauna. In most cases, the reason for counting specimens at a site is to obtain estimates that will best reflect the number of taxa and their individual abundances in the fossil assemblage. Additional considerations arise in inferring the abundances of taxa in the original living fauna (see "Inferring Population Abundances in Original Faunas from Fossil Assemblages").

The best way to count individuals in the fossil assemblage depends on the circumstances under which fossils were concentrated in the assemblage (Badgley, 1986a). No single method is appropriate for all circumstances. Initially, it is necessary to decide upon the level of taxonomic resolution and a method of quantification to use in estimating the number of individuals. The desirable level of taxonomic resolution depends upon the nature of the fossil material and the identifiability of specimens. Most vertebrate fossils are readily identifiable to species level from even fragmentary dental, gnathic, or cranial fossils but cannot be identified beyond higher taxonomic levels based on postcranial remains. The level of taxonomic resolution need not be the same for all taxa considered in a faunal analysis; for example, a mixed set of species-, genus-, and family-level taxa may be quite suitable in most studies. However, it is essential that fossils be unambiguously assigned to one taxon. For example, if jaws from a taxon are assigned to species level, but postcranials from that taxon can only be assigned to family level, then the quantification process for that taxon must focus on the family level. In other words, the number of individuals belonging to the family, rather than each of its component species, must be counted.

Three main protocols are used to quantify faunal abundances from bonebed deposits. These protocols involve estimates for each taxon of (1) the number of identifiable specimens (NISP), (2) the minimum number of skeletal elements (MNE), and (3) the minimum number of individuals (MNI) (Badgley, 1986a). For NISP, every separate specimen that can be identified to the specified level of taxonomic resolution is counted as a separate individual; specimens may include fragments with matched breaks that can be fitted together. For MNE, bone fragments for the same element are grouped together even if matched breaks are not present, and an estimate is made of the minimum number of skeletal elements that could account for the material present. This estimate accounts for left and right specimens, juvenile and adult specimens, and other relevant information. Each element recognized is counted as one individual. For MNI, the minimum number of individuals necessary to account for all of the elements at a site is evaluated. One way to achieve this estimate is to count the most common element present for each taxon (e.g., the number of left humeri). The common element need not be the same for every taxon. Additional information in the remaining fossil material for each taxon may require the estimate to be increased, based upon the size, life stages, and preservational state of the common element in relation to the remaining fossil material. If the remaining fossil material has elements signifying a body size, life stage (e.g., juvenile), or preservational state not represented by the most common element, then the estimate of MNI can be revised as appropriate. For example, suppose that the most common element in a sample of 28 bovid fossils is the left dentary. If seven left dentaries are present, then seven individuals would be counted for the bovids. If all seven dentaries are of adult individuals, and the postcranial remains contain one long bone with an unfused epiphysis, then the count would be revised to eight individuals in order to include the juvenile animal.

The primary criterion for choosing among these methods of counting is the probability that fossil specimens of a particular taxon belong to the same individual animal (Badgley, 1986a). This probability of association, or p(A), is estimated qualitatively from the taphonomic information at a locality, including its depositional context. Taphonomic features relevant to the evaluation of p(A) for a bonebed include the general depositional environment, the distribution of bones within the sedimentary source unit, the presence or absence of articulated or contiguous skeletal material of the same taxon, the preservational state of fossils, and other taphonomic information that reflects the circumstances of accumulation of skeletal material at the site. The underlying rationale is that some circumstances of accumulation disassociate skeletal elements from their original skeletons

Table 6.2. Guidelines and examples of quantification (estimation of the number of individuals) for vertebrate fossil assemblages

A. Guidelines for quantification

| Probability of Association | High | | Low |
|---|---|---|---|
| Example | Articulated skeletons | Dispersed, fragmented | Dispersed, fluvial transport |
| Context | Natural trap | Floodplain surface | Channel lag |
| Counting Method | Minimum no. of individuals MNI | Minimum no. of elements MNE | No. of identified specimens NISP |

*Note:* Guidelines for assessing the probability of association among skeletal elements and the quantification method suited to different probabilities of association.
*Source:* Modified from Badgley (1986a).

B. Quantification of surface-collected mammalian assemblages from Miocene of Pakistan

| Taxonomic Group | Channel Lag NISP* | Concentration on Floodplain MNE | MNI* |
|---|---|---|---|
| Hominoidea | 6 | 5 | 1 |
| Carnivora indet. | 29 | 7 | ? |
| Mustelidae | 4 | 11 | 1 |
| Viverridae | 1 | 38 | 4 |
| Hyaenidae | 9 | 43 | 5 |
| Felidae | 3 | 1 | 1 |
| Proboscidea indet. | 16 | 5 | 2 |
| Gomphotheriidae | 12 | 6 | 4 |
| Deinotheriidae | 1 | — | — |
| Equidae | 166 | 23 | 7 |
| Chalicotheriidae | 3 | — | — |
| Rhinocerotidae | 54 | 9 | 3 |
| Suidae | 40 | 51 | 7 |
| Anthracotheriidae | 18 | — | — |
| Tragulidae | 39 | 2 | 2 |
| Giraffidae | 41 | 5 | 3 |
| Bovidae | 378 | 80 | 16 |

*Note:* Example from Siwalik, late Miocene mammalian assemblages from Pakistan.
Note that the different quantification methods may result in substantially different estimates of the number of individuals per taxon.
*Most appropriate quantification method for the depositional context.
*Source:* Modified from Badgley (1986a).

and aggregate elements from different individuals, whereas other circumstances of accumulation tend to keep whole or partial skeletons together until burial (Rogers and Kidwell, 2000). Table 6.2A illustrates three examples from the spectrum of probabilities of association.

If the site evaluation leads to the inference that p(A) is very low and little fragmentation is present, then NISP will provide the best estimate of the number of individuals represented in a bonebed. For example, NISP is appropriate for an assemblage of largely intact but disassociated elements from a conglomerate lens representing a channel-lag deposit. In contrast, if the fossils at a site show considerable fragmentation, such as in screen-washed microfossil collections, then NISP is likely to inflate species counts. In such cases, MNE gives a better estimate of faunal abundances, as it avoids overcounting highly fragmented fossils. However, p(A) should still be assessed as low for the fossil elements recovered from a site, if MNE is to be applied. If, on the other hand, p(A) is inferred to be high, then MNI should be used. MNI is appropriate for mass-mortality sites, such as natural traps or assemblages of drowned carcasses; it is also appropriate for carnivore dens. The results of different quantification methods may differ greatly or little when applied to the same assemblage, depending upon the characteristics of the fossil assemblage (Table 6.2B).

Since the taphonomic histories of individual fossil assemblages may vary widely, even for localities from a single depositional basin, the question of which quantification method to use should be evaluated for every fossil assemblage. Usually, in a stratigraphic sequence with many fossil localities, fossil assemblages recur in a few depositional contexts, each context with consistent taphonomic features. Thus, the main analytical decision may be which quantification method to select for fossil assemblages with similar taphonomic histories. It is perfectly legitimate to use a different quantification method for different fossil assemblages from the same sequence. For the Miocene Siwalik sequence of northern Pakistan, for example, Badgley (1986a, 1986b) used NISP for fluvially transported vertebrate assemblages and MNI for contemporaneous assemblages interpreted as carnivore accumulations or long-term predation sites.

### Taxon-Specific Modifications of Counting Protocols

In some cases, taxon-specific modifications to quantification protocols may help to limit overcounting or undercounting of particular groups of animals that could bias faunal-abundance estimates for a locality. Bone-fragmentation patterns can vary considerably among taxa, and failure to account for these differences could confound efforts to census bonebeds in which fragmented specimens are common, such as screen-washed microvertebrate localities. To eliminate the chance of counting multiple fragments from the same fossil element (which would inflate counts for taxa with highly fragmented fossils), counts can be restricted to include only bones more than half complete in faunal censuses (Blob and Fiorillo, 1996). However, exclusive use of this protocol could bias counts against taxa typically represented in fossil assemblages by fragile elements that are rarely recovered half complete. Such a pattern could reflect predation or scavenging rather than original abundance.

To counter such biases, taxon-specific adjustments can be applied to the standard of completeness required for a specimen to be considered a countable "element" for faunal censuses. Salamanders, for instance, are commonly represented in Mesozoic and Cenozoic microvertebrate assemblages by fossil vertebrae. However, salamander vertebrae typically have severely constricted midcentra and, in collections made by screen washing, these fossils are usually recovered as broken half vertebrae (Brinkman, 1990; Blob and Fiorillo, 1996). Under a standard MNE counting protocol, these fossils would not be considered sufficiently complete for inclusion in faunal counts. To avoid drastically undercounting salamanders in faunal censuses of microvertebrate bonebeds, every half centrum could be considered a discrete element for the purpose of specimen counts (Brinkman, 1990; Blob and Fiorillo, 1996). Adjustments to standards of specimen completeness also may be warranted for turtle shell bones included in censuses of bonebeds preserving highly fragmented fossils. The platelike bones of turtle shells often break into large numbers of fragments that cannot be assigned to specific anatomical elements (Hutchison and Archibald, 1986), suggesting an abundance of turtles that likely would be undercounted if MNE protocols were followed strictly, but potentially overcounted if the NISP protocol were adopted. To moderate potential under- or overcounting of turtles in bonebeds, the standards for including specimens in faunal censuses could be modified in a variety of ways. In addition to counting elements more than half complete, for example, shell fragments indicating individuals of body sizes unrepresented among more complete specimens also could be tallied, using shell-thickness disparities to indicate body-size differences (Hutchison and Archibald, 1986; Fiorillo, 1989; Blob and Fiorillo, 1996). These examples from two common vertebrate taxa emphasize that taxon-specific anatomical or preservational vagaries should be a standard consideration when counting individuals from vertebrate assemblages, particularly for multitaxic bonebeds in which different faunal components may be affected differently.

The number of diagnostic elements that different taxa are likely to contribute to bonebeds also can vary, a consideration that is particularly relevant for faunal analyses of bonebeds composed of disarticulated elements from a wide variety of taxonomic groups. For example, although the internal skeletons of osteichthyan fishes and tetrapods possess roughly similar numbers of bones, some armored fishes such as gars (e.g., *Atractosteus, Lepisosteus*) can contribute several hundred durable, diagnostic armored scales per individual to a bonebed (Bryant, 1989). Even for bonebeds assembled through fluvial transport, in which the p(A) of individual skeletal elements is near zero, it seems likely that NISP and MNE could overestimate the abundance of such fishes relative to other taxa. Adjustments to NISP or MNE protocols that could account for superabundant, nonunique elements among certain taxa have not been proposed. However, alternative counting methods such as MNI are likely to drastically underestimate abundances for taxa possessing numerous nonunique elements in assemblages of fossils with a low p(A) (Badgley, 1986a; Bryant, 1989), suggesting that NISP or MNE protocols are still the best options for counting such taxa if the sedimentological setting is appropriate.

For some assemblages, however, it may be possible to quantify likely differences in fossil production among taxa and to use those evaluations to refine faunal counts. For example, in Mesozoic microvertebrate bonebeds, isolated teeth are among the most common fossils preserved for taxa ranging from mammals to crocodilians to a wide variety of dinosaurs, including theropods, hadrosaurs, and ceratopsians (e.g., Dodson, 1987; Brinkman et al., 1998). However, these lineages not only possessed different numbers of teeth, but also shed and replaced their teeth at different rates. For instance, individual hadrosaurid dinosaurs possessed hundreds of functioning teeth simultaneously (Weishampel, 1984) and may have shed and replaced most of these teeth as often as four to eight times per year (Erickson, 1996), whereas theropod jaws held approximately 50 teeth that were replaced once every one to two years (Erickson, 1996). These data suggest that a single hadrosaur could have produced as many as 10 times more isolated teeth than a single theropod during the same time span. Thus, in comparisons of hadrosaur and theropod abundance based on counts of isolated teeth (e.g., Brinkman et al., 1998), tooth abundance and replacement data could be used to adjust counts and reduce the likely inflation of hadrosaur abundance. Depending on the estimated replacement rate and number of functional teeth for a particular hadrosaur taxon, it may be appropriate to divide hadrosaur NISP counts by a factor between 4 and 16, if those counts are to be compared to counts for theropods. Such adjustments are less straightforward to apply to faunal counts based on wider ranges of taxa and anatomical elements but could help to refine a variety of specific comparisons.

## TAPHONOMIC EQUIVALENCE AND COMPARISONS OF BONEBED FAUNAS

### Taphonomic Equivalence: Conceptual Framework

Comparisons of fossiliferous sequences are fundamental to paleobiological research. For many comparative purposes, and especially paleoecological ones, it is important to compare and analyze multiple fossil assemblages from either the same or different sequences. The effects of taphonomic and depositional factors must be considered in such comparisons in order to avoid performing inappropriate analyses and to limit the risk of drawing erroneous conclusions. If environments of deposition or other taphonomic attributes differ among analysis units, then differences in faunal composition may reflect different habitats or processes of accumulation rather than differences in the original living faunas. However, if the depositional environments of bonebeds are similar, then at least some of the taphonomic processes that affected the fossils preserved at those sites likely also were similar, reducing the chance that faunal differences among those bonebeds are mainly taphonomic artifacts (Eberth, 1990; Badgley and Behrensmeyer, 1995). Behrensmeyer (1991) proposed the term "isotaphonomic" for fossil assemblages that have similar general taphonomic histories and argued that taphonomically equivalent sites can be usefully compared over long stretches of geologic time or great distances.

Several kinds of data are useful for assessing the taphonomic equivalence of fossil sites (Table 6.3). Lithofacies at the scale of the formation and the fossil assemblage indicate the general depositional environment (e.g., fluvial system) and the location of the fossil assemblage within it (e.g., distal floodplain), respectively. The spatial arrangement of fossils in the sedimentary matrix provides information about association among skeletal elements, preferred orientation of bones, depositional and postdepositional processes influencing the fossil assemblage, and the duration of accumulation. The spatial density of fossils reflects the rate of accumulation of skeletal remains in relation to the sediment accumulation rate. Bone modifications, including abrasion, rounding, weathering, punctures, scratches, and breakage patterns, indicate physical or biological postmortem processes. The size range of fossil material is informative about the agent(s)

Table 6.3. Quantitative and semiquantitative taphonomic attributes that can be compared among vertebrate assemblages.

| Attribute | Possible Manifestations for an Assemblage | | | |
|---|---|---|---|---|
| **Assemblage data** | | | | |
| Sample size | —10———-100———-1000———-10,000— | | | |
| No. of individuals | 1———-10———-100———-1000——— | | | |
| No. of species | 1———-10———-100——— | | | |
| Relative abundance | Equal frequencies———-Unequal frequencies | | | |
| Body-size range | Small species missing—All sizes present—Large species missing | | | |
| Skeletal-element dist. | Unsorted———-Sorted———-One element only | | | |
| Articulation | Articulated—Disarticulated but—Associated and—Isolated and | | | |
| | | associated | dispersed | dispersed |
| Age distribution | Juveniles only———-Mixed———-Adults only | | | |
| | | | | |
| **Quarry data** | | | | |
| Area of accumulation | —1 m² ———-100 m²——— 1000 m²——— | | | |
| Density | —0.1/m²———-1/m²———-10/m²———-100/m²— | | | |
| Spatial arrangement | | | | |
| Plan view | Random azimuths———-Preferred orientation | | | |
| Profile view | High dips———-Horizontal | | | |
| Patchiness | Evenly distributed———-Uneven———-Clumped | | | |
| | | | | |
| **Bone modification** | | | | |
| Breakage | Complete elements—Broken—Fragments | | | |
| Weathering stage* | Stage 0———-Stage 1———-Stages 0–3 present | | | |
| Abrasion/polish | Unabraded———-Abraded———-Highly abraded | | | |
| Surface marks | None———-Present———-Extensive | | | |

*Note:*\*Refers to weathering stages of Behrensmeyer (1978).
*Source:* Modified from Behrensmeyer (1991: Table 1).

of accumulation. Hydraulic equivalence indicates whether the processes that delivered clastic sediments to the site could also have transported the observed skeletal elements. The body-size range of animals represented by fossils, in relation to the size range of vertebrates known to be present at the relevant time, indicates whether gross size biases affected the fossil assemblage. The range and distribution of individuals by ontogenetic age also are potentially informative about the circumstances of mortality and agent(s) of accumulation. Finally, the skeletal-element composition of individual taxa provides evidence of the degree of sorting and winnowing that affected the assemblage. These data are the basis for inferring environment of deposition and processes and agents of accumulation of the fossils. In turn, these inferences form the basis for assessing taphonomic equivalence.

Ideal taphonomic equivalence occurs when the general and specific environments of deposition and the processes and agents of accumulation are the same. An example is lag concentrations in conglomerate lenses of fluvial channels in different foreland-basin settings, although even in these settings the amount of sourcing from preexisting channel accumulations during channel migration can vary. Ideal taphonomic equivalence is rarely achieved since fossil preservation and accumulation involve stochastic processes; hence, stochastic variation is present even among fossil assemblages produced under very similar circumstances in the same basin. Also, some important attributes of fossil assemblages, such as the time span represented in a fossil assemblage, are difficult to estimate even qualitatively (e.g., Kidwell and Behrensmeyer, 1993). Nonetheless, systematic comparison of taphonomic attributes, such as those in Table 6.3, can steer the investigator away from attributing differences in the composition of fossil assemblages to evolution or environmental change, if the differences are more likely due to taphonomic processes alone. Taphonomic comparisons of localities can serve as an informal "null model" in assessing the factors that caused changes in faunal composition over time or space. The null hypothesis is that differences between two (or more) localities are due to different taphonomic histories alone (Westrop, 1986; Blob and Fiorillo, 1996). If taphonomic attributes of the localities are similar, then it becomes more difficult to explain differences in faunal composition by the null hypothesis. Examples of isotaphonomic vertebrate assemblages include the fluvial channel-lag and fluvial channel-fill "modes" compared by Behrensmeyer (1988) for a Permian sequence in Texas and the Miocene Siwaliks of Pakistan. Each depositional setting showed a characteristic set of taphonomic attributes. The channel-lag setting preserved bones from a large part of the floodbasin, whereas the channel-fill (abandoned channel) setting preserved mainly local accumulations. The pedogenically overprinted, floodplain mudstones of ancient fluvial systems provide another example of isotaphonomic settings (Fig. 6.1). Vertebrate assemblages in this context also show a characteristic set of taphonomic features, despite large differences in age, location, and faunal composition (e.g., Retallack, 1983; Bown and Kraus, 1993; Badgley et al., 1995).

A

B

*Figure 6.1.* Eocene and Miocene fluvial sequences, illustrating similar depositional environments that have preserved abundant vertebrate assemblages. A. Early Eocene Willwood Formation of Clark's Fork Basin, Wyoming. B. Late Miocene Nagri Formation, Potwar Plateau, northern Pakistan. Photographs by C. Badgley.

### Testing Taphonomic Equivalence: Quantitative Methods

To test for differences in the type and intensity of taphonomic processes that acted among sites, large samples of individual fossils from each bone-bed can be evaluated with respect to the taphonomic parameters outlined in Table 6.3 (e.g., size, shape, weathering, abrasion, breakage, source-taxon size, and skeletal-element distribution). From these data, each site can be described quantitatively with respect to each taphonomic variable through the calculation of mean values (e.g., mean fossil size), proportion of fossils belonging to different categories (e.g., percentage of teeth, percentage of limb bones, etc.), or frequency distributions (e.g., numbers of fossils displaying weathering stages 0, 1, 2, etc.). Statistical comparisons of these quantitative profiles then can be performed to evaluate the taphonomic equivalence of bonebeds with respect to specific variables and processes.

*Comparisons among Sites: Continuous and Categorical Variables*

The type of quantitative description applied to a taphonomic variable will help to determine the most appropriate statistical method for comparing that variable among sites. For some continuous variables, means and standard deviations might be sufficient to characterize localities, and simple parametric comparisons (e.g. *t*-tests, analyses of variance) or nonparametric comparisons (e.g., Mann-Whitney *U*-tests) could be used to evaluate taphonomic equivalence. However, many taphonomic variables are described by categorical data (i.e., discrete classes such as weathering or shape categories) that must be compared using a different set of statistical tests. For some categorical variables, there is no natural sequence to the various categories. When comparing the proportions of fossils of different shapes between sites, for example, there is no inherent reason to order plate-shaped fossils before rod-shaped or cone-shaped fossils. Similarly, the frequencies of various anatomical elements at a site do not reflect an underlying continuous variable. For such nonordered, categorical variables, methods such as *z*-tests can be used to compare the proportion of fossils in each category and test for significant differences between sites (Moore and McCabe, 1993). These tests can help to indicate whether different taphonomic processes influenced the fossils that were preserved among several localities, which, if true, would suggest that comparisons of their faunas should be made cautiously.

To illustrate the *z*-test, consider bonebeds 1 and 2, with respective total samples of $n_1$ and $n_2$ fossils, of which $X_1$ and $X_2$ fossils, respectively, are cone shaped. To test the null hypothesis that these two bonebeds contain the same proportions of cone-shaped fossils, the test statistic $z$ can be calculated as

$$z = [(X_1/n_1) - (X_2/n_2)]/s_p \qquad (6.1)$$

where

$$s_p = \{p(1 - p)[(1/n_1) + (1/n_2)]\}^{0.5} \qquad (6.2)$$

and

$$p = (X_1 + X_2)/(n_1 + n_2) \qquad (6.3)$$

(Moore and McCabe, 1993). The test statistic $z$ is simply the difference between the sample proportions at the two localities, divided by their standard deviation. Standard normal probability tables (e.g., Moore and McCabe, 1993, Table A) can be used to obtain a $P$-value for this calculation of $z$ (which should be doubled if a two-tailed test is desired). The test is based on a normal approximation to the binomial distribution and performs best when the absolute frequency of specimens in the tested category, as well as the number of remaining specimens that do not belong to the tested category, is at least 5 in both samples (bonebeds) compared. If multiple comparisons of fossil proportions are performed between two bonebeds, it may be appropriate to adjust $P$-values using techniques such as the sequential Bonferroni method in order to limit the probability that a correct null hypothesis will be rejected (Rice, 1989).

An alternative type of categorical variable is actually a simplified representation of a continuous variable in which the categories can be ordered in a natural sequence. For instance, the continuous progress of bone weathering and abrasion is described using a scale of discrete stages from 0 to 5 (Behrensmeyer, 1978; 0 to 3 for fossil bones [Fiorillo, 1988]). Similarly, data pertaining to continuous measured variables like bone lengths could be compiled into incremental bins, and the count of fossils in each bin reported (e.g., 0–1 cm, 1–2 cm [Blob and Fiorillo, 1996; Brinkman et al., 2004]). Explicitly quantitative descriptions of fossil shape could be organized in the same fashion. For such ordinal variables, an ordered frequency distribution can be constructed from the frequencies of fossils in each category. Nonparametric Kolmogorov-Smirnov two-sample tests then can be used to test for differences in frequency distributions between bonebeds. This test can indicate whether differences in the intensity of taphonomic processes likely contributed to any faunal contrasts observed between localities. In the two-sample test, cumulative frequencies of specimens are calculated for each sample (i.e., bonebed), starting with the first increment of the categorical variable and proceeding in sequence through the last increment of the distribution (Sokal and Rohlf, 1995). Cumulative frequencies for each sample then are divided by the total sample size for that sample, in order to calculate the cumulative proportion of specimens through all the increments of the variable for each sample. The difference between the cumulative proportions of specimens at each site then is calculated for each increment of the variable. The largest absolute difference between the cumulative proportions of the two samples becomes the test statistic $D$, which can be compared to critical values to evaluate whether the two distributions differ significantly. For large samples of specimens ($n_1 > 40$ and $n_2 > 40$), an approximate test is performed, in which, for a specified level of significance $\alpha$, the critical value $D_\alpha$ can be calculated as

$$D_\alpha = (K_\alpha[(n_1 + n_2) + (n_1 n_2)]^{0.5} \qquad (6.4)$$

where

$$K_\alpha = \{0.5[-ln(\alpha/2)]\}^{0.5} \qquad (6.5)$$

(Sokal and Rohlf, 1995).

It may be necessary to apply a combination of these tests to clarify the likely influence of taphonomic factors on faunal composition. For example, the microvertebrate localities EAQ and BBR from the Upper Cretaceous Judith River Formation of south-central Montana are both preserved in fine-grained channel sandstones in close geographic and stratigraphic proximity, but the localities show a number of differences in faunal composition (Blob and Fiorillo, 1996). One of the most striking differences is the high proportion of turtles at EAQ, but the virtual absence of turtles from BBR (Fig. 6.2A and 6.2B). The absence of turtles from a channel locality seems unusual—is it possible that this faunal distinction, and perhaps more subtle ones as well, might have taphonomic, rather than ecological origins?

It is likely that differences in the sizes of fossils preserved at EAQ and BBR contributed to the contrasts in the faunas recovered from these sites (Blob and Fiorillo, 1996). Even though both sites occur in fine-grained channel deposits, at BBR, 98% of the fossils are under 1 cm in maximum length, whereas at EAQ only about 70% of the fossils are under 1 cm (Fig. 6.2C). A nonparametric Kolmogorov-Smirnov test shows that these size distributions differ significantly. However, turtle fossils from both localities are nearly all (99.2%) greater than 1 cm in maximum length (Fig. 6.2D). Thus, if very few fossils over 1 cm long are preserved at BBR, turtles would likely not have been preserved there, whether they lived in the area or not.

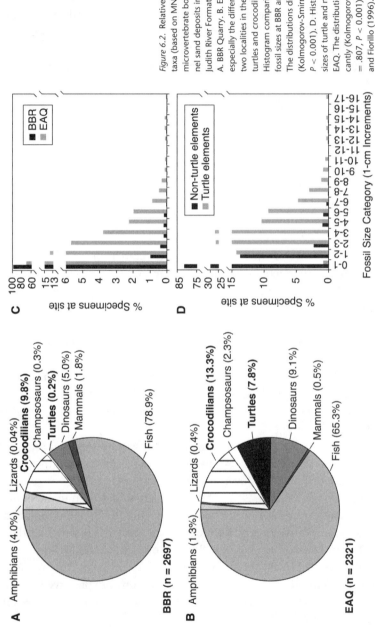

*Figure 6.2.* Relative abundances of taxa (based on MNE counts) at two microvertebrate bonebeds from channel sand deposits in the Cretaceous Judith River Formation of Montana. A. BBR Quarry. B. EAQ Quarry. Note especially the differences between the two localities in the abundances of turtles and crocodilians (boldface). C. Histogram comparing distributions of fossil sizes at BBR and EAQ localities. The distributions differ significantly (Kolmogorov-Smirnov test: $D = .275$, $P < 0.001$). D. Histogram comparing sizes of turtle and nonturtle fossils from EAQ. The distributions differ significantly (Kolmogorov-Smirnov test: $D_{max} = .807$, $P < 0.001$). Adapted from Blob and Fiorillo (1996).

Differences in the shapes of fossils preserved at EAQ and BBR also may have contributed to faunal contrasts between these sites, particularly with respect to crocodilians. Crocodilian fossils are over 30% more abundant at EAQ than at BBR (Blob and Fiorillo, 1996). More than 80% of crocodilian fossils at both sites are teeth and scutes under 1 cm in length; thus, although intensive size sorting may have excluded some larger crocodilian fossils from BBR, differential size sorting between the sites seems unlikely to have affected crocodilian abundance to the extent that it affected turtle abundance. However, two-tailed z-tests show that significantly greater proportions of plate-shaped and cone-shaped fossils were concentrated at EAQ, whereas a significantly greater proportion of approximately equidimensional microfossils were concentrated at BBR (Blob and Fiorillo, 1996). Because the most common crocodilian fossils, scutes and teeth, are, respectively, plate shaped and cone shaped, it is possible that differences in shape sorting also contributed to the faunal differences between EAQ and BBR. The mechanism responsible for the differences in sorting between these sites remains to be determined, but it is clear that taphonomic causes cannot be disregarded for the contrasting paleofaunas recovered from these sites.

One complication in evaluating the extent of differential size or shape sorting among bonebeds is that some taxa typically are represented by fossils of a restricted range of sizes or shapes (e.g., turtles by plate-shaped shell elements). As a result, it may be difficult to establish whether preferential preservation of a fossil size or shape class may have inflated the abundance of a particular taxon, or whether the original abundance of a taxon at a site led to the preferential preservation of specific fossil sizes or shapes (Blob and Fiorillo, 1996). Two approaches have been suggested to help deal with this potentially confounding issue. First, comparisons of taxonomic abundance might be focused on particular faunal groups, rather than performed across entire faunas preserved in bonebeds (Blob and Fiorillo, 1996). For example, the teeth of most Cretaceous mammals are compact elements much smaller than 1 cm in all dimensions. Such elements probably would not have been affected differently by processes leading to fossil size or shape sorting, so comparisons of the relative dominance of different mammalian taxa among Cretaceous bonebeds would be very likely to produce ecologically meaningful results (Blob and Fiorillo, 1996). Second, fossil size and shape profiles used to evaluate the isotaphonomy of bonebeds might be derived from nondiagnostic specimens (i.e., specimens that cannot be identified taxonomically), either in place of or in addition to profiles derived from diagnostic specimens with a potentially limited range of characteristic sizes and shapes (Blob and

Fiorillo, 1996; Brinkman et al., 2004). If diagnostic and nondiagnostic fossils from a bonebed showed similar size and shape profiles, those profiles could be considered more likely to typify the locality as a whole (Blob and Fiorillo, 1996).

*Evaluating Deviations from Unbiased Preservation in Bonebeds*

For some taphonomic variables, it is possible to evaluate the relative severity of preservation bias among bonebeds by comparing the degree to which each site deviates from an unbiased pattern of fossil preservation. For example, contrasts in the anatomical elements preserved at different bonebeds could suggest that the intensity of taphonomic factors differed among those localities, or that distinct taphonomic processes acted during the formation of each site. As several authors have demonstrated, skeletal elements differ in their tendencies to disperse in flowing water (Voorhies, 1969; Dodson, 1973; Behrensmeyer, 1975; Blob, 1997; Trapani, 1998). Voorhies (1969) and Behrensmeyer (1975) divided mammalian skeletal elements into three broad categories (termed Voorhies groups) based on their dispersal tendencies. Early dispersers such as ribs and vertebrae are classified in Voorhies group 1, intermediate dispersers such as many long bones are classified in Voorhies group 2, and late dispersers such as skulls and teeth are in Voorhies group 3. If preserved elements belong mostly to group 1 (i.e., early-dispersing elements are most common), then the assemblage may have accumulated as several elements were transported to the site over time; in contrast, if preserved elements belong mostly to group 3 (i.e., late-dispersing elements are most common), then light elements may have been winnowed away from the site (Voorhies, 1969; Behrensmeyer, 1975). Lag (group 3) assemblages are more likely than transported (group 1) assemblages to reflect a single, original local fauna, but some faunal components may have been selectively removed from lag deposits (Behrensmeyer, 1975, 1991). Thus, faunal comparisons of bonebeds displaying different patterns of Voorhies group preservation should be approached cautiously, because differences in the taxa preserved at such sites may stem from taphonomic factors rather than (or as well as) differences in original local faunas.

Before contrasts in element frequencies among bonebeds are interpreted, it is important to evaluate not only whether particular Voorhies groups are common, but also whether they are more common than would be expected in an unbiased assemblage. In many vertebrate taxa including mammals, group 1 elements (e.g., ribs and vertebrae) are naturally more abundant than later-dispersing group 2 elements (e.g., limb bones), so an abundance of group 1 elements could indicate the absence of fluvial winnowing rather than a transported assemblage. To interpret the frequencies of anatomical elements among sites, the proportion of each anatomical element or Voorhies group can be compared to the proportion expected for that element or group in an intact skeleton (Badgley, 1986b; Hunt, 1990; Fiorillo, 1991; Henrici and Fiorillo, 1993; Lyman, 1994; Varricchio, 1995; Coombs and Coombs, 1997). For example, in the Careless Creek locality from the Upper Cretaceous Judith River Formation of Montana, Fiorillo (1991) found that the absolute numbers of dinosaur elements belonging to Voorhies groups 1 and 2 differed by almost an order of magnitude, but that the ratio of group 1 elements to group 2 elements closely matched expectations for intact dinosaur skeletons. Thus, fluvial winnowing probably played little role in shaping the fauna preserved at this locality.

In another analysis of a Cretaceous dinosaur bonebed, Varricchio (1995) applied $\chi^2$-tests to evaluate whether counts of anatomical elements collected from Jack's Birthday Site (Two Medicine Formation of Montana) deviated significantly from the counts that would be expected in intact dinosaur skeletons, in order to judge whether hydrodynamic processes might have selectively removed fossils from the site. Evaluation of expected element frequencies is required to calculate the test statistic $X^2$ as

$$X^2 = \sum^{a} [(f_o - f_e)^2 / f_e] \qquad (6.6)$$

where $a$ is the number of classes that have been counted, $f_o$ is the observed count of specimens for that class, and $f_e$ is the count of specimens expected for that class (Sokal and Rohlf, 1995). This statistic is compared to the $\chi^2$ distribution for $(a - 1)$ degrees of freedom to test whether observed frequencies differ significantly from expected frequencies. Varricchio (1995) found that some easily transportable (group 1) elements, such as ribs and caudal chevrons, were significantly less abundant than expected, but that other group 1 elements, such as vertebrae, were significantly more abundant than expected. Thus, hydrodynamic winnowing seems unlikely to have been the agent of selective element removal that acted during the formation of this bonebed.

For multitaxic bonebeds preserving taxonomically diverse species or species of different sizes, it may be appropriate to analyze anatomical representation separately for each species, because different species can yield different numbers of elements in each Voorhies group, and homologous elements of different species may fall into different sorting categories (Blob, 1997; Coombs and Coombs, 1997). For example, in their analysis of the Morava Ranch Quarry from the Miocene of Nebraska, Coombs and Coombs (1997) found that all Voorhies groups were well represented among the total sample of bones collected from the quarry. However, for the chalicothere mammal *Moropus*, slightly more cranial elements and slightly fewer ribs, phalanges, and vertebrae were collected than would be expected based on the proportions of these elements in an intact *Moropus* skeleton. Species-specific analyses of anatomical representation indicated moderate fluvial winnowing that was not apparent based on pooled analyses of the elements collected from this bonebed (Coombs and Coombs, 1997).

Counts of anatomical-element frequencies from bonebeds also can be used to evaluate whether (or which) predators might have contributed to the formation of an assemblage. For such evaluations, frequencies of elements recovered from a deposit can be compared to typical frequency profiles for a variety of bone-accumulating predators (e.g., canids, mustelids, owls, or hawks). For example, Pratt (1989) sought to evaluate the likelihood that particular predator species might have concentrated bones of microvertebrates (rodents and frogs) at Unit 15 from the Miocene Thomas Farm locality in Florida. She compared the frequencies of rodent and frog elements to expectations for complete skeletons, calculating the proportion of representation for each element. She then compared the profiles of element representation for rodents and frogs to the profiles that would be expected for several different mammalian and avian predators, using rank-order correlation and $r \times c$ tests of association (Pratt, 1989).

Rank-order correlation is a nonparametric technique in which a coefficient of rank correlation is calculated to evaluate whether the ranks of abundance of different variables in a sample (e.g., which skeletal element is most abundant, which is second most abundant) show significantly similar patterns among samples (Sokal and Rohlf, 1995). Parametric analyses assume that data conform to a particular distribution (e.g., the normal distribution for analysis of variance) and generally have greater power to detect deviations from null hypotheses if their assumptions about data distributions are met. However, nonparametric methods are free from such assumptions and, thus, are preferable if data clearly are not normally distributed. The goal of correlation analyses is to evaluate the association between variables through the calculation of a correlation coefficient. Correlations calculated from rank orders are appropriate for inherently ordinal data, as well as for data that are related in a nonlinear fashion (Sokal and Rohlf, 1995). The use of rank-order methods also can facilitate the recognition of variation among rare data classes that might be obscured by variation in abundance counts among a few dominant classes

(Brinkman, 1990). One common coefficient (especially appropriate when the reliability of close ranks is uncertain) is Spearman's coefficient of rank correlation, $R_s$, which can be calculated as the Pearson product-moment correlation ($R$) between two columns of ranks in most statistical software packages. Pratt (1989) found that calculations of $R_s$ indicated significant similarity between the element representation of fossil rodents in Unit 15 from the Thomas Farm locality and the representation profiles of elements recovered from the scat of small mammalian predators.

In $r \times c$ tests, a contingency table is constructed in which counts for each class of a sample are assigned to a new row, and each sample being compared is assigned to a separate column (Sokal and Rohlf, 1995). Thus, in Pratt's (1989) study, each table row included a count of a different anatomical element, with the first column reporting counts for the Thomas Farm locality, and the second column reporting counts for a specific predator. The counts in the cells of the contingency table then are used to calculate the statistic $G$ (see Sokal and Rohlf, 1995, p. 738 for calculation details), which is tested for significance by comparing its value to a $\chi^2$-distribution with degrees of freedom equal to the product of (columns − 1) and (rows − 1). This approach is similar to that applied commonly in traditional $\chi^2$ tests but allows expansion of the test to a contingency table with more than 2 × 2 cells. The $r \times c$ tests indicated that the representation of frog elements at Thomas Farm did not differ significantly from the representation profiles of elements concentrated by avian predators, such as owls. Thus, different types of predators may have contributed different components of the Thomas Farm microvertebrate fauna.

Profiles of element completeness (i.e., breakage) for bonebeds also can be compared to the profiles typical for different predator species to evaluate potential contributors to an assemblage. In her study of Thomas Farm, Pratt (1989) counted the frequencies of complete and broken limb elements from the heteromyid rodent *Proheteromys floridanus* preserved in Unit 15. She used $\chi^2$-tests to compare these frequency distributions to the patterns that would be expected for assemblages produced by a variety of modern mammalian and avian predators. The frequencies of broken fossil elements differed significantly from those produced by modern birds but were consistent with those in accumulations produced by small mammals, just as the element-representation profiles had indicated (Pratt, 1989).

Quantitative analyses of features such as fossil sizes, fossil shapes, skeletal-element representation, and breakage among bonebeds, such as the comparisons outlined here, can substantially clarify the likely extent of taphonomic influence on specific contrasts among fossil assemblages. Although differences in the taphonomic profiles of localities do not necessarily preclude paleoecological comparisons of bonebeds, such differences do establish the context within which such comparisons can be made. As bonebeds depart further from isotaphonomic preservation, analyses of their faunas should be approached with greater caution.

## SAMPLING METHOD AND SAMPLE SIZE

### Sampling Method, Effort, and Size

Further issues that can potentially confound the evaluation and comparison of faunas are differences in sampling method, sampling effort, and sample size among bonebed localities. It is crucial to recognize the ways in which these features can affect faunal censuses before comparing bonebeds with the goal of drawing conclusions about faunal change and evolution.

*Sampling Method*

Vertebrate fossils are generally collected by any of three methods: surface collecting, excavation (quarrying), and screen washing. These methods tend to be appropriate under rather different circumstances, and each method has advantages and disadvantages. Surface collecting is appropriate when numerous fossils have weathered out of their source strata and lie scattered on the surface of the modern landscape. The advantage of this method is relatively rapid retrieval of fossils. Often it is possible to determine the strata from which fossils were derived, and clustering or association among specimens can be recorded. The disadvantages of surface collecting are that remains of smaller animals tend to be missed and that the provenance of some specimens is often uncertain. A structured variant of surface collecting involves marking a grid system (of equal-area squares) or transect and searching meticulously within the specified area; this spatial structure usually ensures a thorough search for fossil remains on the surface.

Excavation is appropriate for specimens that are still encased in the sedimentary matrix. Excavation may involve the use of a structured grid system for recording the identity, location, and orientation of each specimen; alternatively, excavation may emphasize fossil retrieval rather than the documentation of specimen depositional context. Fossil collection through excavation improves the potential to retrieve complete or delicate material, allows detailed spatial relationships and associations among specimens to be documented, and allows a careful search of a relatively small volume of sediment to be performed. Though these advantages can be substantial, disadvantages of excavation include a slow pace of work and, unless excavation occurs in a preparatory lab, a frequent inability to recover all remains of very small species.

Screen washing involves gathering bulk sediment from fossiliferous strata and processing that sediment by dry- or wet-screening methods. The main advantage of this method is that it greatly improves the recovery of fossils from small vertebrates (<1 kg in original body weight) relative to other collecting methods. Disadvantages of screen washing include the intensive labor required to screen sediment and sort fossils, and the possible fragmentation of fossils during these processes. This method is not well suited to recovering remains of larger vertebrates (≥1 kg).

Each sampling method imposes a size bias on the material collected, and each method samples a different volume of sediment. Thus, each method tends to be most appropriate for a different taphonomic setting. Furthermore, in quantitative analyses of bonebed faunas, it is critical to keep track of differences in sampling method among sites in order to avoid drawing spurious conclusions about faunal contrasts. Contemporaneous samples recovered by different methods are often most appropriately treated as complementary samples from the same original fauna, with surface-collected and excavated samples documenting larger species and screen-washed samples documenting smaller species.

*Sampling Effort*

Effort is a measure of the person-hours or volume of sediment involved in collecting samples. Samples are rarely standardized for sampling effort, but there may be circumstances in which it is desirable to do so, such as when surface prospecting is conducted over large areas of outcrop. Changes in sample size in relation to sampling effort indicate changes in fossil productivity within a sedimentary sequence. The stabilization of faunal composition (for taxa in the same size range) in relation to sampling effort may help to establish confidence in the first and last appearances of taxa within a stratigraphic sequence. Surveys standardized by sampling effort over narrow stratigraphic intervals of the Miocene Siwalik record of Pakistan documented biostratigraphic and taphonomic trends that complemented information from individual localities (Behrensmeyer and Barry, 2005). For example, an increase over time in the frequency of teeth versus axial skeletal elements tracked a facies shift toward a more mountain-proximal fluvial system. The surveys also increased the stratigraphic resolution of the first appearance datum of equids and documented their rapid increase in frequency relative to ungulates of similar body size.

The size of samples being compared strongly influences the richness and composition of the samples and the significance of statistical tests. For modern ecosystems and paleontological samples, there is a positive correlation between sample size and species richness. This relationship holds for sets of samples from different faunas and for successively larger samples from the same fauna. The form of the relationship varies according to the species-abundance distribution of the fauna being sampled. Because most faunas are not sampled exhaustively, it is important to evaluate the effects of sample size (or "sampling") on taxonomic richness and faunal composition.

Sample size affects the interpretation of all quantitative aspects of fossil assemblages. Here we discuss its effects on interpretations of taxonomic richness, specifically species richness, but the same principles apply to data for higher taxa. Changes in faunal composition over time are a primary focus of research on long fossiliferous sequences. Such data often are reported in the form of a stratigraphic-range chart or a graph of species richness versus time. It is important to note the changes in sample size for the units of analysis over the stratigraphic range being analyzed. Three kinds of graph are especially revealing (Fig. 6.3). One is a graph of species richness versus sample size for each unit of analysis (e.g., locality or sampling interval), along with the Pearson's correlation coefficient between these two variables (Fig. 6.3A). Such a scatterplot may indicate a nonlinear relationship between species richness and sample size. However, a square-root transformation of sample size, which is appropriate when the underlying sampling model corresponds to Poisson sampling (Snedecor and Cochran, 1980), can improve the linearity of the relationship. The magnitude of the correlation coefficient indicates the strength of the linear relationship, whereas the evenness of the species-abundance distribution from which the samples were drawn affects the slope of the relationship. A second type of informative plot illustrates how both sample size and the number of species per interval change through the stratigraphic sequence (Fig. 6.3B). The third useful type of graph depicts the number of taxonomic

---

*Figure 6.3.* Mammalian faunal data from Wasatchian localities, Clark's Fork Basin, to illustrate the effects of sampling on species richness and faunal turnover. A. Number of catalogued mammal specimens versus number of species per stratigraphic interval. B. Number of catalogued specimens and number of species per interval in relation to stratigraphic position. Note the close correspondence between increases and decreases in these two curves. C. Number of catalogued specimens, number of appearances, and number of disappearances per interval in relation to stratigraphic position to illustrate the potentially confounding effect of large changes in sample size on (apparent) faunal turnover. Cf, Clarkforkian; Wa, Wasatchian. Modified from Badgley and Gingerich (1988).

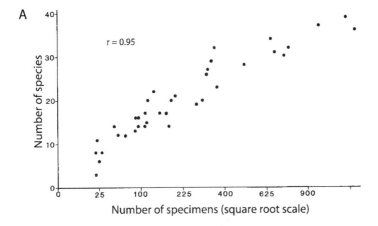

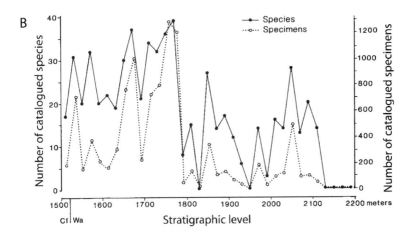

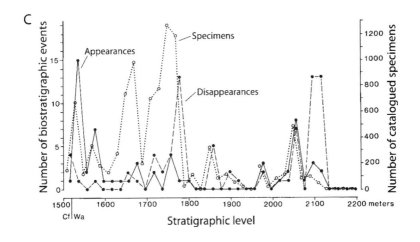

appearances, the number of taxonomic disappearances, and the sample size for each stratigraphic interval (Fig. 6.3C). This graph indicates whether marked changes in sample size coincide with changes in faunal composition. Sampling theory predicts that species richness should increase when sample size increases substantially and should decrease when sample size drops sharply, other things being equal. If such patterns emerge in paleontological data, then changes in faunal composition may be more apparent than real.

An analysis of Wasatchian faunas from the Clark's Fork Basin, Wyoming (Badgley and Gingerich, 1988) illustrates these issues (Fig. 6.3). The Clark's Fork Basin has a rich vertebrate record covering much of the Paleocene and early Eocene. The record of mammalian faunal change from early Wasatchian (early Eocene) localities has peaks in faunal turnover, with certain narrow stratigraphic intervals showing high numbers of first appearances or last appearances (Fig. 6.3C). Some of these peaks were correlated with faunal-turnover events (biohorizons) documented in the adjacent Bighorn Basin (Gingerich, 1983; Bown and Kraus, 1993). These data from the Clark's Fork Basin showed a strong correlation ($r = 0.95$) between the number of catalogued specimens and the number of mammalian species. For parts of the record, the number of appearances rose when the number of specimens per stratigraphic interval increased (e.g., at 1520–1540 m), and the number of disappearances rose when the number of specimens per interval fell from high values to very low values (between 1760 and 1800 m). These patterns suggested that the fluctuations in sample size concealed the real pattern of faunal change, and that some of the faunal turnovers were artifacts of changes in sample size.

A subsequent study evaluated further the likelihood that species disappearances (at 1760–1800 m in Fig. 6.3C) were a consequence of decrease in sample size (Badgley, 1990). For each species present in the subjacent stratigraphic intervals of large sample size, Badgley calculated the likelihood ratio of two hypotheses: $H_a$, the species is absent, and $H_p$, the species is present but not represented by fossils. The likelihood ratio for these hypotheses is

$$P(D/H_a)/P(D/H_p) \qquad (6.7)$$

where $P$ stands for probability and $D$ stands for the documented frequency of fossils from the stratigraphic interval in question. Under $H_a$, the probability of observing zero fossils is 1.0. Under $H_p$, the probability of observing zero fossils depends on the frequency of each species in the original fauna and the sample size of the fossil assemblage. This probability was determined by Monte Carlo simulation and by a Poisson model, methods that gave similar results. These methods indicated not only how many species were likely to show apparent disappearances when sample size declined, but also which species were most susceptible to apparent disappearance and which were most likely to show a genuine last appearance. The results indicated that about 45% to 80% (depending upon the sample size under consideration) of the observed disappearances likely resulted from the decrease in sample size and not from actual disappearance. The cautionary note from these two studies of faunal change in the Clark's Fork Basin is that large changes in sample size through a stratigraphic section are likely to create apparent faunal turnovers. Fortunately, apparent faunal turnovers can be recognized readily if sampling effects are monitored through the sequence.

### Rarefaction

One way to limit the confounding effects of sampling is to restrict comparisons to sites from which similar numbers of fossils have been collected. Alternatively, faunal comparisons could be restricted to bonebeds for which diversity can be considered "completely" sampled, although this raises the issue of how sampling completeness should be evaluated for a concentration of fossils. If sampling completeness is recognized as a potential focus of study from the outset of a bonebed excavation, an effort could be made during the excavation to document the number of specimens that are collected before each new taxon is identified. Plotting the cumulative number of specimens collected against the cumulative number of taxa represented results in a taxon-accumulation curve, which could provide a reasonable indication of the completeness of taxonomic sampling at a locality. This curve will reach a plateau as the yield of new taxa diminishes (Wolff, 1975; Pratt, 1989; Colwell and Coddington, 1994; Jamniczky et al., 2003). If taxon abundances are recorded during excavation, these also can also provide insight into the completeness of locality sampling. For example, if few taxa are represented by single specimens (i.e., are rare) at a bonebed, it suggests that few new taxa remain to be collected from the site (Colwell and Coddington, 1994; Alroy, 2000). Colwell and Coddington (1994) review a variety of asymptotic and nonasymptotic methods for evaluating sample completeness and for predicting improvements in sampling completeness for a specified level of additional collection effort.

In many cases, however, on-site evaluation of sampling complete-ness may be impractical; further, analyses of faunal patterns often require comparisons among bonebeds that differ in sample size. In such cases, rarefaction techniques can address the question of whether differences in diversity among bonebeds are simply consequences of their differing sam-ple sizes (Raup, 1975, Tipper, 1979; Jamniczky et al., 2003). An estimate of the number of species, $E(S_m)$, that a collection of $N$ specimens would contain if it were restricted to $m$ ($<N$) specimens can be derived from the equation

$$E(S_m) = \sum_{i=1}^{s} \left[ 1 - \binom{N - N_i}{m} \middle/ \binom{N}{m} \right] \qquad (6.8)$$

where $S$ is the number of species in the original collection, $N_i$ is the number of individuals in the $i$th species, $N$ is the total number of speci-mens in the original collection, and $m$ is the number of individuals in the subsample (Tipper, 1979). Rarefaction curves for each assemblage can be constructed by calculating $E(S_m)$ for several values of $m$. To facilitate com-parisons among assemblage curves, $E(S_m)$ calculations are recommended for a minimum of $(S + 1)$ values of $m$, where $S$ is the number of species in the richest sample. The variance $Var(S_m)$ of each estimate $E(S_m)$ can be evaluated as

$$Var(S_m) = \sum_{i=1}^{s} \left\{ \left[ \binom{N - N_i}{m} \middle/ \binom{N}{m} \right] \cdot \left[ 1 - \binom{N - N_i}{m} \middle/ \binom{N}{m} \right] \right\}$$
$$+ 2 \sum_{j=2}^{s} \sum_{i=1}^{j-1} \left\{ \binom{N - N_i - N_j}{m} \middle/ \binom{N}{m} \right.$$
$$\left. - \left[ \binom{N - N_i}{m} \binom{N - N_j}{m} \right] \middle/ \left[ \binom{N}{m} \binom{N}{m} \right] \right\}$$
$$(6.9)$$

With these variance evaluations, the significance of differences in rarefac-tion curves for two bonebeds can be evaluated on a point-by-point basis, using tests outlined by Tipper (1979). It is only valid to use rarefaction curves to interpolate what the diversity of a bonebed would be if a smaller sample of specimens had been collected from the site. Unlike empirical species-accumulation curves, rarefaction curves cannot extrapolate taxo-nomic diversity for hypothetical samples containing more specimens than the original collection (Raup, 1975; Tipper, 1979; Miller and Foote, 1996).

Jamniczky and colleagues (2003) demonstrated the application of rarefaction techniques in analyses of bonebeds through an evaluation of the vertebrate paleofauna preserved at the Bonebed 105 microfossil locality from the Upper Cretaceous Oldman Formation of Dinosaur Provincial Park, Alberta. They first divided screen-washed sediment excavated from the site into several subsamples of equal weight and sorted the vertebrate fossils from each subsample. They then constructed an additive empirical diversity curve for the site by counting the total number of taxa recognized after specimens from each subsample were tallied. These data were then rarefied, and the resulting curve was evaluated to determine the point at which additional sampling ceased to yield a substantial number of new taxa. From an initial set of 61 subsamples, the authors determined that taxonomic stability had been achieved after the first 48 subsamples (Jamniczky et al., 2003). These data provided a basis for comparisons of faunal diversity between Bonebed 105 and other sites from which fewer specimens had been collected.

Rarefaction methods also can be applied to diversity data from bonebeds in broad analyses of changes in regional taxonomic diversity through time. Diversity for specific time intervals can be estimated from lists of the faunas recovered from localities in each interval. However, intervals across the time period studied often are not sampled with equal density, once again raising the question of whether trends in raw diversity data merely reflect changes in sampling intensity. To control for variation in sampling intensity, Alroy (1996, 1998, 2000) developed several modifications to the classical rarefaction procedures outlined above. In one approach (Alroy, 2000), the time series under study is divided into bins of equal duration, and the faunal list from each locality in the sample is assigned to a bin. Within each bin, entire faunal lists are sampled without replacement until the sum of the number of taxa recorded from each sampled list reaches a specified quota. For example, if two lists were selected that included 5 and 10 species, respectively, then 15 taxonomic occurrences would have been sampled toward the quota. The quota of taxonomic occurrences per bin that is appropriate for a study can be evaluated by plotting a range of possible quota values against the number of time intervals (i.e., bins) that fail to supply that quota of taxonomic occurrences (Alroy, 1998). The inflection point of this curve, where its slope rises above zero, will indicate the maximum suitable occurrence quota for an analysis. After a full quota is obtained for each bin, age ranges are calculated for each taxon. Taxonomic diversity then can be calculated at the boundary between each pair of neighboring bins (i.e., temporal intervals) by counting the number of taxa that survive across each boundary.

Plotting these counts versus time will produce a diversity curve that is standardized for temporal variations in sampling. These steps should be iterated several times so that an average, sampling-standardized diversity curve can be calculated. Additional adjustments to these procedures can be made to depict more refined relationships between sample size and species richness, and to account for variation in (1) the number of individual fossils collected from different localities, and (2) locality-level (as opposed to interval-level) species richness through time (Alroy, 2000).

Some procedural cautions should be taken in the application of rarefaction methods. For example, in a recent summary of rarefaction approaches, Gotelli and Colwell (2001) cautioned against using category-subcategory taxon ratios (such as genus/species) to compare taxonomic richness among times or places, because taxon-sampling curves for the two categories have different shapes, resulting in nonlinear relationships between the number of taxa in one category and the number in the subcategory. Nonetheless, the application of rarefaction techniques to analyses of data from vertebrate assemblages has increased considerably in recent years (e.g., Alroy, 1996, 2000; Clyde and Gingerich, 1998; Markwick, 1998) and promises to generate insight into a variety of questions about broad-scale faunal change.

### Inferring population abundances in original faunas from fossil assemblages

Once a satisfactory estimate of abundance has been achieved for each taxon recovered from a fossil assemblage, there is a tendency to treat these estimates as though they represented taxonomic abundances in the original, living fauna. However, the relationship between abundances in fossil and living faunas is rarely so simple. Several processes filter information about the original living fauna as some of its individuals become fossilized, and most of these filtering processes do not act in an equivalent manner on all taxa or on all individuals of the same taxon. In fact, living populations can be linked to fossil samples by four discrete, conceptual stages (although the processes underlying these links are continuous).

**Stage 1:** *The living fauna* consists of multiple species, each with characteristic ecological attributes and population size. Living populations are distributed heterogeneously over a geographic region. Species vary in birth rates, death rates, and lifespan; the death rate, along with population size, determines the number of animals that can potentially contribute to the fossil record during a given time interval. Although these population attributes vary stochastically in nature, documented birth and death rates for some animal groups vary systematically with body size, so it is sometimes feasible to estimate population turnover for ancient vertebrates (e.g., Western, 1980).

**Stage 2:** *The death assemblage* is the set of carcasses (or parts of carcasses) present on the landscape as fossils in the making. Depending upon the agents of mortality, many individuals may never contribute to the death assemblage at all. For example, some prey animals are consumed entirely by their predators, with little left to fossilize even as part of a coprolite. Thus, two filtering processes separate the living assemblage and the death assemblage: species-specific turnover rates and processes of mortality.

**Stage 3:** *The fossil assemblage* is the set of vertebrate remains preserved in a depositional unit. Between the death assemblage and the fossil assemblage are disarticulation and scattering, weathering, dissolution, transport by animals or physical agents, burial, and diagenesis. The time interval over which skeletal material accumulates is also relevant to inferring abundance in the living fauna. Case studies of particular taxa or ecosystems have enhanced our understanding of these processes (e.g., Weigelt, 1989), but they are not well documented for vertebrates generally. These taphonomic processes do not act in an equivalent manner on the remains of different species. For instance, some processes vary in their effects according to the size and shape of skeletal elements (Behrensmeyer, 1978; Hanson, 1980; Hill and Behrensmeyer, 1984).

**Stage 4:** The fourth stage is *the sample*, the set of fossils that the paleontologist retrieves. The sample is usually a subset of an entire fossil assemblage, which itself is a taphonomically altered subset of the death assemblage, which in turn is a spatially restricted product of the turnover rates of the populations in the living fauna. The samples typically vary in their composition and taxonomic abundance, reflecting the combined effects of all the processes linking the original living fauna to samples taken from different points on the landscape. The point of this conceptual exercise is to demonstrate why the relative abundances of taxa in the sample of a fossil assemblage are unlikely to reflect closely the relative abundances of taxa in the living fauna.

How, then, does one reconstruct taxonomic abundances in the original living fauna, based on data sampled from fossil assemblages? One approach is to model the data from the fossil sample as the result of multiple transformations from the living source fauna, providing quantitative "transfer functions." For example, information about birth rates and death rates for extant populations can be modeled as regression equations of population turnover rate (which equals the supply of carcasses per unit time) in relation to body size for a specified duration (Western, 1980; Badgley, 1982). Also, information about taphonomic processes immediately following death can be used to estimate the probability of preservation of vertebrate remains for different taxa or for taxa in different size classes. Such taphonomic models may incorporate all taphonomic effects, or focus on individual taphonomic processes (such as weathering rate, effects of fluvial sorting, loss from diagenesis, etc.) separately. It is useful to create only as many distinct transformations as can be supported by empirical studies (unless the objective of the analysis is modeling for the sake of modeling). For example, in reconstructing Miocene Siwalik mammalian faunas, Badgley (1982, 1984, 1986b) proposed a range of plausible population sizes for each taxon, turnover rates that varied in relation to taxon body size, probabilities of preservation that varied in relation to size class of mammalian taxa, and estimates of duration of accumulation based on sedimentation rates in modern fluvial environments. She used Bayesian likelihood estimation to infer the source population size that resulted in the observed abundance of each taxon in samples from fossil assemblages of similar taphonomic history. Since both turnover rates and probabilities of preservation varied among taxa, the relative abundance of taxa in the fossil assemblages differed from the relative abundance of taxa in the inferred living fauna (Fig. 6.4). Taxa with a low probability of preservation and high numbers of fossils became even more abundant in the inferred living fauna (e.g., bovids), while taxa with a high probability of preservation and high numbers of fossils were less abundant in the living fauna (e.g., equids).

In general, estimates of living population abundances based on data from fossil assemblages are only as reliable as the information that supports the transfer functions themselves. For vertebrates, considerable study of the taphonomic processes affecting different taxa in different settings remains to be performed before the transfer functions can be considered more than reasonable guesses. In contrast, transfer functions between fossil pollen assemblages and the abundances of plant populations that produced the fossil pollen have been well studied for many different biomes (e.g., Birks and Gordon, 1985).

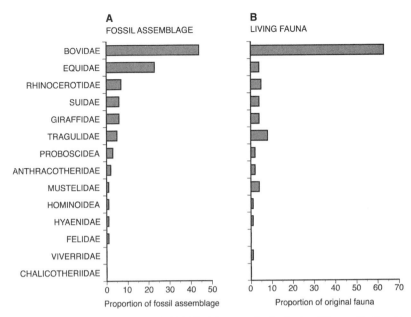

*Figure 6.4.* Relative abundance (percent) of individuals in mammalian families from Miocene, Siwalik rocks of Pakistan. A. Pooled data for fossil assemblages from channel-lag deposits. B. Reconstructed composition of the original living fauna, based on a maximum likelihood model of source population sizes that contributed to the fossil assemblage in (A). Modified from Badgley (1982, 1984).

## QUANTITATIVE COMPARISONS OF BONEBED FAUNAS: TECHNIQUES AND CASE STUDIES

The numerical analyses we have described to this point help to ensure that faunal counts for sites are appropriate given their depositional context, and help to evaluate the taphonomic factors that may have influenced those counts. Often, however, these are only preliminary steps toward the ultimate goal of comparing faunas collected from many localities. In this section, we present several techniques that can be used to quantitatively compare faunas or to relate properties of faunas to other variables, such as environmental factors. We also provide examples of studies that have used these methods. Bivariate methods are particularly useful for examining patterns of change in specific faunal components with respect to temporal, geographic, or environmental gradients. However, because comparisons of the diversity and abundance of multiple taxa among multiple localities are inherently multivariate endeavors, multivariate statistical techniques are used extensively in analyses of bonebed faunas. We introduce basic aspects of some of the most common analytical methods applied to bonebed faunas in the sections that follow. Texts that provide more detailed information on the procedures required to implement these methods include works by Neff and Marcus (1980), Gauch (1982), Greenacre (1984), Pielou (1984), Digby and Kempton (1987), Krzanowski (1988), Hair et al. (1992), Moore and McCabe, (1993), Jongman et al. (1995), and Sokal and Rohlf (1995).

### Analysis of Variance

The analysis of variance (ANOVA) is widely used in biology and paleontology to test for differences among groups of data (Simpson et al., 1960; Sokal and Rohlf, 1995). Most statistical software packages can perform ANOVAs (e.g., Statview, JMP), and several statistics texts provide excellent descriptions of the theory and calculations required (e.g., Moore and McCabe, 1993; Sokal and Rohlf, 1995). ANOVA is a method for evaluating the hypothesis that sample means do or do not differ significantly because they were drawn from different groups or categories of data. The groups represent circumstances that have a possible effect on the variable being measured. In the example presented below, the groups are different lithologies from which fossil assemblages were retrieved, and the variables are numbers of specimens of small-mammal families from screen-washed samples. The hypothesis under evaluation is that lithology (facies) has an effect on the taxonomic composition of the sample from a locality—in other words, that for each taxon, the frequency of specimens differs in assemblages from different lithologies. The ANOVA partitions the variance of samples into the part that can ascribed to the main effect (differences among groups) and the part that can be ascribed to random or unmeasured effects (variation within groups). There are several kinds of ANOVA. Here we describe the one-way analysis of variance (or single-classification ANOVA) with fixed effects, in which the data are grouped according to one class of nonrandom effects (in our example, three kinds of lithology). Other types of ANOVA, entailing more structured analyses, include hierarchical (or nested) ANOVA, two-way ANOVA, and multifactorial ANOVA (Sokal and Rohlf, 1995).

Computationally, ANOVA evaluates three kinds of variance in a one-way classification (Sokal and Rohlf, 1995). (1) The total variance is the mean square of observations in relation to the grand mean of the entire data set. (2) The variance within groups is the average variance of samples (the mean of several variances), representing the variation of data within groups due to measurement error, unmeasured effects, and individual variation. (3) The variance among groups is the variance among means of the different samples (the variance of several means); here the emphasis is on variation due to different effects or treatments. ANOVA partitions the total variance of a data set into among-group variance, representing the main effect or treatment, and within-group variance, representing error (noneffect). An $F$-test evaluates whether or not the among-group variance equals the within-group variance. The null hypothesis is that these two kinds of variance are equal (i.e., the effect is not significant). The alternative hypothesis may be one tailed (variance among groups > variance within groups) or two tailed (variance among groups $\neq$ variance within groups). Each kind of variance has its own degrees of freedom (df), and the critical value of the $F$-statistic depends upon the df of among-group variance and the df of within-group variance. The analysis is conceptually straightforward but algebraically elaborate, especially for hierarchical, two-way, and multifactorial ANOVAs.

ANOVA can test a variety of hypotheses in comparisons of bonebed faunas. In a taphonomic study of small-mammal assemblages preserved in different depositional settings from Middle Miocene, Siwalik deposits of Pakistan, Badgley et al. (1998) used ANOVA to test whether differences in lithology affected the faunal composition of the localities. The localities were preserved in one of three types of source sediments (Table 6.4A): (1) siltstone or claystone with variable sand content, representing deposition of suspended sediments; (2) sandstone to fine conglomerate, representing sheet-wash deposits in the context of abandoned channels; and (3) fine- and coarse-grained sediments, deposited in distinct thin layers in abandoned channels (and collected together because it was impractical to separate them). Could such differences in microhabitat or in fluvial processes have resulted in differences in faunal composition? Table 6.4A shows the number of identifiable teeth (sample size = 5 to 256) recovered from screen-washed samples from 10 different strata. Some samples were from different lithologies at the same locality. Eleven higher taxa were represented over all the samples, but most specimens belonged to five families: the Erinaceidae (hedgehogs), Sciuridae (squirrels), Cricetidae (hamsters), Rhizomyidae (bamboo rats), and Muridae (true mice and rats). Results of the ANOVA (Table 6.4B) showed that lithology did not significantly affect the frequency of specimens for any of the five families. For each family, the mean square effect (the among-group variance due to differences in lithology) was less than the mean square error (the within-group variance due to unmeasured effects or sampling error), resulting in a low $F$-ratio and a failure to reject the null hypothesis at $P = 0.05$. The variance due to effect must be much greater than the variance due to error ($F$-ratio greater than 4.74 for (2, 7) degrees of freedom in a one-tailed test) in order for the *F*-test to reject the null hypothesis. The results of this ANOVA are consistent with the interpretation that these Siwalik small-mammal concentrations resulted from one mechanism of accumulation, despite differences in source lithology (Badgley et al., 1998).

Table 6.4. Example of ANOVA applied to evaluate the effects of lithology on faunal composition for Middle Miocene, Siwalik bonebed localities from Pakistan

A. Faunal data

| Site | 41 | 59 | 59a | 430 | 640b | 640c | 641a | 641c | 651 | 660 |
|---|---|---|---|---|---|---|---|---|---|---|
| Lithology | F | C | F | M | M | M | M | C | F | F |
| Family | | | | | | | | | | |
| Erinaceidae | 0 | 12 | 9 | 6 | 7 | 2 | 21 | 0 | 7 | 2 |
| Soricidae | 0 | 3 | 0 | 1 | 1 | 0 | 1 | 1 | 0 | 0 |
| Chiroptera | 1 | 1 | 2 | 0 | 0 | 0 | 0 | 0 | 0 | 0 |
| Adapidae | 0 | 1 | 0 | 0 | 0 | 0 | 1 | 0 | 1 | 0 |
| Sciuridae | 0 | 14 | 4 | 11 | 5 | 1 | 7 | 1 | 9 | 0 |
| Gliridae | 0 | 5 | 1 | 1 | 1 | 0 | 1 | 0 | 0 | 1 |
| Cricetidae | 10 | 43 | 31 | 152 | 25 | 4 | 41 | 2 | 67 | 2 |
| Rhizomyidae | 1 | 15 | 7 | 36 | 12 | 2 | 5 | 1 | 39 | 0 |
| Muridae | 4 | 17 | 12 | 42 | 11 | 1 | 27 | 0 | 23 | 0 |
| Ctenodactylidae | 0 | 1 | 0 | 6 | 0 | 0 | 0 | 0 | 3 | 0 |
| Thryonomyidae | 0 | 0 | 1 | 1 | 0 | 0 | 1 | 0 | 1 | 1 |

*Note:* Number of identifiable teeth for families of small mammals from 10 contemporaneous localities. C, coarse (sandstone or conglomerate); F, fine (siltstone or claystone), M, mixed coarse and fine.
*Source:* Modified from Badgley et al. (1998).

B. ANOVA results

| Faunal Variable | Mean Square Effect | Mean Square Error | *F*-ratio | *P*-level |
|---|---|---|---|---|
| Erinaceidae | 20.70 | 47.29 | 0.44 | 0.66 |
| Sciuridae | 14.18 | 27.32 | 0.52 | 0.62 |
| Cricetidae | 1072.80 | 2353.50 | 0.46 | 0.65 |
| Rhizomyidae | 22.05 | 261.36 | 0.08 | 0.92 |
| Muridae | 144.05 | 204.00 | 0.71 | 0.53 |

*Note:* Results from analysis of variance for effect of lithology (coarse, fine, mixed) on frequency of specimens in each of five families. (Six families omitted because of low overall frequencies.) The degrees of freedom for main effect are 2 and for error are 7; the critical value of $F(2,7)$ is 4.74 for a one-tailed test at $P = 0.05$. Since the variance due to sampling error is greater than the variance due to the main effect, the null hypothesis (there is no effect of lithology on faunal composition) cannot be rejected.
*Source:* Modified from Badgley et al. (1998).

Richard W. Blob and Catherine Badgley

## Linear Regression

Regression is another method used widely in quantitative studies. In linear regression, the relationship between two or more variables is modeled with a linear equation between a dependent variable and one or more independent variables (or transformations of them). Even though the equation for a line is conceptually and algebraically simple and familiar, the use of random variables adds many procedural details and subtleties. The regression equation expresses the nature of the linear relationship between dependent and independent variables based on the variance of the dependent variable $(Y - \overline{Y})^2$ and the variance of the independent variables $(X - \overline{X})^2$, where $\overline{Y}$ and $\overline{X}$ are the respective means of the dependent and independent variables. As in ANOVA, the regression model partitions the variance in the dependent variable into the variation due to the independent variable (the effect) and the variation due to unmeasured effects or error. The model can be used to quantify the effect of the independent variable on the dependent variable and also to predict the dependent variable for new observations of the independent variable(s). Bivariate regression refers to a model with one independent variable; multiple regression refers to a model with two or more independent variables. The variable(s) may be categorical, ordinal, or continuous, although special kinds of regression analysis are required to deal with some of these combinations (Gujarati, 1995; Sokal and Rohlf, 1995).

The result of a regression analysis includes the regression equation, several statistics about it, and the residuals. The linear equation itself consists of an estimated intercept and an estimated slope coefficient for each independent variable:

$$Y = b_0 + b_1 X_1 + b_2 X_2 + \ldots + b_j X_j + e \qquad (6.10)$$

where the $b$-values are coefficients for each of $j$ variables, $b_0$ is the intercept, and $e$ is error. The significance of each coefficient can be evaluated from a $t$-test for a predetermined level of significance, commonly $P = 0.05$. The significance of the slope coefficient(s) is especially important; if the estimate fails the significance test (i.e., the $P$-value is greater than the chosen level of significance), then the hypothesis that the real slope coefficient $= 0$ cannot be rejected. The coefficient of determination, $R^2$, expresses the linearity of the relationship (higher values indicate greater linearity); it also measures the proportion of the total variance in $Y$, the dependent variable, accounted for by variation in the independent variable, $X$. However, it does not indicate how well the regression model estimates the slope coefficients and is not a measure of goodness-of-fit for a regression model. $R^2$ is often confused with $R$, the correlation coefficient for two variables. While the square root of $R^2$ is indeed $R$, the appropriate statistic to report in regression analysis is $R^2$, not $R$. For a good discussion of the conceptual differences between correlation analysis and regression analysis, see work by Sokal and Rohlf (1995). The standard error of the estimate, $S_e$, is the standard error of the error (or residual) variance in the linear model. Lower values indicate better prediction of the dependent variable by the model; thus, $S_e$ measures goodness of fit (Hamilton, 1992). As with ANOVA, the overall regression model is evaluated with an $F$-test. In this test, the ratio of the variance from effect to the variance from error is compared to the appropriate $F$-value, chosen on the basis of the degrees of freedom in the analysis. The residuals of the regression (predicted value – observed value for each case) can be examined for heteroscedasticity (residual variance changing systematically over different values of $X$) and for autocorrelation. If either condition is present, it can sometimes be adjusted by transformations of the independent variables (e.g., logarithms, square root, addition of extra terms). In multiple regression, it is also important to examine the degree of correlation among the independent variables (multicollinearity or tolerance). The best measure of multicollinearity is the auxiliary $R^2$ or $R^2_j$, which is the amount of variation in one $X$-variable explained by the other independent variables. The quantity $(1 - R^2_j)$ is the tolerance for the focal $X$-variable. Multiple regression models should aim for $R^2_j$ values below 0.80; if any auxiliary $R^2$ value is higher, then excessive multicollinearity is present (Hamilton, 1992). This means that at least one independent variable is highly redundant with others present and that the slope coefficients are not well estimated.

Commonly only $R^2$ is reported to indicate the strength of a regression equation. But it is not the only or even the best indicator of a good regression model. A good regression model has (1) a significant slope coefficient (meaning that its independent variable has an effect in the linear equation, or that the slope of the line is nonzero) for each independent variable; (2) a low standard error of regression (meaning that predicted values are close to observed values of the dependent variable); (3) a high $R^2$, signifying a strongly linear relationship; and (4) homoscedastically distributed residuals (i.e., residual error is similar over the entire spectrum of the independent variable) (Gujarati, 1995). For multiple regression, low multicollinearity among the independent variables also is important.

We illustrate the use of regression analysis with two examples. The first comes from a body of literature on the estimation of body size from fossil

Table 6.5. Least-squares linear regression models of body weight (in kg) on three linear skeletal measures for mammalian carnivores

| Skeletal Feature | Group ($n$) | Slope | Intercept | $R^2$ | $S_e$ | %PE |
|---|---|---|---|---|---|---|
| Skull length | Canidae ($n = 27$) | 2.86 | −5.21 | 0.86 | 0.117 | 21 |
| (cm) | Ursidae ($n = 15$) | 2.02 | −2.80 | 0.49 | 0.193 | 39 |
| | Mustelidae ($n = 37$) | 3.39 | −6.03 | 0.90 | 0.199 | 40 |
| | Felidae ($n = 30$) | 3.11 | −5.38 | 0.85 | 0.196 | 38 |
| | Total ($n = 109$) | 3.13 | −5.59 | 0.90 | 0.220 | 47 |
| Occiput- | Canidae ($n = 27$) | 3.08 | −5.03 | 0.88 | 0.114 | 22 |
| orbit | Ursidae ($n = 15$) | 1.98 | −2.38 | 0.41 | 0.207 | 42 |
| length (mm) | Mustelidae ($n = 37$) | 3.29 | −5.53 | 0.86 | 0.220 | 47 |
| | Felidae ($n = 30$) | 3.54 | −5.86 | 0.85 | 0.196 | 37 |
| | Total ($n = 109$) | 3.44 | −5.74 | 0.90 | 0.207 | 42 |
| $M_1$ length | Canidae ($n = 27$) | 1.82 | −1.22 | 0.76 | 0.158 | 27 |
| (mm) | Ursidae ($n = 15$) | 0.49 | 1.26 | 0.18 | 0.250 | 46 |
| | Mustelidae ($n = 37$) | 3.48 | −3.04 | 0.86 | 0.220 | 45 |
| | Felidae ($n = 30$) | 3.05 | −2.15 | 0.90 | 0.613 | 28 |
| | Total ($n = 109$) | 2.97 | −2.27 | 0.69 | 0.377 | 97 |

*Note:* Each row contains the information from a regression model of the form body weight = Intercept + (slope x skeletal feature) + error. These regression models are the basis for inferring body mass from fossil carnivores.

$n$ refers to the number of species in each group, with males and females counted separately because of the presence of size dimorphism. $R^2$ is the coefficient of determination; $S_e$ is the standard error of the estimate; %PE is the percent prediction error.

*Source:* Modified from Van Valkenburgh (1990).

vertebrate remains. Since many skeletal measurements are highly correlated with overall measures of individual body size (such as body length or body weight), these relationships have been quantified using regression analysis in order to generate predictive equations for use with different vertebrate groups. The general equation takes the form *body weight* = $b_0$ + $b_1$(*skeletal feature*) + *error*, where $b_0$ is the intercept and $b_1$ is the slope coefficient, as in Equation 6.10.

Van Valkenburgh (1990) analyzed skeletal measurements in relation to body weight for four groups of mammalian carnivores (Table 6.5). Skull length, occiput-orbit length, length of the first lower molar ($M_1$), and body weight (kg) were recorded for males and females of 57 species. Data for males and females, when available, were kept separate because of the presence of substantial size dimorphism for some groups. The four groups were Canidae (wolves, foxes, jackals, 14 species, 27 entries by sex),

Ursidae (bears, 8 species, 15 entries by sex), Mustelidae (weasels, badgers, otters, 19 species, 37 entries by sex), and Felidae (cats, 16 species, 30 entries by sex). All data were $\log_{10}$-transformed. In Table 6.5, results of 15 regression models are reported, including the slope and intercept of the regression equation, the coefficient of determination ($R^2$), the standard error of the estimate ($S_e$), and the percent prediction error (%PE). The percent prediction error is the percent difference between the observed and predicted body weights, a version of the residuals. The mean of the absolute values of the %PE (reported in Table 6.5) provides a measure of predictive accuracy that can be compared among regression equations (Smith, 1981; Van Valkenburgh, 1990).

Table 6.5 shows that regression models differ among carnivore groups and among craniodental variables. The percent prediction errors are lower for Canidae than for the other groups for all three variables, whereas Mustelidae and Ursidae have high percent prediction errors for all variables. For skull length and $M_1$ length, the regression models for each taxonomic group perform better (lower %PE) than the regression model for all species (Total). But for occiput-orbit length, the Total model has a lower percent prediction error than does the equation for Mustelidae. Depending upon what fossil carnivore material is available, Table 6.5 offers different options for inferring the body weight of a fossil carnivore. These approaches are relevant to studies of fossil bonebeds because body-weight estimation is a precursor to many kinds of taphonomic and paleoecological analyses. For example, the distribution of species by body weight, or size structure, of a vertebrate fauna is an important aspect of its ecological characteristics.

In our second example of regression analysis, multiple-regression techniques were applied to study the density of mammalian species (number of species per quadrat) in relation to climate and topography for 388 equal-area quadrats of North America (Badgley and Fox, 2000). Although this study focused on the geographic ranges of extant mammal species, the approaches used could easily be extended to fossil taxa collected from bonebed localities. In addition, this example illustrates the procedure for determining a good regression model when strong correlations exist among the independent variables.

Badgley and Fox (2000) examined the geographic ranges of 721 species of extant mammals and nine environmental variables and used multiple regression to model species density as a function of the environmental variables (Table 6.6). All variables were log-transformed before analysis, so the regression equations express proportional change in species density as a function of proportional change in environmental variables. The full model (Table 6.6A), using all nine environmental variables, has a high $R^2$ and a low standard error, but some of the coefficients are non-significant (at $P = 0.05$) and there is high multicollinearity for several variables. To determine a more satisfactory regression model with fewer independent variables, several reduced regression models were compared to the full model via a nested $F$-test. The nested $F$-test compares the significance of different sets of independent variables with the same dependent variable in related regression models. The test works by comparing the sum of squared errors (SSE), appropriately adjusted for degrees of freedom, between two regression models. If model 1 contains a full set ($v$) of

Table 6.6. Results of standard, linear, multiple regression of species density on physiographic and climatic variables

A. Full regression model

### Regression Model

Species density $= -61.04 -1.84$(annual range $T$) $+ 1.80$(annual min. $T$) $+ 11.04$(annual max. $T$) $+ 0.03$(Frostfree pd) $- 0.06$(annual precipitation) $- 0.09$(annual PET) $+ 0.32$(AET) $+ 0.07$(Relief) $+ 0.18$(Elevation) $+$ error

$n = 388$

### Regression Diagnostics

$R^2 = 0.87$

Adjusted $R^2 = 0.87$, $P = 0.000000$

Standard error of regression $= 0.186$

SSE $= 13.07$, df $= 378$

| Independent Variable | Coefficient | Standard Error of Coefficient | P-level |
|---|---|---|---|
| Annual range T | −1.844 | 1.155 | 0.1111 |
| Annual minimum T | 1.799 | 0.645 | 0.0055 |
| Annual maximum T | 11.038 | 1.871 | 0.0000 |
| Frostfree period | 0.026 | 0.042 | 0.5326 |
| Annual precipitation | −0.064 | 0.037 | 0.0839 |
| Annual PET | −0.088 | 0.069 | 0.1998 |
| Annual AET | 0.322 | 0.048 | 0.0000 |
| Relief | 0.065 | 0.016 | 0.0001 |
| Elevation | 0.184 | 0.020 | 0.0000 |

(*Continued*)

B. Best regression model (passes nested $F$-tests, all coefficients significant at $P = 0.05$, low multicollinearity)

---

**Regression Model**

Species density $= -69.25 + 2.53$(annual min. $T$) $+ 9.88$ (annual max. $T$) $+ 0.24$(AET) $+ 0.06$(Relief) $+ 0.18$(Elevation) $+$ error
$n = 386$(2 outliers omitted)

**Regression Diagnostics**
$R^2 = 0.88$
Adjusted $R^2 = 0.87$, $P = 0.000000$
Standard error of regression $= 0.181$

| Independent Variable | Coefficient | Standard Error of Coefficient | $P$-Level |
|---|---|---|---|
| Annual minimum $T$ | 2.532 | 0.200 | 0.0000 |
| Annual maximum $T$ | 9.876 | 0.822 | 0.0000 |
| Annual AET | 0.236 | 0.025 | 0.0000 |
| Relief | 0.065 | 0.014 | 0.0000 |
| Elevation | 0.175 | 0.017 | 0.0000 |

*Note:* All variables were $\log_e$-transformed before analysis.

$T$, temperature; PET, potential evapotranspiration; AET, actual evapotranspiration.

*Source:* From Badgley and Fox (2000).

possibly useful independent variables, and model 2 contains a reduced set $(v - w)$ of useful independent variables, then the quantity

$$\frac{(\text{SSE}_2 - \text{SSE}_1)/w}{\text{SSE}_1/\text{df}_1} \quad (6.11)$$

where $\text{df} =$ degrees of freedom and $w = (\text{df}_1 - \text{df}_2)$, provides the test statistic for an $F$-test with $(w, \text{df}_1)$ degrees of freedom (Gujarati, 1995).

The purpose of employing this test was to construct a regression model with a subset of the nine environmental variables that had no significant reduction in SSE relative to the full model. The process involved three stages to arrive at the "best" model that was both ecologically and statistically sound. First, the authors tested whether all the temperature variables, all the moisture variables, and both topography variables were each significant as a group to the regression. Each group was significant: its omission caused a significant increase in the SSE (greater prediction errors) relative to the full model. Next, within each group of environmental variables, the auxiliary $R^2$ was calculated for each variable with the other variables in the group. For variables with a high auxiliary $R^2$, one or more variables were omitted from the group and the auxiliary $R^2$ was recalculated with the remaining variables. From each group of variables, a meaningful set of variables that were not highly correlated with other variables was identified. This process resulted in several possible reduced regression models. Each of these then was compared to the full model using the nested $F$-test (see Badgley and Fox, 2000, for specific test results). From these reduced models, one was selected (Table 6.6B) that best met the criteria for a good regression (i.e., significant slope coefficients, low standard error, low multicollinearity). Inspection of residuals for this regression (observed – predicted values of species density) revealed two outliers. The model improved slightly after these quadrats were omitted and the regression was recalculated (results in Table 6.6B reflect omission of these outliers). The residuals did not show heteroscedasticity in relation to either the independent variables or predicted species density. From an ecological perspective, the regression model that ultimately resulted from this analysis (Table 6.6B) demonstrated that species density can be predicted as a positive function of proportional change in annual minimum temperature (winter temperature), annual maximum temperature (summer temperature), mean annual actual evapotranspiration, relief, and elevation. Each environmental variable had a positive effect as the effects of the other independent variables were held constant.

**Cluster Analysis**

Cluster analysis is a common multivariate classification technique that can be applied with a wide range of variations depending on the goals of an analysis. Classification techniques focus on identifying discontinuities in data sets and use those discontinuities as a basis for recognizing discrete groups that, in paleontological examples, may represent distinct faunal assemblages, or localities formed under similar environmental conditions. Because these methods force samples into discrete groups, they can be especially useful for the analysis of highly heterogeneous data and can help to focus subsequent analyses if applied early in a study; however, these methods also may impose misleading associations among samples (Shi, 1993; Etter, 1999). *Nonhierarchical* cluster analyses assign each sample to a cluster in multivariate space without clarifying the relationships among clusters, whereas *hierarchical* cluster analyses evaluate the interrelationships among clusters of samples (Gauch, 1980; Shi, 1993). Hierarchical analyses, therefore, are able to address a wide range of paleoecological questions and are used widely to explore the correspondence between patterns of faunal similarity among localities and discrete biogeographic regions, periods of time, or environments of preservation (Shi, 1993; Etter, 1999). *Monothetic* analyses cluster samples on the basis of a single attribute. Although straightforward to apply, the single attribute used to define monothetic classifications must be chosen with care to avoid undue separation of samples that are similar in most respects (Williams, 1971; Shi, 1993). In contrast, *polythetic* analyses base the clustering of samples on dissimilarity measures applied to all the attributes quantified in the study, so that samples clustered together are the ones that, on average, resemble each other the most (Williams, 1971). A further division of clustering techniques is between *divisive* methods that successively divide a pool of samples into a hierarchy of smaller clusters, and *agglomerative* methods that group individual samples into a hierarchy of successively larger clusters (Williams, 1971; Shi, 1993; Etter, 1999). Each method is strongest at its earliest stages of computation; thus, divisive methods are best at distinguishing the initial, upper-level divisions of the entire data set into groups, whereas agglomerative methods are best at the initial, lower-level fusions of individual samples into groups. Computational difficulties historically impeded the development and application of polythetic, divisive cluster analyses, but software to perform such analyses is now available, and such methods may be especially useful for very large data sets (Etter, 1999). In contrast, agglomerative cluster analysis can be performed in a wide variety of software packages, and is especially appropriate for smaller data sets.

The hierarchical, polythetic, agglomerative cluster analysis on which we focus discussion (hereafter called "cluster analysis") requires two steps. First, a matrix of distance, or dissimilarity, values must be calculated for the samples in the study. A wide variety of distance and dissimilarity indices, many of which are first calculated as similarity values which then are subtracted from 1, have been proposed for both binary (e.g., presence/absence) data and quantitative (e.g., abundance) data. Shi (1993) discussed the relative merits of 39 different binary similarity indices and advocated use of the Jaccard coefficient because it is not strongly affected by faunal sample sizes and because it emphasizes the shared presence, rather than the shared absence of taxa. The Jaccard coefficient ($J_s$) ranges between 0 (no similarity) and 1 (identical samples) and can be calculated between two faunas B and C as

$$J_s = a/(a + b + c) \qquad (6.12)$$

where *a* is the number of species shared by both faunas B and C, *b* is the number of species found only in fauna B, and *c* is the number of species found only in fauna C (Shi, 1993; Etter, 1999). Other similarity coefficients (e.g., the Dice coefficient) often differ by weighting some of these components, such as the number of shared species, more heavily than the others. For quantitative abundance data, a similarity matrix (and, subsequently, a dissimilarity matrix) can be calculated using standard correlation coefficients such as Pearson's product-moment coefficient ($R$) or Spearman's coefficient of rank correlation ($R_s$) (Sokal and Rohlf, 1995). However, correlation coefficients can be affected strongly by the sample sizes of the faunas compared and should not be used for data matrices that contain a large number of zeros (Etter, 1999). Etter (1999) presented equations for several distance coefficients that can be applied to data with such complications.

The second step in cluster analysis is to apply a clustering algorithm to the distance or dissimilarity matrix in order to group the samples into larger and larger hierarchical groups. The various algorithms that have been devised to assemble clusters differ mainly in how the distances between clusters are evaluated. In the nearest-neighbor (single-linkage) algorithm, clusters are formed sequentially between units (individual samples or clusters of samples) with the lowest dissimilarity, with comparisons involving clusters evaluated according to two rules (Etter, 1999). First, dissimilarity between a single sample and a cluster equals the lowest possible dissimilarity between the sample and any of the samples comprising the cluster. Second, dissimilarity between two clusters equals the lowest dissimilarity between any member of one cluster and any member of the second cluster. An alternative algorithm, the farthest-neighbor (complete-linkage) method, alters these two rules so that dissimilarity between a single sample and a cluster equals the highest possible dissimilarity between the sample and any of the samples comprising the cluster, and dissimilarity between two clusters equals the highest dissimilarity between any member of one cluster and any member of the second cluster (Shi, 1993; Etter, 1999). These two methods tend to arrive at extreme clustering solutions, and nearest-neighbor clustering in particular tends to produce chaining clusters that obscure the recognition of groups (Shi, 1993). One common solution to these difficulties is to calculate an average dissimilarity between clusters, rather than the maximum or minimum, to evaluate intercluster distances (Shi, 1993; Etter, 1999). For example, in the UPGMA average linkage method (unweighted pair-group method using arithmetic averages), every sample in a cluster is assigned equal weight, so that the weight of a cluster in a calculation of average dissimilarity is proportional to the number of samples it includes (Pielou, 1984). A variety of methods have been advocated to evaluate the significance of associations identified during cluster analyses (Strauss, 1982; Nemec and Brinkhurst, 1988; Shi, 1993), but objections have been raised to many of these methods and unambiguous evaluations of cluster significance can be difficult to achieve (Etter, 1999).

A number of studies have applied clustering techniques to evaluate hierarchical associations among bonebed faunas. For example, Alberdí et al. (2001) sought to evaluate similarities in taxonomic composition between Plio-Pleistocene assemblages of fossil mammals from the Gaudix-Baza Basin in Spain and mammalian faunas from the Miocene Dhok Pathan deposits of the Pakistan Siwaliks (Badgley, 1986b) that were deposited in a variety of fluvial environments. From the faunal-abundance data, Alberdí et al. (2001) calculated Spearman's coefficient of rank correlation between each pair of sites and then performed a UPGMA cluster analysis on the matrix derived from these values. Despite the great difference in age and geographic location between these two groups of localities, one of the three Spanish paleofaunas clustered with the Siwalik fluvial assemblages. This led Alberdí et al. (2001) to conclude that fluvial processes strongly influenced the fossils that were preserved at that Spanish locality. At a broader scale, Shubin and Sues (1991) used a variety of cluster analyses to evaluate the similarities of continental tetrapod faunas from several Triassic and Early Jurassic formations worldwide, drawing on bonebed analyses to compile faunal lists for each formation. Using presence/absence data for components of the tetrapod fauna, they calculated three different similarity indices between all the pairs of formations and applied five different clustering algorithms to the resulting similarity matrices. Regardless of the similarity index or clustering algorithm they used, the clusters formed primarily linked faunas of similar age, rather than of close geographic proximity. This result suggested that continental biotic interchange was blocked by few barriers on the Pangaean supercontinent of the Early Mesozoic (Shubin and Sues, 1991).

Depending on the goals of a study, cluster analyses can be performed to evaluate associations among the taxa preserved at different localities, in addition to associations among localities. For instance, using faunal abundance data collected from 19 taphonomically equivalent vertebrate microfossil localities in the lower Judith River Group (Campanian) of Alberta, Brinkman et al. (2004) performed separate average linkage cluster analyses of taxa and localities to identify patterns of association at each of these scales. Localities clustered into three groups strongly correlated with stratigraphic position in the section (low, middle, and high), and taxa showed three clusters correlated with patterns of change in abundance through the section (i.e., a cluster of taxa that increased in abundance through the section, a cluster of taxa that decreased in abundance through the section, and a cluster with fairly constant abundance through the section). The patterns of association identified in these complementary analyses led Brinkman et al. (2004) to conclude that ecological factors related to stratigraphic position, particularly the increased distance from a regressing shoreline for sites higher in the section, were the primary factors determining paleocommunity composition in the region through this period of time.

### Principal Components Analysis

Principal components analysis (PCA) is a multivariate ordination method that summarizes the variation among samples in a smaller number of new variables (the principal components, or PCs) that are weighted functions of the original variables (Marcus, 1990). In contrast to classification methods, ordinations assume an underlying continuity in data and seek to order samples (taxa or localities) along gradients that reflect the major directions of variation in the data. Natural clusters of samples in a data set (for example, bonebed assemblages with similar faunas) may be revealed in scatterplots of the samples on the new axes (Shi, 1993; Etter, 1999). In the context of stratigraphic, environmental, or taphonomic data, PCA can be a powerful tool for clarifying patterns of faunal variation among localities, as well as the factors contributing to those patterns.

In analyses of bonebed faunas, samples are typically fossil localities, and variables are typically the taxa preserved, with the value of each variable for a locality equaling the number of individuals of each taxon preserved at that site. Thus, counts of several taxa for each of several bonebeds comprise a data matrix of columns (taxa) and rows (localities) that can be depicted as representing a cloud of points in multidimensional space. Each dimension (i.e., axis) in this space represents a taxon, each point represents a bonebed, and the position of each bonebed point in the space is determined by the abundances of each taxon at that site. PCA seeks to define a new set of axes for the sample of points by rigidly rotating the original axes to a new orientation, such that one axis is aligned with the direction of maximum variation in the data, and each subsequent axis is aligned with the largest possible amount of remaining variation in the data (Campbell and Atchley, 1981; Marcus, 1990; Hair et al., 1992; Etter, 1999). These new axes are the principal components, with the first principal component (PC1) explaining the greatest fraction of variance, PC2 explaining the next greatest variance, and so on. With rigid rotation of the original axes, the new axes remain orthogonal, so that the new variables remain mutually independent. As this geometric description indicates, PCA calculates the same number of axes as variables (taxa) that were present in the original data (Marcus, 1990). Simplification is achieved because the first few principal component axes typically account for the majority of variation in the data, reducing the number of dimensions that must be analyzed to identify major patterns. PCA can be performed in a variety of statistical software packages (e.g., StatView, JMP), and details of the calculations required are described in several texts on multivariate data analysis (Pielou, 1984; Krzanowski, 1988; Hair et al., 1992; Reyment and Jöreskog, 1993). Our discussion of this method focuses on a variety of considerations that arise during application of PCA.

In PCA, the original taxonomic-abundance data matrix is usually "centered" by subtracting, for each variable, the mean value of the variable from the value of that variable for each locality (Pielou, 1984). This process produces bipolar axes, along which some data points have positive scores and some data points have negative scores, and is appropriate when the data project appreciably (i.e., can be plotted extending out from zero) along all axes. However, if many species are not shared among groups of localities, then points (localities) may tend to cluster such that each cluster has low scores on different sets of axes. For such data sets, centering will tend to force PC1 through some of these clusters, obscuring distinctions among groups that would be apparent if the data were left uncentered. The data matrix also can be standardized prior to PCA by dividing each element for a species by the standard deviation of all the locality-abundances for that species (Pielou, 1984). This process reduces the likelihood that analyses will be dominated by the most abundant species because of their typically greater variances but may overestimate the importance of rare species. However, standardization is critical for data sets in which variables have been measured in different units (e.g. biomass vs. counts of individuals).

After decisions about centering and standardization have been made, PCA is performed on either a correlation matrix or a covariance matrix calculated from the taxonomic-abundance data matrix (Neff and Marcus, 1980; Reyment and Jöreskog [1993] discuss considerations for data in the form of proportions, rather than counts). Correlation and covariance matrices are generally calculated in the software used to perform PCA, but an informed choice must be made between these alternatives because they produce differing results and each is appropriate for different types of data (Neff and Marcus, 1980; Marcus, 1990). A PCA performed on a covariance matrix will give greater weight to the original variables with high-variance data. This is not true for a correlation matrix PCA, because the correlation matrix is derived by standardizing the data so that all variables have an identical mean and standard deviation. If all of the variables in a PCA are measured in the same units, then the different amounts of variation among variables may be informative, and the PCA should be performed on the covariance matrix. Because data sets consisting of faunal abundance counts generally meet this criterion, PCAs based on the covariance matrix are appropriate for most analyses of bonebed faunas.

The next step in PCA is to compute eigenvectors and corresponding eigenvalues from the covariance or correlation matrix (Shi, 1993). The coefficients of the eigenvectors indicate the relative contribution of each original variable to each of the principal components calculated from them (Marcus, 1990). The eigenvalue of each eigenvector indicates the amount of total sample variance that is explained by the associated principal component (Neff and Marcus, 1980). In some cases, eigenvector coefficients for a principal component (termed "loadings" by some authors) can be similar for most of the original variables, leaving the relative importance of their contributions to the new axis ambiguous (Hair et al., 1992). In such cases (for example, if all variables load highly on the first component), interpretation of the analysis can be aided by a rigid rotation of the orthogonal components so that variable loadings on each axis are as close to 1 (the maximum) or 0 (the minimum) as possible. This procedure is called VARIMAX rotation (Hair et al., 1992) and is available as an option in most statistical packages that perform PCA. Scores for each sample (i.e., bonebed) on each of the new rotated (or unrotated) principal components then can be calculated.

Interpretation of PCA results focuses on evaluating the scores of samples on each PC axis in the context of the original variables that contribute to the axis (as indicated by the variable loadings). For example, if a site shows a high positive score on PC1, it would indicate that the site has high values of the variables that contribute to PC1. PC scores also can be used to depict each locality in scatterplots based on the new PC axes, facilitating the interpretation of analysis results. This type of analysis, in which new variables are calculated from the original variables (simplifying the number of columns in the data set) and the rows of the data matrix (localities) are ordinated, is the standard form of PCA and is widely referred to as "R-mode" analysis (Reyment, 1990; Shi, 1993). Gabriel (1971) has described additional methods for ordinating the original variables ("Q-mode analyses") and plotting the results on the same principal component axes as the locality ordination. In this biplot, the original variables are generally plotted as vectors radiating from the origin, with the length of each vector reflecting the importance of its contribution to the new principal components (Gabriel, 1971; Ter Braak, 1983). Biplots can be constructed in some statistical packages (e.g., JMP, CANOCO) and can provide a useful means of evaluating which original variables (i.e., taxa) contribute most strongly to associations among bonebed faunas.

An issue that must be resolved before the results of a PCA can be interpreted is how high the loading of an original variable on a principal component must be in order for the variable to be considered to contribute significantly to the new component. One common criterion for evaluating loading significance is based on the fact that a loading represents the correlation between an original variable and a principal component (Hair et al., 1992). As such, loading significance can be interpreted in a fashion similar to that applied to correlation coefficients, generating, for example, calculations of minimum loadings of 0.26 and 0.19 for significance at the 5% and 1% significance levels for a sample of 100 localities. The loading magnitude necessary for significance will decrease as more variables are analyzed, but this approach but may not be sufficiently conservative for principal components beyond the first (Hair et al., 1992).

Another issue that must be resolved prior to PCA interpretation is the number of principal components that should be interpreted. Three main criteria have been applied to evaluate the significance of components (Hair et al., 1992). The "latent root" criterion states that only components with eigenvalues (i.e., latent roots) greater than 1 are significant. This criterion is based on the rationale that any new component should account for at least the same variance as one of the original variables to justify its interpretation. The "latent root" criterion tends to be highly conservative in analyses involving less than 20 variables, but insufficiently conservative in analyses involving more than 50 variables. Alternatively, a "percentage of variance" criterion can be applied, in which components are judged significant until the cumulative variance explained reaches 95%, or until an individual component explains less than 5% of the data variance. A third alternative is the scree test, in which the eigenvalue of each component is plotted against the number of the component. The point at which this curve begins to flatten out indicates the last significant component. Jackson (1993) describes additional methods for evaluating the number of principal components that should be considered significant.

PC scores calculated for data from fossil localities are commonly used for two purposes. First, similarities among localities can be evaluated based on the clustering of sites when their scores are graphed in a scatterplot based on the new PC axes (Shi, 1993; Etter, 1999). Faunal similarities among sites are commonly examined, but similarities in other features, such as depositional environment, also can be evaluated if the analysis includes sedimentological or taphonomic variables in addition to faunal variables. The correspondence of clusters to predefined groups (e.g., sites from particular geographic regions or environments) also can be compared. However, PC scores also can be evaluated directly in the context of variables extrinsic to the original analysis. For example, PC scores could be plotted against stratigraphic position, or distance from a paleoshoreline. These examples illustrate the potential for PC scores to be used as variables in studies of faunal change.

PCA has been applied to a wide range of studies of bonebed faunas. Wilson (1980), for example, used PCA to evaluate similarities among the fish faunas of 22 lacustrine localities from the Middle Eocene of western North America. He used the abundances of different fish genera, counts of articulated and disarticulated specimens, and counts of bone-rich and bone-poor coprolites as variables in order to evaluate taphonomic distinctions among the localities. PCA produced new axes that ordinated sites based on features that suggested a gradient from shallow, nearshore environments to deep, offshore environments. Sites with high negative scores on PC1, for example, typically preserved large numbers of articulated specimens (mainly from a single catostomid genus, *Amyzon*), whereas sites with high positive scores on PC1 preserved mainly disarticulated specimens from a variety of taxa. Because low diversity is generally consistent with deeper water, and disarticulation tends to increase in nearshore environments where water is warmer, more aerobic, and more turbulent than in deeper water, locality scores on PC1 appeared to reflect an environmental gradient from deep to shallow habitats. Similar patterns emerged from analyses of faunas collected from different stratigraphic horizons at a single locality, suggesting changes in the depth of the lake through time. Thus, by plotting localities on the new principal component axes, Wilson (1980) could infer which bonebeds preserved faunas typical of shallow, intermediate, and deep lake environments.

PCA has been used to examine faunal change through time as well as across geographic regions. Gingerich (1989), for example, used PCA to analyze changes in taxonomic structure among mammalian faunas from the Late Paleocene through the Early Eocene in the Clark's Fork Basin of Wyoming. He compiled counts of the minimum number of individuals for each of 33 taxa from 49 Clark's Fork Basin localities that were ordered in stratigraphic sequence. He then performed a PCA on this taxonomic abundance matrix and plotted the score of each locality on the first principal component (PC1, accounting for 71% of analysis variance) against the stratigraphic position of the locality (Fig. 6.5). Based on the correlations of the original variables with the new axes, high positive scores on

CLARK'S FORK BASIN

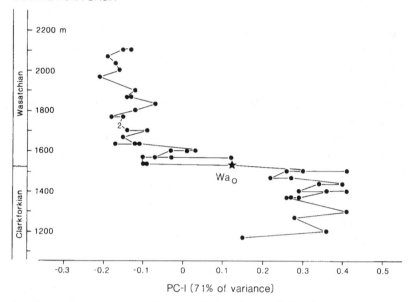

Figure 6.5. Principal component scores for mammalian faunas from the Clark's Fork Basin of Wyoming graphed versus the stratigraphic positions of the localities. From Gingerich (1989).

PC1 indicated faunas dominated by phenacodontid condylarths, whereas high negative scores indicated faunas dominated by hyopsodontid condylarths, adapid primates, and equid perissodactyls. Gingerich (1989) found a sharp drop in PC1 scores at the Clarkforkian/Wasatchian boundary, signifying a sudden shift from dominance by phenacodontids to dominance by hyopsodontids, adapids, and equids.

**Correspondence Analysis**

Correspondence analysis (CA), also known as reciprocal averaging, is a multivariate ordination method closely related to PCA that also can be used to explore the structure of faunal-abundance data from multiple localities. Like PCA, CA simplifies the original data set by calculating a new set of orthogonal axes (each accounting for the maximum possible variance in the data) that can be used to ordinate localities and taxa. Points representing both localities and taxa can be plotted simultaneously on the new axes in CAs (Pielou, 1984; Reyment and Jöreskog, 1993; Jongman et al., 1995), just as in PCA biplots (Gabriel, 1971; Ter Braak, 1983). If points for particular variables (taxa) plot near clusters of locality points, it suggests that those localities are similar in their abundances of those taxa. However, PCA and CA are based on different assumptions and, thus, are appropriate under different circumstances (Jongman et al., 1995). PCA is related to linear multiple regression: the method constructs theoretical explanatory (latent) variables by fitting straight surfaces (in multiple dimensions) to faunal data, minimizing the residual sum of squares. Thus, PCA assumes that the faunal variable (e.g., the abundance of a taxon or number of species in an ecological guild) increases or decreases linearly in relation to values on the PCA axis. In contrast, CA is a simplification of a maximum likelihood algorithm, often computed by a weighted-averaging procedure in which faunal variable scores are weighted averages of site scores (and vice versa). CA assumes a bell-shaped or unimodal response of the faunal variable to the latent variable, a response in which the faunal variable peaks over a limited range of values of the latent variables. Thus, the final faunal variable scores are estimates of the optima (the height of the bell curve) of each variable. CAs can be performed by some statistical software packages (e.g., JMP, CANOCO), and Tian and colleagues (1993) have published code for a routine to perform CA in the Matlab programming environment.

Locality and taxon scores on the new axes of CA can be obtained through a series of iterative calculations involving the original data matrix (Hill, 1974; Pielou, 1984). However, identical results can be obtained through an eigenanalysis similar to that performed during PCA (Pielou, 1984), the approach that we focus on here to compare the two methods. One further difference between PCA and CA occurs in the transformations that each applies to the data matrix before analysis. In PCA, the data can be (but need not be) centered and/or standardized, but in CA, the data matrix is not centered and each datum is divided by the square root of its row total and its column total, generating a new matrix **M** (Pielou, 1984). The matrix **M** then is used to generate eigenvectors and corresponding eigenvalues for both the localities (R-mode) and taxa (Q-mode). As in PCA, the eigenvectors indicate the relative contributions of each of the original localities and taxa to the new axes, and the eigenvalue for each eigenvector indicates the amount of variance that each axis accounts for. Details of the required calculations are provided in several texts (e.g., Pielou, 1984; Krzanowski, 1988). Locality and taxon scores for the CA axes then can be calculated based on the eigenvectors derived from these analyses.

One potential difficulty with the interpretation of CA ordinations is a tendency for samples to become arrayed as an arch (Gauch et al., 1977; Hill and Gauch, 1980). This tendency, known as the horseshoe effect, also can occur in PCA and is especially prevalent among analyses in which many localities have no pairs of species in common (Digby and Kempton, 1987; Etter, 1999). Several authors view this pattern as a mathematical artifact, and methods to correct for the horseshoe effect have been developed (e.g., detrended correspondence analysis [Hill and Gauch, 1980]). However, by using detrending methods, an investigator runs the risk of obscuring some of the actual structure of the data (Wartenberg et al., 1987; Reyment and Jöreskog, 1993).

CA is appropriate for many of the same types of examinations of faunal data structure as PCA. For example, de Bonis and colleagues (1992) applied CA to evaluate the similarities in large-mammal faunas between Neogene fossil localities from Europe, Asia, and Africa and recent African and Asian localities from different environments. In addition to analyzing the entire data set together (20 taxa, 20 recent localities, and 51 fossil localities), they also performed separate CAs on subsets of the data restricted to just the fossil localities and to fossil localities of a specific age. Similar patterns of separation emerged from most of these analyses for both localities and taxa. Middle Miocene localities (the oldest included in the analyses) separated from all other sites, fossil and recent, largely due to the presence of lineages such as creodonts and anthracotheres that were extinct when younger assemblages were fossilized. Post-Middle-Miocene localities tended to separate from recent localities due to a greater abundance of primates in recent faunas and a greater abundance of equids, rhinos, and giraffes in the fossil localities. However, a second axis of separation distinguished faunas from open versus forested environments (regardless of geographic location) for both recent and fossil localities, with open-environment faunas dominated by bovids, and forest faunas dominated by deer, primates, and tapirs (de Bonis et al., 1992).

Blob and colleagues (Blob and Dodson, 1993; Blob et al., 1997) used CA to examine faunal variation across Judithian microvertebrate bonebeds from Alberta and Montana. The three locations in which collecting has concentrated are Dinosaur Provincial Park in the north (Brinkman, 1990), the Academy of Natural Sciences of Philadelphia collecting areas in the south (Fiorillo, 1989; Blob and Fiorillo, 1996), and the Judithian type area between these two, sampled by Sahni (1972) and by teams from the University of Chicago (Carrano et al., 1995, 1997; Rogers, 1995, 1998; Blob et al., 1997, 2001) (Fig. 6.6A). One of the primary goals of the recent collections from the type area was to evaluate whether a north-south gradient could be detected in the Judith River faunas of these three regions. Faunal abundances from Sahni's type area collection differed strongly from those of Judith River deposits to the north and south (Blob and Dodson,

1993), suggesting that Judithian abundance patterns might show uni-modal variation across the geographic region. This pattern was explored by performing a CA on faunal abundance counts from Judithian microver-tebrate localities, including additional sites from the type area. The results indicated that Sahni's (1972) Clambank Hollow locality preserved a fauna that was highly atypical even among other sites from the type area, in which mammals and dinosaurs were much more abundant than at other localities (Fig. 6.6B). Thus, characterizing a "Judithian" fauna strictly on the basis of Sahni's type locality collection could produce misleading in-terpretations of faunal change across broader time scales. Other faunas from the Judithian type area of central Montana plotted amid the cluster of sites from Alberta, indicating considerable faunal similarity between these regions. However, assemblages from southern Montana tended to differ from those of sites further to the north due to an abundance of gar and a paucity of amphibians (Fig. 6.6B). These geographic faunal distinc-tions suggest a focus for future paleoecological studies that might identify environmental correlates of regional differences in Judithian taxonomic abundances.

## SUMMARY

Quantitative analyses are an effective means of identifying and evalu-ating patterns among the faunas preserved in vertebrate bonebeds. We have outlined a variety of quantitative methods that can facilitate broad-scale analyses of temporal, geographic, and environmental patterns of faunal change, as well as a number of techniques that can help to eval-uate whether the sites contributing to broad-scale analyses are valid to compare. To conclude this chapter, we would like to reemphasize three critical points regarding quantitative analyses of bonebed faunas. First, counting specimens from bonebeds is not a trivial issue. Because many ecological, biogeographic, and evolutionary interpretations are based on specimen counts, approaches to censusing bonebed faunas should be con-sidered carefully. Second, comparisons among bonebed faunas also should be chosen with taphonomic and sampling effects in mind. If a particular comparison appears likely to be confounded by taphonomic factors, an attempt should be made to correct for the effects of those factors. If this is not possible, the taphonomic complications should be acknowledged, or the comparison should be avoided entirely. Evaluating and correcting for sampling effects should be standard practice. Finally, a wide variety of analytical techniques can provide insights into the biology of the animals

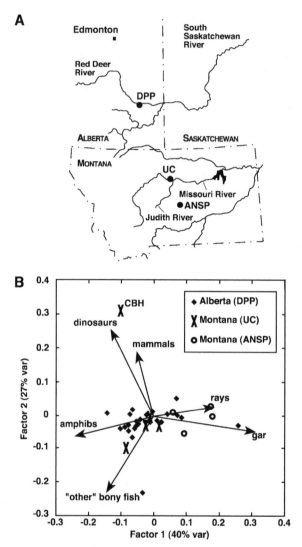

*Figure 6.6.* A. Map of Alberta and Montana, showing locations from which collections have been made from Judithian microvertebrate bonebeds. DPP, Dinosaur Provincial Park; UC, University of Chicago (Judith River Formation Type Section); ANSP, Academy of Natural Sciences, Philadelphia. B. Results of correspondence analysis performed on faunal abundance data from Judithian localities. Ordinations on factor 1 (accounting for 40% of total variance) versus factor 2 (accounting for 27% total variance) are plotted for taxa and localities. Points (diamonds, X's, and circles) indicate individual localities. The Clambank Hollow locality from the type section is labeled (CBH). Vectors radiating from the origin indicate the taxa most responsible for ordination along a particular axis. Faunal data were derived from Brinkman (1990) for DPP localities; Blob and Dodson (1993) and Blob and Fiorillo (1996) for ANSP localities; Sahni (1972) for CBH; and Blob et al. (1997) and Carrano et al. (1995, 1997) for UC localities.

preserved in bonebeds. Although quantitative analyses have the potential to clarify predictions, and even facilitate the formation of process-based models, these methods can also be used sloppily or erroneously. It is important to understand the principles and assumptions of a method before choosing it—simply "trying everything in the book until a pattern emerges" is a strategy that yields dubious results. However, use of appropriate techniques can yield novel insights into paleofaunal questions. Thus, we urge students of fossil bonebeds to gain experience with different quantitative techniques and to explore the complementary perspectives provided by these approaches as they use bonebeds to analyze patterns of faunal and environmental change.

## ACKNOWLEDGMENTS

We extend sincere thanks to the numerous colleagues who contributed to portions of the studies that have been described in this chapter, and to Ray Rogers, Dave Eberth, and Tony Fiorillo for organizing the symposium upon which the contributions to this book are based. Jeff Walker and Pete Wagner provided helpful literature, and Bonnie Miljour assisted with preparation of Figures 6.1, 6.3, and 6.4.

## REFERENCES

Alberdí, M.T., M.A. Alanso, B. Azanza, M. Hoyos, and J. Morales. 2001. Vertebrate taphonomy in circum-lake environments: Three cases in the Guadix-Baza Basin (Granada, Spain). Palaeogeography, Palaeoclimatology, Palaeoecology 165:1–26.

Alroy, J. 1996. Constant extinction, constrained diversification, and uncoordinated stasis in North American mammals. Palaeogeography, Palaeoclimatology, Palaeoecology 127:285–311.

Alroy, J. 1998. Equilibrial diversity dynamics in North American mammals. Pp. 232–287 *in* Biodiversity dynamics: Turnover of populations, taxa and communities. M.L. McKinney and J. Drake, eds. Columbia University Press, New York.

Alroy, J. 2000. New methods for quantifying macroevolutionary patterns and processes. Paleobiology 26:707–733.

Badgley, C. 1982. How much time is represented in the present? The development of time-averaged modern assemblages as models for the fossil record. Third North American Paleontological Convention Proceedings 1:23–28.

Badgley, C. 1984. The paleoenvironment of South Asian Miocene hominoids. Pp. 796–811 *in* The evolution of the East Asian environment, vol. 2. R.O. Whyte, ed. Centre of Asian Studies, University of Hong Kong, Hong Kong.

Badgley, C. 1986a. Counting individuals in mammalian fossil assemblages from fluvial environments. Palaios 1:328–338.

Badgley, C. 1986b. Taphonomy of mammalian fossil remains from Siwalik rocks of Pakistan. Paleobiology 12:119–142.

Badgley, C. 1990. A statistical assessment of last appearances in the Eocene record of mammals. Pp. 153–167 in Dawn of the Age of Mammals in the northern part of the Rocky Mountain Interior, North America. Special paper 243. T.M. Bown and K.D. Rose, eds. Geological Society of America, Boulder, Colorado.

Badgley, C., and A.K. Behrensmeyer. 1995. Preservational, paleoecological and evolutionary patterns in Paleogene of Wyoming-Montana and the Neogene of Pakistan. Palaeogeography, Palaeoclimatology, Palaeoecology 115:319–340.

Badgley, C., and D.L. Fox. 2000. Ecological biogeography of North American mammals: Species density and ecological structure in relation to environmental gradients. Journal of Biogeography 27:1437–1467.

Badgley, C., and P.D. Gingerich. 1988. Sampling and faunal turnover in early Eocene mammals. Palaeogeography, Palaeoclimatology, Palaeoecology 63:141–157.

Badgley, C., W.S. Bartels, M.E. Morgan, A.K. Behrensmeyer, and S.M. Raza. 1995. Taphonomy of vertebrate assemblages from the Paleogene of northwestern Wyoming and the Neogene of northern Pakistan. Palaeogeography, Palaeoclimatology, Palaeoecology 115:157–180.

Badgley, C., W. Downs, and L.J. Flynn. 1998. Taphonomy of small-mammal fossil assemblages from the middle Miocene Chinji formation, Siwalik group, Pakistan. Pp. 145–166 in Advances in vertebrate paleontology and geochronology. Y. Tomida, L.J. Flynn, and L.L. Jacobs, eds. National Science Museum Monographs, Tokyo.

Barry, J.C., M.E. Morgan, L.J. Flynn, D. Pilbeam, L.L. Jacobs, E.H. Lindsay, S.M. Raza, and N. Solounias. 1995. Patterns of faunal turnover and diversity in the Neogene Siwaliks of northern Pakistan. Palaeogeography, Palaeoclimatology, Palaeoecology 115:209–226.

Barry, J.C., M.E. Morgan, L.J. Flynn, D. Pilbeam, A.K. Behrensmeyer, S.M. Raza, I.A. Khan, C. Badgley, J. Hicks, and J. Kelley. 2002. Faunal and environmental change in the Late Miocene Siwaliks of Northern Pakistan. Paleobiology Memoirs 3:1–72.

Behrensmeyer, A.K. 1975. The taphonomy and paleoecology of Plio-Pleistocene vertebrate assemblages of Lake Rudolf, Kenya. Bulletin of the Museum of Comparative Zoology 146:473–578.

Behrensmeyer, A.K. 1978. Taphonomic and ecologic information from bone weathering. Paleobiology 8:211–227.

Behrensmeyer, A.K. 1988. Vertebrate preservation in fluvial channels. Palaeogeography, Palaeoclimatology, Palaeoecology 63:183–199.

Behrensmeyer, A.K. 1991. Terrestrial vertebrate accumulations. Pp. 291–335 in Taphonomy: Releasing the data locked in the fossil record. P.A. Allison and D.E.G. Briggs, eds. Plenum Press, New York.

Behrensmyer, A.K., and J.C. Barry. 2005. Biostratigraphic surveys in the Siwaliks of Pakistan: A method for standardized surface sampling of the vertebrate fossil record. Paleontologia Electronica 8.1.15A. http://palaeo-electronica.org/2005_1/behrens15/issue1_05.htm

Birks, H.J.B., and A.D. Gordon. 1985. Numerical methods in quaternary pollen analysis. Academic Press, London.

Blob, R.W. 1997. Relative hydrodynamic dispersal potentials of soft-shelled turtle elements: Implications for interpreting skeletal sorting in assemblages of non-mammalian terrestrial vertebrates. Palaios 12:151–164.

Blob, R.W., and P. Dodson. 1993. Quantitative analysis of faunal patterns among Late Cretaceous microfossil localities from Montana and Alberta. Journal of Vertebrate Paleontology 13:27A.

Blob, R.W., and A.R. Fiorillo. 1996. The significance of vertebrate microfossil size and shape distributions for faunal abundance reconstructions: A Late Cretaceous example. Paleobiology 22:422–435.

Blob, R.W., M.T. Carrano, R.R. Rogers, C.A. Forster, and N.R. Espinoza. 1997. New taxonomic and taphonomic data from the herpetofauna of the Judith River Formation (Campanian), Montana. Journal of Vertebrate Paleontology 17:32A–33A.

Blob, R.W., M.T. Carrano, R.R. Rogers, C.A. Forster, and N.R. Espinoza. 2001. A new fossil frog from the Upper Cretaceous Judith River Formation of Montana. Journal of Vertebrate Paleontology 21:190–194.

Bown, T.M., and M.J. Kraus. 1993. Time-stratigraphic reconstruction and integration of paleopedologic, sedimentologic, and biotic events (Willwood Formation, Lower Eocene, northwest Wyoming, U.S.A.). Palaios 8:68–80.

Brinkman, D.B. 1990. Paleoecology of the Judith River Formation (Campanian) of Dinosaur Provincial Park, Alberta, Canada: Evidence from vertebrate microfossil localities. Palaeogeography, Palaeoclimatology, Palaeoecology 78:37–54.

Brinkman, D.B., M.J. Ryan, and D.A. Eberth. 1998. The paleogeographic and stratigraphic distribution of ceratopsids (Ornithischia) in the Upper Judith River Group of western Canada. Palaios 13:160–169.

Brinkman, D.B., A.P. Russell, D.A. Eberth, and J. Peng. 2004. Vertebrate paleocommunities of the lower Judith River Group (Campanian) of southeastern Alberta, Canada, as interpreted from vertebrate microfossil assemblages. Palaeogeography, Palaeoclimatology, Palaeoecology 213:295–313.

Brinkman, D.B., D.A. Eberth, and P.J. Currie. This volume. From bonebeds to paleobiology: Applications of bonebed data. Chapter 4 *in* Bonebeds: Genesis, analysis, and paleobiological significance. R.R. Rogers, D.A. Eberth, and A.R. Fiorillo, eds. University of Chicago Press, Chicago.

Bryant, L.J. 1989. Non-dinosaurian lower vertebrates across the Cretaceous-Tertiary Boundary in northeastern Montana. University of California Publications in Geological Sciences 134:1–107.

Campbell, N.A., and W.R. Atchley. 1981. The geometry of canonical variate analysis. Systematic Zoology 30:268–280.

Carrano, M.T., R.W. Blob, J.J. Flynn, C.A. Forster, and R.R. Rogers. 1995. Additions to the fauna of the Judith River Formation (Campanian) type area, north-central Montana, with possible range extensions of two genera of eutherian mammals. Journal of Vertebrate Paleontology 15:21A–22A.

Carrano, M.T., R.W. Blob, J.J. Flynn, R.R. Rogers, and C.A. Forster. 1997. The mammalian fauna of the Judith River Formation type area (Campanian, central Montana) revisited. Journal of Vertebrate Paleontology 17:36A.

Clyde, W.C., and P.D. Gingerich. 1998. Mammalian community response to the latest Paleocene thermal maximum: An isotaphonomic study in the northern Bighorn Basin, Wyoming. Geology 26:1011–1014.

Colwell, R.K., and J.A. Coddington. 1994. Estimating terrestrial biodiversity through extrapolation. Philosophical Transactions of the Royal Society of London B 345:101–118.

Coombs, M.C., and W.P. Coombs Jr. 1997. Analysis of the geology, fauna, and taphonomy of Morava Ranch Quarry, Early Miocene of northwest Nebraska. Palaios 12:165–187.

Currie, P.J., and P. Dodson. 1984. Mass death of a herd of ceratopsian dinosaurs. Pp. 61–66 *in* Third symposium on Mesozoic terrestrial ecosystems, short papers. W.E. Reif and F. Westphal, eds. Attempto Verlag, Tübingen.

de Bonis, L., G. Bouvrain, D. Geraads, and G. Koufos. 1992. Multivariate study of late Cenozoic mammalian faunal compositions and paleoecology. Paleontologia I Evolució 24–25:93–101.

Digby, P.G.N., and R.A. Kempton. 1987. Multivariate analysis of ecological communities. Chapman and Hall, London.

Dodson, P. 1973. The significance of small bones in paleoecological interpretation. Contributions to Geology, University of Wyoming 12:15–19.

Dodson, P. 1987. Microfaunal studies of dinosaur paleoecology, Judith River Formation of southern Alberta. Pp. 70–75 *in* Fourth symposium on Mesozoic terrestrial ecosystems, short papers. P.J. Currie and E. Koster, eds. Royal Tyrrell Museum of Palaeontology, Drumheller.

Eberth, D.A. 1990. Stratigraphy and sedimentology of vertebrate microfossil sites in the uppermost Judith River Formation (Campanian), Dinosaur Provincial Park, Alberta, Canada. Palaeogeography, Palaeoclimatology, Palaeoecology 78:1–36.

Erickson, G.M. 1996. Incremental lines of von Ebner in dinosaurs and the assessment of tooth replacement rates using growth line counts. Proceedings of the National Academy of Sciences, USA 93:14623–14627.

Etter, W. 1999. Community analysis. Pp. 285–359 *in* Numerical paleobiology: Computer-based modelling and analysis of fossils and their distributions. D.A.T. Harper, ed. John Wiley and Sons, Chichester.

Fiorillo, A.R. 1988. Taphonomy of Hazard Homestead Quarry (Ogallala Group), Hitchcock County, Nebraska. Contributions to Geology, University of Wyoming 26:57–97.

Fiorillo, A.R. 1989. The vertebrate fauna from the Judith River Formation (Upper Cretaceous) of Wheatland and Golden Valley Counties, Montana. The Mosasaur 4:127–142.

Fiorillo, A.R. 1991. Taphonomy and depositional setting of Careless Creek Quarry (Judith River Formation), Wheatland County, Montana, U.S.A. Palaeogeography, Palaeoclimatology, Palaeoecology 81:281–311.

Gabriel, K.R. 1971. The biplot graphic display of matrices with application to principal component analysis. Biometrika 58:453–467.

Gauch, H.G. 1980. Rapid initial clustering of large data sets. Vegetatio 42:103–111.

Gauch, H.G. 1982. Multivariate analysis in community ecology. Cambridge University Press, Cambridge.

Gauch, H.G., R.H. Whittaker, and T.R. Wentworth. 1977. A comparative study of reciprocal averaging and other ordination techniques. Journal of Ecology 65:157–174.

Gingerich, P.D. 1983. Paleocene-Eocene faunal zones and a preliminary analysis of

Laramide structural deformation in the Clark's Fork Basin, Wyoming. Pp. 185–195 *in* 34th annual field conference guidebook. Wyoming Geological Association, Casper.

Gingerich, P.D. 1989. New earliest Wasatchian mammalian fauna from the Eocene of northwestern Wyoming: Composition and diversity in a rarely sampled high-floodplain assemblage. University of Michigan Papers on Paleontology 28:1–97.

Gotelli, N.J., and R.K. Colwell. 2001. Quantifying biodiversity: Procedures and pitfalls in the measurement and comparison of species richness. Ecology Letters 4:379–391.

Greenacre, M.J. 1984. Theory and application of correspondence analysis. Academic Press, London.

Gujarati, D.N. 1995. Basic econometrics, 3rd ed. McGraw-Hill, New York

Hair, J.F. Jr., R.E. Anderson, R.L. Tatham, and W.C. Black. 1992. Multivariate data analysis with readings, 3rd ed. Macmillan, New York.

Hamilton, L.C. 1992. Regression with graphics: A second course in applied statistics. Duxbury Press, Belmont, California.

Hanson, C.B. 1980. Fluvial taphonomic processes: Models and experiments. Pp. 156–181 *in* Fossils in the making. A.K. Behrensmeyer and A. Hill, eds. University of Chicago Press, Chicago.

Henrici, A.C., and A.R. Fiorillo. 1993. Catastrophic death assemblage of *Chelomophrynus bayi* (Anura, Rhinophrynidae) from the Middle Eocene Wagon Bed Formation of Central Wyoming. Journal of Paleontology 67:1016–1026.

Hill, A., and A.K. Behrensmeyer. 1984. Disarticulation patterns of some modern East African mammals. Paleobiology 10:366–376.

Hill, M.O. 1974. Reciprocal averaging: An eigenvector method of ordination. Journal of Ecology 61:237–249.

Hill, M.O., and H.G. Gauch Jr. 1980. Detrended correspondence analysis: An improved ordination technique. Vegetatio 42:47–58.

Horner, J.R. 1982. Evidence of colonial nesting and "site fidelity" among ornithischian dinosaurs. Nature 297:575–576.

Horner, J.R., and R. Makela. 1979. Nest of juveniles provides evidence of family structure among dinosaurs. Nature 282:296–298.

Hunt, R.M. Jr. 1990. Taphomony and sedimentology of Arikaree (lower Miocene) fluvial, eolian, and lacustrine paleoenvironments, Nebraska and Wyoming; A paleobiota entombed in fine-grained volcaniclastic rocks. Pp. 69–111 *in* Volcanism and fossil biotas. Special paper 244. M.J. Lockley and A. Rice, eds. Geological Society of America, Boulder, Colorado.

Hutchison, J.H., and J.D. Archibald. 1986. Diversity of turtles across the Cretaceous/Tertiary boundary in northeastern Montana. Palaeogeography, Palaeoclimatology, Palaeoecology 55:1–22.

Jackson, D.A. 1993. Stopping rules in principal components analysis: A comparison of heuristical and statistical approaches. Ecology 74:2204–2214.

Jamniczky, H.A., D.B. Brinkman, and A.P. Russell. 2003. Vertebrate microsite sampling: How much is enough? Journal of Vertebrate Paleontology 23:725–734.

Jongman, R.H.G, C.J.F. Ter Braak, and O.F.R. Tongeren. 1995. Data analysis in community and landscape ecology. Cambridge University Press, Cambridge.

Kidwell, S.M., and A.K. Behrensmeyer. 1993. Taphonomic approaches to time resolution in fossil assemblages. Short courses in paleontology 6. Paleontological Society, Knoxville, Tennessee.

Krause, D.W. 1986. Competitive exclusion and taxonomic displacement in the fossil record: The case of rodents and multituberculates in North America. Contributions to Geology, University of Wyoming, Special Paper 3:95–117.

Krzanowski, W.J. 1988. Principles of multivariate analysis. Clarendon Press, Oxford.

Lillegraven, J.A., and J.J. Eberle. 1999. Vertebrate faunal changes through Lancian and Puercan time in southern Wyoming. Journal of Paleontology 73:691–710.

Lyman, R.L. 1994. Relative abundances of skeletal specimens and taphonomic analysis of vertebrate remains. Palaios 9:288–298.

Maas, M.C., D.W. Krause, and S.G. Strait. 1988. The decline and extinction of Plesiadapiformes (Mammalia: ?Primates) in North America: Displacement or replacement? Paleobiology 14:410–431.

Marcus, L.F. 1990. Traditional morphometrics. Pp. 77–122 in Proceedings of the Michigan morphometrics workshop. special publication 2. F.J. Rohlf and F.L. Bookstein, eds. University of Michigan Museum of Zoology, Ann Arbor.

Markwick, P.J. 1998. Crocodilian diversity in space and time: The role of climate in paleoecology and its implication for understanding K/T extinctions. Paleobiology 24:470–497.

Miller, A.I., and M. Foote. 1996. Calibrating the Ordovician Radiation of marine life: Implications for Phanerozoic diversity trends. Paleobiology 22:304–309.

Moore, D.S., and G.P. McCabe. 1993. Introduction to the practice of statistics, 2nd ed. W. H. Freeman, New York.

Neff, N.A., and L.F. Marcus. 1980. A survey of multivariate methods for systematics. Privately published, New York.

Nemec, A.F.L., and R.O. Brinkhurst. 1988. Using the bootstrap to assess statistical significance in the cluster analysis of species abundance data. Canadian Journal of Fisheries and Aquatic Sciences 45:965–970.

Pielou, E.C. 1984. The interpretation of ecological data: A primer on classification and ordination. John Wiley and Sons, New York.

Pratt, A.E. 1989. Taphonomy of the microvertebrate fauna from the early Miocene Thomas Farm locality, Florida (U.S.A.). Palaeogeography, Palaeoclimatology, Palaeoecology 76:125–151.

Raup, D.M. 1975. Taxonomic diversity estimation using rarefaction. Paleobiology 1:333–342.

Retallack, G.J. 1983. Late Eocene and Oligocene paleosols from Badlands National Park, South Dakota. Special paper 193. Geological Society of America, Boulder, Colorado.

Reyment, R.A. 1990. Reification of classical multivariate statistical analysis in morphometry. Pp. 123–144 in Proceedings of the Michigan morphometrics workshop. Special publication 2. F.J. Rohlf and F.L. Bookstein, eds. University of Michigan Museum of Zoology, Ann Arbor.

Reyment, R.A., and K.G. Jöreskog. 1993. Applied factor analysis in the natural sciences. Cambridge University Press, Cambridge.

Rice, W.R. 1989. Analyzing tables of statistical tests. Evolution 43:223–225.

Rogers, R.R. 1995. Sequence stratigraphy and vertebrate taphonomy of the Upper Cretaceous Two Medicine and Judith River Formations, Montana. Ph. D. dissertation, University of Chicago, Chicago.

Rogers, R.R. 1998. Sequence analysis of the Upper Cretaceous Two Medicine and Judith River Formations, Montana: Nonmarine response to the Claggett and Bearpaw marine cycles. Journal of Sedimentary Research 68:615–631.

Rogers, R.R., and S.M. Kidwell. 2000. Associations of vertebrate skeletal concentrations and discontinuity surfaces in terrestrial and shallow marine records: A test in the Cretaceous of Montana. Journal of Geology 108:131–154.

Sahni, A. 1972. The vertebrate fauna of the Judith River Formation, Montana. Bulletin of the American Museum of Natural History 147:323–412.

Shi, G.R. 1993. Multivariate data analysis in paleoecology and paleobiogeography—a review. Palaeogeography, Palaeoclimatology, Palaeoecology 105:199–234.

Shubin, N.H., and H.-D. Sues. 1991. Biogeography of early Mesozoic continental tetrapods: Patterns and implications. Paleobiology 17:214–230.

Simpson, G.G., A. Roe, and R.C. Lewontin. 1960. Quantitative zoology. Harcourt, Brace, New York.

Smith, R. J. 1981. Interpretation of correlations in intraspecific and interspecific allometry. Growth 45:291–297.

Snedecor, G.W., and W.G. Cochran. 1980. Statistical methods, 7th ed. Iowa State University Press, Ames.

Sokal, R.R. and F.J. Rohlf. 1995. Biometry: The principles and practice of statistics in biological research, 3rd ed. W. H. Freeman, San Francisco.

Strauss, R.E. 1982. Statistical significance of species clusters in association analysis. Ecology 63:634–639.

Ter Braak, C.J.F. 1983. Principal components biplots and alpha and beta diversity. Ecology 64:454–462.

Tian, D., S. Sorooshian, and D.E. Myers. 1993. Correspondence analysis with Matlab. Computers and Geosciences 19:1007–1022.

Tipper, J.C. 1979. Rarefaction and rarefiction—the use and abuse of a method in paleoecology. Paleobiology 5:423–434.

Trapani, J. 1998. Hydrodynamic sorting of avian skeletal remains. Journal of Archaeological Science 25:477–487.

Van Valkenburgh, B. 1990. Skeletal and dental predictors of body mass in carnivores. Pp. 181–205 *in* Body size in mammalian paleobiology: Estimation and biological implications. J. Damuth and B.J. MacFadden, eds. Cambridge University Press, Cambridge.

Varricchio, D.J. 1995. Taphonomy of Jack's Birthday Site, a diverse dinosaur bonebed from the Upper Cretaceous Two Medicine Formation of Montana. Palaeogeography, Palaeoclimatology, Palaeoecology 114:297–323.

Voorhies, M.R. 1969. Taphonomy and population dynamics of an Early Pliocene Vertebrate Fauna, Knox County, Nebraska. Contributions to Geology, University of Wyoming, Special Paper 1:1–69.

Wartenberg, D., S. Ferson, and F.J. Rohlf. 1987. Putting things in order: A critique of detrended correspondence analysis. American Naturalist 129:434–448.

Weigelt, J. 1989. Recent vertebrate carcasses and their paleobiological implications. J. Schaefer, transl. University of Chicago Press, Chicago.

Weishampel, D.B. 1984. The evolution of jaw mechanisms in ornithopod dinosaurs. Advances in Anatomy, Embryology and Cell Biology 87:1–110.

Western, D. 1980. Linking the ecology of past and present mammal communities. Pp. 41–54 *in* Fossils in the making. A. K. Behrensmeyer and A. Hill, eds. University of Chicago Press, Chicago.

Westrop, S.R. 1986. Taphonomic versus ecologic controls on taxonomic relative abundance patterns in tempestites. Lethaia 19:123–132.

Williams, W.T. 1971. Principles of clustering. Annual Review of Ecology and Systematics 2:303–326.

Wilson, M.V.H. 1980. Eocene lake environments: Depth and distance-from-shore variation in fish, insect, and plant assemblages. Palaeogeography, Palaeoclimatology, Palaeoecology 32:21–44.

Wolff, R.G. 1975. Sampling and sample size in ecological analyses of fossil mammals. Paleobiology 1:195–204.

# CHAPTER 7

# Trace Element Geochemistry of Bonebeds

*Clive Trueman*

## INTRODUCTION

Fossil bones and teeth are the last physical remains of once living vertebrates, and the chemistry of these fossils at least in part reflects the biochemistry of the individual. If the mineral phase of bone remains unaltered
during diagenesis and fossilization, then the bone will contain a biogenic
signal that can be used to infer directly physiological, dietary, and climatic
information about the animal and its environment. Once removed from
body fluids, however, bones will react rapidly, either disintegrating or al-
tering to become "fossilized." After death, bones acquire trace elements
from the burial environment, and these trace elements form natural environmental tags that provide a direct record of the postmortem history
of a bone.

Geochemical analyses of trace elements are particularly well suited
for the study of bonebeds. Trace element analyses can be used to assess the extent of mixing within a bonebed, to determine the origin of
reworked bones within a mixed assemblage, and to help to distinguish between catastrophic and attritional modes of accumulation. Trace element
analyses have the additional advantage that although they are generally
destructive, they are independent of taxonomy and can be performed
on small indeterminate fragments of bone. Furthermore, trace element
analyses are relatively rapid, and the instrumentation required is now

commonly available in many research institutions. Finally, trace element analyses provide paleoenvironemental as well as taphonomic information. For all these reasons, geochemical techniques are powerful weapons in the taphonomist's and paleontologist's arsenal.

## DIAGENESIS: THE KEY TO GEOCHEMICAL TAPHONOMY

Diagenesis or lithification of one sort or another has featured in many literary and mythological texts. For example, Perseus used the decapitated head of Medusa to defeat the Kraken, Lot's wife was turned into a pillar of salt, and trolls rather conveniently turn to stone in daylight. Many of these literary references to diagenesis demonstrate its usefulness and value (although Lot's views on the subject are not exhaustively recorded). Diagenesis has endured a much more hostile press in the paleontological literature, where it is almost always portrayed as a process that destroys information and renders potentially useful techniques ambiguous at best. However, while diagenetic change in one form or another is inevitable when dealing with fossil bone, these changes can be viewed in a positive light. To understand how diagenetic change can be used as a taphonomic tool, we need to understand what happens to bones during fossilization, and also how chemical elements behave on landscape surfaces. Both of these questions are worthy of books in themselves, and the purpose of this chapter is not to provide an exhaustive (and exhausting) review of bone fossilization or environmental aqueous geochemistry. This chapter is divided into two main sections; the first section provides a basic understanding of early diagenetic trace metal patterns in fossil bones, and the second section outlines the appropriate procedures for conducting analyses of trace metals in fossil bone. Any reader who would like to apply these techniques is strongly encouraged to obtain a more thorough appreciation of the subtleties of bone–trace element interaction through the references provided. Finally, a small section outlining good practice in the practical analysis of trace element geochemistry in fossil bone is provided.

### Fossilization: A Brief Review

Bone is a composite material, composed of crystallites of carbonated calcium phosphate (bioapatite) set in a collagenous matrix. Fresh bone crystallites will dissolve slowly in most phosphorus-undersaturated pore waters and are sparingly soluble in seawater (Nriagu, 1983, Karkanas et al.,

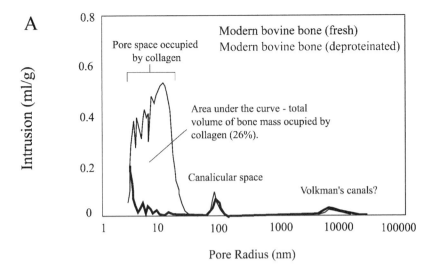

A

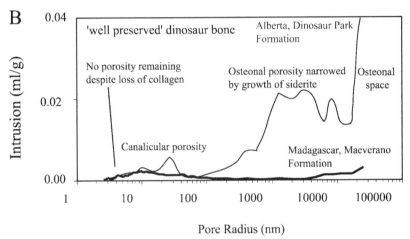

B

Figure 7.1. Porosimetry measurements for fresh and deproteinated modern bovine bone (A) and dinosaur bone from the Cretaceous Dinosaur Park Formation, Alberta, Canada, and Maevarano Formation, Madagascar (B). The porosity of modern cow bone increases by ~26% after collagen is removed. Well-preserved or unpermineralized dinosaur bones that contain essentially no collagen and no minerals other than apatite do not preserve the porosity originally occupied by collagen.

2000; Berna et al., 2004). Consequently, bone crystals are unstable once they are removed from body fluids. Bone crystallites are contained within a matrix of the protein collagen, essentially producing mineralized collagen fibrils. Approximately 30–40% by volume of the substance of bone (not including vascular spaces) is made up of this collagen (Fig. 7.1)

(Nielsen-Marsh et al., 2000), and the vulnerable crystal surfaces are therefore removed from contact with any fluids as long as their protective collagen coating remains.

Fossil bones contain at best trace amounts of collagen (e.g., Wykoff, 1972). Evidently collagen is lost during diagenesis, probably through hydrolysis to its soluble hydrated derivative gelatin (see Collins et al. [1995], Trueman and Martill [2002], and Collins et al. [2002] for a discussion of collagen loss and long-term survival of bone). Hydrolysis of collagen would theoretically increase bone porosity and expose apatite crystallite surfaces and should, therefore, lead to rapid destruction of the bone. In fossil bones, however, the space left by collagen does not remain as pore space but is filled principally by apatite (Fig. 7.1). This suggests that fossilization (or stabilization) of bone occurs when the size of exposed apatite crystals increases, reducing solubility and porosity (Hubert et al., 1996; Trueman and Tuross, 2002). This "recrystallization" or growth of new apatite must occur relatively rapidly, as exposed crystallites would soon dissolve or be exploited biologically in most pore-water environments. It is likely that recrystallization proceeds together with hydrolysis of collagen, although the relative rates of collagen loss and mineral growth are undoubtedly environment specific and are currently unknown. Changes in crystal size and perfection can be monitored optically (Hubert et al., 1996; Pfretzschner, 2000; Tütken et al., 2004), by x-ray diffraction (XRD) (e.g., Tuross et al., 1989; Hiller et al., 2004), or by Fourier transform infrared (FTIR) spectroscopy (e.g., Weiner et al., 1993; Trueman et al., 2004), and bone crystal growth has been shown to occur contemporaneously with degradation of collagen (e.g., Tuross et al., 1989; Roberts et al., 2002; Trueman et al., 2004). Bones exposed on savannah grasslands in Kenya show marked changes in apatite crystal size and collagen content within 5 years postmortem (Tuross et al.1989; Trueman et al., 2004). However, analyses of large numbers of bones from archaeological sites with known ages suggest that it may take from less than a thousand to hundreds of thousands of years for fresh bone to achieve the levels of recrystallization seen in fossil bones. It should be noted that archaeological bone samples do not necessarily provide a good analogue for the processes that led to preservation of most fossil bones, as many archaeological samples are derived from human burials, and the taphonomy of deliberate burial is different from that of natural accumulation and burial.

The exact nature of bone recrystallization, and the effect of recrystallization on the chemistry (particularly the isotopic composition) of bone apatite are currently debated (e.g., Tuross et al., 1989; Sillen, 1990; Barrick et al., 1996; Reiche et al., 2002; Trueman and Tuross, 2002; Trueman et al.,

**(1) Fresh bone** - bioapatite crystallites are arranged parallel to collagen fibres - intra-crystalline porosity is c. 30% of the volume of the bone

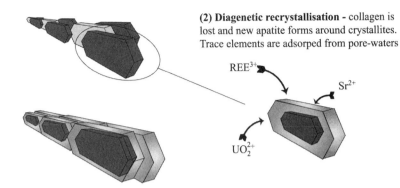

**(2) Diagenetic recrystallisation** - collagen is lost and new apatite forms around crystallites. Trace elements are adsorped from pore-waters

$REE^{3+}$

$Sr^{2+}$

$UO_2^{2+}$

**(3) Fossil bone** - recrystallisation ceases when intra-crystalline porosity is closed by growth of new apatite. Little further change in trace element composition

*Figure 7.2.* Schematic model of bone recrystallization by addition of authigenic apatite. (*Note:* Original bone crystals are on the order of 30 × 30 × 5 nm and are mostly poorly formed plate-shaped crystals of bioapatite.)

2003a; Kohn and Law, 2006; Fricke, Chapter 8 in this volume). Many authors argue that diagenetic alteration is an inevitable consequence of the addition of authigenic apatite into bone, but these authors are unable to demonstrate a source for the extra phosphate needed to recrystallize bone. This is troubling, as only 15% of phosphorus in vertebrates is contained in their soft tissues (an insufficient source to fill bone pore spaces), and any phosphorus contained in sediments is generally bound up by clay minerals or biologic processes (Lucas and Prévôt, 1984). Nonetheless, recrystallization of bone apatite does occur, and this recrystallization results in a reduction of porosity of at least 20–30% by volume (Trueman and Tuross, 2002). In many cases this reduction in porosity is achieved with no significant addition of other authigenic minerals and therefore must be achieved by growth of apatite alone. One view of bone fossilization (Fig. 7.2) is that recrystallization occurs by growth of authigenic apatite within pore spaces (e.g., Hubert et al., 1996; Kolodny et al., 1997). A consequence of this model is that the trace metal and isotopic composition of bone would be altered, with the extent of any alteration depending on the relative amount of biogenic versus authigenic apatite occurring in any single bone (Trueman et al., 2003a).

Apatite has a strong affinity for many elements that are only sparingly present in body fluids, such as U, Th, and the rare earth elements (REE). Fresh bone contains low levels of these elements, as they are present in very low concentrations in the body fluids from which bone forms. Soil pore waters often contain significant concentrations of these elements, however, and once bone is removed from body fluids and exposed to soil pore waters, the concentration of many metals increases dramatically in a relatively short span of time (e.g., Bernat, 1975; Tuross et al., 1989; Shinomiya et al., 1998; Trueman et al., 2004). Any diagenetically altered fossil bone will, thus, contain a geochemical signal related to the environment of fossilization (e.g., Henderson et al., 1983; Wright et al., 1984; Trueman, 1999; Martin et al., 2005). All fossil bones show increases in trace element content that are unequivocally diagenetic in origin, and increases in the trace metal composition of bone postmortem are intimately associated with changes in apatite chemistry and mineralogy (e.g., Tuross et al., 1989; Reynard et al., 1999; Trueman et al., 2004).

The trace element composition of a fossil bone is controlled by a number of variables, including the concentration of trace elements in the pore water and burial sediments, the apatite-fluid partition coefficient, the chemistry of the microenvironment of burial, the hydrology of the microenvironment of burial, the bone microstructure, and the length of exposure (Trueman, 1999). This highlights the fact that the final trace element composition of a bone is a function of many variables, none of which can be known with any certainty. The most important control on the trace element composition of a fossil bone is the trace element composition and chemistry of the pore water in the burial environment. This in turn is controlled by a number of environmental variables, such as (1) source rock, (2) weathering rates, (3) hydrology, (4) climate, (5) redox conditions, and (6) suspended particles. Given these various controls, the trace element composition of fossil bone can vary in response to relatively minor changes in land surface and subsurface conditions. Changes in the chemistry of fossil bones either between or within assemblages may therefore be interpreted in the context of differences in depositional environments. Most geochemical studies of fossil bones employ such a comparative approach.

The total geochemical variation within any assemblage (i.e., the variation in trace element composition between individual bones within an assemblage) can be used as a taphonomic character (e.g., Trueman et al.,

2003b; Metzger et al., 2004). Geochemical variation can be expressed as a function of the amount of mixing:

$$V_{as} = f(M, V_{be}, t_i) \qquad (7.1)$$

where $V_{as}$ is the geochemical variation within an assemblage, $M$ is a measure of mixing (time and space averaging), $V_{be}$ is the variation in early depositional and diagenetic environments, and $t_i$ is the amount of time that each bone spends in different early depositional environments (related to the rate of reworking). $V_{be}$ and $t_i$ are environmental variables that can be controlled by sampling bone assemblages from similar sedimentary facies (ideally assemblages within the same sedimentary system). $V_{as}$ may then be used as a comparative measure of temporal and spatial averaging (Trueman, 1999; Trueman et al., 2003b; Metzger et al., 2004) (Fig. 7.3).

### The Rare Earth Elements

The rare earth elements (REE) are a group of elements that are particularly useful to geochemists. Despite their name, the REE are in fact relatively common in the earth's crust. For example, several of the REE are more abundant in the crust than more familiar elements such as lead. In this overview, the REE are defined as the elements La ($Z = 57$) through Lu ($Z = 71$). The term "REE" generally includes La, Y, and Sc; however, in geological studies, the term often refers to the entire 15 elements in the f-block of the periodic table, with Sc and Y considered with the REE but not within the REE (Gill, 1996). The REE form trivalent ions (with the exception of Eu and Ce); and the ionic radius of the $REE^{3+}$ ion decreases smoothly as atomic number increases (Fig. 7.4). This is known as the "lanthanide contraction," and as the geochemical behavior of elements is governed to a large extent by the ionic radius and ion charge, the relative abundances of the REE in natural systems generally are controlled by ion size considerations.

To be of use to the taphonomist, trace elements, including the REE, must fulfill a number of criteria:

**Criterion 1:** *They must vary significantly between environments.*
In natural waters the REE are typically complexed to a range of charged molecules forming for instance carbonate, phosphate, or humic complexes (e.g., Wood, 1990; Byrne and Li, 1995; Johannesson et al., 1996a, 1996b). The

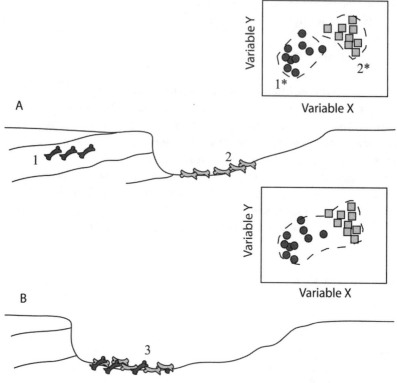

*Figure 7.3.* Principles of geochemical taphonomy. A. Two separate vertebrate assemblages (1, 2) form in different environments, and the geochemical composition of bones from each assemblage is distinct (1*, 2*). B. Reworking of bones from assemblage 1 into assemblage 2 produces an attritional assemblage (3). The geochemical composition of bones within this mixed assemblage reflects both original sources. The variation in geochemical composition of bones within the attritional assemblage increases as more bones are reworked from different initial depositional environments. Reworked bones within the attritional assemblage can be identified and assigned to their original burial associations on the basis of their geochemical composition. Note that schematic bivariate plots are drawn here for convenience, but many geochemical variables can be determined. Multivariate statistical techniques such as MANOVA, discriminate analysis, cluster analysis, and/or PCA should be applied to analyze geochemical data (see Blob and Badgley, Chapter 6 in this volume).

strength of these REE complexes varies, and REE will be more easily removed from solution if the dissolved species is weakly bound. Thus, where REE are held as weak complexes, they will be readily sorbed onto particle surfaces and removed from solution. The lighter REE (known as the LREE) form relatively large ions, whereas the heavier REE (known as the HREE) form relatively small ions (Fig. 7.4). Due to these differences in ion size, the proportion of each of the REE complexed to a particular species varies. In waters with near neutral pH and low organic content, for instance, a greater proportion of the HREE

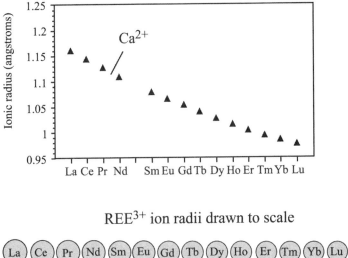

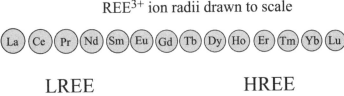

REE³⁺ ion radii drawn to scale

$$La \quad Ce \quad Pr \quad Nd \quad Sm \quad Eu \quad Gd \quad Tb \quad Dy \quad Ho \quad Er \quad Tm \quad Yb \quad Lu$$

## LREE                                        HREE

*Figure 7.4.* A. The "lanthanide effect." The REE form trivalent ions whose radius decreases smoothly with atomic number. This variation in ion radius explains much of the aqueous chemical behavior of the REE.

are held as the strong carbonate complex compared to the LREE, which are preferentially removed from solution

The strength of REE complexes also varies across the REE series. HREE-carbonate complexes are more stable than LREE-carbonate complexes, so even if all REE are held as carbonate complexes, LREE will still be preferentially removed from solution. In alkaline soils, the lighter REE are preferentially sorped onto mineral surfaces and removed from solution, whereas the heavier REE form more stable aqueous complexes and groundwaters (and their associated bones) are relatively enriched in HREE (e.g., Roaldset, 1973; Duddy, 1980; Wood, 1990; Dupré et al., 1996; Morey and Stetterholm, 1997; Braun et al., 1998; Dia et al., 2000). In circum-neutral or acidic waters, REE are held as free ions or commonly as complexes with humic substances (Tang and Johannesson, 2003). The strength of these complexes varies much less across the REE series (Byrne and Li, 1995), so preferential removal of either LREE or HREE is less likely, and waters develop relatively flat REE patterns relative to shale. Subtle changes in pore-water chemistry or REE transport, thus, result in fractionation of the REE (i.e., changes in the relative concentrations of each REE). The direction of any observed fractionation, such as an increase in relative enrichment of HREE, can be used to infer the underlying cause and thus give paleoenvironmental information (Fig. 7.5).

**Criterion 2:** *They must not be common (present) in living tissue.*
The REE have no known physiological function, and as a consequence are present in low levels in living bone (typical levels are ~0.1 ppm total REE). Furthermore, as the REE are not concentrated significantly in any major food source, there are no known taxomonic effects on the REE composition of bone.

**Criterion 3:** *They must be incorporated rapidly and easily into bone postmortem.*
Ichthyoliths (ancient phosphatic fish debris) provide dramatic evidence for rapid postmortem uptake of REE in bones and teeth. Modern fish teeth and bones typically contain ~1 ppm REE (Bernat, 1975; Shaw and Wasserburg, 1985; Elderfield and Pagett, 1986). However, teeth recovered from the uppermost 4 cm of sediment cores in the deep Atlantic and Pacific Oceans (and therefore less than 10,000 years old) contain ~1000 ppm REE (Goldberg et al., 1963; Bernat, 1975; Elderfield and Pagett, 1986). Furthermore, there is no systematic increase in the REE concentrations in ichthyoliths with depth, suggesting that the majority of the REE are incorporated within the first few thousand years postmortem (Elderfield and Pagett, 1986; Martin and Haley, 2000). Fossil bones from all depositional locations typically contain similar REE concentrations, although REE concentrations in some terrestrial localities can reach several weight-percent (Denys et al., 1996). Finally, the REE composition of ichthyoliths largely mirrors the bottomwater composition (Goldberg et al., 1963; Bernat, 1975; Elderfield and Pagett, 1986; Shaw and Wasserburg, 1985; Lécuyer et al., 2004; but see Shields and Webb [2004] for a contrasting opinion). Ichthyoliths recovered from the deep Atlantic have different REE patterns than ichthyoliths recovered from the deep Pacific, and both are different from ichthyoliths recovered from shallow shelf locations (Elderfield and Pagett, 1986). This observation suggests that bones inherit their REE directly from overlying (or pore) waters with no fractionation (Grandjean et al., 1987; Grandjean-Lecuyer et al., 1993; Girard and Albarède, 1996; Holser, 1997) and, thus, record the chemistry of the waters within which they were buried (Fig. 7.6). Further work has suggested that there may be some other factors that control the final trace element composition of a fossil bone (e.g., Reynard et al., 1999; Lécuyer et al., 2003; Shields and Webb et al., 2004), but these late diagenetic processes do not appear to be common in the fossil record.

**Criterion 4:** *They must not be susceptible to appreciable fractionation after initial incorporation into bone.*
Bone is a chemically open system, with high intracrystalline porosity, and is therefore susceptible to diagenetic exchange with pore waters. During

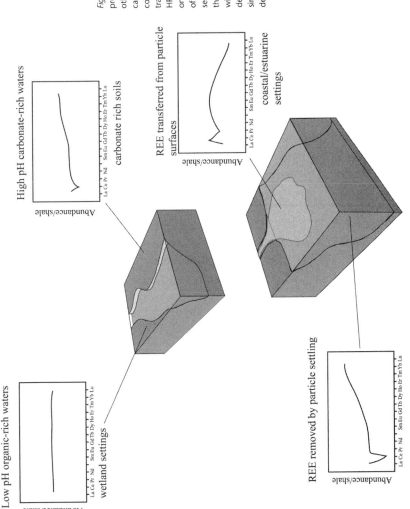

*Figure 7.5.* Cartoon summarizing some important processes that fractionate the REE from one another during transport and weathering. The REE can be transported as dissolved ions, complexes, or colloids or adsorbed onto particle surfaces. These transport mechanisms may favor either LREE or HREE, and so the REE may be fractionated from one another depending on the dominant method of REE transport. All plots are schematic representations of shale-normalized REE abundances that might be developed in bones in contact with differing pore-water composition. Typical depositional settings are indicated. Note that no single REE pattern is definitive for any particular depositional setting. Compare with Figure 7.7.

High pH carbonate-rich waters
carbonate rich soils

Abundance/shale

La Ce Pr Nd   Sm Eu Gd Tb Dy Ho Er Tm Yb Lu

REE transferred from particle surfaces

coastal/estuarine settings

Abundance/shale

La Ce Pr Nd   Sm Eu Gd Tb Dy Ho Er Tm Yb Lu

Low pH organic-rich waters

wetland settings

Abundance/shale

La Ce Pr Nd   Sm Eu Gd Tb Dy Ho Er Tm Yb Lu

REE removed by particle settling

open marine settings

Abundance/shale

La Ce Pr Nd   Sm Eu Gd Tb Dy Ho Er Tm Yb Lu

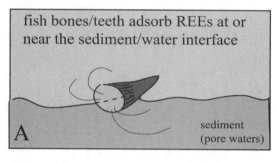

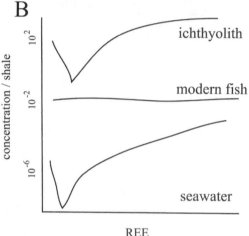

*Figure 7.6.* REE uptake in ichthyoliths. A. Fish debris lying on the ocean floor adsorb REE rapidly from the overlying seawater or the uppermost pore waters. B. Schematic REE patterns developed in seawater, modern fish teeth, and ichthyoliths. The concentrations of the REE in ichthyoliths buried for <4 Ka are much higher than those in modern seawater or modern fish teeth, but the relative abundance resembles that of the overlying seawater indicating that REE are incorporated into ichthyoliths postmortem directly from overlying waters with little fractionation (Bernat, 1975; Elderfield and Pagett, 1986).

recrystallization this porosity is closed, however, reducing the potential for ion exchange (e.g., Millard and Hedges, 1999). Once recrystallization is complete, further exchange with pore waters is limited to solid-state diffusion, and the fossil bone is a closed system, suffering little or no late diagenetic alteration (e.g., Trueman and Tuross 2002). Several studies have demonstrated successfully that the trace element and isotopic composition of bone developed during early diagenesis is preserved throughout later diagenesis. For example, Keto and Jacobsen (1987) and Scher and Martin (2006) were able to distinguish paleoceanic water masses on the basis of Nd signatures in Cambrian and Ordovician conodonts and Eocene fish teeth, respectively. Wright et al. (1984) used cerium anomalies in conodonts and ichthyoliths to reconstruct redox conditions in Paleozoic oceans, and in a series of studies, Lécuyer and colleagues have used

REE profiles in ichthyoliths to reconstruct ocean chemistry through time (e.g., Grandjean et al., 1987; Picard et al., 2002; Lécuyer et al., 2004). Trueman and Benton (1997) and Staron et al. (2000) showed that REE signals in bone survive reworking into contrasting sedimentary environments, and, in the terrestrial realm, Williams et al. (1997) used REE patterns in Pleistocene bones to distinguish between oxidizing and reducing burial environments in Olduvai Gorge. In contrast, Reynard et al. (1999) suggested that later diagenetic recrystallisation could have severe effects on the REE patterns of ichthyoliths, but that these changes should be recognizable. However, Armstrong et al. (2001) showed that relatively severe levels of thermal metamorphism (heating to temperatures of $\sim$350–500°C) where needed to alter the REE pattern developed in conodont elements. There is thus good theoretical and empirical evidence to suggest that that the REE signals developed so readily in bones during early diagenesis are retained through their later diagenetic history.

In summation, the rare earth elements fulfill all of the criteria required for use in geochemical taphonomy. Additionally, the REE are relatively simple to measure with modern analytical equipment (see the section on good practice for a recommended analytical protocol). The REE are not the only trace elements that could be used (or have been used) to reconstruct paleoenvironments or taphonomy from fossil bones. However, many other elements either have a physiological function or are very insoluble and consequently rare in many porewater systems. Additionally, much less is known about the behavior of other trace elements in the weathering-transport-bone system.

## APPLICATIONS IN THE FOSSIL RECORD

### Determining Provenance

Taphonomists would dearly like to be able to group any bones sharing a common initial burial location, and if possible to identify that particular location. This would then allow a quantitative assessment of the extent of mixing within attritional assemblages and would allow bones of disputed provenance to be assigned correctly their stratigraphic and/or geographic origin. Geochemical provenance techniques offer this possibility.

Provenance studies are familiar to geologists, archaeologists, and forensic chemists who have used geochemical fingerprints to (1) link lavas and volcanic ash layers to their magma chambers, (2) thwart unscrupulous dealers attempting to pass off cheap wines as wines from the great vineyards, and (3) trace toxic spills in rivers back to their source. Geochemical provenance techniques have also been used to police the illegal trade in animal products (e.g., Van der Merwe et al., 1990 Amin et al., 2003) and to establish the identity or geographic origin of unknown human remains (e.g., Beard and Johnson, 2000; Hoogewerff et al., 2001; Pye, 2004). From a taphonomic perspective, many attritional deposits are formed by the gradual concentration of bones and teeth over long periods of time, and bonebeds may contain bones from more than one early diagenetic environment (Fig. 7.3). Geochemical tests of provenance can thus be applied to bone assemblages to (1) determine whether an assemblage contains bones derived from more than one early diagenetic environment, and (2) identify the geographic and stratigraphic origin of reworked bones within an assemblage. In the case of vertebrate assemblages, the trace element chemistry of bones from a particular assemblage is determined and analyzed statistically to test whether the bones form one continuous population or several discrete populations. If the bones can be split into several populations on the basis of their trace element signals it is likely that the assemblage was derived from more than one discrete source (Fig. 7.3). In this case the trace element composition of bones from each of these separate populations could be compared to that of bones recovered from stratigraphically or geographically adjacent assemblages.

A clear demonstration of provenance analysis using fossil bone trace element geochemistry was provided by Plummer et al. (1994). This study addressed a common problem in paleontology, namely, assigning surface collected or otherwise poorly assigned fossil bones to their correct stratigraphic level. Plummer et al. (1994) investigated surface-collected fossils from the early to late Pleistocene Kanjera Formation of western Kenya. This formation presented a particularly significant problem, as several anatomically modern hominid fossils were recovered from surface collection within the Kanjera Formation and assigned to a middle Pleistocene age. If true, this would represent some of the oldest anatomically modern fossils known (Plummer et al., 1994). However, the Kanjera Formation may span 1 million years, and the provenance of the hominids has remained controversial. To resolve this problem, Plummer et al. (1994) used X-ray fluorescence techniques to analyze the trace element composition of a suite of 261 bones and 40 teeth collected in situ from six major stratigraphic units. The concentrations of seven elements (U, Rb, Rn, Zr, Th, Sr, Y) were determined, and although the authors did not record their reasons for choosing this suite of elements, it is likely that these elements were the easiest to detect with their analytical techniques. The data set was then subjected to principal components analysis (PCA), which resulted in the differentiation of two main groups.

A sample of 289 bones and teeth of unknown provenance (including the disputed hominid fossils) was then analyzed using the same analytical techniques and equipment. Fossils were assigned to specific beds based on the similarity in their PCA scores to those of the control suite of samples with known provenance. The results of these analyses were very clear: 74.4% of the samples were assigned to a specific bed based on their PCA scores, and there was no overlap between the fossil hominid samples and the rest of the Kanjera Formation samples. The hominid fossils largely overlapped the range described by bones from overlying Holocene sediments. By contrast, a large sample suite of *Theropithecus* bones grouped together in a very well-defined cluster that itself characterized a particular bed. This study clearly shows the potential for trace element geochemistry to discriminate between different early burial environments, and thus to assess the provenance of bones of unknown or disputed origin. It should be noted however, that much of the variation seen between beds in the Kanjera Formation is described by variation in the ratio of uranium to other elements—essentially a function of redox conditions. Using a more comprehensive (or different) suite of trace elements may have provided more detailed discrimination between sample lithologies, but this would have required a different analytical technique.

The REE are particularly well suited to provenance analyses, as minor differences in pore-water chemistry result in fractionation within the REE series. As discussed above, the REE pattern developed in a fossil bone is controlled by the chemistry and chemical evolution of the pore water in the burial environment. In waters with high pH, $CO_3^{2-}$ complexes dominate, and the LREE are preferentially removed onto particle surfaces. These waters and any bones recrystallizing in contact with them become enriched in HREE. In waters with neutral or acidic pH, or very high organic contents, REE patterns are relatively flat compared to shale. Where REE are supplied by particle surfaces, pore waters and their associated bones inherit LREE-enriched patterns. This is perhaps most easily demonstrated by plotting the shale-normalized REE composition of a large number of fossil bones recovered from a range of depositional environments (Fig. 7.7). Fossil bones from different geological settings are grouped according to their depositional setting, so that bones from coastal marine environments are relatively enriched in LREE and mostly (but not exclusively) plot in different areas to those from deep marine or terrestrial settings, which are relatively enriched in HREE. Bones from estuarine settings plot intermediate between soil and coastal environments, and bones from aeolian and desert environments, where REE are supplied through particle

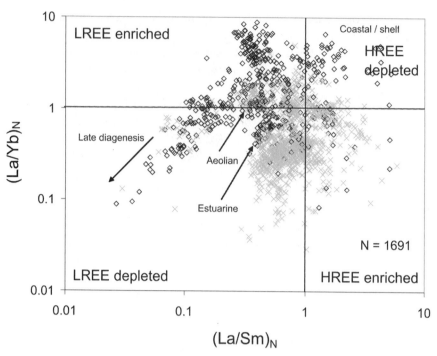

*Figure 7.7.* Relative abundance of REE in >1500 bones, teeth and conodont elements from a range of depositional settings and ages, compiled from published analyses. Open symbols, bones from marine environments; crossed symbols, bones from terrestrial settings. REE patterns are often displayed in bivariate plots of ratios of representative light, middle, and heavy REE (e.g., Reynard et al., 1999; Trueman and Tuross, 2002) or as ternary diagrams (e.g., Patrick et al., 2002). Bones, teeth and conodont elements from coastal marine settings group toward the top left side in this REE ratio plot, reflecting patterns enriched in light and middle REE. Bones from terrestrial settings plot toward the lower portions of the diagram, reflecting relatively low levels of LREE. Bones from carbonate-rich terrestrial soils are particularly enriched in HREE and commonly plot toward the right-hand side of the diagram. These differences reflect differing levels of fractionation of REE from one another during aqueous transport in pore waters of varied composition. After Trueman and Tuross [2002] and Trueman et al. [2006]. Compare with Figure 7.5.

surfaces, are relatively enriched in LREE compared to most other terrestrial environments. According to the general model presented above, bones fossilized in carbonate-rich overbank and paleosol facies should yield bones that are enriched in HREE compared to bones from facies with less alkaline pore waters. This is indeed the case in bones from the Campanian Two Medicine and Dinosaur Park formations of Montana and Alberta, respectively (Fig. 7.8). Within paleosols and associated sediments from the Oligocene Brule Formation of South Dakota, Metzger et al. (2004) demonstrated that bones recovered from oxbow lake facies could be distinguished from those recovered from overbank and paleosol facies through their REE signals (presumably due to differences in pH and organic content of the

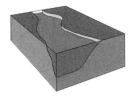

Dinosaur Park Formation (DPP) - fluvially dominated - bones removed rapidly into fluvial channels

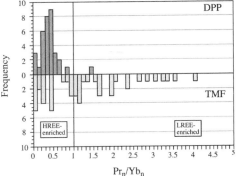

Two Medicine Formation (TMF) - soil dominated - bones remain in soil profiles

Figure 7.8. REE uptake in bone in terrestrial environments. Bones from carbonate-rich soil facies are relatively enriched in HREE (e.g., Yb) compared to bones from fluvial facies. Data from Trueman (1999).

associated pore waters). The possibility of using the REE composition of bones to address the problem of identifying the source location of bones removed illegally from protected lands was also explored by Patrick et al. (2002), Metzger et al. (2004), and Trueman (2004).

### Recognizing Reworking

Resistant skeletal elements such as teeth may survive multiple cycles of sediment transport, deposition, and erosion, finally ending up in sediments as mixtures of fossils with varying ages. Clearly such mixtures of fossils may give misleading paleontological and paleoecological information, particularly where the stratigraphic or paleontologic range of a particular taxon is extended (e.g., within a transgressive lag deposit). Fluvial and shoreline systems are particularly susceptible to this kind of reworking (e.g., Rogers and Kidwell, 2000). Geochemical taphonomic techniques are ideally suited to identifying allochthonous bones or teeth, as by definition they are derived from different depositional settings (note that the terms "allochthonous" and "autochthonous" here refer to where bones were originally buried with respect to a fossil assemblage, not necessarily where the animals lived).

The critical advantage offered by geochemical taphonomic analysis lies in the ability to test the provenance of individual bones within an assemblage. An example of the use of REE taphonomy to test the stratigraphic integrity of a controversial vertebrate assemblage was provided by Trueman et al. (2005). Late quaternary extinctions of large vertebrate taxa (megafauna) are widespread and broadly coincide with human colonization, leading to hypotheses suggesting that extinctions are directly attributable to human activity. In Australia, the most extreme version of these hypotheses (the blitzkrieg or overkill hypothesis) suggests that megafaunal extinctions were caused by hunting or possibly habitat alteration by humans and were complete on a continental scale within a few thousand years of human arrival. Determining the role of humans in the Australian Pleistocene extinction requires an accurate and high-resolution chronology of megafaunal and human remains. Unfortunately the late Pleistocene sedimentary record in Australia is limited, and most vertebrate remains are found in attritional deposits. In some cases megafaunal remains have been found in sediments dated considerably younger than the proposed extinction date of $\sim$46 Ka. Most notable is the site of Cuddie Springs, southeastern Australia, where megafaunal remains have been found in association with cultural artifacts in sediments that significantly postdate 46 Ka. However, as the vertebrate assemblages in these deposits are largely disarticulated, proponents of the blitzkrieg hypothesis suggest that all megafaunal bones found in sediments significantly younger than 46 Ka must have been reworked. To test this, Trueman et al. (2005) measured the REE, U, and Th composition in bones from extinct and extant vertebrates from a sequence of attritional deposits from Cuddie Springs, spanning the proposed 46 Ka extinction date. The trace element composition of bones from each sampled stratigraphic layer contained a distinct geochemical signature. Based on multiple ANOVA analyses, there was a <0.1% probability that bones (including remains of megafauna) recovered from sediments dated at $\sim$36–30 Ka were reworked from underlying deposits. These results demonstrate that the megafauna persisted at least in some regions of south-central Australia for at least 10 Ka after the arrival of humans and argue against the blitzkrieg hypothesis.

In their study of bones from Kanjera, Kenya, Plummer et al. (1994) found two geochemically distinct groups of bones recovered from above an erosive unconformity. One group was unique to stratigraphic layers above the unconformity, and one group had trace element compositions similar to bones from layers immediately below the unconformity. Plummer et al. (1994) concluded that the presence of bones with anomalous geochemical signals above the unconformity was due to reworking. This conclusion was important as it proposed a definitive test for reworking within complex assemblages. A slightly different approach was taken by Trueman and Benton (1997), who used the REE patterns in bones from Triassic coastal marine bonebeds in the southwest of England to test taphonomic pathways proposed on the basis of sedimentological and paleogeographic evidence. This study showed that it is possible to infer a parent deposit for reworked bones when the original parent deposit has been lost.

The same technique was used by Staron et al. (2001) to locate the K/T boundary in marine sediments. The marine K/T transition in Sewel, New Jersey, is embedded in two units, namely, the Navesink and overlying Hornerstown formations. Within the Hornerstown Formation, approximately 10 cm above the basal contact, there is a bonebed known as the main fossiliferous layer (MFL). This bonebed contains abundant fossils that are exclusively Cretaceous in age, such as mosasaurs and the turtle *Peritresius ornatus*. The location of the K/T boundary is controversial in this succession, with some workers placing the boundary at the Navesink-Hornerstown contact, and others placing the boundary within or above the MFL. Traditional methods of locating the K/T boundary, such as microfossil stratigraphy and chemostratigraphy, have yielded inconclusive results. If the K/T boundary is placed at the Navesink-Hornerstown boundary, then all fossils within the MFL must have been reworked from the underlying Navesink Formation. Staron et al. (2001) analyzed the REE composition of disarticulated mosasaur bones and partially articulated turtle and crocodile bones from the Navesink and Hornerstown formations and the MFL bonebed. The results were subjected to discriminant and hierarchical clustering statistical techniques in an attempt to determine whether the bones from the MFL grouped with the Navesink Formation, which would indicate reworking. Discriminant analyses successfully assigned 18 out of 22 fossils to their respective lithological unit. Ten out of 11 fossils recovered from the MFL form a well-defined discrete group that does not overlap with either the Navesink or Hornerstown assemblages. The single bone from the MFL that could not be assigned by discriminant analysis was anomalously rounded with no cortical bone preserved, suggesting that it may have been reworked. Hierarchical clustering of the data produced similar results, again grouping all MFL fossils. In this study, the

REE data clearly show that the MFL does not contain bones reworked from the underlying Navesink Formation. Instead the MFL assemblage appears to be in place and represents a relatively unique depositional event. The K/T boundary should accordingly be placed either above the MFL within the Hornerstown Formation, or within the MFL.

All geochemical techniques designed to test for provenance or reworking require that bones must (1) be derived from geochemical environments that are significantly different from one another, and (2) remain in their original depositional environment long enough to acquire a stable, unique depositional signal. The time scales of uptake undoubtedly vary from location to location, and work is ongoing to determine rates of uptake in different environments. Of particular note is the fact that the rate of recrystallization of bone can be inferred by determining the concentrations of elements such as uranium in profiles through bone cortices, and comparing the slope of these concentration profiles with mathematical models describing the rate of diffusion and adsorption of elements through bone (e.g., Millard and Hedges, 1999; Pike et al., 2002; Trueman and Tuross, 2002). Such studies show that many fossil bones were completely recrystallized within 1–10 Kyr, indicating that this is the limit of temporal resolution that could be achieved using trace element techniques. This time scale is in agreement with the observation that ichthyoliths recovered from shallow depths within deep sea sediment cores already contain REE concentrations similar to those seen in fossil fish teeth (Bernat, 1975; Elderfield and Pagett, 1986; Martin and Haley, 2000), and with recent estimates of the rates of recrystallization based on the stable isotope composition of fossil bone (Kohn and Laws, 2006). The high temporal resolution suggested by these rapid rates of recrystallization is supported by field studies. In Pleistocene paleosol sediments from the Olorgesailie Formation, Kenya, Trueman et al. (2006) successfully used the REE composition of bone to differentiate between bones from single excavations separated in time by ~1000 years.

The limits of spatial resolution available to REE geochemical studies are more difficult to quantify. The limit of spatial resolution available is a function of the rate of recrsytallization of bone, the amount of post-depositional movement of bone, and the original variation in landscape chemistry. Where conditions are favorable, taphonomic techniques employing bone chemistry may achieve extremely high spatial resolution. For example, in their study of bones from Pleistocene paleosol sediments, Trueman et al. (2006) were able to resolve geochemical differences in bones separated by less than 10 m laterally. The sampling strategy employed will also have a large effect on the spatial resolution that can be obtained. In almost all studies known to the author, significant differences in the REE composition of bones between depositional localities and stratigraphic horizons have been found.

### Using Trace Element Variation to Characterize an Assemblage

Attritional bonebeds represent temporally or spatially averaged samples that potentially mix faunal elements that did not coexist in the life or in the death assemblage. Recognizing this character of attritional bonebeds is extremely important, as the diversity and faunal composition of bonebeds is a function of the method and rate of accumulation (time and space averaging). Assemblages that mix bones from a wide temporal and spatial range are more likely to sample rare faunal elements or mix temporally successive species and therefore will tend to record high faunal diversities compared to snapshot or census samples. It is important, therefore, to quantify the amount of time and space represented in each attritional assemblage so that characters such as diversity are only compared among taphonomically equivalent (isotaphonomic) assemblages. Existing taphonomic methods such as analyses of bone surface textures and the hydrodynamic classes to which recovered bones belong can indicate whether an assemblage is attritional and likely contains reworked elements (e.g., Hill, 1980; Behrensmeyer, 1978, 1990; Fiorillo, 1988; Lyman, 1994), and sedimentological analyses may constrain the maximum temporal span of mixing in any attritional bonebed (Behrensmeyer et al., 1992). These techniques cannot, however, be used to compare the *relative* amount of mixing between attritional vertebrate assemblages.

The trace element geochemistry of fossil bones provides a method for assessing the relative degree of mixing between vertebrate accumulations. It has already been shown that allochthonous bones may often be identified in vertebrate assemblages, as they have a unique trace element signal. In other words, adding a derived bone to an assemblage increases the overall variation in trace element geochemistry, and the total variation in geochemical signals within a bonebed is a function of the number of early burial environments sampled (Fig. 7.3). The total variation in the trace element composition of bones from a bonebed can thus be used as a quantitative character related to the amount of potential mixing (time and space averaging).

Before this is illustrated with an example, it is important to remember that the variation in trace element chemistry of bones in an assemblage is also related to the heterogeneity of the local diagenetic environment (see

Equation 7.1). Because the deep ocean and large lakes are chemically homogenous environments, their bone assemblages are likely to be relatively homogenous geochemically when compared to terrestrial assemblages, irrespective of the amount of taphonomic mixing. Along these lines, Patrick et al. (2004) showed that the REE composition of mosasaur bones recovered from discrete stratigraphic horizons within the upper Cretaceous Pierre Shale of South Dakota is uniform over lateral distances of ~250 km. Similarly, Martin et al. (2005) showed that the REE compositions of bones recovered from single depositional units within Pleistocene lake sediments from south-central Oregon are uniform over lateral distances of up to 16 km. Variation in bone chemistry should only be used, therefore, to compare mixing between two assemblages where there is a good geological reason to believe that the depositional and diagenetic environments of the assemblages were equivalent (Trueman et al., 2003b; Metzger et al., 2004). It is also important to use randomized or indeterminate samples for these geochemical taphonomic tests; any attempt to select bones by taxonomy or surface condition could bias the results significantly.

To test whether variation in geochemical signals could be used as a measure of averaging, Trueman et al. (2003b) compared REE compositions in previously analyzed conodonts from single depositional horizons with those from artificially averaged assemblages. Simple analysis of variance based on the F-test showed that artificially averaged assemblages indeed showed significantly more chemical variation than assemblages from single depositional horizons. This approach was then extended to field samples. Marine sediments of the Triassic Westbury Formation in southwest England (e.g., MacQuaker, 1994) provide an excellent test for geochemical taphonomic techniques. The Westbury Formation contains many attritional bonebeds whose character varies dramatically both laterally and temporally. Two of the best-studied bonebeds (e.g., MacQuaker, 1994; Storrs, 1994; Benton et al., 2002) are found at Aust Cliff and Westbury Garden Cliff, on opposing sides of the River Severn. These two bonebeds are roughly contemporaneous and are separated laterally by less than 32 km, but each has a very different taphonomic history (Trueman and Benton, 1997). The basal bonebed developed at Aust Cliff is an event deposit formed by storm-driven reworking of preexisting attritional accumulations. By contrast, the basal bonebed at Westbury Garden Cliff is a restricted condensation deposit formed by low-energy winnowing in a relatively shallow coastal setting (e.g., Rogers and Kidwell, 1992). As predicted from the sedimentological and physical taphonomic evidence, the REE compositions of bones from the Aust Cliff assemblage are significantly more varied geochemically than those from the Westbury Garden Cliff assemblage (Trueman et al., 2003b).

A similar approach can be applied to terrestrial bone accumulations. Compared to marine environments, terrestrial landscapes are typically composed of a mosaic of discrete microenvironments, leading to localized differences in burial geochemistry and greater potential for recognizing postdepositional movement and mixing of bones. In their study of paleosols from the Brule Formation, South Dakota, Metzger et al. (2004) also sampled bones from floodplain, channel, and oxbow lake environments for REE composition. Their data confirmed what was suggested by the sedimentological evidence: fossils from channel lags showed relatively high levels of geochemical variation equivalent to large amounts of time and space averaging; fossils from overbank paleosols (particularly rapidly buried paleosols) showed modest levels of geochemical variation reflective of lower levels of mixing; and fossils from abandoned channel–oxbow lake settings showed the lowest geochemical variation indicative of the lowest levels of mixing. Significantly, bones from two distinct overbank facies exhibited similar levels of variance in their REE composition, suggesting that these represent isotaphonomic accumulations.

A similar example is provided by the attritional assemblages of Cuddie Springs, Australia (Trueman et al., 2005). In this sequence of claypan sediments, a single bone-rich conglomerate shows sedimentological features indicating relatively high energy deposition, presumably reflecting deposition after a local flooding event. In contrast, all other assemblages (including assemblages containing megafaunal remains in association with sediments dated at 36–30 Ka and cultural artifacts) are hosted in fine-grained sediments reflecting low-energy deposition within a lake or claypan environment. Bone samples from the conglomerate show relatively high levels of geochemical variation (and thus, time averaging) compared to other bone assemblages within the sequence.

A final example of the use of variance in geochemical signals as an indication of the amount of taphonomic mixing can be drawn from the study of Pleistocene paleosol deposits of the Olorgesailie Formation, Kenya (Potts et al., 1999). Bones were sampled from multiple localities within a single paleosol, estimated as representing 1000 years in time. Bones recovered from small channel bodies show significantly greater geochemical variation than bones recovered from adjacent excavations separated laterally from the channel facies by less than 10 m (Trueman et al., 2006).

## Catastrophic or Attritional Accumulation?

Bone accumulations can form extremely rapidly, especially if they result from a single killing agent. These catastrophic vertebrate accumulations (CVAs) may pass relatively unaltered into the fossil record, resulting in bonebeds with very low temporal and spatial mixing. Identifying bone accumulations that may have formed as CVAs is important, as such event beds provide a unique sample of a population at a particular point in time. However, recognizing CVAs is not always simple, particularly in the case of scattered disarticulated remains. In many cases a catastrophic origin for a bone accumulation is inferred because of monotaxic accumulations, or association with clearly catastrophic sedimentary events such as airfall tuffs or floods. Catastrophic events do not, however, always trap monotaxic populations and often do not result in immediate burial. The fossilized remains of catastrophic accumulations that are not immediately buried may resemble time-averaged deposits due to in situ trampling, bioturbation, phytoturbation, and differential weathering. Furthermore, monotaxic or low-diversity assemblages could form over long periods of time under some ecological scenarios (Rogers and Kidwell, Chapter 1 in this volume). Geochemical taphonomic methods can be used to compliment sedimentological and physical taphonomic analyses to provide independent evidence for the rate of accumulation of a bonebed assemblage.

The *Maiasaura* hadrosaur bonebed in the Two Medicine Formation of Montana is essentially monotaxic ("monodominant" in the terminology of Eberth et al. [Chapter 3 in this volume]) and is a reasonable candidate for a catastrophic death assemblage (Hooker, 1987; Horner and Gorman, 1988). Bones from this bonebed are disarticulated, and the bed itself is interpreted as a mass-flow deposit (Schmitt et al., 1998). Because bones were introduced into the deposit from a preexisting land surface, it is unclear whether they were derived from a preexisting high concentration of bones on that land surface (perhaps after a mass-death event) or if they represent a secondary concentration of bones during erosive weathering and reworking. Isolated bones recovered at various intervals above and below the bonebed have contrasting REE profiles, and it is therefore possible to constrain the amount of reworking and mixing in the bonebed itself. If bones were introduced into the bonebed by extensive reworking of prefossilized underlying deposits, the geochemistry of bones from the bonebed and the lower deposits should be similar. If, on the other hand, the bonebed preserves only bones from contemporaneous units, then the geochemistry of bones in the bonebed should resemble those from adjacent deposits. Alternatively, the bonebed may have sampled material from a unique burial environment, in which case the bones from the bonebed may not correspond to either population.

The REE composition of bones from the Willow Creek Anticline bonebed fall entirely within the range of bones from contemporaneous levels and partially overlap the range of bones from lower levels (Trueman, 1999). The variation in REE composition in bones from the bonebed is, however, significantly less than that of bones recovered from both the contemporaneous and earlier deposits, and none of the bones sampled from the bonebed fall in the range of bones sampled from channel sands in contemporaneous levels. Thus, the REE evidence suggests that the bonebed represents a restricted sample. The REE evidence adds weight to the argument that this bonebed formed as a mudflow swept over a land surface littered with bones shortly after a mass-death event.

A similar approach was used in an analysis of bonebeds from the Dinosaur Park Formation of Alberta. Both monodominant and multitaxic bonebeds are found within the Dinosaur Park Formation, and it is difficult to tell whether differences in taxonomic diversity of the bonebeds reflect taphonomic differences (rate of accumulation, time averaging) or paleoecologic differences (low-diversity faunal associations). In a pilot study (Trueman, 1999), bones from a multitaxic bonebed were shown to be significantly more varied in REE composition than bones from a monodominant bonebed. If one accepts the assumption that the range of early diagenetic environments in the local landscape was equivalent between the two assemblages, then this data suggests that the monodominant bonebeds are less averaged than the multitaxic bonebeds and may represent catastrophic mass-death events, whereas multitaxic bonebeds probably do not. Again, the variation in REE chemistry provides a method to identify isotaphonomic bonebeds, which can then be used to study diversity and faunal associations through time.

### Paleoenvironmental Reconstruction

The REE composition of fossil biominerals was first investigated as a potential proxy for ancient seawater composition, and a large literature exists describing REE patterns in ichthyoliths (fish debris) and conodonts. For example, Wright et al. (1984) used the cerium chemistry of conodonts to infer changes in ocean redox state, and Lécuyer and colleagues have inferred major changes in seawater chemistry over the Phanerozoic based on REE patterns in ichthyoliths (e.g., Lécuyer et al., 2004). The extent to which ichthyoliths and conodonts record seawater rather than sediment pore-water chemistry has been questioned (German and Elderfield, 1990; Shields and Webb, 2004). If fossilization is rapid with respect to deposition, then the REE composition of the ichthyolith may be controlled by the overlying seawater. If, however, deposition rates are rapid with respect to fossilization, then the REE composition of the ichthyolith will reflect the REE composition of sediment pore waters (e.g., Kemp and Trueman, 2003; Martin et al., 2005). In principle, as the mobility of the REE and their availability for uptake into bone is controlled by the composition and chemistry (particularly pH) of the ambient fluid, variations in REE patterns in fossil bones either laterally or stratigraphically within a depositional sequence can be used to infer changes in pore-water or groundwater chemistry.

A good example of the use of fossil bone chemistry to infer paleoenvironmental changes is provided by Martin et al. (2005), who measured the REE composition of bones and teeth from the Pleistocene Fossil Lake area, south-central Oregon. By comparing REE patterns in fossil biominerals and a suite of natural waters, Martin et al. (2005) inferred that lake waters became progressively more alkaline and saline with time, presumably reflecting increasing aridity. This interpretation was supported by decreasing taxonomic diversity and an increase in the abundance of species indicative of arid conditions.

Numerous workers have derived paleoenvironmental information related to aridity, salinity, and water mass circulation from the REE composition of fossil bones and teeth, including Samoilov and Benjamini (1996), Picard et al. (2002), Kemp and Trueman (2003), Patrick et al. (2004), and Anderson et al. (2006). The geochemical composition of fossil bones and teeth can clearly be used to derive both taphonomic and paleoenvironmental information from vertebrate assemblages.

## SUMMARY

Many trace metals are rapidly introduced into postmortem bones, and these metals are retained during later diagenesis. The trace metal compositions of fossil bones within vertebrate assemblages reflects the early diagenetic environment and can be used as an effective taphonomic indicator. No other method of analysis can test whether specific bones within an assemblage are reworked (allochthonous), assess the provenance of bones of disputed origin, or compare quantitatively the relative amount of mixing within attritional assemblages. Geochemical taphonomic techniques should be considered as complimentary to existing taphonomic techniques based largely on surface modifications, sedimentology, and hydrodynamic considerations. Geochemical and sedimentological taphonomic investigations are mutually independent, so inferences drawn from coupled geochemical and sedimentological analyses are more robust than those drawn from any single approach. The environmental causes of changes in the REE composition of groundwaters and bones are predictable, and measurements of bone chemistry made for taphonomic investigations can therefore also be used to infer paleonenvironmental information, and vice versa—at no extra cost!

It is hoped that this contribution will spur other workers to delve more deeply into their bonebeds, to explore the variety of issues that only geochemical techniques can resolve, and to test the limits of this powerful method of analysis.

## GOOD PRACTICE: A PRIMER ON THE ANALYSIS OF TRACE METALS IN FOSSIL BONE

The analytical practicalities of determining trace element concentrations in bones are relatively simple, and it is possible to go from a pile of 100 dusty bone fragments to a table of their REE concentrations in a few days. The following section provides a basic guide to the analyses of the trace element geochemistry of fossil bone. It is not meant as a comprehensive or definitive instruction book, as the practicalities of each site and scientific question will force revisions and modifications, but hopefully it will be useful to those taking a first step into the field of geochemical paleontology. Good luck!

The trace element chemistry of fossil bone is complex and potentially controlled by many variables. There is little to be gained from adopting a "scattergun" or "suck-it-and-see" approach to sampling for trace element analyses, as it is virtually impossible to identify the cause of geochemical variation. Perhaps the most effective use of trace element composition of fossil bone in the study of bonebeds is to test hypotheses regarding mixing and provenance constructed from sedimentological, stratigraphic, taxonomic, and physical taphonomic analyses. These hypotheses will dictate the sampling protocol necessary and sample size required. This section outlines the basic protocol for the analysis of trace metals in fossil bone.

## Sedimentology

There is no point in spending time and money collecting geochemical data unless the sedimentary context of the bones is well constrained (unless of course one is trying to determine the provenance of a subset of bones). The first step in any geochemical taphonomic analysis is, therefore, good field work. Without detailed sample location and lithological information (Eberth et al., Chapter 5 in this volume) it is often impossible to interpret the fine detail of trace element analyses.

## Sample Size

As trace element taphonomic techniques rely on statistical analyses, it is important that sample sizes are sufficiently large to justify the statistical tests employed. Furthermore, the geochemistry of terrestrial pore waters can vary significantly over very short distances, so that relatively large sample sizes are needed to ensure that the full range of "natural" variation in the target population is known. Ultimately, the sample size required for each study will vary depending on the demands of the study and the geochemical complexity of the setting. In other words, collect more samples than you think you can analyze! Initially, an absolute minimum of 20 specimens should be selected from any single locality. As the geographical area of the sample locality increases, so the potential for variation in burial environments also increases, and larger samples will be needed to characterize the depositional environment. A pilot study may be used to define statistically the minimum number of samples required to test whether assemblages are geochemically similar.

As trace element studies may be performed on indeterminate bone fragments, relatively large sample populations can often be obtained for destructive analyses with no significant loss to other paleontological aspects of the study. Before analyzing samples, it is important to record as much information as possible; at least record the bone type (compact, cancellous, or trabecular bone) and physical taphonomic condition (e.g., weathering abrasion, shape category) and for each locality obtain some petrographic thin sections and perform XRD or FTIR analyses on a subset of the bones. These thin sections and mineralogical analyses are essential, as bones are usually permineralized, and the nature of the permineralizing material should be assessed before proceeding with other bulk chemical analyses. Techniques of sample preparation may need to be altered to ensure that any chemical analyses of bone apatite are not contaminated by other authigenic minerals.

## Field Collection

It is always preferable to collect specimens directly from the field, specifically for trace element analyses, as most other collection strategies will bias sample populations. For instance, collections may specifically target intact, identifiable bone at the expense of fragmentary bone or may preferentially target bones that look "interesting" taphonomically. Such bones are almost by definition unrepresentative of the assemblage as a whole. It is good practice to collect and record a representative sample of bone fragments from any study site, even if you have no immediate intention to do any geochemistry. A quantitative assessment of taphonomic mixing might become critical to a scientific argument at a later date; also, who knows what might be analytically possible in the next few years?

Most importantly, as with all geochemical analyses, all efforts must be made to avoid treatment of bone with any preservatives or glue. Use of these solvents renders bone effectively useless for any subsequent analysis, whether based on trace elements, stable isotopes, or organic molecules. Preservatives should be therefore be used with great reluctance in the field!

## Target Bone Structure

Cortical bone should be targeted for analyses. Williams (1988) demonstrated that in a transect through cortical bone, the concentrations of REE, U, and Th fall dramatically from both the periosteal and endosteal surfaces toward a minimum within the dense cortical bone. This observation has been subsequently confirmed in a wide range of fossil bones. The relative passage of different trace metals through bone cortex is in part a function of how easily they are accommodated into the bone lattice, and therefore trace metals will be fractionated with depth in a bone cortex. The amount of this fractionation probably depends upon the rate of water flow though the bone and around the bone and may vary with soil type. Within-cortex fractionation means that variations in the thickness of bone cortex, and position of any sample within cortical bone, will affect not only the absolute abundance of trace metals measured in a bone, but also the relative abundance. To minimize this effect, small fragments of bone cortex should be analyzed, with a cortical thickness between 1 and 10 mm. Ideally, bones with similar cortical thickness (e.g., midshaft cylindrical bones of similar size) should be sampled, although this may not be possible in practice in many cases, and most cited studies have not limited sampling to equivalent bone sizes or types. Bones with intact outer (periosteal) surfaces should be targeted, and the full thickness of the cortex should be sampled and homogenized during crushing. This will help to ensure that all REE contained in the cortical bone section are sampled and will help to minimize the effects of sampling different depths of cortical bone. If possible, a subsample of the sample population should be microsampled or analyzed using in situ techniques such as laser-ablation Inductively Coupled Plasma-Mass Spectrometry to determine the extent of trace metal fractionation within individual bone cortices.

## Taxonomy

The REE are largely absent in fresh bone and have no known physiological function, so there is no significant variation in REE composition with taxonomy. This may not be true for all trace elements, and care should be taken when selecting the suite of elements that is going to be analyzed. Bones should be chosen with no regard to taxonomy (unless the point of the anlaysis is to test hypotheses that are based on taxonomy). Using indeterminate bone fragments largely circumvents this problem.

## Measuring Techniques

Trace elements are relatively easily measured in fossil bones, and the most appropriate method of analysis will vary according to the trace elements of interest and their concentrations within fossil bones. Major, minor, and trace elements with concentrations higher than ~1 ppm may be analyzed by a range of techniques including Inductively Coupled Plasma Optical Emission Spectroscopy, X-Ray Fluorescence, and possibly Electron Probe-Micro Analysis or even X-ray microanalysis if concentrations of the selected trace elements are >~0.5 wt%. However, while these techniques are suited to relatively rapid analyses of a wide range of elements, they are not ideally suited to analyses of REE, U, and Th contents in fossil bones, as the practical limits for accurate analyses of these elements are often close to the concentrations in bone. Most analyses of REE in fossil bone are performed using ICP-MS techniques. Most ICP-MS machines analyze trace metals in solution, and while modern machines can potentially measure a wide range in analyte concentrations, the optimum concentrations for the analytes in the sample solution range from 0.1 to 100 ppb. In any case the total dissolved solids in a sample solution should not exceed 0.05%. Evidently bone samples must be dissolved and diluted prior to analysis.

Laser ablation (LA) offers an alternative approach. In LA-ICP-MS analysis, a pulse of laser light is fired onto a flat surface of the sample, and the heating ablates fragments of the sample, which can be introduced directly into the mass spectrometer. This technique allows in situ spatially resolved analyses with a spot size of $\sim$20 $\mu$m, and minimal sample preparation. Significant problems exist, however, regarding calibration and standardization with LA-ICP-MS studies. LA-ICP-MS techniques are ideally suited to determining concentration profiles within bone cortices but are less suitable for comparing trace element geochemistry between bones from different assemblages, where an average signal from the bone cortex is more effective.

## Sample Dissolution

Fossil bones contain apatite and a range of secondary diagenetic and detrital materials. Ideally, then, apatite should be separated from all diagenetic minerals prior to analysis. In practice this is extremely difficult, especially when one is dealing with large sample sizes, but several simple steps will minimize contamination (or dilution). Fossil bone apatite usually contains several orders of magnitude higher concentrations of REE than any associated authigenic or detrital grains. In most cases therefore the inclusion of minor (<5%) amounts of detrital or authigenic minerals within analyzed sample solutions will have little effect on the measured REE content of fossil bone other than a mild dilution. Most geochemical taphonomic analyses are based on the relative rather than absolute concentrations of trace elements, so mild dilution effects are generally not significant. The sample procedure outlined below is designed to provide a relatively quick and simple methodology for determining REE compositions of large sample suites of fossil bones. It is certainly not perfect, and the suggested multiple weighing cycles may introduce relatively large errors into any analytical measurement. However, these errors will affect all analyzed elements equally and so should have no effect on the relative abundance of REEs within a sample.

- Firstly, and obviously, all bone samples should be cleaned thoroughly with dental picks or drills to remove adhering sediment or encrusting authigenic minerals.
- Samples should be crushed in agate pestle and mortar and dried in an oven at 100°C (or freeze-dried) overnight. This step removes any trapped water prior to weighing.

- Approximately 0.01 g bone powder should be accurately weighed into a clean heat-resistant sample tube, and the mass of the sample and tube recorded.
- One to 2 ml sodium or ammonium acetate solution adjusted to pH 5 with acetic acid should be added to the bone powder and left for several hours at room temperature. This step removes authigenic calcite. Apatite crystal surfaces may begin to dissolve and may release some surface-bound REEs, but this effect will be minor compared to the effect of analyzing authigenic calcite together with apatite.
- Buffer solutions should be decanted and the samples dried and reweighed. Approximately 2 ml 2N $HNO_3$ is then added. Dissolution of apatite will occur rapidly at room temperature, and no additional heating is needed. Detrital or authigenic minerals such as quartz, clay minerals, and siderite, and also adsorbed organics such as humic acids, will not dissolve, but surface-bound REEs may be released to the solution.
- The solution is then taken up with a syringe and filtered through at least a 0.45 $\mu$m filter into a sample bottle. If clays or other insoluble materials form a large portion of the original bone sample, it may be necessary to weigh any residue trapped on the filter paper and therefore determine the proportion of apatite within the fossil bone sample.
- The analyte solution should then be made up to 50 ml by weight with 1–2% $HNO_3$ to form a sample solution that may be stored prior to analyses.
- Immediately prior to analysis, 1 ml of sample solution is taken, together with 1 ml of 100 ppb internal standard (e.g., Re/Ru) and made up to 10 ml with 1–2% $HNO_3$.
- The dissolution procedure outlined above is intended as a general simple digestion protocol suitable for analyzing bones with total REE concentrations of ~1000–5000 ppm. Other conditions may require weaker or stronger dilutions, additional steps, or other modifications. Most importantly, all samples should be analyzed together with internal laboratory standards and international rock standards (such as NIST 120c – Florida Phosphate) for quality control.

### ICP-MS Running Conditions

Ba has several naturally occurring isotopes which, together with the LREE, readily form oxides within the ICP-MS injection process. These oxides interfere with middle and heavy REE—particularly Eu (BaO), Tm, and

Gd—and therefore can lead to errors. Ba also forms isobaric overlaps with La and may therefore lead to falsely high measured La concentrations (e.g., Trueman, 1999). Bones often contain high Ba and relatively high LREE concentrations, therefore it is important to optimize the operational conditions of the ICP-MS to minimize oxide formation and to correct for isobaric overlaps and oxide formation within the analysis software on a daily basis.

The REE and other trace metal composition of fossil bone can be measured using other analytical techniques such as ICP-OES. However, as these machines have higher detection limits, the REE may need to be preconcentrated prior to analysis. Preconcentration is usually achieved via cation exchange, and careful laboratory work can then yield results comparable to other analytical techniques. However, cation exchange methods are relatively time consuming and can prove expensive, needing large volumes of trace metal–grade acids. Such procedures are therefore not ideally suited to analyzing very large sample populations.

## REFERENCES

Amin, J., M. Bramer, and R. Emslie. 2003. Intelligent data analysis for conservation: Experiments with rhino horn fingerprint identification. Knowledge-Based Systems 16:329–336.

Anderson, P.E., M.J. Benton, C.N. Trueman, B. Paterson, and G. Cuny. 2007. Palaeoenvironments of vertebrates on the southern shore of Tethys: The nonmarine Early Cretaceous of Tunisia. Palaeogeography, Palaeoclimatology, Palaeoecology 243:118–131.

Armstrong, H.A., D.G. Pearson, and M. Griselin. 2001. Thermal effects on rare earth element and strontium isotope chemistry in single conodont elements. Geochimica et Cosmochimica Acta 65:435–441.

Barrick, R.E., W.J. Showers, and A.G. Fischer. 1996. Comparison of thermoregulation of four ornithischian dinosaurs and a varanid lizard from the Cretaceous Two Medicine Formation: Evidence from oxygen isotopes. Palaios 11:295–305.

Beard, B.L., and C.M. Johnson. 2000. Strontium isotope composition of skeletal material can determine the birth place and geographic mobility of humans and animals. Journal of Forensic Sciences 45:1049–1061.

Behrensmeyer, A.K. 1978. Taphonomic and ecological information from bone weathering. Paleobiology 4:150–162.

Behrensmeyer, A.K. 1990. Transport-hydrodynamics: Bones. Pp. 291–335 *in* Paleobiology: A synthesis. D.E.G. Briggs, and P.R. Crowther, eds. Blackwell Scientific Publications, Oxford.

Behrensmeyer, A.K., R.S. Hook, C.E. Badgley, J.A. Boy, R.E. Chapman, P. Dodson, R.A. Gastaldo, R.W. Graham, L.D. Martin, P.E. Olsen, R.A. Spicer, R.E. Taggart, and M.V.H. Wilson. 1992. Paleoenvironmental contexts and taphonomic modes. Pp. 15–136 *in* Terrestrial ecosystems through time: The evolutionary paleoecology of terrestrial plants and animals. A.K. Behrensmeyer, J.D. Damuth, W.A. DiMichelle, R. Potts, H.-D. Sues, and S.L. Wing, eds. University of Chicago Press, Chicago.

Benton, M.J., E. Cook, and P. Turner. 2002. Permian and Triassic red beds and the Penarth Group of Great Britain. Geological Conservation Review Series, no 24. Joint Nature Conservation Committee, Peterborough, UK.

Berna, F., A. Matthews, and S. Weiner. 2004. Solubilities of bone mineral from archaeological sites: The recrystallization window. Journal of Archaeological Science 31:867–882.

Bernat, M. 1975. Les isotopes de l'uranium et du thorium et les terres rares dans l'environment marin. Cahiers OSTROM Series Geologie 7:65–83.

Braun, J.-J., J. Viers, B. Dupré, M. Polve, J. Ndam, and J.-P. Muller. 1998. Solid/liquid fractionation in the lateritic system of Goyoum, East Cameroon: The implication for present dynamics of the soil covers of the humid tropical regions. Geochimica et Cosmochimica Acta 62:273–299.

Byrne, R.H., and B.Q. Li. 1995. Comparative complexation behavior of the rare-earths. Geochimica et Cosmochimica Acta 59:4575–4589.

Collins, M.J., M.S. Riley, A.M. Child, and G. Turner-Walker. 1995. A basic mathematical simulation of the chemical degredation of ancient collagen. Journal of Archaeological Science 22:175–183.

Collins, M.J., C.M. Nielsen-Marsh, J. Hiller, C.I. Smith, J.P. Roberts, R.V. Prigodich, T.J. Wess, J, Csapò, A.R. Millard, and G. Turner-Walker. 2002. The survival of organic matter in bone: A review. Archeometry 44:383–394.

Denys, C., C.T. William, Y. Dauphin, P. Andrews, and Y. Fernandez-Jalvo. 1996. Diagenetical changes in Pleistocene small mammal bones from Olduvai Bed-1. Palaeogeography, Palaeoclimatology, Palaeoecology 126:121–134.

Dia, A., G. Gruau, G. Olivié-Lauquet, C. Riou, J. Molénat, and P. Curmi. 2000. The distribution of rare earth elements in groundwaters: Assessing the role of source-rock composition, redox changes and colloidal particles. Geochimica et Cosmochimica Acta 64:4131–4151.

Duddy, I.R. 1980. Redistribution and fractionation of rare-earth and other elements in a weathering profile. Chemical Geology 30:363–381.

Dupré, B., J. Gaillardet, D. Rousseau, and C.J. Allègre. 1996. Major and trace elements of river-bourne material: The Congo Basin. Geochimica et Cosmochimica Acta 60:1301–1321.

Eberth, D.A., R.R. Rogers, and A.R. Fiorillo. This volume. A practical approach to the study of bonebeds. Chapter 5 *in* Bonebeds: Genesis, analysis, and paleobiological significance. R.R. Rogers, D.A. Eberth, and A.R. Fiorillo, eds. University of Chicago Press, Chicago.

Elderfield, H., and R. Pagett. 1986. Rare earth elements in ichthyoliths: Variations with redox conditions and depositional environment. The Science of the Total Environment 49:175–197.

Fiorillo, A.R. 1988. Taphonomy of Hazard Homestead Quarry (Ongalla Group), Hitchcock County, Nebraska. Contributions to Geology, University of Wyoming 26:57–97.

Fricke, H. This volume. Stable isotope geochemistry of bonebed fossils: Reconstructing paleoenvironments, paleoecology, and paleobiology. Chapter 8 *in* Bonebeds:

Genesis, analysis, and paleobiological significance. R.R. Rogers, D.A. Eberth, and A.R. Fiorillo, eds. University of Chicago Press, Chicago.

German, C.R., and H. Elderfield. 1990. Application of the Ce-anomaly as a paleoredox indicator: The ground rules. Paleoceanography 5:823–833.

Gill, R. 1996. Chemical fundamentals of geology. Chapman and Hall, London.

Girard, C., and F. Albarède. 1996. Trace elements in conodont phosphates from the Frasnian/Famennian boundary. Palaeogeography, Palaeoclimatology, Palaeoecology 126:195–209.

Goldberg, E.D., M. Koide, R.A. Schmitt, and J. Smith. 1963. Rare earth distributions in the marine environment. Journal of Geophysical Research 68:4204–4217.

Grandjean, P., H. Cappetta, A. Michard, and F. Albarède. 1987. The assessment of REE patterns and 143Nd/144Nd ratios in fish remains. Earth and Planetary Science Letters 84:181–196.

Grandjean-Lécuyer, P., R. Feist, and F. Albarède. 1993. Rare earth elements in old biogenic apatites. Geochimica et Cosmochimica Acta 57:2507–2514.

Henderson, P., C.A. Marlow, T.I. Molleson, and C.T. Williams. 1983. Patterns of chemical change during bone fossilization. Nature 306:358–360.

Hill, A.P. 1980. Early postmortem damage to the remains of some contemporary east African mammals. Pp. 131–152 in Fossils in the making. A.K. Behrensmeyer and A.P. Hill, eds. University of Chicago Press, Chicago.

Hiller, J. C., M.J. Collins, A.T. Chamberlain, and T.J. Wess. 2004. Small-angle X-ray scattering: A high-throughput technique for investigating archaeological bone preservation. Journal of Archaeological Science 31:1349–1359.

Holser, W.T. 1997. Evaluation of the application of rare-earth elements to paleoceanography. Palaeogeography, Palaeoclimatology, Palaeoecology 132:309–323.

Hoogewerff, J., W. Papesch, M. Kralik, M. Berner, P. Vroon, H. Meisbauer, O. Graber, K.-H. Künzel, and J. Kleinjans. 2001. The last domicile of the iceman from Hauslabjoch: A geochemical approach using Sr, C and O isotopes and trace element signatures. Journal of Archaeological Science 28:983–989.

Hooker, J.S. 1987. Late Cretaceous ashfall and the demise of a hadrosauran "herd." Geological Society of America, Rocky Mountain Section, Abstracts with Programs 19:284.

Horner, J.R., and J. Gorman. 1988. Digging dinosaurs. Workman Publishing Company, New York.

Hubert, J.F., P.T. Panish, D.J. Chure, and K.S. Prostak. 1996. Chemistry, microstructure, petrology, and diagenetic model of Jurassic dinosaur bones, Dinosaur National Monument, Utah. Journal of Sedimentary Research 66:531–547.

Johannesson, K., W.B. Lyons, M.A. Yelken, H.E. Gaudette, and K.J. Stetzenbach. 1996a. Geochemistry of rare-earth elements in hypersaline and dilute acidic terrestrial waters: Complexation behavior and middle rare-earth enrichment. Chemical Geology 133:124–144.

Johannesson, K., K.J. Stetzenbach, V.F. Hodge, and W.B. Lyons. 1996b. Rare earth element complexation behavior in circumneutral pH groundwaters: Assessing the role of carbonate and phosphate ions. Earth and Planetary Science Letters 139:305–319.

Karkanas, P., O. Bar-Yosef, P. Goldberg, and S. Weiner. 2000. Diagenesis in prehistoric caves: The use of minerals that form in situ to assess the completeness of the archaeological record. Journal of Archaeological Science 27:915–929.

Kemp, R.A., and C.N. Trueman. 2003. Rare earth elements in Solnhofen biogenic apatite: Geochemical clues to the palaeoenvironment. Sedimentary Geology 155:109–127.

Keto, L.S., and S.B. Jacobsen. 1987. Nd and Sr isotopic variations of Early Paleozoic oceans. Earth and Planetary Science Letters 84:27–41.

Kohn, M.J., and J.M. Law. 2006. The stable isotope composition of fossil bone as a new palaeoclimate indicator. Geochimica et Cosmochimica Acta 70:931–946.

Kolodny, Y., B. Luz, M. Sander, and W.A. Clemens. 1997. Dinosaur bones: Fossils or pseudomorphs? The pitfalls of physiology reconstruction from apatitic fossils. Palaeogeography, Palaeoclimatology, Palaeoecology 126:161–171.

Lécuyer, C., C. Bogey, J.-P. Garcia, P. Grandjean, J.-A. Barrat, M. Floquet, N. Bardet, and X. Pereda-Superbiola. 2003. Stable isotope composition and rare earth element content of vertebrate remains from the Late Cretaceous of northern Spain (Laño): Did the environmental record survive? Palaeogeography, Palaeoclimatology, Palaeoecology 193:457–471.

Lécuyer, C., B. Reynard, and P. Grandjean. 2004. Rare earth element evolution of Phanerozoic seawater recorded in biogenic apatites. Chemical Geology 204:63–102.

Lucas, J., and L. Prévôt. 1984. Synthèse de l'apatite par voie bactérienne à partir de matière organique phosphates et divers carbonates de calcium dans des eaux douce et marine naturelles. Chemical Geology 42:104–118.

Lyman, R.L. 1994. Vertebrate taphonomy. Cambridge manuals in archaeology. Cambridge University Press, Cambridge.

MacQuaker, J.H.S. 1994. Palaeoenvironmental significance of "bone-beds" in organic-rich mudstone successions: An example from the Upper Triassic of south-west Britain. Zoological Journal of the Linnean Society 112:285–308.

Martin, E.E., and B.A. Haley. 2000. Fossil fish teeth as proxies for seawater Sr and Nd isotopes. Geochimica et Cosmochimica Acta 64:835–847.

Martin, J.E., D. Patrick, A.J. Kihm, F.F. Foit Jr., and D.E. Grandstaff. 2005. Lithostratigraphy, tephrochronology, and rare earth element geochemistry of fossils at the classical Pleistocene Fossil Lake area, south central Oregon. Journal of Geology 113:139–155.

Metzger, C.A., D.O. Terry, and D.E. Grandstaff. 2004. Effect of paleosol formation on rare earth element signatures in fossil bone. Geology 32:467–500.

Millard A.R., and R.E.M. Hedges. 1999. A diffusion-adsorption model of uranium uptake by archaeological bone. Geochimica et Cosmochimica Acta 60:2139–2152.

Morey, G.B., and D.R. Stetterholm.1997. Rare earth elements in weathering profiles and sediments of Minnesota: Implications for provenance studies. Journal of Sedimentary Research 67:105–115.

Nielsen-Marsh, C.N., and R.E.M. Hedges. 2000. Patterns of diagenesis in bone. I The effects of site environments. Journal of Archaeological Science 27:1139–1150.

Nriagu, J.O. 1983. Rapid decomposition of fish bones in Lake Erie sediments. Hydrobiologica 106:217–222.

Patrick, D., J.E. Martin, D.C. Parris, and D.E. Grandstaff. 2002. Rare earth element signatures of fossil vertebrates compared with lithostratigraphic subdivisions of the Upper Cretaceous Pierre Shale, central South Dakota. Proceedings of the South Dakota Academy of Science 81:161–179.

Patrick, D., J.E. Martin, D.C. Parris, and D.E. Grandstaff. 2004. Paleoenvironmental interpretations of rare earth element signatures in mosasaurs (Reptilia) from the upper Cretaceous Pierre Shale, central South Dakota, USA. Palaeogeography, Palaeoclimatology, Palaeoecology 212:277–294.

Pfretzscnher, H.-U. 2000. Microcracks and fossilization of haversian bone. Neues Jahrbuch Für Geologie und Palaontologie-abhandlungen 216:413–432.

Picard, S., C. Lécuyer, J.A. Barrat, J.-P. Garcia, G. Drommart, and S.M.F. Sheppard. 2002. Rare earth elements contents of Jurassic fish and reptile teeth and their potential relation to seawater composition (Anglo-Paris Basin, France and England). Chemical Geology 186:1–16.

Pike, A.W.G., R.E.M. Hedges, and P. Van Calsteren. 2002. U-series dating of bone using the diffusion-adsorption model. Geochimica et Cosmochimica Acta 66:4273–4286.

Plummer, T.W., A.M. Kinuyua, and R. Potts. 1994. Provenancing of hominid and mammalian fossils from Kanjera, Kenya, using EDXRF. Journal of Archaeological Science 21:553–563.

Potts, R., A.K. Behrensmeyer, and P. Ditchfield, P. 1999. Paleolandscape variation and Early Pleistocene hominid activities: Members 1 and 7, Oloregsailie Formation, Kenya. Journal of Human Evolution 37:747–788.

Pye, K. 2004. Isotope and trace element analysis of human teeth and bones for forensic purposes. Pp. 215–236 in Forensic geoscience: Principles, techniques and applications. Special publications 232. K. Pye, and D.J. Croft, eds. Geological Society of London, London.

Reiche I., C. Vignaud, and M. Menu. 2002. The crystallinity of ancient bone and dentine: New insights by transmission electron microscopy. Archeometry 44:447–460.

Reynard, B., C. Lécuyer, and P. Grandjean. 1999. Crystal chemical controls on rare-earth element concentrations in fossil biogenic apatites and implications for paleoenvironmental reconstructions. Chemical Geology 155:233–241.

Roaldset, E. 1973. Rare earth elements in Quaternary clays of the Numedal area, southern Norway. Lithos 6:349–372.

Roberts, S.J., C.I. Smith, A. Millard, and M.J. Collins. 2002. The taphonomy of cooked bone: Characterizing boiling and its physico-chemical effects. Archaeometry 44:485–494.

Rogers, R.R., and S.M. Kidwell. 2000. Associations of vertebrate skeletal concentrations and discontinuity surfaces in terrestrial and shallow marine records: A test in the cretaceous of Montana. Journal of Geology 108:131–154.

Samoilov, V.S., and C. Benjamini. 1996. Geochemical features of dinosaur remains from the Gobi Desert, south Mongolia. Palaios 11:519–531.

Scher, H., and E.E. Martin. 2006. Timing and climatic consequences of the opening of Drake Passage. Science 312:428–430.

Schmitt, J.G., J.R. Horner, R.R. Laws, and F. Jackson. 1998. Debris-flow deposition of a hadrosaur-bearing bone bed, Upper Cretaceous Two Medicine Formation, Northwest Montana. Journal of Vertebrate Paleontology 18(Supplement to 3): 76A

Shaw, H.F., and G.J. Wasserburg. 1985. Sm-Nd in marine carbonates and phosphates: Implications for Nd isotopes in seawater and crustal ages. Geochimica et Cosmochimica Acta 49:503–518.

Shields, G.A., and G.E. Webb. 2004. Has the REE composition of seawater changed over geological time? Chemical Geology 204:103–107.

Shinomiya, T., K. Shinomiya, C. Orimoto, T. Minami, Y. Tohno, and M. Yamada. 1998. In- and out-flows of elements in bones embedded in reference soils. Forensic Science International 98:109–118.

Sholkovitz, E.R., W.M. Landing, and B.L. Lewis. 1994. Ocean particle chemistry—the fractionation of rare-earth elements between suspended particles and seawater. Geochimica et Cosmochimica Acta 58:1567–1579.

Sillen, A. 1990. Response to N. Tuross, A. K. Behrensmeyer and E.D. Eanes. Journal of Archaeological Science 17:595–596.

Staron, R.M., B.S. Grandstaff, W.B. Gallagher, and D.E. Grandstaff. 2001. REE signatures in vertebrate fossils from Sewel, NJ: Implications for location of the K-T boundary. Palaios 16:255–265.

Storrs, G.W. 1994. Fossil vertebrate faunas of the British Rhaetian (latest Triassic). Zoological Journal of the Linnean Society 112:217–259.

Sykes, J.H. 1977. British Rhaetian bone-beds. Mercian Geologist 5:39–48.

Tang J.W., and K.H. Johannesson. 2003. Speciation of rare earth elements in natural terrestrial waters: Assessing the role of dissolved organic matter from the modeling approach. Geochimica et Cosmochimica Acta 67:2321–2339.

Trueman, C.N. 1999. Rare earth element geochemistry and taphonomy of terrestrial vertebrate assemblages. Palaios 14:555–566.

Trueman, C.N. 2004. Forensic geology of bone mineral: Geochemical tracers for postmortem movement of bone remains. Pp. 249–256 in Forensic geoscience, principles, techniques and applications. Special publications 232. K. Pye, and D.J. Croft, eds. Geological Society of London, London.

Trueman, C.N., and M.J. Benton. 1997. A geochemical method to trace the taphonomic history of reworked bones in sedimentary settings. Geology 27:263–265.

Trueman, C.N., and D. Martill. 2002. The long term preservation of bone: The role of microbial attack. Archaeometry 44:371–382.

Trueman, C.N., and N. Tuross. 2002. Trace elements in modern and ancient bone. Pp. 489–522 in Phosphates: Geochemical, geobiological and materials importance. Reviews in Mineralogy and Geochemistry 48. M.J. Kohn, J. Rakovan, and J.M. Hughes, eds. Mineralogical Society of America, Washington, DC.

Trueman C., C. Chenery, D.A. Eberth, and B. Spiro. 2003a. Diagenetic effects on the oxygen isotope composition of dinosaurs and other vertebrates recovered from terrestrial and marine sediments. Journal of the Geological Society of London 160:895–901.

Trueman, C.N., M.J. Benton, and M.R. Palmer. 2003b. Geochemical taphonomy of shallow marine vertebrate assemblages. Palaeogeography, Palaeoclimatology, Palaeoecology 197:151–169.

Trueman, C.N., A.K. Behrensmeyer, N. Tuross, and S. Weiner. 2004. Mineralogical and compositional changes in bones exposed on soil surfaces in Amboseli National Park, Kenya: Diagenetic mechanisms and the role of sediment pore fluids. Journal of Archaeological Science 31:721–739.

Trueman, C.N.G., J. Field, S. Wroe, B. Charles, and J. Dortch. 2005. Prolonged coexistence of humans and megafauna in Pleistocene Australia. Proceedings of the National Academy of Sciences of the U.S.A. 102:8381–8385.

Trueman, C.N., A.K. Behrensmeyer, R. Potts, and N. Tuross. 2006. High-resolution records of location and stratigraphic provenance from the rare earth element composition of fossil bones. Geochimica et Cosmochimica Acta 70:4343–4355.

Tuross, N., A.K. Behrensmeyer, and E.D. Eanes. 1989. Strontium increases and crystallinity changes in taphonomic and archaeological bone. Journal of Archaeological Science 16:661–672.

Tütken, T., H.-U. Pfretzschner, T.W. Vennemann, G. Sun, and Y.D. Wang. 2004. Paleobiology and skeletal chronology of Jurassic dinosaurs: Implications from the histology and oxygen isotope compositions of bones. Palaeogeography, Palaeoclimatology, Palaeoecology 206:217–238.

Van der Merwe, N.J., J.A. Lee-Thorp, J.F. Thackeray, A. Hall-Martin, F.J. Kruger, H. Coetzee, R.H.V. Bell, and M. Lindeque. 1990. Source-area determination of elephant ivory by isotopic analysis. Nature 346:744–746.

Weiner, S., P. Goldberg, and O. Bar-Josef. 1993. Bone preservation in Kebara Cave, Israel, using on-site Fourier transform infrared spectrometry. Journal of Archaeological Science 20:613–627.

Williams, C.T. 1988. Alteration of chemical composition of fossil bones by soil processes and groundwater. Pp. 27–40 *in* Trace elements in environmental history. G. Grupe, and B. Herrmann, eds. Springer-Verlag, Berlin.

Williams, C.T., P. Henderson, C.A. Marlow, and T.I. Molleson. 1997. The environment of deposition indicated by the distribution of rare earth elements in fossil bones from Olduvai Gorge, Tanzania. Applied Geochemistry 12:537–547.

Wood, S.A. 1990. The aqueous geochemistry of the rare-earth elements and yttrium. 1. Review of available low-temperature data for inorganic complexes and the inorganic REE speciation of natural waters. Chemical Geology 82:159–186.

Wright, J., R.S. Seymour, and H.F. Shaw. 1984. REE and Nd isotopes in conodont apatite: Variations with geological age and depositional environment. Pp. 325–340 *in* Conodont biofacies and provincialism. Special paper. D.L. Clark, ed. Geological Society of America, Boulder, Colorado.

Wykoff, R.W.G. 1972. The biochemistry of animal fossils. Scientechnica, Bristol.

# CHAPTER 8

# Stable Isotope Geochemistry of Bonebed Fossils: Reconstructing Paleoenvironments, Paleoecology, and Paleobiology

*Henry Fricke*

## INTRODUCTION

The tissues of living animals contain carbon, oxygen, and nitrogen, which are sourced from the food, water, and air that animals ingest and respire. This linkage holds true for vertebrate hardparts such as bones and teeth, which consist of a matrix of organic molecules (mostly collagen) surrounded by crystals of bioapatite $[Ca_5(PO_4, CO_3)_3(OH, CO_3)]$. For example, carbon found in bioapatite is related to ingested organic material, such as plants in the case of herbivores and flesh in the case of carnivores. Similarly, oxygen in apatite is obtained from the atmosphere and from water ingested from streams, ponds, lakes, and leaves.

These links have proven very important to scientists interested in earth history because the ratio of stable isotopes of carbon, oxygen and nitrogen, in plants and surface waters can vary a great deal. Carbon isotope ratios of plants change in response to environmental conditions and to the type of photosynthetic pathways utilized by the plant; oxygen isotope ratios of waters in streams, lakes, and leaves vary significantly in response to environmental factors such as temperature and aridity, and in relation to the hydrological "history" of air masses that supply precipitation to these surface water reservoirs. As a result, stable isotope analyses of fossils that record these stable isotope variations can be used to address diverse topics, ranging from the nature of ancient environmental conditions, to differences in the behavior, dietary choices, and preferred habitats of extinct animals. Vertebrate fossils concentrated in bonebeds offer a particularly exciting opportunity to take advantage of these isotopic relations.

The goals of this chapter are to (1) outline the different ways to undertake stable isotope research using fossils, and (2) explore the different types of questions that can be addressed using stable isotope data. The chapter begins with a review that focuses on isotopic variability in plants and surface waters, isotopic relations between these ingested materials and vertebrate remains, and the preservation of primary isotopic information over time. Many excellent reviews of these topics are already available and, thus, only a brief introduction is provided here. The review is followed by a general consideration of study design in relation to bonebed type, and a summary of the critical assumptions and unknowns that may impact study design. The chapter concludes with specific examples of paleoenvironmental and paleoecological reconstructions using stable isotope data, with the hope that they may provide a template and incentive for future research.

This chapter has two important limitations. First, it focuses only on what can be learned from analyzing bioapatite, and does not discuss organic molecules such as protein and collagen, cholesterols, and lipids, which are also an integral part of vertebrate skeletons. The primary reason for limiting the discussion to bioapatite is because organic remains are very susceptible to decomposition via enzymatic, microbial, and geochemical processes, and whereas bioapatite can be preserved for millions of years, organic molecules are rarely found preserved in fossil remains older than ~75 ky. Second, most of the studies described here center on terrestrial mammal remains, which have received the greatest amount of stable isotope research to date. However, much of what is covered in this chapter is theoretically applicable to vertebrates in general, and researchers are encouraged to apply these approaches to other groups.

## STABLE ISOTOPES AND ISOTOPIC VARIABILITY OF BONEBED REMAINS

The primary reason that stable isotope ratios of bonebed fossils can be used to investigate environmental conditions or ecological relations of the past is that plants and surface waters distributed across terrestrial landscapes are often characterized by different stable isotope ratios that can be passed on to animal consumers. The underlying causes for isotopic variability are the small differences in atomic mass between isotopes that, in turn, affect rates of reaction and the strength of bonds (O'Neil, 1986). For example, isotopes of oxygen, carbon, and hydrogen are differentiated from each other by the number of neutrons in the atomic nucleus (e.g., $^{18}O$ and $^{16}O$,

$^{13}C$ and $^{12}C$, $^{2}H$ and H) and these differences often result in a separation—or fractionation—of heavy and light isotopes during physical, chemical, and biological processes. Differences in the relative abundance of heavy to light isotopes (the isotope ratio) in various materials can be quite large, particularly when reactions are ongoing (e.g., most biochemical reactions, evaporation of large water bodies). In such cases, products are often enriched in the light isotope.

The relative abundance of heavy to light isotopes in most materials is on the order of parts per thousand, and stable isotope ratios in any material are reported relative to a standard using the $\delta$ notation, where $\delta = (R_{sample}/R_{standard} - 1)1000\%$o (rather than %). Isotope standards are often derived from common materials and include standard mean ocean water (VSMOW) for oxygen, and marine carbonate (VPDB) for carbon. Hydrogen isotope ratios of bioapatite are not commonly analyzed, do not appear to be preserved well over time (Kohn et al., 1999), and are not discussed further.

### Carbon Isotope Variability of Plants

Plants living in terrestrial environments may exhibit a large range in $\delta^{13}C$, from approximately −36‰ to −10‰, even across a relatively small geographic area. As illustrated schematically in Figure 8.1, plants growing in open, dry, saline, or nutrient-poor settings generally have higher $\delta^{13}C$ than those living in closed, wet, nutrient-rich environments. Fully aquatic plants, however, often have higher carbon isotope ratios than fully terrestrial plants. Such carbon isotope variability can be key in helping to differentiate between fossil herbivores that routinely ate different plants, or ingested water from different parts of a large paleoecosystem.

Causes of carbon isotope variability in plants have been the focus of several excellent reviews (O'Leary, 1988; Farquhar et al., 1989; O'Leary et al., 1992; Kohn and Cerling, 2002) and are briefly summarized here. One very important factor is the photosynthetic pathway utilized by the plant. Most carbon in plant organic material is derived from the reduction of atmospheric $CO_2$ during photosynthesis. Stable isotopes of carbon are fractionated from one another during this process, and the extent of this fractionation (i.e., carbon isotope discrimination between organic material and the atmosphere) varies depending on which specific photosynthetic pathway is utilized by a given plant. For example, plants using the C3 (Calvin) pathway are characterized by a large isotopic discrimination and have $\delta^{13}C$ values that range from approximately −36‰ to −21‰

*Figure 8.1.* Schematic illustration of relative $\delta^{13}C$ values that may be found for C3 plants occupying different parts of a coastal/terrestrial ecosystem. The latter include: A, fully marine plants; B, closed-canopy forest; C, open forest, water available; D, open grassland; E, mixed forest/grassland; F, coastal marsh/mangrove forest; G, high-elevation areas. Using area E as a baseline, plants living in areas A, D, F, and G are likely to have higher $\delta^{13}C$, while those living in areas B and C are likely to have lower $\delta^{13}C$. Reasons for relatively higher $\delta^{13}C$ values include slow rates of diffusion of $CO_2$ through water (A); enhanced aridity and thus less moisture availability in open environments (D); osmotic stress in brackish waters along with the possibility of evaporative tidal flat settings (F); lower concentrations of $CO_2$ in the atmosphere at higher elevations (G). Reasons for relatively lower $\delta^{13}C$ values include recycling of $CO_2$ in the understory of dense, closed-canopy forests (B), and the abundant availability of fresh water (C). C4 plants living in any of these areas can greatly increase the average $\delta^{13}C$ of plants living in them. Many C4 plants are grasses common to open, dry environments (such as D) or warm tropical regions. Because combinations of different factors such as regional climate, hydrology, and plant type may all act to modify the general pattern presented here, this figure alone should not be used to interpret isotope data from bonebed remains.

compared to $\delta^{13}C$ of about $-8‰$ for modern atmospheric $CO_2$. In contrast, C4 (Hatch-Slack pathway) plants are characterized by less isotopic discrimination and variability, with $\delta^{13}C$ values generally ranging from $-14$ to $-10‰$. Less common CAM (Crassulacean acid metabolism pathway) plants utilize a combination of C3 and C4 photosynthetic processes and have $\delta^{13}C$ values intermediate to those of C3 and C4 plants.

Local environmental conditions also play an important role in influencing carbon isotope ratios of plants, particularly in the case of plants utilizing the C3 photosynthetic pathway. These plants are very sensitive to the amount of $CO_2$ in a leaf cell, and when there is an excess of $CO_2$, $\delta^{13}C$ values in synthesized tissues will be lower, reflecting greater selectivity for $^{12}C$. In contrast, when $CO_2$ concentrations are limited, a greater proportion of all C atoms will be utilized for tissue generation, and the associated isotopic discrimination between $^{12}C$ and $^{13}C$ will be reduced (Farqhar et al., 1989). In turn, concentrations of $CO_2$ in a leaf cell are influenced a great deal by the opening and closing of leaf stomata, which controls the flux of $CO_2$ into the plant. Stomata are more likely to remain closed when

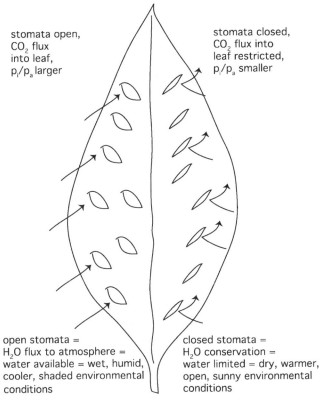

$p_a$=partial pressure of $CO_2$ in the atmosphere,
$p_i$ =partial pressure of $CO_2$ in leaf cells

$\delta^{13}C$ plant = $\delta^{13}C$ atm - 4.4 -22 *($p_i/p_a$), so if
$p_i/p_a$ smaller, $\delta^{13}C$ plant higher,
$p_i/p_a$ higher, $\delta^{13}C$ plant lower

stomata open,
$CO_2$ flux
into leaf,
$p_i/p_a$ larger

stomata closed,
$CO_2$ flux into
leaf restricted,
$p_i/p_a$ smaller

open stomata =
$H_2O$ flux to atmosphere =
water available = wet, humid,
cooler, shaded environmental
conditions

closed stomata =
$H_2O$ conservation =
water limited = dry, warmer,
open, sunny environmental
conditions

*Figure 8.2.* Influences on the $\delta^{13}C$ of C3 plants can be represented by the following equation: $\delta^{13}C_{plant} = \delta^{13}C_{atmosphere} - a - (b - a)*(p_i/p_a)$, where "a" represents the isotopic fractionation associated with the diffusion of $CO_2$ into a leaf and is typically about 4.4‰, "b" is the isotopic fractionation associated with carbon fixation into organic matter and is typically about 29‰, $p_i$ is the partial pressure of $CO_2$ inside a leaf cell, and $p_a$ is the partial pressure of $CO_2$ in the atmosphere (after Farquhar et al., 1989).

environmental factors such as temperature, water availability, and light intensity are such that water needs to be conserved (Fig. 8.2) (O'Leary, 1988; Farqhar et al., 1989; Tieszen, 1991; O'Leary et al., 1992). For example, under arid, high-temperature, or high-light conditions, stomata remain closed to minimize water loss. As a result, $CO_2$ concentrations in leaf cells decrease, and $\delta^{13}C$ increases. Conversely, plants growing in wetter, shaded, cooler areas keep their stomata open for a longer period, $CO_2$ concentrations in the leaf are higher, and $\delta^{13}C$ values are lower.

Summaries of these environmental effects are also provided by Heaton (1999) and Arens et al. (2000).

$\delta^{13}$C of plant material may also be affected by the "canopy effect." In this situation, $CO_2$ under the canopy of a closed forest exhibits lower carbon isotope ratios than the open atmosphere due to plant respiration and decomposition on or near the forest floor. When incorporated into understory plants during photosynthesis, the lower $\delta^{13}$C values stand out in comparison to the same plants living in open canopy settings (van der Merve and Medina, 1991; Cerling et al., 2004). Because the canopy effect is due to local changes in $\delta^{13}$C of $CO_2$, it is independent of photosynthetic pathways or specific environmental conditions.

Lastly, there may be taxon-specific differences in $\delta^{13}$C of C3 plants living in any one place (Tieszen, 1991; Heaton, 1999; Arens et al., 2000, Codron et al., 2005). However, they have not been studied systematically for many modern and ancient environments and are not discussed further. $\delta^{13}$C of atmospheric $CO_2$ can also change over time in response to perturbations in the global carbon cycle, and absolute $\delta^{13}$C of all plant types will change in a similar direction (Koch et al., 1992), although not necessarily to the same degree (Bowen et al., 2004). It should also be noted that most of the paleobotanical fossil record prior to the late Cenozoic is dominated by C3 plants. There is evidence that plants utilizing both C4 and CAM photosynthetic pathways may have evolved several times during the Mesozoic; their occurrence however, appears to have been limited in number and restricted to only certain kinds of terrestrial settings (Wright and Vanstone, 1991; Kuypers, et al., 1999; Krull and Retallack, 2000; Edwards et al., 2004). It is not until the Miocene, ~8 Ma, that the global expansion of C4 grasslands occurred, resulting in modern terrestrial ecosystems (Cerling et al., 1997a). The C4 pathway is well adapted to high-light, high-temperature, and arid conditions, thus the recognition of these kinds of plants in the fossil record can be used to indicate plant biology as well as broad paleoenvironmental conditions.

### Oxygen Isotope Variability of Surface Waters

Oxygen isotope systematics of surface waters are more complicated than those of carbon in plants because large regional and global scale patterns in oxygen isotope ratios exist (Fig. 8.3), and any local variations in oxygen isotope ratios of surface waters that may occur over a small area (e.g., Fig. 8.4) are then superimposed on the larger-scale patterns. Despite this additional complexity, oxygen isotope ratios in the hardparts of vertebrates

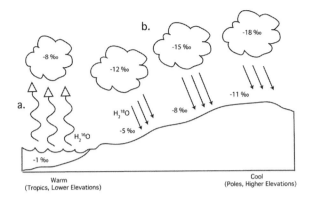

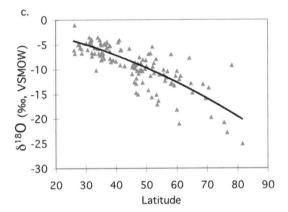

*Figure 8.3.* Schematic illustration of the hydrologic processes which effect oxygen isotope ratios of precipitation (and surface waters) at global and regional scales. A. Because the lighter isotope is preferentially incorporated into the vapor phase during evaporation, water in air masses over the ocean have lower $\delta^{18}O$ values than ocean water. B. Condensation takes place as these air masses move over land or up elevation. The heavier isotope is preferentially incorporated into the liquid or solid phase so that precipitation has a higher $\delta^{18}O$ than the remaining water vapor. Ongoing cooling and, thus, distillation of air masses results in progressively lower $\delta^{18}O$ values of both vapor and precipitation. C. As latitude increases, air-mass distillation generally results in precipitation with a decreased $\delta^{18}O$ (after Rozanski et al., 1993).

that drink surface water can record oxygen isotope variability and, in turn, can be used to infer both paleoenvironmental and paleoecological conditions. The following is a brief summary of causes of oxygen isotopic variability in surface waters on global and local scales.

At present, $\delta^{18}O$ of global precipitation ranges from approximately 0‰ to –30‰ (Dansgaard, 1964; Rozanski et al., 1993). The primary cause of isotopic variability in precipitation is the preferential incorporation of $^{18}O$ into condensate as water is precipitated and removed from cooling air masses (Fig. 8.3). As more precipitation is removed from an air mass,

*Figure 8.4.* Schematic illustration of relative $\delta^{18}O$ values that may be found for surface water (both standing and leaf water) located in different parts of a coastal to terrestrial ecosystem. A. Freshwater swamps. B. Zone of marine and fresh water mixing. C. Forest canopy. D. Streams with precipitation from high elevation. E. River and river margin. F. Ponds and lakes. G. Open grassland. H. Mixed forest/grassland. I. Trunk river. Using area H as a baseline, water from A, B, and G are likely to have higher $\delta^{18}O$, while that from C, D, and E are likely to have lower $\delta^{18}O$. Reasons for relatively higher $\delta^{18}O$ values include evaporation of standing water, particularly in open vegetation or arid settings (A); mixing of high $\delta^{18}O$ ocean water with freshwater sourced in precipitation (B); enhanced evaporation of leaf water in sunnier, windier open settings (G). Reasons for relatively lower $\delta^{18}O$ values include reduced evaporation of leaf water in shady, still understory settings (C); collection of precipitation from higher elevations having lower $\delta^{18}O$ values (D); more humid air and saturated soils (E). $\delta^{18}O$ values of ponds and lakes are variable but are likely to be higher than $\delta^{18}O$ of local precipitation, except when very small or when located in very humid settings. $\delta^{18}O$ of a trunk river (I) will depend on the source and hydrologic history of all the waters collected and mixed into it at any given point. Because combinations of different factors such as regional climate, hydrology, and plant type may all act to modify the general pattern presented here, this figure alone should not be used to interpret isotope data from bonebed remains.

$\delta^{18}O$ of the remaining vapor becomes progressively lower. The resulting patterns in $\delta^{18}O$ of precipitation ($\delta^{18}O_{pt}$) include a regular decrease in $\delta^{18}O_{pt}$ as air masses cool while rising over mountains, moving away from coastal areas, or moving from tropical source areas to polar sinks (Epstein and Meyada, 1953; Dansgaard, 1964; Rozanski et al., 1993; Gat, 1996). In tropical regions, where vertical convection results in cooling, a correlation also occurs between the amount of precipitation and $\delta^{18}O_{pt}$ (Dansgaard, 1964; Araguas-Araguas et al., 1998).

Although regional temperatures and rainout patterns play a major role in determining the oxygen isotope ratio of precipitation at any given locality, terrestrial vertebrates generally do not ingest precipitation directly. Instead, they ingest water from surface water reservoirs such as streams, lakes, and leaves. In turn, these reservoirs may have oxygen isotope ratios that differ significantly from those of local precipitation due to a variety of local hydrological processes (Fig. 8.4). For example, small ponds and streams in humid areas may hold local precipitation with little isotopic modification, but larger lakes and soil waters, especially in arid regions, may undergo evaporation that modifies their oxygen isotope ratio via the preferential incorporation of $^{16}O$ into the vapor phase. Similarly, $\delta^{18}O$ values of leaf water ingested by herbivorous animals can be shifted to higher values relative to precipitation as a result of evaporation at the surface of a leaf, particularly in less-humid environments (Sternberg, 1989). Lastly, precipitation from large areas and over long periods of time can be mixed together during the formation of lakes, soil water, groundwater, and larger rivers.

## Bioapatite as a Record of Isotope Variability over an Ancient Landscape

Animals record the isotopic characteristics of ancient landscapes when they ingest organic material and drink from surface water reservoirs. After being eaten by an herbivore, organic compounds in plants are metabolized and carbon is incorporated into a number of different phases, including dissolved $CO_2$ and $HCO_3^-$, which ultimately come to reside in the herbivore's bioapatite (Koch et al., 1994; Koch, 1998; Cerling and Harris, 1999; Passey et al., 2005). Carbon isotope fractionations associated with these processes result in $\delta^{13}C$ values of both bioapatite carbonate and organic materials that are significantly higher than those of ingested plant matter. For large wild mammals, the offset between $\delta^{13}C$ of bioapatite and bulk diet is approximately 12–15‰ (Koch, 1998; Cerling and Harris, 1999; Kohn and Cerling, 2002; Hoppe et al., 2004a; Passey et al., 2005), for rodents the offset is approximately 10‰ (DeNiro and Epstein, 1978; Ambrose and Norr, 1993), for large birds the offset between $\delta^{13}C$ of eggshell carbonate and bulk diet is about 14–16‰ (Schaffner and Swart, 1991; Johnson et al., 1998), and for dinosaurs the offset between $\delta^{13}C$ of bioapatite and presumed bulk diet is estimated to be roughly 18‰ (Fricke et al., in review). The reason for these different amounts of offset is not clear, although they may be related to differences among taxa regarding (1) the biogeochemical processes that take place as carbon from plants is incorporated into bioapatite and organic materials of herbivores, or (2) which organic compounds in a plant (i.e., proteins, carbohydrates, lipids) are actually utilized by the animal (DeNiro and Epstein, 1978; Krueger and Sullivan, 1984; Lee-Thorp et al., 1989; Ambrose and Norr, 1993; Gannes et al., 1998; Koch, 1998; Hedges, 2003; Jim et al., 2004; Passey et al., 2005).

Carbon isotope ratios of modern terrestrial carnivores have also been studied (both bioapatite and bone proteins; Ambrose and DeNiro, 1986;

Lee-Thorp et al., 1989; Hilderbrand et al., 1999; Roth and Hobson, 2000; Roth, 2002; Bocherens and Drucker, 2003; Kohn et al., 2005; Fox-Dobbs et al., 2006), but not as extensively as in herbivores. Paleoecological and dietary interpretations of bioapatite isotope data from carnivores are difficult to make because there is a lack of empirical data regarding the isotopic offset between $\delta^{13}C_{herbivore}$ and $\delta^{13}C_{carnivore}$, and an incomplete understanding of why and how these offsets occur. Lee-Thorpe et al. (1989) calculated isotopic offsets between carnivores and other organisms from a South African ecosystem where C4 plants were presumed to be absent. They observed that $\delta^{13}C$ of herbivore meat is ~2.5‰ higher than that of consumed plants and calculated an offset of ~9‰ between carnivore bioapatite and herbivore meat. Accordingly, they calculated that the carbon isotope offset between carnivore bioapatite and local plants is ~11.5‰, a value similar to that observed between herbivore bioapatite and plants living in the same area. In contrast, Fox-Dobbs et al. (2006) analyzed remains of coexisting wolves and large herbivores from C3 ecosystems of North America, and observed that carnivore $\delta^{13}C$ was consistently ~1‰ lower than herbivore $\delta^{13}C$.

In order to understand these offsets and whether they should be expected to remain constant in the face of taxonomic, dietary, or environmental changes, researchers have used experimental approaches (Ambrose and Norr, 1993; Tieszen and Fagre, 1993; Balasse et al., 1999; Passey et al., 2005; Zazzo et al., 2005) and modeling (Hedges, 2003). If the results of Lee-Thorpe et al. (1989) or Fox-Dobbs et al. (2006) are confirmed by future research, then $\delta^{13}C_{herbivore} - \delta^{13}C_{carnivore}$ values may be useful in identifying the dietary choices of ancient carnivores and, thus, ancient trophic relationships.

Oxygen in vertebrate bioapatite has sources primarily in ingested water and atmospheric oxygen that contribute to blood/metabolic water (Longinelli 1984; Luz and Kolodny, 1985; Bryant and Froelich, 1995; Kohn, 1996; Kohn and Cerling, 2002). $\delta^{18}O$ of atmospheric oxygen has remained relatively constant over time and space with a value of ~23‰ (Kohn, 1996). Thus, it probably does not influence oxygen isotope variations in bioapatite of vertebrates living in different places or drinking different waters (Fig. 8.4).

The isotopic offset between ingested surface water and both phosphate and carbonate phases that are present in biogenic apatite is controlled by (1) body temperature, which determines the isotopic fractionation between apatite and body water, and (2) fractionations that occur during the formation of body water from ingested water. Where body temperature is known and constant (i.e., homeothermic mammals and birds), both of these factors can be considered together using physiological models that account for the fluxes of oxygen into and out of the body, and the oxygen isotope fractionations associated with each metabolic process (Bryant and Froelich, 1995; Kohn, 1996). Alternatively, direct empirical relations between $\delta^{18}O$ of bioapatite and $\delta^{18}O$ of meteoric water may be used for different kinds of animals. In general, both methods of determining the isotopic offset between $\delta^{18}O$ of bioapatite (or carbonate in the case of eggshell) and $\delta^{18}O$ of ingested waters provide similar results for large mammals and birds that do not rely on ingested leaf water or rely on unique water conservation strategies (Schaffner and Swart, 1991; Johnson et al., 1998; Koch, 1998; Iacumin and Longinelli, 2002; Kohn and Cerling, 2002; Hoppe et al., 2004a; Hoppe, 2006).

Determining the offset between $\delta^{18}O$ of bioapatite and $\delta^{18}O$ of ingested waters is not straightforward if body temperature is unknown (e.g., dinosaurs). In such cases, only relative differences in $\delta^{18}O$ of ingested waters may be inferred. Alternatively, if body temperature varies in response to environmental conditions (e.g., ectothermic reptiles, amphibians, fish), then environmental temperature, which is used to estimate the $\delta^{18}O$ of ingested waters, must be determined independently.

### Bones, Teeth, and Scales

Bioapatite is a major component of several different skeletal elements found in vertebrates, including teeth, tusks, bones, and body scales. Both teeth and tusks consist of two main kinds of materials: enamel and dentine. The former forms a hard shell around the softer dentine core. Body scales of some fish are analogous to teeth in that a carapace of enamel-like bioapatite (ganoine) covers dentine-like material. Teeth and body scales grow incrementally for a limited period early in the life of an animal, whereas tusks continue to grow until an animal dies. In all three cases, once bioapatite is deposited it is no longer open to chemical modification via biological or physiological processes. Thus, teeth, tusks and scales "lock in" the isotopic ratios that existed during bioapatite deposition. Bone, however, can be remodeled over the entire life of an organism, and stable isotope ratios in its bioapatite may simply reflect conditions prior to death.

Kohn and Cerling (2002) provide an excellent overview of the mineralogy and biogeochemistry of these materials, and only key factors are repeated here. In living vertebrates, enamel, dentine, and bone differ primarily in (1) the size of apatite crystals and (2) the amount of organic collagen originally present. Bone has very small apatite crystals tens of nanometers in length, and a framework of organic collagen that makes up $\sim$35% of unaltered bone. Tooth dentine is characterized by similar crystal

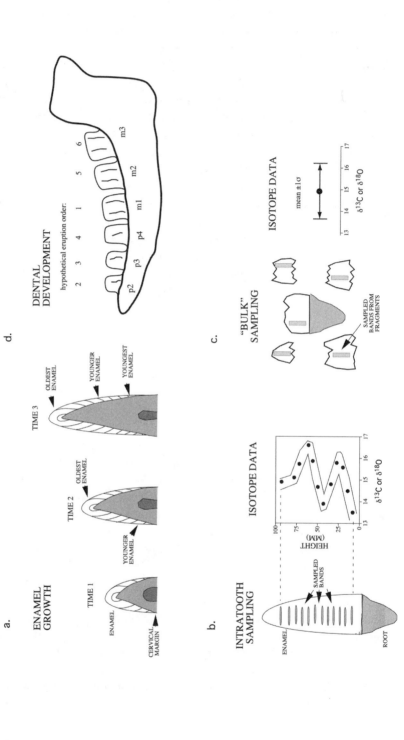

a.
ENAMEL
GROWTH

TIME 1

ENAMEL

CERVICAL
MARGIN

TIME 2

OLDEST
ENAMEL

YOUNGER
ENAMEL

TIME 3

OLDEST
ENAMEL

YOUNGER
ENAMEL

YOUNGEST
ENAMEL

b.
INTRATOOTH
SAMPLING

ENAMEL

SAMPLED
BANDS

ROOT

ISOTOPE DATA

HEIGHT
(MM)

100

75

50

25

0

$\delta^{13}C$ or $\delta^{18}O$

13    14    15    16    17

c.
"BULK"
SAMPLING

SAMPLED
BANDS FROM
FRAGMENTS

ISOTOPE DATA

mean ±1σ

$\delta^{13}C$ or $\delta^{18}O$

13    14    15    16    17

d.
DENTAL
DEVELOPMENT

hypothetical eruption order:

2    3    4    1    5    6

p2    p3    p4    m1    m2    m3

sizes, but less collagen (~20%). In contrast, enamel is made up of larger apatite crystals hundreds of nanometers in length and only contains <5% original organic material. These differences influence porosity and, thus, the quality of preservation of isotopic information over time (see below).

One final detail to consider is that teeth of many vertebrates grow incrementally over the span of months to years (Fig. 8.5a; Hillson, 1986) and, thus, tooth enamel and tusks can capture short-term isotopic variations in carbon and oxygen isotope ratios of ingested food and water (Fig. 8.5b; Koch et al., 1989; Fricke and O'Neil, 1996; Stuart-Williams and Schwarcz, 1997; Fricke et al., 1998a; Kohn et al., 1998; Sharp and Cerling, 1998; Wiedemann et al., 1999; Gadbury et al., 2000; Fox and Fisher, 2001, 2004; Passey and Cerling., 2002a; Feranec 2004a; Higgins and MacFadden, 2004; Hoppe et al., 2004b; Rinaldi and Cole, 2004; Straight et al., 2004; Nelson, 2005; Zazzo et al., 2005). In fact, both carbon isotope ratios of plants and oxygen isotope ratios of surface water can vary over the course of a year in response to changes in environmental conditions such as temperature, humidity, and precipitation. Furthermore, vertebrate herbivores may occupy different parts of an ecosystem and consume different plants and waters depending on the time of year. Thus, the intratooth isotope variability that results must be quantified in order to effectively compare data between and among populations of animals. Ultimately, such variability may be useful in studying seasonal changes in climatic variables or seasonal shifts in animal behavior (see below).

## STABLE ISOTOPE RATIOS AND DIAGENESIS

In order for stable isotope ratios of vertebrate fossils from bonebeds to provide useful information regarding environments or ecological relations

---

*Figure 8.5.* A. Enamel and dentine formation occurs incrementally as a tooth erupts past the cervical margin. Thus, a continuous time series of isotopic information is preserved. Note that bands of enamel form at an angle to the enamel and dentine contact. Because it is difficult to measure the isotopic composition of only a single time band, some mixing, or dampening, of isotopic signals may occur. B. Isotopic data ($\delta^{18}O$ of water and $\delta^{13}C$ of plants) from sampled bands of enamel along the direction of growth provide information about the magnitude of seasonal variations in environmental conditions. Using these data it is also possible to estimate rates and season of tooth growth in an individual. C. Seasonal variations in isotope ratios can also be assessed by sampling enamel from isolated tooth fragments. Assuming no major climatic changes over time, average $\delta^{18}O/\delta^{13}C$ values for a population of fragments approximates the annual mean isotope ratio, and the standard deviation approximates the seasonal variability (Clementz and Koch, 2001). D. Teeth in a jaw do not all erupt through the cervical margin at the same time, and the eruption rate may vary from tooth to tooth. It may be possible to quantify these rates of dental development by conducting intratooth sampling of all the teeth in a jaw and comparing their isotopic patterns (e.g., Kohn et al., 1998; Gadbury et al., 2000; Nelson, 2005).

of the past, primary isotopic information must be obtained. Therefore, the topic of diagenesis (postdepositional alteration) of original isotope ratios is of tremendous importance. There is now broad consensus on a number of diagenetic issues. First, enamel, dentine, and bone have different suscepti-bilities to diagenetic alteration, with enamel regarded as the least suscep-tible and, thus, the most preferred for analysis. Secondly, mechanisms of geochemical alteration may be different for stable isotopes than for trace elements, including rare earth elements (REE), and therefore a decoupling of isotopic and REE alteration is possible. Lastly, although there is no unambiguous way to determine whether diagenetic alteration of primary isotope ratios has occurred, it is still possible to obtain useful paleoen-vironmental and paleoecological information using stable isotope ratios as long as the potential for some degree of alteration is recognized and materials suitable for analysis are identified.

### Differing Susceptibilities to Alteration

Biogenic apatite that makes up hardparts of vertebrate fossils is far more common in the geologic record than organic material, and this fact reflects its general resistance to decomposition over a wide range of geochemical conditions. Nevertheless, there are both physical and chemical variations within bioapatite that may result in differing susceptibilities to alteration. For example, the size of apatite crystals and the original amount of organic matter present are not uniform among skeletal elements. Furthermore, the isotopic ratios of two chemical components found in bioapatite, $PO_4$ and $CO_3$, are commonly studied, and these may be affected differently even when exposed to the same diagenetic conditions.

In general, diagenetic alteration of bioapatite is thought to occur by two end-member processes: (1) isotopic exchange between biogenic ap-atite and surrounding fluids containing $H_2O$, $HCO_3^-$, $CO_2$, $CH_4$, and (2) dissolution and/or addition of secondary apatite and carbonate (Zazzo et al., 2004). The former requires that P-O or C-O bonds of anionic com-plexes within apatite are broken and then reformed so that isotope ex-change may occur. In order for the isotope ratio of carbonate or phosphate in recrystallized apatite to differ significantly from initial ratios (and thus be recognized as resulting from diagenesis), the temperature of this di-agenetic process must be significantly different than that of the original bioapatite formation, or the isotopic exchange must occur in the presence of C and O from external sources that have an isotope ratio much different than that found in the primary phosphate or carbonate complex.

In the case of secondary mineral precipitation, biogenic apatite may retain its original isotope ratios but the primary signal can be overwhelmed by that of secondary precipitates. In these cases, the degree of isotopic alteration observed will depend on the percentage of secondary mineral present. Secondary carbonate and apatite minerals are often precipitated from groundwater that is isotopically different than body water and at temperatures that are much different than those in the body of an animal. Accordingly, secondary minerals often have an isotope ratio that is significantly different from that of original (and unaltered) biogenic apatite.

A basic understanding of these processes is important because skeletal remains with high porosities can be subjected to great fluxes of exogenous fluids, while those with small apatite crystals will have much more surface area available to undergo isotopic exchange and more volume available for precipitation of secondary phosphates and carbonates. Of the common skeletal materials, bone has very small apatite crystals and a high porosity potential due to the high percentage of organic collagen that is likely to be altered or removed soon after burial. In contrast, enamel is made up of much larger crystals and has much less porosity potential. Thus, although bone is most likely to be susceptible to all diagenetic processes (Nelson et al., 1986; Kolodny et al., 1996; Kohn and Cerling 2002; Trueman et al., 2003), tooth dentine is less likely to be affected, and tooth enamel is the least likely to be affected. For this reason, it is strongly recommended that bone be avoided in favor of tooth enamel as a substrate for isotopic analysis.

In the case of chemical differences within enamel bioapatite, P-O bonds are stronger and more resistant to breaking during inorganic reactions than C-O bonds (Tudge, 1960; Zazzo et al., 2004). In general, therefore, phosphate oxygen in enamel is considered better suited for isotopic research than carbonate carbon and oxygen in enamel (see review by Kohn and Cerling, 2002), but ratios from both should be considered suspect unless there is convincing evidence that alteration did not occur.

There are also important exceptions to these generalities. In the presence of microbial activity, enzyme-mediated reactions may break P-O bonds (Blake et al., 1997; Lécuyer et al., 1999), and may do so to an even greater extent than C-O bonds (Zazzo et al., 2004). In contrast, skeletal remains buried quickly in clay-rich sediment, or buried in very arid settings, may be effectively sealed off from exposure to groundwater and microbial activity. For this reason, alteration needs to be considered on a case-by-case basis, and it is necessary to have testable means of identifying whether primary isotope information is preserved.

It must also be stressed that the mechanisms of stable isotope alteration are not the same as those that cause changes in REE signatures of bioapatite (Trueman, Chapter 7 in this volume). Alteration of REE values is due primarily to cation substitution of elements for Ca in the crystal lattice and/or the adsorption of these elements onto surfaces of small apatite crystals, the effects of which may be enhanced by recrystallization of apatite (Kohn et al., 1999; Trueman and Tuross, 2002; Trueman, Chapter 7 in this volume). Neither cation substitution nor apatite recrystallization result in the breaking of C-O and P-O bonds that make up anionic complexes in this mineral, thus neither process will necessarily be associated with diagenetic alteration of stable isotope ratios of enamel bioapatite (Trueman et al., 2003).

Because there is no direct geochemical link between REE and stable isotope diagenesis, the occurrence of each must be addressed independently. It is quite possible that REE signatures of vertebrate fossils will reflect geochemical conditions of early depositional environments due to the ease with which cation substitution occurs during diagenesis (Kohn et al., 1999; Trueman and Tuross, 2002; Trueman, Chapter 7 in this volume), while at the same time primary stable isotope ratios related to animal behavior and biology will be preserved. In such an ideal case, combined stable isotope and REE analysis of bonebed remains can be amazingly powerful, having the potential to provide insight into both environmental conditions and ecological relations while the animals were living (stable isotopes; see below) and information regarding depositional environments and taphonomic processes after their deaths (REE).

### Identifying Effects of Isotopic Alteration

Although reasonable arguments can made that certain kinds of vertebrate remains are more or less susceptible to diagenetic alteration, it should never be assumed that original stable isotope information is or is not preserved for any given material from any bonebed. Instead, attempts should be made to resolve whether a primary signal remains. Many different methods of doing so have been proposed, but none provide unambiguous results (Sharp et al., 2000; Kohn and Cerling, 2002; Fricke et al., in review). Therefore, it is suggested here that several different tests for diagenetic alteration be applied with the goal of obtaining consistent results. Tests emphasized here involve comparing isotope data from bioapatite in enamel with isotope data from (1) associated authigenic carbonates and phosphates and (2) associated dentine or bone. A third test involves comparing isotope data from bioapatite in enamel or biocarbonate of shell among different taxa.

The first approach is based on the premise that authigenic minerals found in sediments hosting vertebrate remains should have isotope ratios that reflect those of diagenetic fluids and temperatures. Thus, any evidence of an isotopic relation between data sets, such as the formation of an isotopic mixing relationship can be used to identify effects of diagenetic overprinting of primary isotope ratios (Fig. 8.6a), whereas lack of mixing trends indicate that alteration is perhaps minimal (Fig. 8.6b; Barrick and Showers, 1994, 1995; Quade et al., 1992). It is difficult to determine, however, if the timing of authigenic mineral formation and the geochemical conditions reflected by authigenesis are, in fact, the same as the diagenetic processes that may be influencing the isotope ratios of bioapatite. In other words, authigenic mineral formation and isotopic alteration may be geochemically decoupled, as are processes of REE alteration and isotope alteration (Trueman and Tuross, 2002; Trueman, Chapter 7 in this volume).

Because of this uncertainty, it is better to compare isotope data between different skeletal components such as enamel, dentine, and bone (Fig. 8.6c; Wang and Cerling, 1994). Here, the assumptions are that (1) alteration of these biogenic materials occurred at the same time and (2) any isotopic difference between enamel, dentine, or bone reflects differences in crystal size and porosity, and thus susceptibility to isotopic exchange and precipitation of secondary minerals during diagenesis. In this test, if isotope ratios of tooth enamel are observed to have significantly different average values and variances than dentine or bone, then it can be concluded that the latter have been altered to some degree by diagenetic processes (Fig. 8.6c). Such a result does not rule out alteration of enamel, but it does provide strong evidence that enamel has not been affected by diagenesis to the same degree as dentine or bone.

Additional support for preservation of primary isotope information in enamel can be gathered by comparing isotope data from fossil remains of different taxa (Fig. 8.6d). Differences in the mean and/or variance for populations of different animals have been observed for a number of taxa from different time periods and localities (Feranec and MacFadden, 2000, 2006; Clementz et al., 2003; Cerling et al., 2004; Kohn et al., 2005), and are expected if the taxa are characterized by different physiologies and/or ecological behaviors. For example, a semiaquatic herbivore that drinks river water and eats plants that have not experienced stress due to limited water availability should have lower $\delta^{18}O$ and $\delta^{13}C$ than herbivores that rely on leaf water and eat stressed plants from drier parts of the local environment. Resulting isotopic offsets among taxa would not exist if isotopic alteration was extensive. In such a case isotopic exchange with groundwaters or secondary precipitation of apatite during diagenesis should result

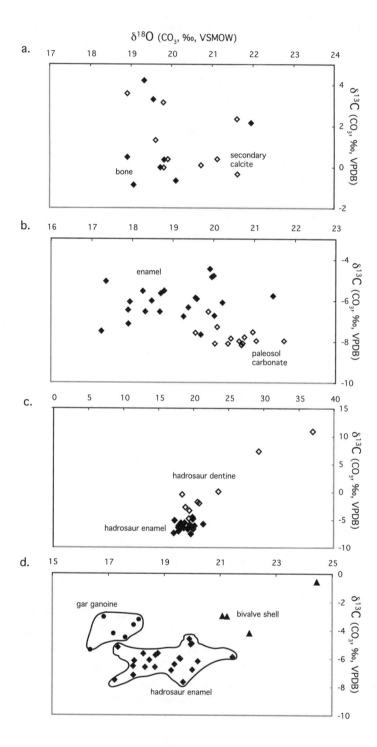

in uniform isotope ratios. It is possible, of course, that isotopic overlap will occur between taxa whose remains are unaltered due to similarities in the isotopic composition of animal diet, drinking water, or physiological processes. Thus, a lack of isotopic offset between fossil taxa does not necessarily imply diagenesis.

A slightly different way of constraining the amount of oxygen isotope alteration that may have occurred involves a comparison of $\delta^{18}O$ values from $CO_3$ and $PO_4$ components from the same skeletal sample. Modern mammal bone and enamel show a consistent offset in $\delta^{18}O$ between these phases of $\sim$9‰ (Bryant et al., 1996b; Iacumin et al., 1996; Iacumin and Longinelli, 2002), which reflects differences in isotopic fractionation associated with carbonate and phosphate precipitation in bioapatite. If the offset in $\delta^{18}O$ values of $CO_3$ and $PO_4$ components of fossil mammal bioapatite vary from this modern norm, it can be concluded that diagenetic alteration of $CO_3$ or $PO_4$ or both, has occurred. However, exactly which component is affected, and to what degree, is difficult to quantify (Zazzo et al., 2004). Furthermore, even among modern samples the offset in $\delta^{18}O$ between $CO_3$ and $PO_4$ may vary by several per mil within a single population of animals. In the absence of a method to differentiate normal variation from that introduced during diagenesis, the usefulness of this approach as an unambiguous means of identifying isotopic alteration must be regarded as limited.

Lastly, studies of bioapatite using infrared spectroscopy (IR) allow for the investigation of elements in specific sites within a crystal lattice, such as phosphate and carbonate groups, and for a characterization of how they are modified during diagenesis. Integration of these methods with isotopic analysis of oxygen and carbon from these groups may ultimately allow changes in the crystallographic structure of bioapatite to be tied directly

---

*Figure 8.6.* There are several possible ways of identifying whether significant diagenetic alteration of carbon and oxygen isotope ratios has occurred. One is a comparison of bioapatite and diagenetic carbonate. A. Complete overlap of isotopic ratios between bone (filled) and secondary calcite (open) indicates that primary bone values have been overprinted and now reflect diagenetic conditions (data from Trueman, personal communication, 2006). B. Limited overlap and a lack of a linear mixing relationship between tooth enamel (filled) and paleosol (i.e., diagenetic; open) carbonate is not consistent with diagenetic alteration of tooth enamel carbonate (data from Fricke et al., in review). Another is comparisons of isotope data from tooth enamel and dentine of the same tooth (C). Increased isotopic variability and different isotope ratios are evidence that one of the bioapatite phases (most likely dentine) have been adversely affected to a larger degree by diagenesis (see text). Lastly, isotopic differences between coexisting taxa can be used (D). Extensive overprinting of primary isotope signals by diagenetic processes should result in uniform isotope ratios that reflect temperatures and fluid conditions of diagenesis. The lack of such uniformity, as shown by the occurrence of taxonomic differences, is evidence that complete isotopic resetting did not occur.

to changes in isotope ratios. Although this type of research may allow for unambiguous interpretations of isotopic preservation to be made in the future, at present, the technique has limited usefulness (e.g., Sponheimer and Lee-Thorpe, 1999a).

Application of some or all of the above approaches to a suite of bone-bed samples may allow for the independent confirmation that bioapatite of some material, most likely enamel, has not been altered significantly during diagenesis. That said, such a result begs the question of how much evidence for or against isotopic alteration is sufficient to render a study viable or impracticable. A large part of the answer depends on the type of questions being asked. If the goal is to test whether different herbivores partitioned food or water resources by eating plants and ingesting waters from different parts of the ecosystem, then the preservation of isotopic offsets in bone—even in the face of shifts in mean values or population variance—may be sufficient. However, if the paleoenvironmental and paleoecological investigations being conducted require more precise estimates of isotopic means and variances for a given population, then analysis of bone will likely be insufficient. Instead, enamel should be the focus of study, and as many methods as possible should be applied to increase confidence in the results.

## APPLICATIONS: WORKING WITH BONEBED SAMPLES

Because stable isotope ratios of vertebrate remains are influenced by environmental conditions, animal behavior, and animal physiology, it is possible to study all of these subjects using stable isotope data from bonebeds. However, the feasibility of addressing questions related to these topics depends on the number of individuals sampled from a bonebed, the temporal relationship of these individuals (are they contemporaneous?), and the degree to which seasonal isotopic variability is captured by the sample. In turn, these factors are strongly influenced by the taphonomic history, taxonomic diversity, and geology of a given bonebed (e.g. Eberth et al., Chapter 5 in this volume). These general issues are discussed below.

### Accounting for Seasonal Isotope Variations

It is critical to identify any seasonal bias that may exist in isotope data sets before comparing isotope data from within or between assemblages at one or more bonebeds. By so doing, any observed isotopic similarities and differences within and between assemblages can be more confidently attributed to biology, behavior, or local environment.

One way to characterize seasonal isotope variability is to undertake intratooth sampling, which involves the collection of multiple samples from along the length of a single tooth. This method works because enamel is added incrementally during tooth growth, creating a time series of isotope data (Fig. 8.5b). Depending on the rate of tooth growth and the amount of tooth wear, this sampling approach can allow median $\delta^{13}$C and $\delta^{18}$O values as well as maxima and minima values to be determined precisely for time periods of up to several years (Fig. 8.5b). However, intratooth sampling of a single tooth provides information for only a one- to two-year period in the life of a single individual. Studies of multiple teeth from more than one individual will more likely provide a representative sample of the environmental conditions experienced by a population of animals that lived and died together (mass-death assemblages) or during a relatively brief period of geologic time (time-averaged assemblages). Among mass-death assemblages, an additional benefit of doing a multitooth multiindividual study is that a comparison of yearly isotope records can provide insight into annual variability in isotope ratios and isotope ranges.

Another approach to characterizing seasonal variations in tooth enamel isotope ratios involves the analysis of many bulk samples of tooth enamel from the same taxon. Such bulk samples, usually obtained from worn or broken tooth fragments, will contain only a small part of the seasonal isotope record, and any single $\delta^{13}$C/$\delta^{18}$O value is unlikely to represent the annual median or seasonal maxima/minima. Statistical models suggest, however, that the range in $\delta^{13}$C/$\delta^{18}$O for a large population of bulk samples will accurately reflect the seasonal isotopic variability experienced by that population of animals, whereas mean $\delta^{13}$C/$\delta^{18}$O values will reflect the annual median isotope ratio (Fig. 8.5c; Clementz and Koch, 2001). In this bulk-sampling approach it is not possible to obtain precise records of seasonal isotope variations for single years.

### Bonebed End-members

The number of specimens available for study, the taxonomic diversity of the assemblage, and the amount of time represented by the specimens are the three most important factors influencing the degree to which isotopic data from bonebeds effectively and accurately reflect original biological and paleoenvironmental conditions. Here, these factors are considered in the context of two bonebed end-members: those that yield low-diversity, mass-death assemblages, and those that yield high-diversity, time-averaged assemblages.

Mass-death assemblages often consist of small to modest numbers of articulated or partially articulated skeletons that were deposited and buried over short periods of time (e.g., Ashfall Fossil Beds [Voorhies, 1985; Rogers and Kidwell, Chapter 1 in this volume]). This type of assemblage is often limited in its taxonomic diversity (monotaxic to monodominant; Eberth et al., Chapter 3 in this volume), making it more difficult to obtain isotope samples for comparative analysis. Moreover, teeth in jaws and other exceptional specimens of entire teeth in collections are usually off limits to the destructive sampling techniques of isotopic analysis. Mass-death assemblages often do not consist of paleofaunas or yield isotopic data that accurately reflect long-term paleoenvironmental or paleoecological trends. However, they do have tremendous potential to provide information about the paleobiology of the most abundant animals present, including habitat preferences and tooth eruption rates and patterns. Also, as indicated above, analyses of these assemblages can provide important insight into seasonal climatic variability.

At the other end of the spectrum are high-diversity (multitaxic) assemblages that include isolated remains that were deposited and possibly reworked over time (e.g., channel lag and ravinement surface deposits [Brinkman, 1990; Eberth, 1990; Rogers and Kidwell, 2000]). In many cases, these types of assemblages preserve large numbers of individuals representing multiple taxa, and many fragmentary specimens that are of little value in morphological analyses. Accordingly, collections from bonebeds that yield these assemblages are frequently available for destructive sampling techniques.

By analyzing remains of different animals in multitaxic, time-averaged assemblages, a larger part of the isotopic variability present in any given ecosystem is likely to be sampled due to differences in dietary choices and drinking behaviors of different taxa and individuals. For example, different species of mammalian herbivores often partition plant resources and have different strategies for obtaining water (e.g., directly from water bodies; indirectly from leaves). Thus, analyses of many mammalian taxa from a bonebed, including carnivores and herbivores, have the potential to record more of the extant carbon and oxygen isotope variability associated with different plants, photosynthetic pathways, and microenvironments that make up any ecosystem (e.g., Fig. 8.7; Bocherens et al., 1996, Cerling and Harris, 1999; Sponheimer and Lee-Thorp, 1999c; Sponheimer et al., 2003; Passey et al., 2005). Similarly, by collecting data from large numbers of individuals, the impact of climatic anomalies, such as unusually cold or warm years, on isotopic means and ranges for a given

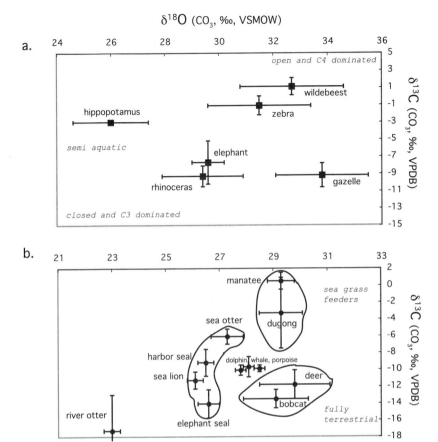

*Figure 8.7.* Carbon and oxygen isotope data from modern mammals illustrate how vertebrates living in the same terrestrial (A) and marine (B) ecosystems can have significantly different isotope ratios. The distribution reflects differences in environmental conditions, water sources, and plant types between microhabitats. For example, terrestrial herbivores that are more likely to obtain water from leaves (e.g., gazelle, deer) are expected to have higher $\delta^{18}O$ values due to evaporation at leaf surface. This evaporative effect will be magnified in more open and arid habitats but may be negligible for semiaquatic herbivores that eat aquatic plants (e.g., hippopotamus) or for carnivores (e.g., bobcat). In the case of carbon, animals eating plants that utilize the C4 pathway (e.g., zebra) or sea grass (e.g., manatee) are expected to have higher $\delta^{13}C$ values. Aquatic taxa are generally characterized by less isotopic variability than terrestrial taxa as fewer microenvironments exist to accommodate these forms (Clementz and Koch, 2001). Data from Bocherens et al. (1996), Clementz and Koch (2001), and Clementz et al. (2003). Symbols represent population means and bars represent one standard deviation away from the mean.

population is decreased. Similarly, large sample sizes decrease the impact of behavioral variation over time or between individuals. In general, multitaxic bonebeds provide greater opportunities to more thoroughly characterize past environments and animal behavior compared to bonebeds containing low-diversity, mass-death assemblages.

Isotopic analyses of vertebrate remains from bonebeds of different age or from different areas offer a number of exciting research possibilities. Data can be used to establish a regional isotope record of paleoenvironmental change through time, which, because they are linked to records of ecological interactions, may allow for the study of how plant-animal and animal-animal interactions responded to changing climatic or hydrologic conditions. Focusing instead on a single time slice, isotope data from multiple bonebeds can be used to establish spatial patterns in isotopic variation. These spatial pattern "maps" can then be used to infer relative differences in paleoenvironmental conditions (e.g., temperature, humidity), explore plant type and animal dietary choices (e.g., C3/C4), and reconstruct regional patterns in air-mass rainout.

Bonebeds must be spaced at semi-regular to regular stratigraphic intervals in order to document and quantify temporal changes in paleoenvironments. Alternatively, bonebeds must occur within narrow stratigraphic limits if valid paleogeographic comparisons are to be made. These limitations place investigators at the mercy of the geologic record and chronostratigraphic data. Because most bonebeds are not closely associated with rocks that yield absolute ages, their chronostratigraphic relationships are usually determined in the context of broadly constrained time intervals based on paleomagnetic, biostratigraphic, and sequence stratigraphic data. Thus, the time intervals within which some bonebeds are compared may represent hundreds of thousands to millions of years. Furthermore, researchers may choose to combine bonebeds from different geologic time intervals in order to increase the number of comparable sites. For example, Fricke and Rogers (2000) lumped together localities from the "Campanian/Maastrichtian boundary interval" that may have spanned several million years. Similarly, Fricke (2003) compared isotope data from Eocene fossils of "Wasatchian" land mammal age, with possible differences in age of several million years. In both cases, climate records from marine sediment cores were used to argue that climate change during these intervals was minimal.

The effect that spatial-temporal lumping of bonebeds may have on the interpretation of stable isotope data must be addressed on a case-by-case basis. If it can be argued, using independent lines of data, that climate did not change significantly during the specified time interval, then spatial differences in isotope ratios can be interpreted as reflecting geographic differences in paleoenvironmental and paleoecological factors.

Bonebeds may be associated with nonbiogenic minerals and other sedimentary material that also contain stable isotope information of paleoenvironmental and paleoecological importance. For example, carbon isotope ratios of soil carbonates may accurately reflect the average $\delta^{13}C$ for all plants that lived in the area as the carbonates were forming (Fox and Koch, 2003, 2004). By comparing $\delta^{13}C$ values of the carbonates to those from associated vertebrate remains, it is theoretically possible to reveal an animal's dietary and ecological preferences for plants with either high or low carbon isotope ratios. In the case of oxygen, evaporation can affect $\delta^{18}O$ of soil waters, particularly in arid settings, and oxygen isotope fractionation during carbonate formation is more temperature dependent than carbon isotope fractionation. Therefore, $\delta^{18}O$ of soil carbonates is not an ideal monitor of $\delta^{18}O$ of surface waters, but it may confirm the occurrence of arid or nonarid conditions.

Sedimentary organic matter such as wood, charcoal, or leaf cuticle, which is sourced from plants that once lived in the area of study, can provide useful isotopic data when collected from a bonebed, particularly if such organic matter is taxonomically identifiable. As carbon isotope ratios of such organic matter are the most direct representation of herbivore diet that can be obtained, they can be used to (1) determine which carbon isotope offsets between diet and bioapatite are reasonable for a given vertebrate, (2) describe variability in isotope ratios of different plants living in an ancient ecosystem, and (3) identify specific kinds of plants that were eaten by ancient vertebrates. Finely comminuted plant material ("coffee grounds") present in a bonebed is different in that it most likely provides a homogenized average $\delta^{13}C$ value reflecting contributions from many plants. Lastly, it is also possible that a preservational bias exists against certain plant remains and organic molecules with high carbon isotope ratios. In such cases, $\delta^{13}C$ of the sedimentary organic matter in a bonebed is generally high relative to that for the actual plant communities in the area (e.g., Fogel and Tuross, 1999; Krull and Retallack, 2000; Wynn et al., 2005).

## BONEBED APPLICATIONS: EXAMPLES

Having considered background issues and how sampling influences study design, it is now time to describe the types of paleoenvironmental, paleoecological, and paleobiological investigations that can be undertaken using isotope data from bonebeds. The selection of research highlighted below includes studies that focus on multiple animals and/or taxa. It is fully recognized that some, but not all, of these studies utilize samples derived from bonebeds. Research topics are presented separately so that the reader can focus on (1) the assumptions involved in each kind of study and (2) the kinds of interpretations that are possible. The following examples do not represent all potential research topics, rather, they reflect the background of the author and are intended to show how stable isotope analyses can be profitably included in a bonebed research program.

For general reviews of applications see Koch et al. (1994), Gröcke (1997), Gannes et al. (1998), Koch (1998), Kohn and Cerling (2002), and Lee-Thorp and Sponheimer (2005). Examples focusing on marine vertebrates include Lécuyer et al. (1993, 1996, 2003), Vennemann and Hegner (1998), Vennemann et al. (2001), and Pucéat et al., (2003). Lastly, the archaeological and paleoanthropological literature is rich with studies aimed at unraveling the diets and ecological preferences of primates and hominids using isotope data (Schwarcz and Schoeninger, 1991; Schoeninger, 1996; Sponheimer and Lee-Thorpe, 1999b; Schoeninger et al., 2001; Lee-Thorpe et al., 2003; Sponheimer et al., 2005).

### Paleoenvironmental Conditions

The successful study of environmental conditions at the scale of a single ecosystem relies on (1) accounting for any seasonal bias in $\delta^{13}C$ and $\delta^{18}O$ of vertebrate remains, (2) a careful consideration of isotopic offsets between vertebrate remains and ingested food and water, and (3) interpreting both the absolute values and ranges of $\delta^{13}C$ and $\delta^{18}O$. Once obtained, isotopic data can be utilized in several different ways.

#### Vegetation Structure

By providing a snapshot of isotopic variability in plants, data from vertebrate remains in bonebeds can be used to document the presence of a variety of vegetation structures, such as closed- and open-canopy forest, open grassland, or some combination of these. However, because such documentation requires a record of all isotopic variability in plants and surface water (e.g., Fig. 8.7), it is essential that $\delta^{13}C$ and $\delta^{18}O$ be determined for as many taxa as possible in a bonebed.

Many studies of this type have focused on Pleistocene landscapes. For example, Kohn et al. (2005) analyzed tooth enamel carbonate from six herbivore taxa that lived in South Carolina 115 ky ago (Fig. 8.8a). Ranges in $\delta^{13}C$ for an individual taxon such as horses ($\sim$6‰) or tapirs ($\sim$1‰) were small, but together $\delta^{13}C$ values from these six common taxa ranged from –16‰ to –1‰ and presented a more complete picture of vegetation structure. In particular, this broader range indicated that C3 plants were affected by a variety of environmental conditions in forested areas, whereas high values pointed to the existence of C4 grasslands. Furthermore, a positive correlation between $\delta^{13}C$ and $\delta^{18}O$ indicated that grassland areas were more open and subject to greater evaporation (Kohn et al., 2005). Other mixed C3–C4 forest-grassland paleoenvironments have been described using multiple-taxa isotope data from the Pleistocene of Florida (Koch et al., 1998) and from the Miocene of Kenya (Cerling et al., 1997b), and these studies demonstrate well how isotope data can complement other means of reconstructing vegetation structure.

In the absence of C4 plants, there is less potential for carbon isotope variability across a landscape, making it more difficult to resolve differences in vegetation type. Nevertheless, such resolution is still possible. In a recent study of a modern closed-canopy C3 forest in Africa, Cerling et al. (2004) observed a range in $\delta^{13}C$ of 12‰ for tooth enamel carbonate from more than a dozen vertebrate taxa living in and below closed-forest canopy. Similar ranges in isotope ratios were also observed in plants due to the canopy affect. These results indicate that isotope data of mammal remains have the ability to capture aspects of the vertical structure of forested areas. To date, applications of isotopic methods to ancient C3 ecosystems have been limited (MacFadden and Higgins, 2004; Botha et al., 2005; Feranac and MacFadden, 2006), but they do reveal this potential. Feranec and MacFadden (2006) observed dietary differences among five herbivorous taxa of Miocene age (Fig. 8.8b). The relatively small range in $\delta^{13}C$ of $\sim$5‰ is not consistent with a closed-canopy C3 forest (Cerling et al., 2004); rather, variability in $\delta^{13}C$ and $\delta^{18}O$ is attributed to animals occupying different parts of a more open C3 forest, some of which were more open or closer to flowing water than others. These insights into the physical structure of ancient forests, such as spacing of trees and occurrence of open areas, and into animal behavior are otherwise difficult to obtain using more traditional paleobotanical or sedimentological techniques.

As a complement to fossil remains, isotope data from paleosol carbonates in some cases can provide an even more complete isotopic picture of ancient landscapes. For example, analysis of paleosol carbonates by Fox and Koch (2003, 2004) indicate the presence of C4 grasses, which suggest a more open landscape. Because this interpretation was not supported by isotope data collected from the tooth enamel of the associated fossil equids,

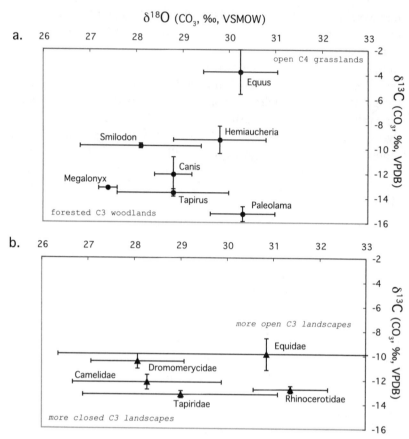

$\delta^{18}O$ (CO$_3$, ‰, VSMOW)

*Figure 8.8.* Carbon and oxygen isotope data from fossils of coexisting mammals can be used to reconstruct vegetation structure and to identify niche and resource partitioning between taxa. Higher $\delta^{13}C$ and $\delta^{18}O$ are generally associated with more open environments and evaporatively modified water sources. A. Samples from a Pleistocene terrestrial locality in South Carolina are interpreted to represent a mix of C3 and C4 plants eaten by herbivores. Isotopic differences suggest that carnivores also hunted in different parts of the ecosystem (Kohn et al., 2005). B. Samples from a Miocene terrestrial locality in Florida where only C3 plants were present (Feranec and MacFadden, 2006). A limited range in $\delta^{13}C$ implies a lack of closed canopy forest, while significant isotopic differences among some taxa illustrate preferences for more open or closed microhabitats. Symbols represent population means and bars represent one standard deviation away from the mean.

they concluded that isotope ratios of vertebrates are not necessarily a direct reflection of all vegetation and water present. Rather, isotope ratios from vertebrates reflect only the plants and water that are actually ingested.

### Climatic Conditions

The link between rainout of moisture from air masses, $\delta^{18}O$ of resulting precipitation and surface water, and $\delta^{18}O$ of vertebrate remains in bonebeds can be used to study several different aspects of paleoclimate. In particular, it may be possible to estimate mean annual temperature (MAT) and the amounts of precipitation. Records of climatic conditions in terrestrial environments are generally sparse, thus studies of bonebed remains with this goal in mind will be of interest to climate modelers and paleontologists alike.

Because of the unreliability in applying modern relations between $\delta 18O$ of precipitation and MAT to the past (Boyle, 1997; Fricke and O'Neil, 1999; Rowley et al., 2001; Fricke and Wing, 2004), this author believes that the best isotopic estimates of MAT are those that rely on the temperature-sensitive fractionation of oxygen isotopes between apatite and water at the scale of local paleoenvironments. This approach uses two or more physiologically distinct, but coexisting, taxa (collected from a single bonebed) that are assumed to have been ingesting the same waters (e.g., a terrestrial mammal with a constant body temperature and a freshwater fish or invertebrate; Fig. 8.9). Because biogenic apatite in the mammal forms at a

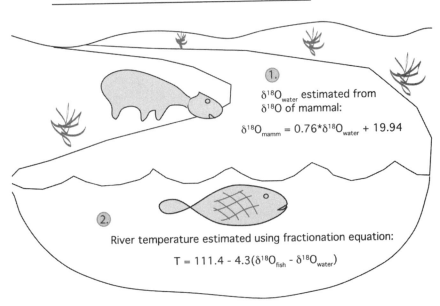

$\delta^{18}O$ from co-existing riverine taxa: mammal and fish

1. $\delta^{18}O_{water}$ estimated from $\delta^{18}O$ of mammal:

$\delta^{18}O_{mamm} = 0.76 * \delta^{18}O_{water} + 19.94$

2. River temperature estimated using fractionation equation:

$T = 111.4 - 4.3(\delta^{18}O_{fish} - \delta^{18}O_{water})$

*Figure 8.9.* The two-taxa approach to estimating temperature in terrestrial environments (after Fricke and Wing, 2004). 1. Biogenic apatite in mammals is formed at a constant temperature. Using a physiological model (Kohn [1996] for biogenic phosphate), $\delta^{18}O$ of river water can be estimated. 2. The value calculated above is then substituted along with measured $\delta^{18}O$ values from fish into carbonate/phosphate-water fractionation equations to estimate river water temperature, which is assumed to be similar to that of the overlying atmosphere.

constant body temperature, its $\delta 18O$ can be used (along with a physiological model; e.g., Kohn, 1996) to estimate $\delta 18O_{water}$. In contrast, $\delta 18O$ of fish scales is dependent on $\delta 18O$ of water and on water temperature (in the case of phosphate, $T_{river} = 111.4 - 4.3 [\delta 18O_{water} - \delta 18O_{bp}]$; Longinelli and Nuti, 1973). $\delta 18O_{water}$ values estimated from mammalian phosphate and $\delta 18O$ values from fish are substituted into this phosphate-water fractionation equation and are used to estimate river water temperature, which is very close to that of the overlying atmosphere at present (Fricke and Wing, 2004).

A test of the "two-taxa approach" was conducted by Fricke and Wing (2004). In that study a comparison of standard paleobotanical and two-taxa isotope methods for estimating MAT in the Paleogene of Wyoming showed similar results, thus indicating that data from large-mammals and fish can be combined to provide useful estimates of ancient MATs. Therefore the two-taxa approach represents a potentially powerful paleoclimate tool (Kolodny et al., 1983; Fricke et al., 1998b; Barrick et al., 1999; Grimes et al., 2003, 2004a, 2005). In the case of Grimes et al. (2003), an oxygen isotope data set was obtained from rodent teeth, fish scales and otoliths, and several aquatic invertebrates of Eocene age from Britain. In turn, these were used to produce an internally consistent set of paleotemperature estimates during periods of skeletal growth for the different taxa. Overall, these results from the Eocene are very encouraging and will hopefully lead to similar studies over a wider range of time and space.

*Seasonality*

The ability of tooth enamel from single teeth to record seasonal variations in isotope ratios (Fig. 8.5a and 8.5b) represents a special climatic application. At this time, however, it appears difficult to use seasonal variations in isotope ratios to study climatic conditions quantitatively. One problem is that the isotopic measure of seasonality preserved in tooth enamel is generally less than that which actually exists in plants and waters ingested by an animal. This discrepancy occurs because body water represents a time-averaged reservoir of carbon and oxygen with a turn over time of weeks to months (Kohn and Cerling, 2002; Passey and Cerling, 2002 et al., 2002a; Balasse et al., 2003; Ayliffe et al., 2004; Hoppe et al., 2004a; Zazzo et al., 2005). Therefore, carbon and oxygen isotope ratios of enamel forming at any one time reflect a homogenized pool of carbon and oxygen in which shorter-term seasonal extremes in isotope ratios of ingested materials are dampened. Furthermore, contemporaneous layers of tooth enamel are deposited at variable angles to the tooth surface (Fig. 8.5a), and when enamel is sampled via drilling or laser ablation, enamel of different ages is combined, resulting in isotopic mixing. Lastly, even if a true reflection of seasonal variations in $\delta^{13}C$ and $\delta^{18}O$ of ingested plants and waters can be reconstructed using tooth enamel data, the fact remains that changes in animal behavior may also occur over the course of a year that *also* impact $\delta^{13}C$ and $\delta^{18}O$ of ingested food and water. Thus, it is not valid to attribute all isotopic variability to climatic factors alone, and unambiguous interpretations of seasonal isotope variations are difficult to make.

Despite these problems, the simple observation that intratooth ranges in isotope ratios do or do not vary over time, from place to place, or among taxa can provide useful qualitative paleoenvironmental and paleoecological information. For example, oxygen isotope data from the teeth of mammals and dinosaurs have also been used to investigate seasonal patterns in the amount of precipitation and in humidity from the Pleistocene to the Cretaceous (Higgins and MacFadden, 2004; Straight et al., 2004). These studies note that surface temperature and the amount of air-mass rainout can both influence $\delta^{18}O$ of precipitation in climatic regimes characterized by marked wet and dry seasonality. Depending on the interplay of these factors, different seasonal patterns in $\delta^{18}O$ of precipitation are produced. By comparing predictions with observed intratooth patterns in $\delta^{18}O$ from theropod teeth, Straight et al. (2004) inferred that the Late Cretaceous in Alberta was often characterized by several months of high rainfall and high humidity during tooth growth. Similarly, Higgins and MacFadden (2004) inferred a rainy season for southwestern North America during the late Pleistocene based on the isotopic analysis of horse and bison teeth. Many other excellent studies have utilized intratooth variations in $\delta^{18}O$ to infer local climatic conditions (Stuart-Williams and Schwarcz, 1997; Gadbury et al., 2000; Fox and Fisher, 2001, 2004; Stanton-Thomas and Carlson, 2003; Nelson, 2005).

*Inferences Drawn from Multiple Bonebeds*

Comparisons of isotope data from multiple bonebeds may underscore environmental changes through time, or paleogeographic variation in paleoenvironmental conditions during a given time interval in the past. The key assumption in these approaches is that isotopic differences are due only to differences in paleoenvironmental conditions, not the basic biology of a given taxon. For this reason, it is best to compare isotope data from closely related taxa, and, preferably, from within a species. Furthermore,

a.

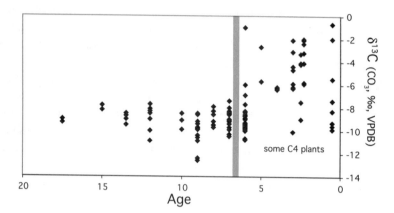

some C4 plants

$\delta^{13}C$ ($CO_3$, ‰, VPDB)

Age

b.

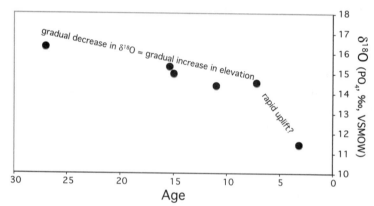

gradual decrease in $\delta^{18}O$ = gradual increase in elevation

rapid uplift?

$\delta^{18}O$ ($PO_4$, ‰, VSMOW)

Age

c.

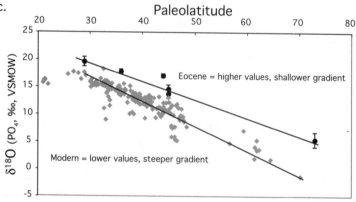

Paleolatitude

Eocene = higher values, shallower gradient

Modern = lower values, steeper gradient

$\delta^{18}O$ ($PO_4$, ‰, VSMOW)

only those taxa that are known to have been widely distributed in time and space will be useful in making intrabonebed comparisons.

The most straightforward application of isotope data from bonebeds of different age is the creation of temporal records of vegetation change or climate change in a given area. One of the best examples of this approach is the study of grassland expansion during the Neogene (Fig. 8.10a; Quade et al., 1992; Cerling et al., 1993, 1997a; Wang et al., 1994; Passey et al., 2002). By analyzing vertebrate remains—generally those of horses—of different ages, an obvious increase in $\delta^{13}C$ at ~8 Ma is visible. In turn, this increase is interpreted to reflect the expansion of C4 plants, particularly grasses, over the landscape. However, because many of these studies rely on the analysis of only one taxon, feeding behavior must be considered; if the studied taxon did not consume C4 grass, then no record of the grassland would be preserved in the taxon's isotope ratios. In the case of climate, multiple bonebeds have been used to document (1) changes in mean annual temperature across the Paleocene-Eocene boundary that capture short-term warming in Wyoming associated with the Paleocene-Eocene Thermal Maximum (Fricke and Wing, 2004; Koch et al., 1995), (2) possible climatic stability across the Eocene-Oligocene boundary (Grimes et al., 2005; Kohn et al., 2004; Bryant et al., 1996a), and (3) changes in the seasonal variation in $\delta^{18}O$ in the Late Cretaceous of Alberta (Straight et al., 2004).

Another application of isotope data from multiple bonebeds has been the study of changes in paleoelevation (and paleoenvironment) through time (Fig. 8.10b). This research takes advantage of the fact that rainout of air masses occurs as they are forced over topographic barriers, with a resulting decrease in $\delta^{18}O$ and the formation of a rain shadow or a monsoonal circulation pattern. For example, Kohn et al. (2002) observed that $\delta^{18}O$ of vertebrate remains decreased over time on the leeward side of the Cascade Mountain and related this decrease to the timing of mountain uplift. Similarly, vertebrate remains from Tibet, China, and Nepal have been used to investigate the uplift history of the Himalayan Mountains

---

*Figure 8.10.* Data from multiple bonebeds can be used to study elevation changes over time, grassland evolution over time, and spatial distributions in isotope ratios for a single time period. A. The occurrence of $\delta^{13}C$ values of equid tooth enamel greater than −8‰ in bonebeds younger than ~8 Ma provides evidence for the expansion of C4 grasslands over North America at this time (Passey et al., 2002). B. $\delta^{18}O$ of tooth enamel from vertebrates living on the leeward side of the Cascade mountains gradually decreases with time in response to increasing elevation and hence air-mass rainout until ~7 Ma when rapid uplift is inferred (Kohn et al., 2002). C. $\delta^{18}O$ of tooth enamel from mammals living over a wide latitudinal range during the Eocene can be used to reconstruct paleohydrologic gradients. In this case, $\delta^{18}O$ of Eocene mammals (black circles) are higher and latitudinal gradients slightly steeper than calculated for mammals drinking modern river water (gray diamonds; from Fricke, 2003).

(Dettman et al., 2001; Wang et al., 2006), and the impact of uplift on average and seasonal variations in temperature and monsoonal precipitation (Dettman et al., 2001).

Other studies have focused on differences in isotope ratios between bonebeds of the same age. Here the goal is to reconstruct paleohydrology, paleoclimate, paleovegetation distributions, and topographic patterns at one instant in geologic time (Fig. 8.10c). In the case of vegetation, Mac-Fadden et al. (1999a) used the distribution of $\delta^{13}C$ in horse tooth enamel to reconstruct the distribution of C3 and C4 grasses across North and South America during the Pleistocene. Because of the relation of $\delta^{18}O$ of precipitation to air-mass rainout, comparisons of $\delta^{18}O$ of mammals over a wide latitudinal range can be used to study the transport of water vapor from tropics to poles under different climatic regimes such as those of the Eocene (Fricke, 2003) and the Late Cretaceous (Amiot et al., 2004). When paired with isotope data from coexisting fish, data from these localities can be used to estimate MAT and, thus, reconstruct latitudinal temperature gradients for these same time intervals (Fricke and Wing, 2004). Lastly, more accurate descriptions of paleoelevation and paleorelief are possible if isotope data are collected from both windward and leeward sides of paleomountains. Such efforts have been undertaken for Eocene age Laramide ranges of western North America (e.g., Fricke, 2003) and the Miocene to recent Sierra Nevada Mountains (Koch and Crowley, 2005).

**Ecological Relations**

The way in which organisms interact with each other and their environments is fundamentally related to their behavior. In the case of extant animals, behavior can be observed, whereas the behavioral activities of extinct animal are only rarely preserved in the fossil record (Brinkman et al., Chapter 4 in this volume) and are much more often inferred by comparing their skeletal and dental morphology and microwear patterns with modern relatives. Even so, such comparisons become increasingly more difficult to undertake and accurately resolve as one moves back through the geologic record. In this context, stable isotope data from fossil remains can provide important complementary information that relates to food choices, sources of water, and habitats. In those cases where different paleoecological niches are characterized by different paleoenvironmental conditions (e.g., water sources, plant types), it may be possible to resolve these differences using stable isotope data. For this reason, stable-isotope-based paleoecological studies focus on isotopic differences between vertebrate taxa, which reflect differences in the kinds of plants and waters ingested.

Before describing paleoecological applications, it is important to clarify two additional aspects of this type of research. First, because isotopic offsets can also result from biological differences between taxa, it is important to carefully consider isotopic offsets between ingested food and water, and vertebrate remains. To do so, it must be known whether the animal was an herbivore, a carnivore, or an omnivore. Secondly, isotopic overlap between fossil taxa does not necessarily imply behavioral similarities; it is possible that animals ate different plants and obtained water from different sources that were isotopically similar. In such cases, isotope data cannot resolve ecological differences.

*Resource Partitioning and Habitat Preferences*

To date, most paleoecological research using stable isotope ratios has centered on mammalian herbivores and interpretations based on isotopic similarities or differences among taxa. At the most basic level, the mere existence of isotopic offsets between herbivores is strong evidence that they were partitioning dietary resources. To the extent that certain kinds of plants live on certain parts of the landscape, these same isotopic offsets can also be used to identify differences in habitats preferred, or at least frequently occupied, by groups of herbivores.

For example, carbon and oxygen isotope data from all common mammals characteristic of a single ecosystem can be described from the perspective of isotopic differences, and systematically higher or lower isotopic ratios between taxonomic pairs can be interpreted appropriately (Ambrose and DeNiro, 1986; Tieszen and Boutton, 1989; Bocherens et al., 1996; Cerling et al., 1997b; Koch et al., 1998; Sponheimer and Lee-Thorpe, 1999c, 2001; Sponheimer et al., 2001, 2003; Palmqvist et al., 2003; Kohn et al., 2005). Studies using carbon and oxygen isotope data to focus on the dietary choices of one or two herbivores reveal similar kinds of isotopic offsets (Quade et al., 1995; McFadden and Cerling, 1996; MacFadden and Shockey, 1997; MacFadden, 1998; Cerling et al., 1999; MacFadden et al., 1999b; Feranec and MacFadden, 2000, 2003; Zazzo et al., 2000; Fox and Fisher, 2001, 2004; Harris and Cerling, 2002). A common interpretation in all these studies is that high $\delta^{13}C$ and $\delta^{18}O$ values are indicative of a diet of C4 grasses and a preference for more open habitats (e.g., North American horses, bison, and mammoths; African elephants). Conversely, lower $\delta^{13}C$

and $\delta^{18}O$ in coexisting herbivores indicate a preference for C3 plants from a denser forest setting (e.g., North American tapirs, mastodons, African giraffes). The lowest $\delta^{18}O$ values are indicative of a preference for riparian habitats and the ingestion of nonevaporated river water (e.g., hippopotamus; Figs. 8.7 and 8.8). Even when C4 plants are not inferred to be present, isotopic differences have revealed dietary preferences for understory plants versus canopy plants or more closed versus more open microenvironments (Fig 8.8b; Cerling et al., 2004; Feranec and MacFadden, 2006). Although an offset between $\delta^{13}C$ of bioapatite and bulk diet of ~12–15‰ is assumed in all these cases, Grimes et al. (2004b) demonstrated that this may not always be true, particularly in the case of rodents, and that paleoecological interpretations depend on this choice.

The dietary preferences of carnivores can also be inferred, although fewer data are usually available. In this approach, the isotope systematics of all common herbivores and potential prey living in an area must be described, and the offset in carbon isotope ratio that corresponds to changes in trophic level (Lee-Thorpe et al., 1989; Fox-Dobbs et al., 2006) must be assumed correct. Feranec (2004a, 2005) and Kohn et al. (2005) used the offset of Lee-Thorpe et al. (1989) and inferred the identity of common prey for carnivores living in several different regions during the Pleistocene. In turn, isotope data from the herbivores in each region were used to determine which part of the landscape (i.e., open areas, dense forests, transitional settings) made up the hunting grounds for these carnivores (Fig. 8.8a; Kohn et al., 2005).

To date, most stable-isotope-based paleoecological research involving vertebrates has focused on Neogene-age terrestrial herbivores, in particular, those of Pleistocene age. There is no a priori reason, however, that material from older bonebeds cannot be studied, and there is now a growing suite of research projects involving bones and teeth from older bonebeds. Barrick et al. (1992), Thewissen et al. (1996), Clementz and Koch (2001), Clementz et al. (2003), and MacFadden et al. (2004) have all used stable isotopes to study preferences of early Cenozoic mammals from coastal, semiaquatic, and fully marine habitats (Fig. 8.11a). Carbon and oxygen isotope offsets between herbivorous dinosaur taxa of the Late Cretaceous have also been observed (Fig. 8.11b) and are interpreted as reflecting dietary differences. It even appears possible to resolve ecological differences among nonmammalian therapsids of Triassic age using stable isotope data (Botha et al., 2005). In fact, stable isotope studies aimed at understanding behaviors and dietary preferences of ancient animals with no anatomically similar modern relatives are arguably the most intriguing, given that few other methods can address such questions.

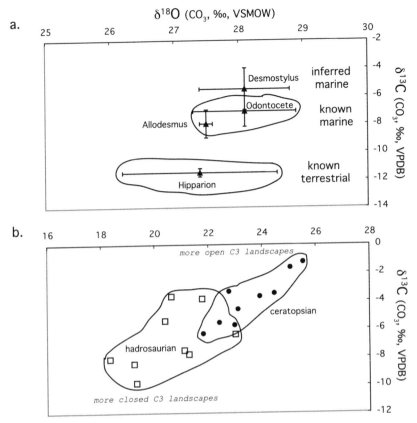

Figure 8.11. Investigations of ecological niche and dietary resource partitioning are not limited to the Neogene or to terrestrial settings. A. Data from early Cenozoic fossils are used to study ecological preferences of different mammals living in coastal settings. Symbols represent population means and bars represent one standard deviation away from the mean. Higher $\delta^{18}O$ and $\delta^{13}C$ values are characteristic of mammals adapted to marine environments, thus making it possible to infer a marine preference for *Desmostylus* (Clementz et al., 2003). B. Higher $\delta^{18}O$ and $\delta^{13}C$ values for herbivorous ceratopsian dinosaurs versus herbivorous hadrosaurian dinosaurs from the Hell Creek Formation, North Dakota, indicate that hadrosaurs utilized forest vegetation while ceratopsians ate a mix of forest plants and plants living in more open, dry or salt-stressed habitats (Fricke, unpublished data).

### Inferences Drawn from Multiple Bonebeds

If remains from multiple individuals of the same or closely related taxa can be studied from a number of contemporaneous bonebeds, then it should be possible to determine if isotopic relations and ecological interactions between taxa are similar from place to place, or if these relations and interactions are dependent on paleoenvironmental conditions. Studies of this type have focused on the dietary habits of Pleistocene mammoths and mastotodons, and other large herbivores from eastern and southern North

America (Koch et al., 1989, 1998; Feranec and MacFadden, 2000, 2003; Feranec, 2004b). The results of these studies suggest that mammoths preferred a diet of grasses, compared to the more mixed plant assemblage preferred by mastodons. Furthermore, the proportion of C3 to C4 grasses ingested by both taxa changed with floral availability and climatic conditions. Similarly, in a study of Late Pleistocene herbivore communities along an east-to-west gradient in southwestern North America, Connin et al. (1998) documented an increase in C3 plant intake by all animals, and attributed this change to differences in rainfall along the gradient.

Similar kinds of isotopic data can be brought to bear on the question of animal movement and migration between bonebed localities, a question that is otherwise difficult to address using fossil morphology alone. For example, if populations of a mobile terrestrial vertebrate taxon are characterized by different isotope ratios at two bonebed sites that are separated by tens to hundreds of kilometers, then a strong argument can be made that these animals did not eat the same food or drink the same water. In turn, it can be concluded that there was no intermingling, or migration, of animals between the sites. Unfortunately, the occurrence of isotopic overlap for two populations of a single taxon found in separate areas is not as easy to interpret; it is possible that two or more populations of a taxon (represented by separate bonebed assemblages) each consumed plants and water from geographically separate localities where isotope ratios happened to be similar to one other. In such a case, the two populations could be misinterpreted as being subsets of a single larger population that migrated between the two areas. In a study of mammoth migration, Hoppe (2004) used a combination of stable and radiogenic isotope data from mammoth tooth enamel to conclude that these animals from the Great Plains did not undertake large migrations of over 600 km but may have migrated distances of several hundred kilometers or less (see also Fig. 8.12). This uncertainty in distance of migration makes it clear that the better the spatial resolution of a suite of contemporaneous bonebeds, the more precise such descriptions of animal movement will be.

Lastly, studies of closely related taxa from a stratigraphic succession of bonebeds can reveal behavioral change through time. MacFadden (2000) provides a review on the topic that covers the entire Cenozoic and illustrates how isotope data can be integrated with morphological information. Other excellent examples of the isotopic approach focus on the evolution of grazing behavior in Neogene herbivores. Because the unique $\delta^{13}C$ values of C4 grasses are easy to track in vertebrate remains, many studies have used the C4 "signal" to document changes in herbivore diet as C4 grasslands expanded during the Miocene, and to study possible relationships

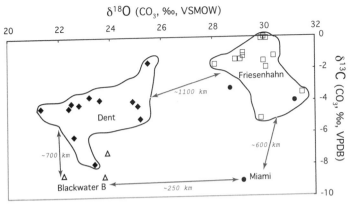

$\delta^{18}O$ (CO$_3$, ‰, VSMOW)

*Figure 8.12.* Isotope data from contemporaneous remains of late Pleistocene mammoths found in four separate areas in western North America along with approximate distances between them (Hoppe, 2004). Because there is no overlap in stable isotope data from mammoths at Dent, Colorado, and Friesenhahn, Texas, it is reasonable to infer that they did not consume the same plants and water. In turn, these data suggest that these groups of mammoths did not migrate between these localities. It is more difficult to make unambiguous interpretations in the case where there is an overlap in data.

between these changes in diet and tooth morphology (Quade et al., 1992; Cerling et al., 1993, 1997a; MacFadden, 1994, 1998; Wang et al., 1994; MacFadden and Cerling, 1996; MacFadden et al., 1999b; Passey et al., 2002; Feranec and MacFadden, 2003). Equally important research has centered on the adaptation of land mammals to semiaquatic and fully marine habitats during the early Cenozoic by identifying changes in diet through time (Barrick et al., 1992; Thewissen et al., 1996; Roe et al., 1998; Clementz and Koch, 2001; Clementz et al., 2003; MacFadden et al., 2004).

**Paleobiology**

Soft tissues of vertebrates can provide a unique window into animal physiology, but they are rarely preserved in the rock record. Hence, detailed knowledge of metabolic processes, body temperatures, and growth rates in extinct taxa is very difficult to obtain. Fortunately, stable isotope ratios of bioapatite may be influenced by biological processes in identifiable ways that provide paleobiological insight. The potential exists to obtain much more paleobiological information if carbon and nitrogen isotope data are also available from organic remains (Gannes et al., 1998, and references therein), however descriptions of these applications are beyond the scope of this chapter.

## Thermoregulation

Body temperatures of vertebrates can be influenced by several factors including (1) internal metabolic processes that result in approximately constant temperatures (endothermy), (2) varying environmental temperatures that are mirrored by the body (ectothermy), (3) behavioral activities such as basking that keep temperatures relatively constant, and (4) large body size, which can result in relatively constant body temperature (inertial homeothermy; gigantothermy). Because the oxygen isotope fractionation that occurs during the precipitation of skeletal apatite from body water is sensitive to body temperature, it is possible to design a study of bonebed remains that can shed some light on thermoregulatory strategies utilized by ancient vertebrates.

In the case of dinosaurs, Barrick and Showers (1994, 1995) and Barrick et al. (1996) measured $\delta^{18}O$ values of bones from body cores and extremities of several dinosaur taxa and observed no statistical differences in the values regardless of the element examined. Based on these results, they argued that dinosaurs were homeothermic and able to maintain a constant temperature throughout their bodies. Using another approach, Fricke and Rogers (2000), and more recently Amiot et al. (2004), compared $\delta^{18}O$ values of coexisting crocodiles, turtles, and theropod dinosaurs from a number of contemporaneous sites across a wide latitudinal range. They observed a steeper latitudinal gradient in $\delta^{18}O$ for dinosaurs, which is consistent with an interpretation of homeothermy for this group, and ectothermy for crocodiles and turtles.

## Rates of Dental Development

Tooth enamel can be used as a temporal marker to measure the rate of tooth growth, the order in which teeth erupt, and overall rates of dental development (Fig. 8.7b and 8.7d). Investigations of bioapatite from tusks may provide similar information as well as insight into season of death (Fisher, 1987). Comprehensive studies of this type require intratooth sampling of all the teeth from one jaw or a serial sampling of an entire tusk. Vertebrates with high-crowned teeth are best suited for this type of approach, whereas low-crowned teeth may not be as well suited due to spatial limitations in sampling and inherent limitations in temporal resolution. In this approach it is best to analyze teeth from the jaws of many individuals in order to confirm growth rates and to document variation.

Several isotope-based studies of the rate of tooth growth and schedule of tooth eruption in modern mammals have been confirmed by direct observation (Balasse et al., 2003; Passey et al., 2005; Zazzo et al., 2005). However, far less is known about extinct animals. Investigations aimed at filling this gap include those of teeth from theropod dinosaurs (growth rate approximately 55 mm/year [Straight et al., 2004]), hadrosaur dinosaurs (growth rate approximately 38 mm/year [Stanton-Thomas and Carlson, 2003]), sabre-toothed cats (growth rate approximately 60–80 mm/year [Feranec et al., 2004a; Feranec, 2005]), Miocene equids (growth rate approximately 47 mm/year [Nelson, 2005]), and beavers (growth rate approximately 270–365 mm/year [Stuart-Williams and Schwarcz, 1997; Rinaldi and Cole, 2004]). In the case of tusks, serial sampling of those from extinct proboscidean has documented rates of growth, and seasons of birth and death (Fisher, 1987; Koch et al., 1989; Fox and Fisher, 2001).

## SUMMARY

The most compelling reason for incorporating stable isotope analyses of vertebrate hardparts into bonebed studies is that these data can shed light on paleoenvironmental conditions and on the behavior and biology of extinct animals. The power of stable isotope ratios of carbon and oxygen to provide insight into these factors lies in their source: carbon from ingested plants and other foods, and oxygen from ingested surface waters. Variation in carbon isotope ratios typically reflect variations in the degree to which leaf stomata are open or closed, which, in turn, is influenced by paleoenvironmental variations across a given landscape. Similarly, oxygen isotope ratios are related to ingested surface waters and typically reflect differences in the amount of evaporation of these waters across a given landscape. In addition to local environmental conditions, animal behavior and physiology play roles in determining stable isotope ratios of vertebrate remains. Given the relation between isotope ratios of vertebrate remains and these factors, isotopic data have the potential to provide insights that are otherwise difficult to obtain via traditional fossil analysis. In turn, these kinds of data may help confirm or refute paleoecological or paleobiological hypotheses based on other types of data or inference (e.g., facies associations, morphological traits, phylogenetic associations).

The challenge associated with such research is making unambiguous interpretations of stable isotope data in the face of multiple environmental, ecological, and biological influences on isotope ratios of vertebrate remains. To overcome this problem and to strengthen interpretations, two things must be done. First, the focus should be less on interpreting absolute isotope ratios and more on interpreting isotopic similarities or differences, regardless of whether comparisons are made between taxa found in a single bonebed, or between taxa found in spatially or temporally separated bonebeds. Second, studies must be designed so that any isotopic similarities or differences observed can be reasonably attributed to environmental, ecological, or biological factors. This is critical because doing so provides a context for the consideration and interpretation of data. For example, if a single herbivorous taxon has an average $\delta^{13}$C value of −10‰, multiple interpretations are possible: this could indicate that all plants available in the ecosystem were water stressed, that this particular taxon preferentially occupied such stressed habitats, or that this animal was eating a mixture of C3 and C4 plants. By comparing these data with data from other herbivorous taxa, however, specific dietary and niche preferences should become apparent.

This example also illustrates the importance of study design. By making the reasonable assumptions that all herbivores are behaviorally similar in respect to diet (i.e., they eat plants) and physiologically similar, then the source of isotopic differences can be assumed to derive mostly from the behavioral—i.e., ecological—differences among them. In contrast, if the research question centers on environmental variability over time or space, then sampling should be restricted to a single taxon. The assumption in this case is that the animals are behaviorally and physiology similar regardless of location, and that any isotopic similarities or differences among bonebeds reflects climatic or hydrological differences. Lastly, to address paleobiological questions, an intertaxonomic comparison of data from an environment where isotopic variability is low (e.g., open C3 forest) may make it reasonable to assume that physiological differences between taxa are the cause of any observed isotopic offsets.

In the end, there are many kinds of questions that a bonebed researcher can strive to answer by measuring stable isotope ratios of vertebrate remains. Keeping in mind site-specific limitations associated with taphonomic factors that control the kinds and numbers of vertebrate remains present, and with due consideration of potential diagenetic alteration of isotopic ratios, here are some suggested starting points. Use oxygen isotope data from mammals and fish to estimate temperatures for a single time period in the past. Make maps of such estimates by analyzing contemporaneous bonebed remains, or alternatively, strive to construct a record of temperature change over time. Such maps and records are of particular interest to climate modelers hoping to test their simulations, and to paleoclimatologists trying to resolve regional versus global climatic change events. If your interests lie in the realm of paleoecology, explore dietary niche partitioning for herbivores, omnivores, and carnivores at the ecosystem scale. This avenue of research is wide open, especially for pre-Pleistocene time periods. A related topic of particular interest to paleontologists is the possibility of intrataxonomic changes in diet in the face of changes in environmental conditions, vegetation structure, or competition, and it is one that can be addressed by comparing isotopic data between spatially or temporally separated bonebeds. Lastly, isotopic data may be useful in characterizing animal migratory patterns, especially if several contemporaneous bonebeds are present over a small geographic range. Questions such as these, which are difficult, if not impossible, to address using traditional approaches, are waiting to be explored using stable isotope data, especially when these data are derived from well-understood bonebed data sets.

## REFERENCES

Ambrose, S.H., and M.J. DeNiro. 1986. The isotopic ecology of East African mammals. Oecologia 69:395–406.

Ambrose, S.H., and L. Norr. 1993. Experimental evidence for the relationship of the carbon isotope ratios of whole diet and dietary protein to those of bone collagen and carbonate. Pp. 1–37 in Prehistoric human bone: Archaeology at the molecular level. J.B. Lambert and G. Grupe, eds. Springer-Verlag, New York.

Amiot, R., C. Lécuyer, E. Buffetaut, F. Fluteau, S. Legendre, and F. Martineau. 2004. Latitudinal temperature gradient during the Cretaceous Upper Campanian—Middle Maastrichtian: $\delta^{18}O$ record of continental vertebrates. Earth and Planetary Science Letters 226:255–272.

Araguas-Araguas, L., K. Froehlich, and K. Rozanski. 1998. Stable isotope composition of precipitation over southeast Asia. Journal of Geophysical Research 103:28721–28742.

Arens, N.C., A.H. Jahren, and R. Amundsen. 2000. Can C3 plants faithfully record the carbon isotopic composition of atmospheric carbon dioxide? Paleobiology 26:137–164.

Ayliffe, L.K., T.E. Cerling, T.F. Robinson, A.G. West, M. Sponheimer, B.H. Passey, J. Hammer, B. Roeder, M.D. Dearing, and J.R. Ehleringer. 2004. Turnover of carbon isotopes in tail hair and breath $CO_2$ of horses fed an isotopically varied diet. Oecologia 139:11–22.

Balasse, M., H. Bocherens, and A. Mariotti. 1999. Intra-bone variability of collagen and apatite isotopic composition used as evidence of a change in diet. Journal of Archaeological Science 26:593–598.

Balasse, M., A.B. Smith, S.H. Ambrose, and S.R. Leigh. 2003. Determining sheep birth seasonality by analysis of tooth enamel oxygen isotope ratios: The Late Stone Age site of Kasteelberg (South Africa). Journal of Archaeological Sciences 30:205–215.

Barrick, R.E., and W.J. Showers. 1994. Thermophysiology of *Tyrannosaurus rex*: Evidence from oxygen isotopes. Nature 265:222–224.

Barrick, R.E., and W.J. Showers. 1995. Oxygen isotope variability in juvenile dinosaurs (*Hypacrosaurus*): Evidence for thermoregulation. Paleobiology 21:552–560.

Barrick, R.E., A.G. Fischer, Y. Kolodny, B. Luz, and D. Bohaska. 1992. Cetacean bone oxygen isotopes as proxies for Miocene ocean composition and glaciation. Palaios 7:521–531.

Barrick, R.E., W.J. Showers, and A.G. Fischer. 1996. Comparison of thermoregulation of four ornithischian dinosaurs and a varanid lizard from the Cretaceous Two Medicine Formation: Evidence from oxygen isotopes. Palaios 11:295–305.

Barrick, R.E., A.G. Fischer, and W.J. Showers. 1999. Oxygen isotopes from turtle bone: Applications for terrestrial paleoclimates? Palaios 14:186–191.

Blake, R.E., J.R. O'Neil, and G.A. Garcia. 1997. Oxygen isotope systematics of biologically mediated reactions of phosphate. I. Microbial degradation of organophosphorus compounds. Geochimica et Cosmochimica Acta 61:4411–4422.

Bocherens, H., and D. Drucker. 2003. Trophic level isotopic enrichment of carbon and nitrogen in bone collagen: Case studies from recent and ancient terrestrial ecosystems. International Journal of Osteoarchaeology 13:46–53.

Bocherens, H., P.L. Koch, A. Mariotti, D. Geraads, and J.J. Jaeger. 1996. Isotopic biogeochemistry ($^{13}C$, $^{18}O$) and mammalian enamel from African Pleistocene hominid sites. Palaios 11:306–318.

Botha, J., J. Lee-Thorp, and A. Chinsamy. 2005. The palaeoecology of the non-mammalian cynodonts *Diademodon* and *Cynognathus* from the Karoo Basin of South Africa, using stable light isotope analysis. Palaeogeography, Palaeoclimatology, Palaeoecology 223:303–316.

Bowen, G.J., D.J. Beerling, P.L. Koch, J.C. Zachos, and T. Quattlebaum. 2004. A humid climate state during the Palaeocene/Eocene thermal maximum. Nature 432:495–499.

Boyle, E.A. 1997. Cool tropical temperatures shift the global $\delta^{18}O$-T relationship: An explanation for the ice core $\delta^{18}O$-borehole thermometry conflict? Geophysical Research Letters 24:273–276.

Brinkman, D.B. 1990. Paleoecology of the Judith River Formation (Campanian) of Dinosaur Provincial Park, Alberta, Canada: Evidence from vertebrate microfossil localities. Palaeogeography, Palaeoclimatology, Palaeoecology 7–8:37–54

Brinkman, D.B., D.A. Eberth, and P.J. Currie. This volume. From bonebeds to paleobiology: Applications of bonebed data. Chapter 4 *in* Bonebeds: Genesis, analysis, and paleobiological significance. R.R. Rogers, D.A. Eberth, and A.R. Fiorillo, eds. University of Chicago Press, Chicago.

Bryant, J.D., and P.N. Froelich. 1995. A model of oxygen isotope fractionation in body water of large mammals. Geochimica et Cosmochimica Acta 59:4523–4537.

Bryant, J.D., P.N. Froelich, W.J. Showers, and B.J. Genna. 1996a. Biologic and climatic signals in the oxygen isotopic composition of Eocene-Oligocene equid enamel phosphate. Palaeogeography, Palaeoclimatology, Palaeoecology 126:75–89.

Bryant, J.D., P.L. Koch, P.N. Froelich, W.J. Showers, and B.J. Genna. 1996b. Oxygen isotope partitioning between phosphate and carbonate in mammalian apatite. Geochimica et Cosmochimica Acta 24:5145–5148.

Cerling, T.E., and J.M. Harris. 1999. Carbon isotope fractionation between diet and bioapatite in ungulate mammals and implications for ecological and paleoecological studies. Oecologia 120:337–363.

Cerling, T.E., Y. Wang, and J. Quade. 1993. Expansion of $C_4$ ecosystems as an indicator of global ecological change in the late Miocene. Nature 361:344–345.

Cerling, T.E., J.M. Harris, B.J. MacFadden, M.G. Leakey, J. Quade, V. Eisenmann, and J.R. Ehleringer. 1997a. Global vegetation change through the Miocene/Pliocene boundary. Nature 389:153–158.

Cerling, T.E., J.M. Harris, S.H. Ambrose, M.G. Leakey, and N. Solounias. 1997b. Dietary and environmental reconstruction with stable isotope analyses of herbivore tooth enamel from the Miocene locality of Fort Ternan, Kenya. Journal of Human Evolution 33:635–650.

Cerling, T.E., J.M. Harris, and M.G. Leakey. 1999. Browsing and grazing in elephants: The isotope record of modern and fossil proboscideans. Oecologia 120:365–374.

Cerling, T.E., J.A. Hart, and T.B. Hart. 2004. Stable isotope ecology in the Ituri Forest. Oecologia 138:5–12.

Clementz, M.T., and P.L. Koch. 2001. Differentiating aquatic mammal habitat and foraging ecology with stable isotopes in tooth enamel. Oecologia 129:461–472.

Clementz, M.T., K.A. Hoppe, and P.L. Koch. 2003. A paleoecological paradox: The habitat and dietary preferences of the extinct tethythere *Desmostylus*, inferred from stable isotope analysis. Paleobiology 29:506–519.

Codron, J., D. Codron, J. Lee-Thorp, M. Sponheimer, W.J. Bond, D. de Ruiter, R. Grant. 2005. Taxonomic, anatomical, and spatio-temporal variations in the stable carbon and nitrogen isotopic compositions of plants from an African savanna. Journal of Archaeological Science 32: 1757–1772.

Connin, S.L., J. Bentancourt, and J. Quade. 1998. Late Pleistocene C4 plant dominance and summer rainfall in the southwestern United States from isotopic study of herbivore teeth. Quaternary Research 50:179–193.

Dansgaard, W. 1964. Stable isotopes in precipitation. Tellus 16:436–468.

DeNiro, M.J., and S. Epstein. 1978. Influence of diet on the distribution of carbon isotopes in animals. Geochimica et Cosmochimica Acta 42:495–506.

Dettman, D.L., M.J. Kohn, J. Quade, F.J. Ryerson, T.P. Ojha, and S. Hamidullah. 2001. Seasonal stable isotope evidence for a strong Asian monsoon throughout the past 10.7 m.y. Geology 29:31–34.

Eberth, D.A. 1990. Stratigraphy and sedimentology of vertebrate microfossil sites in the uppermost Judith River Formation (Campanian), Dinosaur Provincial Park, Alberta, Canada. Palaeogeography, Palaeoclimatology, Palaeoecology 7–8:1–36.

Eberth, D.A., R.R. Rogers, and A.R. Fiorillo. This volume. A practical approach to the study of bonebeds. Chpater 5 *in* Bonebeds: Genesis, analysis, and paleobiological significance. R.R. Rogers, D.A. Eberth, and A.R. Fiorillo, eds. University of Chicago Press, Chicago.

Edwards, G.E., V.R. Franceschi, and E.V. Voznesenskaya. 2004. Single-cell $C_4$ photosynthesis versus the dual-cell (Kranz) paradigm. Annual Review of Plant Biology 55:173–96.

Epstein, S., and T. Mayeda. 1953. Variations in the 18-O content of waters from natural sources. Geochimica et Cosmochimica Acta 4:213–224.

Farquhar, G.D., J.R. Ehrlinger, and K.T. Hubick. 1989. Carbon isotope discrimination and photosynthesis. Annual Review of Plant Physiology and Plant Molecular Biology 40:503–537.

Feranec, R.S. 2004a. Isotopic evidence of saber-tooth development, growth rate, and diet from the adult canine of *Smilodon fatalis* from Rancho La Brea. Palaeogeography, Palaeoclimatology, Palaeoecology 206:303–310.

Feranec, R.S. 2004b. Geographic variation in the diet of hypsodont herbivores from the Rancholarean of Florida. Palaeogeography, Palaeoclimatology, Palaeoecology 207:359–369.

Feranec, R.S. 2005. Growth rate and duration of growth in the adult canine of *Smilodon gracilis* and inferences on diet through stable isotope analysis. Bulletin of the Florida Museum of Natural History 45:369–377.

Feranec, R.S., and B.J. MacFadden. 2000. Evolution of the grazing niche in Pleistocene mammals from Florida: Evidence from stable isotopes. Palaeogeography, Palaeoclimatology, Palaeoecology 162:155–169.

Feranec, R.S., and B.J. MacFadden. 2003. Stable isotopes, hypsodonty, and the paleodiet of *Hemiauchenia* (Mammalia, Camelidae), a morphological specialization creating ecological generalization. Paleobiology 29:230–242.

Feranec, R.S., and B.J. MacFadden. 2006. Isotopic discrimination of resource partitioning among ungulates in C3-dominated communities from the Miocene of Florida and California. Paleobiology 32:191–205.

Fisher, D.C. 1987. Mastodont procurement by paleoindians of the Great Lakes region: Hunting or scavenging? Pp. 309–421 *in* The evolution of human hunting. D.C. Fisher, ed. Plenum, New York.

Fogel, M.L., and N. Tuross. 1999. Transformation of plant biochemicals to geological macromolecules during early diagenesis. Oecologia 120:336–346.

Fox, D.L., and D.C. Fisher. 2001. Stable isotope ecology of a late Miocene population of *Gomphotherium productus* (Mammalia, Proboscidea) from Port of Entry Pit, Oklahoma, USA. Palaios 16:279–293.

Fox, D.L., and D.C. Fisher. 2004. Dietary reconstruction of Miocene *Gomphotherium* (Mammalia, Proboscidea) from the Great Plains region, U.S.A., based on the carbon isotope composition of tusk and molar enamel. Palaeogeography, Palaeoclimatology, Palaeoecology 206:311–335.

Fox, D.L., and P.L. Koch. 2003. Tertiary history of $C_4$ biomass in the Great Plains, U.S.A. Geology 31:809–812.

Fox, D.L., and P.L. Koch. 2004. Carbon and oxygen isotopic variability in Neogene paleosol carbonates: Constraints on the evolution of the C4-grasslands of the Great Plains, USA. Palaeogeography, Palaeoclimatology, Palaeoecology 207:305–329.

Fox-Dobbs, K., P.V. Wheatley, and P.L. Koch. 2006. Carnivore-specific bone bioapatite and collagen carbon isotope fractionations: Case studies of modern and fossil grey wolf populations. Eos Transactions of the American Geophysical Union 87(52): Abstract B53C-0366.

Fricke, H.C. 2003. $\delta^{18}O$ of geographically widespread mammal remains as a means of studying water vapor transport and mountain building during the early Eocene. Geological Society of America Bulletin 115:1088–1096.

Fricke, H.C., and J.R. O'Neil. 1996. Inter- and intra-tooth variations in the oxygen isotope composition of mammalian tooth enamel: Some implications for paleoclimatological and paleobiological research. Palaeogeography, Palaeoclimatology, Palaeoecology 126:91–99.

Fricke, H.C., and J.R. O'Neil. 1999. The correlation between $^{18}O/^{16}O$ of meteoric water and surface temperature: Its use in investigating terrestrial climate change over geologic time. Earth and Planetary Science Letters 170:181–196.

Fricke, H.C., and R.R. Rogers. 2000. Multiple taxon–multiple locality approach to providing oxygen isotope evidence for warm-blooded theropod dinosaurs. Geology 28:799–802.

Fricke, H.C., and S.L. Wing. 2004. Oxygen isotope and paleobotanical estimates of temperature and $\delta^{18}O$-Latitude gradients over North America during the early Eocene. American Journal of Science 304:612–635.

Fricke, H.C., W.C. Clyde, J.R. O'Neil, and P.D. Gingerich. 1998a. Intra-tooth variation in $\delta^{18}O$ of mammalian tooth enamel as a record of seasonal changes in continental climate variables. Geochimica et Cosmochimica Acta 62:1839–1851.

Fricke, H.C., W.C. Clyde, J.R. O'Neil, and P.D. Gingerich. 1998b. Evidence for rapid climate change in North America during the latest Paleocene thermal maximum: Oxygen isotope compositions of biogenic phosphate from the Bighorn Basin (Wyoming). Earth and Planetary Science Letters 160:193–208.

Fricke, H.C., R.R. Rogers, R. Backlund, S. Echt, and C.N. Dwyer. In review. Preservation of primary stable isotope signals in dinosaur remains, and environmental gradients of the Late Cretaceous of Montana and Alberta. Palaeogeography, Palaeoclimatology, Palaeoecology.

Gadbury, C., L.C. Todd, A.H. Jahren, and R. Amundsen. 2000. Spatial and temporal variations in the isotope composition of bison tooth enamel from the early Holocene Hudsen-Meng Bonebed. Palaeogeography, Palaeoclimatology, Palaeoecology 157:79–93.

Gannes, L.Z., C.M.D. Rio, and P.L. Koch. 1998. Natural abundance variations in stable isotopes and their potential uses in animal physiological ecology. Comprehensive Biochemical Physiology 119A(3):725–737.

Gat, J.R. 1996. Oxygen and hydrogen isotopes in the hydrologic cycle. Annual Review of Earth and Planetary Sciences 24:225–262.

Grimes, S.T., D.P. Mattey, J.J. Hooker, and M.E. Collinson. 2003. Paleogene paleoclimate reconstruction using oxygen isotopes from land and freshwater organisms: The use of multiple paleoproxies. Geochimica et Cosmochimica Acta 67:4033–4047.

Grimes, S.T., D.P. Mattey, M.E. Collinson, and J.J. Hooker. 2004a. Using mammal tooth phosphate with freshwater carbonate and phosphate palaeoproxies to obtain mean paleotemperatures. Quaternary Science Reviews 23:967–976.

Grimes, S. T., M.E. Collinson, J.J. Hooker, D.P. Mattey, N.V. Grassineau, and D. Lowry. 2004b. Distinguishing the diets of coexisting fossil theridomyid and glirid rodents using carbon isotopes. Palaeogeography, Palaeoclimatology, Palaeoecology 208:103–119.

Grimes, S.T., J.J. Hooker, M.E. Collinson and D.P. Mattey. 2005. Summer temperatures of late Eocene to early Oligocene freshwaters. Geology 33: 189–192.

Gröcke, D.R. 1997. Stable-isotope studies on the collagen and hydroxylapatite components of fossils: Palaeoecological implications. Lethaia 30:65–78.

Harris, J.M., and T.E. Cerling. 2002. Dietary adaptations of extant and Neogene African suids. Journal of Zoology 256:45–54.

Heaton, T.H.E. 1999. Spatial, species, and temporal variations in the $^{13}C/^{12}C$ ratios of $C_3$ plants: Implications for paleodiet studies. Journal of Archaeological Sciences 26:637–649.

Hedges, R.E.M. 2003. On bone collagen: Apatite-carbonate isotopic relationships. International Journal of Osteoarchaeology 13:66–79.

Higgins, P., and B.J. MacFadden. 2004. "Amount effect" recorded in oxygen isotopes of Late Glacial horse (*Equus*) and bison (*Bison*) teeth from the Sonoran and Chihuahuan deserts, southwestern United States. Palaeogeography, Palaeoclimatology, Palaeoecology 206:337–353.

Hilderbrand, G.V., T.A. Hanley, C.T. Robbins, and C.C. Schwartz. 1999. Role of brown bears (*Ursus arctos*) in the flow of marine nitrogen into a terrestrial ecosystem. Oecologia 121:546–530.

Hillson, S. 1986. Teeth. Cambridge University Press, Cambridge.

Hoppe, K.A. 2004. Late Pleistocene mammoth herd structure migration patterns and Clovis hunting strategies inferred from isotopic analyses of multiple death assemblages. Paleobiology 30:129–145.

Hoppe, K.A. 2006. Correlation between the oxygen isotope ratio of North American bison teeth and local waters: Implication for paleoclimatic reconstructions. Earth and Planetary Science Letters 244:408–417.

Hoppe, K.A., R.G. Amundson, M. Vavra, M.P. McClaran, and D.L. Anderson. 2004a. Isotopic analysis of tooth enamel carbonate from modern North American feral horses: Implications for paleoenvironmental reconstructions. Palaeogeography, Palaeoclimatology, Palaeoecology 203:299–311.

Hoppe, K.A., S.M. Stover, J.R. Pascoe, and R. Amundson. 2004b. Tooth enamel biomineralization in extant horses: Implications for isotopic microsampling. Palaeogeography, Palaeoclimatology, Palaeoecology 206:355–365.

Iacumin, P., and A. Longinelli. 2002. Relationship between $\delta^{18}O$ values for skeletal apatite from reindeer and foxes and yearly mean $\delta^{18}O$ values of environmental water. Earth and Planetary Science Letters 6228:1–7.

Iacumin, P., H. Boecherens, A. Mariotti, and A. Longinelli. 1996. Oxygen isotope analyses of co-existing carbonate and phosphate in biogenic apatite: A way to monitor diagenetic alteration of bone phosphate? Earth and Planetary Science Letters 142:1–6.

Jim, S., S. Ambrose, and R. Evershed. 2004. Stable carbon isotopic evidence for differences in the dietary origin of bone cholesterol, collagen and apatite: Implications for their use in palaeodietary reconstruction. Geochimica et Cosmochimica Acta 68:61–72.

Johnson, B.J., M. Fogel, and G.H. Miller. 1998. Stable isotopes in modern ostrich eggshell: A calibration for paleoenvironmental applications in semi-arid regions of southern Africa. Geochimica et Cosmochimica Acta 62:2451–2461.

Koch, P.L. 1998. Isotopic reconstruction of past continental environments. Annual Review of Earth and Planetary Sciences 26:573–613.

Koch, P.L., and B.E. Crowley. 2005. The isotopic rain shadow and elevation history of the Sierra Nevada mountains, California. Eos Transactions American Geophysical Union 86:52.

Koch, P.L., D.C. Fisher, and D. Dettman. 1989. Oxygen isotope variation in the tusks of extinct proboscideans: A measure of season of death and seasonality. Geology 17:515–519.

Koch, P.L., J.C. Zachos, and P.D. Gingerich. 1992. Correlation between isotope records in marine and continental carbon reservoirs near the Paleocene/Eocene boundary. Nature 358:319–322.

Koch, P. L., M. Fogel, and N. Tuross. 1994. Tracing the diet of fossil animals using stable isotopes. Pp. 63–94 in Stable isotopes in ecology and environmental science. K. Lajtha and R.H. Michener, eds. Blackwell Scientific, Oxford.

Koch, P.L., J.C. Zachos, and D.L. Dettman. 1995. Stable isotope stratigraphy and paleoclimatology of the Paleogene Bighorn Basin (Wyoming, USA). Palaeogeography, Palaeoclimatology, Palaeoecology 115:61–89.

Koch, P.L., K.A. Hoppe, and S.D. Webb. 1998. The isotopic ecology of late Pleistocene mammals in North America, Part I. Chemical Geology 152:119–138.

Kohn, M.J. 1996. Predicting animal $\delta^{18}O$: Accounting for diet and physiological adaptation. Geochimica et Cosmochimica Acta 60:4811–4829.

Kohn, M.J., and T.E. Cerling. 2002. Stable isotope compositions of biological apatite. Reviews in Mineralogy and Geochemistry 48:455–488.

Kohn, M.J., M.J. Schoeninger, and J.W. Valley. 1998. Variability in herbivore tooth oxygen isotope compositions: Reflections of seasonality or developmental physiology. Chemical Geology 152:97–112.

Kohn, M.J., M.J. Schoeninger, and W.W. Barker. 1999. Altered states: Effects of diagenesis on fossil tooth chemistry. Geochimica et Cosmochimica Acta 63:2737–2747.

Kohn, M.J., J.L. Miselis, and T.J. Fremd. 2002. Oxygen isotope evidence for progressive uplift of the Cascade Range, Oregon. Earth and Planetary Science Letters 204:151–165.

Kohn, M.J., J.A. Josef, R. Madden, R.F. Kay, G. Vucetich, and A.A. Carlini. 2004. Climate stability across the Eocene-Oligocene transition, southern Argentina. Geology 32:621–624.

Kohn, M.J., M.P. McKay, and J.L. Knight. 2005. Dining in the Pleistocene—Who's on the menu? Geology 33:649–652.

Kolodny, Y., B. Luz, and O. Navon. 1983. Oxygen isotope variations in phosphate of biogenic apatite. I. Fish bone apatite. Earth Planetary Science Letters 64:398–404.

Kolodny, Y., B. Luz, M. Sander, and W.A. Clemens. 1996. Dinosaur bones: Fossils of pseudomorphs? The pitfalls of physiology reconstruction from apatitic fossils. Palaeogeography, Palaeoclimatology, Palaeoecology 126:161–171.

Krueger, H.W., and C.H. Sullivan. 1984. Models for carbon isotope fractionation between diet and bone. Pp. 205–220 in Stable isotopes in nutrition. Symposium series. J.R. Turnland, and P.E. Johnson, eds. American Chemical Society, Washington, DC.

Krull, E.S., and G.H. Retallack. 2000. $\delta^{13}C$ depth profiles from paleosols across the Permian-Triassic boundary: Evidence for methane release. Geological Society of America Bulletin 112:1459–1472.

Kuypers, M.M.M., R.D. Pancost, and J.S.S. Damste. 1999. A large and abrupt fall in atmospheric $CO_2$ concentration during Cretaceous times. Nature 399:342–345.

Lécuyer C., P. Grandjean, J.R. O'Neil, H. Cappetta, and F. Martineau. 1993. Thermal excursions in the ocean at the Cretaceous-Tertiary boundary (northern Morocco): $\delta^{18}O$

record of phosphatic fish debris. Palaeogeography, Palaeoclimatology, Palaeoecology 105:235–243.

Lécuyer C., P. Grandjean, F. Paris, M. Robardet, and D. Robineau. 1996. Deciphering "temperature" and "salinity" from biogenic phosphates: The $\delta^{18}O$ of coexisting fishes and mammals of the Miocene sea of Western France. Palaeogeography, Palaeoclimatology, Palaeoecology 126:61–74.

Lécuyer, C., P. Grandjean, and M.F. Sheppard. 1999. Oxygen isotope exchange between dissolved phosphate and water at temperatures ≤135° C: Inorganic versus biological fractionations. Geochimica et Cosmochimica Acta 63:855–862.

Lécuyer, C., S. Picard, J.-P. Garcia, S.M.F. Sheppard, P. Grandjean, and G. Dromart. 2003. Thermal evolution of Tethyan surface waters during the Middle-Late Jurassic: Evidence from $\delta^{18}O$ values of marine fish teeth. Paleoceanography18:1076, doi:10.1029/2002PA000863.

Lee-Thorp, J.A., and M. Sponheimer. 2005. Opportunities and constraints for reconstructing palaeoenvironments from stable light isotope ratios in fossils. Geological Quarterly 49:195–204.

Lee-Thorp, J.A., J.C. Sealy, and N.J. van der Merwe. 1989. Stable carbon isotope differences between bone collagen and bone apatite, and their relationship to diet. Journal of Archaeological Science 16:585–599.

Lee-Thorp, J.A., M. Sponheimer, and N.J. van der Merwe. 2003. What do stable isotopes tell us about hominid dietary and ecological niches in the Pliocene? International Journal of Osteoarchaeology 13:104–113.

Longinelli, A. 1984. Oxygen isotopes in mammal bone phosphate: A new tool for paleohydrological and paleoclimatological research? Geochimica et Cosmochimica Acta 48:385–390.

Longinelli, A., and S. Nuti. 1973. Revised phosphate-water isotopic temperature scale. Earth and Planetary Science Letters 19:373–376.

Luz, B., and Y. Kolodny. 1985. Oxygen isotope variations in phosphates of biogenic apatites. IV. Mammal teeth and bones. Earth and Planetary Science Letters 75:29–36.

MacFadden, B.J. 1994. South American fossil mammals and carbon isotopes: A 25 million-year sequence from the Bolivian Andes. Palaeogeography, Palaeoclimatology, Palaeoecology 107:257–268.

MacFadden, B.J. 1998. Tale of two rhinos: Isotopic ecology, paleodiet, and niche differentiation of *Aphelops* and *Teleoceras* from the Florida Neogene. Paleobiology 24:274–286.

MacFadden, B.J. 2000. Cenozoic mammalian herbivores from the Americas: Reconstructing ancient diets and terrestrial communities. Annual Review of Ecology and Systematics 31:33–59.

MacFadden, B.J., and T.E. Cerling. 1996. Mammalian herbivore communities, ancient feeding ecology, and carbon isotopes: A 10-million-year sequence from the Neogene of Florida. Journal of Vertebrate Paleontology 16:103–115.

MacFadden, B.J., and P. Higgins. 2004. Ancient ecology of 15-million-year-old browsing mammals within C3 plant communities from Panama. Oecologia 140:169–182.

MacFadden, B.J., and B.J. Shockey. 1997. Ancient feeding ecology and niche differentiation of Pleistocene mammalian herbivores from Tarija, Bolivia: Morphological and isotopic evidence. Paleobiology 23:77–100.

MacFadden, B.J., T.E. Cerling, J.M. Harris, and J. Prado. 1999a. Ancient latitudinal gradients of $C_3/C_4$ grasses interpreted from stable isotopes of New World Pleistocene horse (*Equus*) teeth. Global Ecology and Biogeography 8:137–149.

MacFadden, B.J., N. Solounias, and T.E. Cerling. 1999b. Ancient diets, ecology, and extinction of 5-million-year-old horses from Florida. Science 283:824–827.

MacFadden, B.J., P. Higgins, M.T. Clementz, and D.S. Jones. 2004. Diets habitat preferences and niche differentiation of Cenozoic sirenians from Florida: Evidence from stable isotopes. Paleobiology 30:297–324.

Nelson, B.K., M.J. DeNiro, and M.J. Schoeninger. 1986. Effects of diagenesis on strontium, carbon, nitrogen, and oxygen concentration and isotopic composition of bone. Geochimica et Cosmochimica Acta 50:1941–1949.

Nelson, S.V. 2005. Paleoseasonality inferred from equid teeth and intra-tooth isotopic variability. Palaeogeography, Palaeoclimatology, Palaeoecology 222:122–144.

O'Leary, M.H. 1988. Carbon isotopes in photosynthesis. Bioscience 38:328–336.

O'Leary, M.H., S. Mahavan, and P. Paneth.1992. Physical and chemical basis of carbon isotope fractionation in plants. Plant, Cell and Environment 15:1099–1104.

O'Neil, J.R. 1986. Theoretical and experimental aspects of isotopic fractionation. Pp. 1–37 *in* Stable isotopes in high temperature geochemical processes. J.W. Valley, H.P. Taylor, and J.R. O'Neil, eds. Reviews in Mineralogy 16:1–37.

Palmqvist, P., D.R. Grocke, A. Arribas, and R.A. Farina. 2003. Paleoecological reconstruction of a lower Pleistocene large mammal community using biogeochemical ($\delta^{13}C$, $\delta^{15}N$, $\delta^{18}O$, Sr:Zn) and ecomorphological approaches. Paleobiology 29:205–229.

Passey, B.H., and T.E. Cerling. 2002. Tooth enamel mineralization in ungulates: Implications for recovering a primary isotopic time-series. Geochimica et Cosmochimica Acta 66:3225–3234.

Passey, B.H., T.E. Cerling, M.E. Perkins, M.R. Voorhies, J.M. Harris, and S.T. Tucker. 2002. Environmental change in the Great Plains: An isotopic record from fossil horses. Journal of Geology 110:123–140.

Passey, B.H., T.F. Robinson, L.K. Ayliffe, T.E. Cerling, M. Sponheimer, M.D. Dearing, B.L. Roeder, and J.R. Ehlinger. 2005. Carbon isotope fractionation between diet, breath $CO_2$, and bioapatite in different mammals. Journal of Archaeological Sciences 32:1459–1470.

Pucéat, E., C. Lécuyer, S.M.F. Sheppard, G. Dromart, S. Reboulet, and P. Grandjean. 2003. Thermal evolution of Cretaceous Tethyan marine waters inferred from oxygen isotope composition of fish tooth enamels. Paleoceanography 18:1029, doi:10.1029/2002PA000823.

Quade, J., T.E. Cerling, J.C. Barry, M.E. Morgan, D.R. Pilbeam, A.R. Chivas, J.A. Lee-Thorp, and N.J. van der Merwe. 1992. A 16-Ma record of paleodiet using carbon and oxygen isotopes in fossil teeth from Pakistan. Chemical Geology, Isotope Geosciences Section 94:183–192.

Quade, J., T.E. Cerling, P. Andrews, and B. Alpagut. 1995. Paleodietary reconstruction of Miocene faunas from Pasalar, Turkey, using stable carbon and oxygen isotopes of fossil tooth enamel. Journal of Human Evolution 28:373–384.

Rinaldi, C., and T.M. Cole III. 2004. Environmental seasonality and incremental growth rates of beaver (*Castor canadensis*) incisors: Implications for paleobiology. Palaeo-geography, Palaeoclimatology, Palaeoecology 206:289–301.

Roe, L.J., J.G.M. Thewissen, J. Quade, J.R. O'Neil, S. Bajpai, A. Sahmi, and S.T. Hussain. 1998. Isotopic approaches to understanding the terrestrial-to-marine transition on the earliest cetaceans. Pp. 399–422 *in* The emergence of whales. J.G.M. Thewissen, ed. Plenum, New York.

Rogers, R.R., and S.M. Kidwell. 2000. Associations of vertebrate skeletal concentrations and discontinuity surfaces in terrestrial and shallow marine records; a test in the Cretaceous of Montana. Journal of Geology 108:131–154.

Rogers, R.R., and S.M. Kidwell. This volume. A conceptual framework for the genesis and analysis of vertebrate skeletal concentrations. Chapter 1 *in* Bonebeds: Genesis, analysis, and paleobiological significance. R.R. Rogers, D.A. Eberth, and A.R. Fiorillo, eds. University of Chicago Press, Chicago.

Roth, J.D. 2002. Temporal variability in arctic fox diet as reflected in stable-carbon isotopes: The importance of sea ice. Oecologia 133:70–77.

Roth, J.D., and K.A. Hobson. 2000. Stable-carbon and nitrogen isotopic fractionization between diet and tissue of captive red fox, *Vulpes vulpes*: Implications for dietary reconstruction. Canadian Journal of Zoology 78:848–852.

Rowley, D.B., R.T. Pierrehumbert, and B. Currie. 2001. A new approach to stable isotope-based paleoaltimetry: Implications for paleoaltimetry and paleohypsometry of the High Himalya since the Late Miocene. Earth Planetary Science Letters 188:253–268.

Rozanski, K., L. Araguás-Araguás, and R. Gonfiantini. 1993. Isotopic patterns in modern global precipitation. Pp. 1–36 *in* Climate change in continental isotope records. P.K. Swart, K.C. Lohmann, J. McKenzie, and S. Savin, eds. American Geophysical Union, Washington, D.C.

Schaffner, F.C., and P.K. Swart. 1991. Influence of diet and environmental water on the carbon and oxygen isotopic signatures of seabird eggshell carbonate. Bulletin of Marine Science 48:23–38.

Schoeninger, M.J. 1996. Stable isotope studies in human evolution. Evolutionary Anthropology 4:83–98.

Schoeninger, M.J., H.T. Bunn, S. Murray, T. Pickering, and J. Moore. 2001. Meat-eating by the fourth African ape. Pp. 179–195 *in* Meat-eating and human evolution. C.B. Stanford, and H.T. Bunn, eds. Oxford University Press, New York.

Schwarcz, H.P., and M.J. Schoeninger. 1991. Stable isotope analyses in human nutritional ecology. Yearbook Physical Anthropology 34:283–321.

Sharp, Z.D., and T.E. Cerling. 1998. Fossil isotope records of seasonal climate and ecology: Straight from the horse's mouth. Geology 26:219–222.

Sponheimer, M., and J.A. Lee-Thorp. 1999a. Alteration of enamel carbonate environments during fossilization. Journal of Archaeological Science 26:143–150.

Sponheimer, M., and J.A. Lee-Thorp. 1999b. Isotopic evidence for the diet of an early hominid, *Australopithecus africanus*. Science 283:368–370.

Sponheimer, M., and J.A. Lee-Thorp. 1999c. Oxygen isotopes in enamel carbonate and their ecological significance. Journal of Archaeological Science 26:723–728.

Sponheimer, M., K. Reed, and J.A. Lee-Thorp. 2001. Isotopic palaeoecology of Makapansgat Limeworks perissodactyla. South African Journal of Science 97:327–329.

Sponheimer, M., J.A. Thorp, D.J. DeRuiter, J.M. Smith, N.J.v.d. Merwe, K. Reed, C.C. Grant, L.K. Ayliffe, T.F. Robinson, C. Heidelberger, and W. Marcus. 2003. Diets of Southern African bovidae: Stable isotope evidence. Journal of Mammalogy 84:471–479.

Sponheimer, M., J. Lee-Thorp, D.D. Ruiter, D. Codron, J. Codron, A.T. Baugh, and F. Thackeray. 2005. Hominids, sedges, and termites: New carbon isotope data from the Sterkfontein valley and Kruger National Park. Journal of Human Evolution 48:301–312.

Stanton-Thomas, K., and S.J. Carlson. 2003. Microscale $\delta^{18}O$ and $\delta^{13}C$ isotopic analysis of an ontogenetic series of the hadrosaurid dinosaur *Edmontosaurus*: Implications for physiology and ecology. Palaeogeography, Palaeoclimatology, Palaeoecology 206:257–287.

Sternberg, L.S.L. 1989. Oxygen and hydrogen isotope ratios in plant cellulose: Mechanisms and applications. Pp. 124–141 *in* Stable isotopes in ecological research. P.W. Rundel, J.R. Ehleringer, and K.A. Nagy, eds. Springer-Verlag, New York.

Straight, W.H., R.E. Barrick, and D.A. Eberth. 2004. Reflections of surface water, seasonality and climate in stable oxygen isotopes from tyrannosaurid tooth enamel. Palaeogeography, Palaeoclimatology, Palaeoecology 206:239–256.

Stuart-Williams, H.L.Q., and H.P. Schwarcz. 1997. Oxygen isotopic determination of climate variation using phosphate from beaver bone, tooth enamel, and dentine. Geochimica et Cosmochimica Acta 61:2539–2550.

Thewissen, J.G.M., L.J. Roe, J.R. O'Neil, S.T. Hussain, A. Sahni, and S. Bajpai. 1996. Evolution of cetacean osmoregulation. Nature 381:379–380.

Tieszen, L.L. 1991. Natural variations in carbon isotope values of plants: Implications for archaeology, ecology, and paleoecology. Journal of Archaeological Science 20:227–248.

Tieszen, L.L., and T.W. Boutton. 1989. Stable carbon isotopes in terrestrial ecosystem research. Pp. 167–195 *in* Stable isotopes in ecological research. P.W. Rundel, J.R. Ehleringer, and K.A. Nagy, eds. Springer-Verlag, New York.

Tieszen, L.L., and T. Fagre. 1993. Effect of diet quality and composition on the isotopic composition of respiratory $CO_2$, bone collagen, bioapatite, and soft tissues. Pp. 123–135 *in* Molecular archaeology of prehistoric human bone. J. Lambert, and G. Grupe, eds. Springer-Verlag, Berlin.

Trueman, C.N., and N. Tuross. 2002. Trace elements in recent and fossil bone apatite. Reviews in Mineralogy and Geochemistry 48:489–521.

Trueman, C.N., C. Chenery, D.A. Eberth, and B. Spiro. 2003. Diagenetic effects on the oxygen isotope composition of dinosaurs and other vertebrates recovered from terrestrial and marine sediments. Journal of the Geological Society of London 160:895–901.

Trueman, C.N. This volume. Trace element geochemistry of bonebeds. Chapter 7 *in* Bonebeds: Genesis, analysis, and paleobiological significance. R.R. Rogers, D.A. Eberth, and A.R. Fiorillo, eds. University of Chicago Press, Chicago.

Tudge, A.P. 1960. A method of analysis of oxygen isotopes in orthophosphate—its use in the measurement of paleotemperatures. Geochimica et Cosmochimica Acta 18:81–93.

van der Merwe, N.J., and E. Medina. 1991. The canopy effect, carbon isotope ratios and foodwebs in Amazonia. Journal of Archaeological Science 18:249–259.

Vennemann, T.W., and E. Hegner. 1998. Oxygen, strontium and neodymium isotope composition of shark teeth as a proxy for the paleooceanography and paleoclimatology of the northern alpine Paratethys. Palaeogeography, Palaeoclimatology, Palaeoecology 142:107–121.

Vennemann, T.W., E. Hegner, G. Cliff, and G.W. Benz. 2001. Isotopic composition of recent shark teeth as a proxy for environmental conditions. Geochimica et Cosmochimica Acta 65:1583–1599.

Voorhies, M.R. 1985. A Miocene rhinoceros herd buried in volcanic ash. National Geographic Research Reports 19:671–688.

Wang, Y., and T.E. Cerling. 1994. A model of fossil tooth and bone diagenesis: Implications for paleodiet reconstruction from stable isotopes: Paleodiet reconstruction from stable isotopes. Palaeogeography, Palaeoclimatology, Palaeoecology 107:281–289.

Wang, Y., T.E. Cerling, and B.J. MacFadden. 1994. Fossil horses and carbon isotopes: New evidence for Cenozoic dietary, habitat, and ecosystem changes in North America. Palaeogeography, Palaeoclimatology, Palaeoecology 107:269–279.

Wang, Y., T. Deng, and D. Biasatti. 2006. Ancient diets indicate significant uplift of southern Tibet after 7 Myr ago. Geology 32:309–312.

Wiedemann, F.B., H. Bocherens, A. Mariotti, A. von den Driesch, and G. Grupe. 1999. Methodological and archaeological implications of intra-tooth isotopic variations ($\delta^{13}$C, $\delta^{18}$O) in herbivores from Ain Ghazal (Jordan, Neolithic). Journal of Archaeological Science 26:697–704.

Wright, V.P., and S.D. Vanstone. 1991. Assessing the carbon dioxide content of ancient atmospheres using palaeo-calcretes; theoretical and empirical constraints. Journal of the Geological Society of London 148:945–947.

Wynn, J.G., M.I. Bird, and V.N.L. Wong. 2005. Rayleigh distillation and the depth profile of $^{13}$C/$^{12}$C ratios of soil organic carbon from soils of disparate texture in Iron Range National Park, Far North Queensland, Australia. Geochimica et Cosmochimica Acta 69: 1961–1973.

Zazzo, A., H. Bocherens, M. Brunet, A. Beauvilain, D. Billiou, H.T. Mackaye, P. Vignaud, and A. Mariotti. 2000. Herbivore paleodiet and paleonvironmental changes in Chad during the Pliocene using stable isotope ratios of tooth enamel carbonate. Paleobiology 26:294–309.

Zazzo, A., C. Lécuyer, and A. Mariotti. 2004. Experimentally-controlled carbon and oxygen isotope exchange between bioapatites and water under inorganic and microbially-mediated conditions. Geochimica et Cosmochimica Acta 68:1–12.

Zazzo, A., M. Balasse, and W.P. Patterson. 2005. High-resolution $\delta^{13}$C intratooth profiles in bovine enamel: Implications for mineralization pattern and isotopic attenuation. Geochimica et Cosmochimica Acta 69:3631–3642.

Chapter 1: A Conceptual Framework for the Genesis and Analysis of Vertebrate Skeletal Concentrations

Raymond R. Rogers
Geology Department
Macalester College
1600 Grand Avenue
Saint Paul, MN 55105
U.S.A.
rogers@macalester.edu

Susan M. Kidwell
Department of Geophysical Sciences
University of Chicago
5734 South Ellis Avenue
Chicago, IL 60637
U.S.A.
skidwell@uchicago.edu

Chapter 2: Bonebeds through Time

Anna K. Behrensmeyer
Department of Paleobiology
MRC 121
National Museum of Natural History
P.O. Box 37012, Smithsonian Institution
Washington, DC 20013-7012
U.S.A.
behrensa@si.edu

## Chapter 3: A Bonebeds Database: Classification, Biases, and Patterns of Occurrence

David A. Eberth
Royal Tyrrell Museum of Palaeontology
Box 7500
Drumheller, Alberta T0J 0Y0
Canada
david.eberth@gov.ab.ca

Matthew Shannon
701-1050 Bidwell Street
Vancouver, British Columbia V6G 2K1
Canada
eryops@hotmail.com

Brent G. Noland
Drumheller, Alberta T0J 0Y1
Canada
brentnoland@hotmail.com

## Chapter 4: From Bonebeds to Paleobiology: Applications of Bonebed Data

Donald B. Brinkman
Royal Tyrrell Museum of Palaeontology
Box 7500
Drumheller, Alberta T0J 0Y0
Canada
don.brinkman@gov.ab.ca

David A. Eberth
See Above

Philip J. Currie
Department of Biological Sciences
University of Alberta,
Edmonton, Alberta T6G 2E9
Canada
philip.currie@ualberta.ca

## Chapter 5: A Practical Approach to the Study of Bonebeds

David A. Eberth
See Above

Raymond R. Rogers
See Above

Anthony R. Fiorillo
Museum of Nature and Science
P.O. Box 151469
Dallas, TX 75315
U.S.A.
tfiorillo@natureandscience.org

Chapter 6: Numerical Methods for Bonebed Analysis

Richard W. Blob
Department of Biological Sciences
Clemson University,
Clemson, SC 29634
U.S.A.
rblob@clemson.edu

Catherine Badgley
Museum of Paleontology
University of Michigan
Ann Arbor, MI 48109
U.S.A.
cbadgley@umich.edu

Chapter 7: Trace Element Geochemistry of Bonebeds

Clive Trueman
School School of Ocean and Earth Science,
National Oceanography Centre
University of Southampton
Waterfront Campus
European Way
Southampton SO14 3ZH
United Kingdom
trueman@noc.soton.ac.uk

Chapter 8: Stable Isotope Geochemistry of Bonebed Fossils: Reconstructing Paleoenvironments, Paleoecology, and Paleobiology

Henry Fricke
Department of Geology
Colorado College
Colorado Springs, CO 80903
U.S.A.
hfricke@coloradocollege.edu

Bonebeds listed by name in the index are mentioned in individual chapters. Many additional bonebeds not listed by name in the index are included in Appendices 2.2, 2.3, and 3.1.

abiotic processes, role in bonebed formation, 86–94. *See also* concentration, of bones/carcasses: abiotic
abrasion, 24, 32–33, 285, 307–8, 342, 346
accumulation, of bones/carcasses. *See* concentration, of bones/carcasses
Agate bonebed, 303. *See also* Agate Fossil Quarry, Agate Spring locality
Agate Fossil Quarry, 271, 274. *See also* Agate bonebed, Agate Spring locality
Agate Spring locality, 3, 148–49, 174–75, 184–85, 190–91. *See also* Agate bonebed, Agate Fossil Quarry
age profile, of fossil assemblage, 289–91
aggregation paleobehavior, 230–41. *See also* herd, herding
agonistic behavior, 9, 12, 250–52
*Albertosaurus* bonebed, 144–45, 150–51, 158–59, 235
Allochthonous assemblages, 20, 32, 296, 413–14, 417, 422
Anderson Quarry, 156–57, 226
Apatite. *See* biopatite
articulation, skeletal, 81–83, 92, 284, 291–94, 342

Ashfall Fossil Beds, 210–11, 227, 458. *See also* Poison Ivy Quarry
association, skeletal, 284, 287–88, 291–94, 336–38
attritional assemblages, 22, 40, 87–89, 229, 404, 419. *See also* concentration, of bones/carcasses: hiatal/passive attritional
authigenic mineralization, 401, 452–453, 461
autochthonous assemblages, 10, 30, 33, 86, 87, 245, 413–14

baseline, 274–76
BBR locality, 347–49
biases, of "bonebeds database," 114–19
bioapatite, 398–401, 437, 445–56
bioerosion, 285, 308, 313–14
biogenic concentrations, of vertebrate remains, 6–19. *See also* concentration, of bones/carcasses: biogenic, extrinsic; biogenic, intrinsic
biotic processes, role in bonebed formation, 86–94. *See also* concentration, of bones/carcasses: biotic
Blitzkrieg hypothesis, 414

block collecting, 69, 271, 273–75
body size, estimation of, 291, 370–71
bone: collecting, 8, 18–19, 85, 87;
  dispersal/sorting, 21–25, 123, 124,
  247–48, 294–96, 342, 349–52;
  modification, 32, 33, 81, 94, 285,
  302–14, 342
bone sands, 5, 98
bonebeds: carnivores vs. herbivores, 79–81;
  characteristics through time, 71–83,
  239–41; classification of, 3–5, 66–67,
  98–99, 106–14, 123–24; definition of,
  3–5, 66–67, 98–99, 106; diversity
  patterns, 116, 119–20; causes of, 5–35,
  83–90, 100–101, 120–21 (see also
  concentration, of bones/carcasses);
  geographic distribution of, 70, 72, 105;
  lithological/paleoenvironmental
  associations of, 71, 74–76, 105, 121–23;
  macrofossil, 3–4, 98, 107–10, 115–19,
  121–23, 272–74, 287, 288, 305, 307;
  microfossil, 4–5, 29–30, 98, 107–10,
  116–23, 251, 278–80, 287–89, 306;
  mixed, 107–10; monodominant, 99,
  108, 109, 112, 113–17, 119–23;
  monogeneric, 111, 114; monospecific,
  76–79, 99, 111, 114; monotaxic, 66, 99,
  108, 109, 111–13, 115–17, 120–22,
  231–33, 333, 420; multidominant, 99,
  108, 109, 112, 113–14, 116, 117,
  119–23; multispecific, 76–78, 86, 87, 89,
  99; multitaxic, 67, 79, 105, 108, 109,
  111–13, 117, 121, 222, 223, 243,
  244–46, 249, 251, 340, 351, 421,
  458–59; paucispecific, 76–79, 87, 99,
  113, 114; paucitaxic, 67, 113, 114, 333;
  Phanerozoic distribution of, 71–73, 91;
  taphonomic patterns (Phanerozoic),
  81–83; taxonomic representation,
  76–79, 111–14. See also names of specific
  bonebeds
borings, in bone, 313–14
breakage, of bone, 33, 305–7, 342, 353
burial, catastrophic, 6, 22, 33–35, 37, 87,
  420. See also obrution
burrows, 33–34, 85, 87, 227, 237, 240

C3 plants (Calvin pathway), 439–42, 463–64,
  470, 472, 474, 478
C4 plants (Crassulacean acid metabolism
  pathway), 440–42, 446, 459, 463–64,
  469–75
canopy effect, 442

Careless Creek locality, 162–63, 351
Centrosaurus bonebeds, 13, 166–67, 196–97,
  226, 233–35, 268–69
chronofauna, 252
Clambank Hollow locality, 387–88
Cleveland-Lloyd Quarry, 140–41, 148–49,
  166–67, 184–85, 186–87, 192–93, 202–3,
  204–5, 251–52
Coelophysis bonebed, 224–25, 239. See also
  Coelophysis Quarry
Coelophysis Quarry, 148–49, 200–1, 275. See
  also Coelophysis bonebed
collagen, 398–401, 437, 438, 447–49, 451
completeness, skeletal, 89, 92, 94, 284, 287,
  294–96, 339, 353
composition of bones, teeth, scales, 398–401,
  447–49
concentration, of bones/carcasses: abiotic,
  86–87, 91–94, 101; biogenic, extrinsic,
  6, 8, 16–19, 40, 100; biogenic, intrinsic,
  6–8, 9–16, 35, 39–40, 100; biotic, 87,
  91–94, 100; catastrophic, 33–34, 85–87,
  100, 120, 124, 224, 246, 283, 289–90,
  420–21; fluvial, hydraulic, 6, 19–25, 37,
  40, 84–87, 90, 101, 121; hiatal/passive
  attritional, 27–30, 86–89, 90–93, 101
  (see also attritional assemblages); mixed,
  89–91, 101; in response to erosion,
  reworking, 6, 20, 22, 25, 31–33, 37, 40,
  86–87, 121, 243, 247, 269, 299, 404,
  409, 413–17; sedimentologic, 6, 22,
  26–35, 37, 40, 86–87, 101; strandline,
  hydraulic, 21, 25–26, 37, 101, 297
corrasion, 285, 308
corrosion, 17, 285, 307–9
counting protocols, for specimens and
  individuals, 287–88, 294, 335–41, 387
Craddock Bonebed, 156–57, 227
Cuddie Springs Bonebed, 132–33, 414, 419

death. See mortality
death assemblage, 363
deflation lag/surface, 37, 121, 270, 284
denning behavior, 33–34, 237, 240. See also
  burrows
depositional system, 104–106, 118, 121, 124,
  282
diagenesis, 6–7, 28, 32, 33, 398–409, 449–56
Diplocaulus bonebed, 226–27
disarticulation, skeletal, 291–94
disease. See mortality: disease/poisoning
drought. See mortality: drought-related
drowning. See mortality: flooding/drowning

EAQ locality, 347–49
*Edmontosaurus* bonebed, 154–55, 176–77, 230, 298
elements vs. specimens, 286–88, 335–41
environmental mitigation, 271–72
etching, 30, 313–14
ETE Bonebed Database, 67–84, 86, 90–95, Appendices 2.2 and 2.3
excavation techniques, 268–80
exhumation, 31–33. *See also* concentration, of bones/carcasses: in response to erosion, reworking

fighting dinosaurs, 250–51
flooding. *See* mortality
flooding surface, 27, 31–32
fluvial hydraulic concentrations. *See* concentration, of bones/carcasses: fluvial, hydraulic
fossilization, 398–409. *See also* diagenesis; prefossilization

geological data, collecting, 280–85, 317
geometry, of bonebeds, 5, 268–70, 281, 283, 284, 296, 297–300
Geraldine Bonebed, 196–97, 246–47
GIS (geographic information system), 276–78
GPS (global positioning system), 267–68

Hagerman Horse Quarry, 192–93, 233
herd, herding, 13–14, 87, 233–36. *See also* aggregation paleobehavior
hiatal concentrations. *See* concentration, of bones/carcasses: hiatal/passive attritional
high diversity vertebrate assemblages, 99, 108–9, 111–14, 116, 244, 458
hydraulic concentrations. *See* concentration, of bones/carcasses: fluvial, hydraulic
hydraulic dispersal. *See* bone dispersal/sorting

ICP-MS running conditions, 428–29
intraspecific variation, 224–28
isotaphonomic, 341–50, 419, 421
isotopes: carbon, 438–42, 463; diagenesis, 449–56; as evidence for climate, 464–70, 478; as evidence for dental development, 476–77; as evidence for diet, 445–46, 461, 470–75; as evidence for grassland expansion, 442, 468–69; as

evidence for habitat preference, 458, 470–75; as evidence for migration, 474; as evidence for paleoelevation, 468–70; as evidence for thermoregulation, 476; as evidence for vegetation structure, 462–64, 467–69; fractionation in bones and teeth, 438–39, 445–47; oxygen, 442–45; seasonal variation, 449, 456–57, 466–67, 469–70; study design, 457–61

Jack's Birthday Site, 208–9, 249, 351
Joffre Bridge Site, 214–15, 232

K/T boundary, 79, 415–16

La Brea Tar Pits, 11, 12, 132–33, 184–85, 186–87, 202–3, 251–52
lag assemblage, 20–22, 30–33, 40, 86, 88, 101, 245, 300, 337–38, 350, 365. *See also* deflation lag/surface
Lamy Metoposaur Quarry, 170–71, 271
Los Chañares locality, 26, 194–95
low diversity vertebrate assemblages, 99, 108–9, 111–14, 116, 119–20, 233–35, 289, 420–21

*Maiasaura* bonebed, 198–99, 420
mapping, 270–71, 276–78, 316
mass mortality. *See* mortality
megafaunal extinction, 414
microsite, 4–5, 109–10. *See also* bonebeds: microfossil
miring. *See* mortality: miring
MNE (minimum number of elements), 284, 287–88, 336–40
MNI (minimum number of individuals), 99, 114, 284, 287–88, 294, 336–40
modern analogs, 315
morphological variation, 229–30
mortality: disease/poisoning, 7, 16, 36, 84, 85–86, 237; drought-related, 7, 15, 36, 84–85, 87, 92, 238–39, 249–50, 252, 315; flooding/drowning, 7, 12–14, 36, 84–85; miring, 7, 12, 36, 85, 87; predation, 8, 9, 12, 14, 16–18, 37, 80, 81–83, 85–86, 87, 94, 100, 249, 250–53, 293, 306–7, 308, 338, 339, 352–53, 363; reproduction-related, 7, 9–11, 36; stranding, 7, 11–12, 13, 36; volcanism-related, 34, 37, 74, 85, 87, 235; weather-related, 7, 14–15, 36, 84, 85, 233; wildfire, 7, 13, 14, 36, 84, 246–47

NISP (number of individual specimens), 98–99, 287–89, 336–41

obrution, 6, 20, 22, 33–35, 37, 40, 87, 101
ontogeny, 222, 228–29, 232, 477
orientation, of bones/carcasses, 300–3, 342
overburden, 272–73

*Pachyrhinosaurus* bonebeds, 230
paleocommunities, 222, 223, 241, 244–54
paleoenvironmental stress, 237–39
paleofaunas, 241–44
paleosols, 28–29, 30, 105, 118, 247, 412, 416, 419, 455, 463
parautochthonous assemblages, 10, 13, 20, 24, 33
Pleistocene Fossil Lake, 422
point counting, vertebrate microfossils, 278
Poison Ivy Quarry, 3, 34, 210–211. *See also* Ashfall Fossil Beds
population abundances, inferring from fossil assemblage, 362–65
predation. *See* mortality: predation
prefossilization, 7, 305. *See also* fossilization

quantitative analyses of bonebed samples: analysis of variance (ANOVA), 366–68; cluster analysis, 375–79; correspondence analysis (CA), 384–88; *F*-test, 367, 368–75; Kolmogorov-Smirnov two-sample test, 346–48; linear regression, 369–75; principal component analysis (PCA), 334, 379–84; rank-order correlation, 352–53; rarefaction, 359–62; $X^2$-test, 351–53; *z*-test, 345–46, 349

rare earth elements (REE), 403–9; analytical methods, 423–29; characterizing bonebeds, 417–19; determining provenance, 409–13, 416; distinguishing catastrophic vs. attritional assemblages, 420–21; paleoenvironmental reconstructions, 421–22; recognizing reworking, 404, 413–17
recrystallization, bone, 400–1, 408, 416
resource partitioning, 464, 471–75

reworking. *See* concentration, of bones/carcasses: in response to erosion, reworking

sample size, 286–87, 424
sampling protocols, 354–65
Scabby Butte bonebed, 180–81, 230
scavenging, 250–52, 293, 306–7
screening, 278–80, 355
sediment starvation. *See* concentration, of bones/carcasses: hiatal/passive attritional
sedimentological concentrations. *See* concentration, of bones/carcasses: sedimentologic
sequence boundary, 31
site assessment, 266–72
size sorting, of bones. *See* bone dispersal/sorting; Voorhies groups/hydraulic transport groups
skeletal concentration. *See* concentration, of bones/carcasses
Smithers Lake, 26
sorting, hydraulic. *See* bone dispersal/sorting
spatial data, 284, 296–302
spawning, 7, 10–11, 232. *See also* mortality: reproduction-related
stable isotopes. *See* isotopes
stereographic projection, 302–3
stranding. *See* mortality: stranding
strandlines, 25–26, 37, 101, 297
surface concentrate, 107, 269–70

taphogram, 286
taphonomic: analysis of vertebrate assemblages, 283–314, 342; equivalence, 95, 334, 341–54 (*see also* isotaphonomic); history of vertebrate assemblages, 35–39, 314–16; mode, 246, 248, 249, 315
taphosystem, 94
taxonomic diversity, 111–14, 115, 288–89, 360, 361
taxonomic resolution, 335–36
teeth, shed, 293
Temeside bonebed, 134–35, 245
thermoregulation. *See* isotopes
Thomas Farm locality, 192–93, 249, 352–53
time averaging, 11, 13, 32, 41, 87, 89, 111, 417–21
tooth marks, 285, 293, 304, 311–13

trace elements, 402–3. *See also* rare earth elements (REE)

trample marks, 185, 310–11

two taxa approach, 465–466

Ukhaa Tolgod locality, 34, 152–53, 189–90

Verdigre Quarry, 23, 210–11

volcanism. *See* mortality: volcanism-related

Voorhies groups/hydraulic transport groups, 22–23, 295–96, 350–52. *See also* bone: dispersal/sorting

weather, severe. *See* mortality, weather-related

weathering, of bones, 28, 285, 308–10, 342

wildfire. *See* mortality: wildfire